FOREWORD

This volume is dedicated to Professor Leslie Fox of Oxford University, whose contribution to numerical analysis is widely recognised. The collection of papers is the outcome of the proceedings of a workshop and symposium on the numerical treatment of integral equations held at Durham University on 19th-29th July, 1982. The list of participants and their topics will reveal much that has been influenced by the Oxford school, promulgated by Leslie Fox and his colleagues.

Sketching the background and origin of the symposium provides me with the perfect opportunity to thank all who contributed to its success. The meeting was the 23rd symposium to be held, with the welcome sponsorship of the London Mathematical Society (LMS), in the series of Durham Symposia. It is representative of a broad outlook that a proportion of such LMS symposia have been in areas of applied or applicable mathematics. I take this opportunity to thank the Durham Symposia Committee of the LMS for its interest and moral support.

The prospects for a meeting were first given encouragement when I was able to talk with Professor John Crank, who was a member of the Durham Symposia Committee. During the long gestation period, the proposal was refereed first by the LMS, for sponsorship, and then by the UK Science and Engineering Research Council (SERC). It gives me great pleasure to thank SERC for its award of visiting fellowships, and to acknowledge the support and advice of L.M.Delves, D. Kershaw, G.F. Miller, F. Ursell and K. Wright who kindly joined me on an organising committee for the preparation and implementation of the proposal. J.R.Whiteman acted courteously and efficiently as SERC assessor. Supplementary support came from the USA European Research Office.

In the straitened circumstances in which British universities currently find themselves, the organiser of a Durham symposium may expect to shoulder the bulk of the administration. Accordingly, I must take this opportunity to thank my colleagues for their forebearance if at times I seemed pre-occupied with this task. The understanding of my family was of invaluable benefit to me.

The City and University of Durham provide an historic and picturesque location for symposia, and I thank the liaison officer, John Coleman, and the staff of Grey College and the Finance

Office for arranging the use of University facilities.

An outcome of the meeting has been the creation of two clubs, the Fredholm 2 Club (co-ordinator K.E. Atkinson) to collect practical examples of Fredholm equations of the second kind, and an Abel-Volterra Club (co-ordinator C.T.H. Baker) to perform a similar task for Abel and Volterra integral equations of all kinds.

A most pleasing enthusiasm and co-operation were evident amongst the participants of this ten-day symposium. I thank all the participants and contributors for their whole-hearted involvement, which made possible the success of the symposium. It is also my wish to thank Geoffrey Miller for acting as co-editor. In arranging for the publication of this volume, we hope to bring to a wider audience (including those who through external circumstances were unable to attend) something of the essence of this meeting.

CHRISTOPHER T. H. BAKER (Reader in Mathematics)
The Victoria University of Manchester

Treatment of Integral Equations by Numerical Methods

Dedicated to Leslie Fox,
Professor of Numerical Analysis and Fellow of
Balliol College, University of Oxford

Treatment of Integral Equations by Numerical Methods

Based on the Proceedings of a Symposium held in Durham from 19-29 July, 1982, organized under the auspices of the London Mathematical Society

Edited by

Christopher T. H. Baker

Department of Mathematics
University of Manchester
Manchester, UK

Geoffrey F. Miller

National Physical Laboratory
Teddington, Middlesex, UK

1982

ACADEMIC PRESS
A Subsidiary of Harcourt Brace Jovanovich, Publishers
London New York
Paris San Diego San Francisco São Paulo
Sydney Tokyo Toronto

ACADEMIC PRESS INC. (LONDON) LTD.
24/28 Oval Road
London NW1

United States Edition published by
ACADEMIC PRESS INC.
111 Fifth Avenue
New York, New York 10003

British Library Cataloguing in Publication Data
Treatment of Integral Equations by Numerical Methods.
1. Integral equations—Numerical solution—
Congresses
I. Baker, C.T.H.
II. Miller, G.F.
515.4'5 QA431

ISBN 0-12-074120-2

Printed in Great Britain by
Whitstable Litho Ltd., Whitstable, Kent

CONTRIBUTORS AND PARTICIPANTS

AHMED, A.H., *University of Newcastle, and Khartoum University.*
AHUÉS, M., *IMAG, University of Grenoble.*
AMINI, S., *Faculty of Technology, Department of Mathematics Statistics and Computing, Plymouth Polytechnic, Plymouth PL4 8AA.*
ATKINSON, K.E., *Mathematics Department, University of Iowa, Iowa City, Iowa 52242, U.S.A.*
BADER, G., *Institut für Angewandte Mathematik Universitat Heidelberg.*
BAMBERGER, L., *Mathematisches Institut, Universität München.*
BAKER, C.T.H., *Mathematics Department, University of Manchester, Manchester M13 9PL.*
BATES, D.M., *Department of Statistics, University of Wisconsin-Madison, U.S.A.*
BELWARD, J.A., *Department of Mathematics, University of Queensland, St. Lucia, Australia.*
BOHL, E., *Fakultät für Mathematik, Universität Konstanz, Postfach 5560, D-7750 Konstanz, West Germany.*
BOWNDS, J.M., *Department of Mathematics, University of Arizona, Tucson, 85721, U.S.A.*
BRUNNER, H., *Institut de mathématiques, Université de Fribourg, CH-1700 Fribourg, Switzerland.*
CABRAL, L.M., *Naval Underwater Systems Center, Newport RI 02840, U.S.A.*
CHATELIN, F., *Mathematical Sciences Division, IBM Research, P.O. Box 218, Yorktown Heights, New York 10598, U.S.A., and IMAG, University of Grenoble, France.*
CHRISTIANSEN, S., *Laboratory of Applied Mathematical Physics, The Technical University of Denmark, Building 303, DK-2800 Lyngby, Denmark.*
CLARYSSE, T.H., *LEUVEN, K.U., Department of Computer Science, Belgium.*
COLLATZ, L., *2 Hamburg 67, Eulenkrugstr 84, Germany.*
COLEMAN, J.P., *Department of Mathematics, The University, Durham.*
D'ALMEIDA, F., *IMAG, University of Grenoble.*
DAVIES, A.R., *Department of Applied Mathematics, University College of Wales, Aberystwyth SY23 3BZ.*

DE HOOG, F., *CSIRO, Canberra, Australia.*
DELVES, L.M., *Department of Computational and Statistical Science, University of Liverpool, Brownlow Hill, Liverpool, L69 3BX.*
DE MEY, G., *Laboratory of Electronics, Ghent State University, Sint Pietersnieuwstraat 41, 9000 Ghent, Belgium.*
EGGERMONT, P.P.B., *Department of Mathematical Sciences, University of Delaware, Newark, De 19711, U.S.A.*
ELLIOTT, D., *Department of Mathematics, University of Tasmania, Box 252C, G.P.O., Hobart, Tasmania, Australia 7001.*
FAIRWEATHER, G., *Mathematics Department, University of Kentucky, Lexington, Kentucky 40506, U.S.A.*
GOLBERG, M.A., *Department of Mathematical Sciences, University of Nevada, Las Vegas 89154, U.S.A.*
GROETSCH, C.W., *Department of Mathematical Sciences - 025, University of Cincinnati, Cincinnati, Ohio 45221, U.S.A.*
HAIRER, E., *Institut für Angewandte Mathematik Universität Heidelberg, D-6900 Heidelberg 1, W. Germany.*
HÄMMERLIN, G., *Mathematisches Institut, Universität München, Theresienstr. 39, D8000 München 2, W. Germany.*
HASHMI, S.M., *Department of Computational and Statistical Science, University of Liverpool, Liverpool, L69 3BX.*
HILL, M., *Department of Mathematics, The University, Reading.*
HOUGH, D.M., *Mathematics Division, South Bank Polytechnic, Wandsworth Road, London.*
VAN DER HOUWEN, P.J., *Mathematical Centre, Kruislaan 413, Amsterdam, Netherlands.*
JAGER, E. *Department of Mathematics, University of Konstanz, Postfach 5560, D-7750 Konstanz, West Germany.*
JASWON, M.A., *Department of Mathematics, The City University, London.*
JEPSON, A., *Stanford University, Palo Alto, California; now at Department of Computer Science, University of Toronto.*
JOHNSTON, R.L., *Department of Computer Science, University of Toronto, Toronto, Ontario, M2J 3H7, Canada.*
KERSHAW, D., *Mathematics Department, University of Lancaster, Bailrigg, Lancaster LA1 4YL.*
KLEINMAN, R., *Institute of Applied Mathematics, University of Delaware, Newark, Delaware, U.S.A.*
LINZ, P., *Department of Mathematics, University of California, Davis, California 95616, U.S.A.*
LUBICH, Ch., *Institut für Angewandte Mathematik, Universität Heidelberg, W. Germany.*
LOINES, J., *Imperial College, London.*
MARTI, J., *Department of Applied Mathematics, ETH - Zürich, CH 8092 Zürich, Switzerland.*
MARTIN, P.A., *Department of Mathematics, University of Manchester, Manchester M13 9PL.*
MASON, J.C., *Department of Mathematics and Ballistics Royal Military College of Science, Shrivenham, Swindon, Wilts.*

MAYERS, D.F., *Computing Laboratory, 19 Parks Road, Oxford, OX1 3PL.*
MCKEE, S., *Oxford University Computing Laboratory, 19 Parks Road, Oxford, OX1 3PL.*
MILLER, G.F., *Division of Information Technology and Computing, National Physical Laboratory, Teddington, Middlesex TW11 OLW.*
MOHAMED, J., *Department of Computational and Statistical Science, University of Liverpool.*
NEVANLINNA, O., *Institute of Mathematics, Helsinki University of Technology, SF-02150, Espoo 15, Finland.*
PAPAMICHAEL, N., *Department of Mathematics, Brunel University, Uxbridge.*
PATTERSON, J. *Department of Mathematics, Dundee University, Scotland.*
PHILLIPS, C., *Department of Computer Studies, The University, Hull.*
RAZALI, M.R., *Faculty of Mathematical Studies, Southampton University, SO9 5NH.*
TE RIELE, H.J.J., *Mathematical Centre, Amsterdam.*
ROACH, G.F., *Department of Mathematics, University of Strathcylde, 26 Richmond Street, Glasgow.*
RUTMAN, R., *College of Engineering, South Eastern Massachusetts University, N. Dartmouth, MA 02747, U.S.A.*
SAYERS, D., *NAG Central Office, 256 Banbury Road, Oxford.*
SCHÄFER, E., *Mathematisches Institut Universität München, Theresienstr 39, D-8000 München 2, W. Germany.*
SLOAN, I.H., *School of Mathematics, University of New South Wales, Kensington, New South Wales 2033, Australia.*
SMITH, R.N.L., *Department of Mathematics and Ballistics, Royal Military College of Science, Shrivenham, Wiltshire.*
SPENCE, A., *School of Mathematics, University of Bath, Claverton Down, Bath BA2 7AY.*
TAYLOR, P.J., *Department of Mathematics, University of Stirling, Stirling, Scotland, FK9 4LA.*
TELIAS, M., *IMAG, University of Grenoble.*
THOMAS, D.P., *Department of Mathematical Sciences, The University, Dundee.*
THOMAS, K.S., *Faculty of Mathematical Studies, The University of Southampton, Southampton SO9 5NH*
URSELL, F., *Department of Mathematics, University of Manchester.*
VOLK, W., *Hahn-Meitner Institut, Bereich D/M, Glienicker Str. 100, D-1000 Berlin 39, Germany.*
WADDINGTON, M., *Electricity Council Research Centre, Capenhurst.*
WADSWORTH, M., *Mathematics Department, Salford University.*
WAHBA, G., *Department of Statistics, University of Wisconsin, 1210 W. Dayton Street, Madison, WI 53705, U.S.A.*
WALSH, J., *Department of Mathematics, University of Manchester.*

WENDLAND, W.L., *Fachbereich Mathematik, Technische Hochschule D-61, Schlossgartenstr. 7, Darmstadt, W. Germany.*

WICKHAM, G.R., *Mathematics Department, University of Manchester, Manchester M13 9PL.*

WILKINSON, J.C., *Headmistress, Stanley Park Comprehensive School, Priory Road, Liverpool, L4 23L.*

WHITEMAN, J.R., *Institute of Computational Mathematics, Brunel University, Uxbridge.*

WOLKENFELT, P.H.M., *Department of Mathematics, Technische Hogeschool Eindhoven, 5600 MB Eindhoven, The Netherlands.*

WRIGHT, K., *University Computing Laboratory, Claremont Tower, Claremont Road, Newcastle-upon-Tyne, NE1 7RU.*

XANTHIS, L.S., *Mathematics and Computing Department, Polytechnic of Central London.*

Poster session contributed *in absentia:*

HANSON, R.J. *Sandia National Laboratories, Div. 1642, Albuquerque New Mexico, 87185, USA.*

PREFACE

This volume contains a set of papers arising out of the ten-day symposium and workshop on the numerical treatment of integral equations, held at Durham in July 1982. In the main, the papers are based upon the lectures contributed during the symposium. Some speakers have submitted their talks for publication elsewhere.

The editors decided that priority should be given to the early publication of the proceedings. However, it was considered nonetheless necessary for the papers contributed to the published proceedings to be refereed, and the editors have acted upon the advice thus received. The editors wish to place on record their sincere appreciation of the co-operation of members of the symposium in agreeing to act as referees, the enthusiasm of the contributors themselves, and the editorial assistance provided by Donald Kershaw. To a large extent, the refereeing process was completed during the symposium; only where major changes were suggested was a second refereeing undertaken. Geoffrey Miller acted as editor for the "applied" papers and Christopher Baker acted for those in "numerical analysis". Papers are arranged in order of receipt of the final copy.

The production of a volume of proceedings differs in some respect from that of a journal, notably in the agreement of a page limit upon each contributor. For an instructional summer school it is appropriate to edit contributions to uniform notational conventions, but here we have followed the practice appropriate to a research symposium (and followed in the journals) in leaving the choice of notation to authors' discretion. Contributors were asked to conform to conventions of layout agreed with Academic Press. Whilst the editors have taken the normal steps to satisfy themselves of the quality of the work, the various authors are responsible for the content of their contributions.

It is our pleasure to thank Academic Press for their willing assistance, Joan Gladwell for typing facilities (required in those cases where the original copy was inadequate for printing), and Helen Baker for arranging the preparation of the index.

Manchester and Teddington

CHRISTOPHER T.H. BAKER and GEOFFREY F. MILLER, October 1982.

CONTENTS

ASYMPTOTIC ANALYSIS OF A FINITE ELEMENT METHOD FOR FREDHOLM EQUATIONS OF THE FIRST KIND

C.W. Groetsch, J.T. King and D. Murio

The University of Cincinnati

1. INTRODUCTION

In this paper we present a rigorous asymptotic error analysis of a finite element method for approximating the minimal norm solution of an equation of the form

$$\int_a^b k(s,t)x(t)dt = g(s) \tag{1.1}$$

where k is a given square integrable kernel. For notational convenience we will phrase (1.1) abstractly as

$$Kx = g \tag{1.2}$$

where K is a compact linear operator from a Hilbert space H_1 into a Hilbert space H_2. We do not assume that (1.2) has a unique solution; by x we mean the solution with minimum norm, that is, the unique solution which is orthogonal to the null-space of K. It is well-known that for nondegenerate kernels equation (1.2) is ill-posed, that is, the minimal norm solution x does not depend continuously on the right hand side g. Recognizing the inherent computational difficulties when g is not precisely known, Tikhonov [8] in 1963 proposed what has become known as the regularization method for solving (1.2). The method involves taking as an approximation to x the minimizer x^α of the so called Tikhonov functional

$$F_\alpha(z) = ||Kz - g||^2 + \alpha||z||^2 \tag{1.3}$$

where α is a positive parameter, the regularization parameter. It is easy to see that x^α is given by

$$x^\alpha = (K^*K + \alpha I)^{-1}K^*g \tag{1.4}$$

where K^* is the adjoint of K.

Of course x^α, being the minimizer of F_α over an infinite dimensional space, is not effectively computable. A computable approximation x_m^α can however be readily obtained by minimizing F_α over a finite dimensional subspace V_m of H_1. Such approximations have been studied previously by Natterer [7], Nashed and Song [6] and Marti [3][4] (Marti's algorithm is presented in geometrical terms; see [2] for its interpretation as a regularized finite element method). Our concern in this note is with relating α to m in such a way that x_m^α converges to x as $m \to \infty$ and with the study of the asymptotic order of convergence of $||x_m^\alpha - x||$.

2. THEORY

Throughout this section we will use the symbols $(\cdot,\cdot)$ and $||\cdot||$ to designate the natural inner product and associated norm on each of the spaces H_1 and H_2; we trust that no confusion will result. We assume that $\{V_m\}$ is a sequence of expanding (i.e., $V_m \subseteq V_{m+1}$) finite dimensional subspaces of H_1 whose union is dense in H_1. Let P_m denote the orthogonal projector of H_1 onto V_m. In the discussion below a prominent role is played by the number γ_m defined by

$$\gamma_m = ||K(I - P_m)||$$

which incorporates properties of the kernel and approximation properties of the subspace V_m. Before proceeding further, we note the following fact, whose simple proof is omitted.

LEMMA 1. $\gamma_m \to 0$ as $m \to \infty$.

Recall that x_m^α represents the minimizer of F_α over the subspace V_m. It follows immediately that x_m^α is characterized as that member of V_m which satisfies

$$(Kx_m^\alpha - g, K\xi) + \alpha(x_m^\alpha, \xi) = 0 \tag{2.1}$$

for all $\xi \in V_m$. Define a new inner product $[\cdot,\cdot]$ on H_1 by

$$[v,w] = (Kv,Kw) + \alpha(v,w)$$

and denote by $|\cdot|$ the norm on H_1 which is induced by this inner product (note that $|\cdot|$ is equivalent to $||\cdot||$). In terms of this new inner product, (2.1) may be more economically expressed as

$$[x_m^\alpha - x, \xi] = -\alpha(x,\xi) \tag{2.2}$$

for all $\xi \in V_m$. The infinite dimensional Tikhonov approximation x^α of course satisfies the same condition for all $\xi \in H_1$ and hence

$$[x_m^\alpha - x^\alpha, \xi] = 0 \tag{2.3}$$

for all $\xi \in V_m$. If we denote by P_m the orthogonal (with respect to $[\cdot,\cdot]$) projector of H_1 onto V_m, then from (2.3) we immediately obtain the following simple characterization of x_m^α.

LEMMA 2. $x_m^\alpha = P_m x^\alpha$

Also, since $x_m^\alpha - P_m x \in V_m$ and since $I - P_m$ projects onto the orthogonal (with respect to $[\cdot,\cdot]$) complement of V_m, we are able to establish the following.

LEMMA 3. $||x^\alpha - x_m^\alpha||^2 \leq (1 + \gamma_m^2/\alpha)||(I - P_m)x^\alpha||^2$.

Proof. By Lemma 2 and the definition of P_m we have

$$|x^\alpha - x_m^\alpha|^2 = |x^\alpha - P_m x^\alpha|^2 \leq |x^\alpha - P_m x^\alpha|^2$$
$$= ||K(I - P_m)^2 x^\alpha||^2 + \alpha||x^\alpha - P_m x^\alpha||^2$$
$$\leq (\gamma_m^2 + \alpha)||(I - P_m)x^\alpha||^2,$$

and since $\alpha||\cdot||^2 \leq |\cdot|^2$, the result follows. #

We will now call on the following three well-known (see e.g. [5]) facts about Tikhonov regularization in infinite dimensional spaces:

$$||x^\alpha - x|| \to 0 \qquad \text{as } \alpha \to 0 \tag{2.4}$$

$$||x^\alpha - x|| = 0(\sqrt{\alpha}) \qquad \text{if } x \in R(K^*) \tag{2.5}$$

$$||x^\alpha - x|| = 0(\alpha) \qquad \text{if } x \in R(K^*K). \tag{2.6}$$

By use of the triangle inequality

$$||x_m^\alpha - x|| \leq ||x_m^\alpha - x^\alpha|| + ||x^\alpha - x|| \tag{2.7}$$

along with Lemma 3 and (2.4-6) we now show that the behaviour displayed in (2.4-6) can be exactly mirrored in the finite element setting. In each of the theorems below α is chosen as a function of m, say $\alpha = \alpha_m$, so that $\alpha_m \to 0$ as $m \to \infty$.

THEOREM 1. If $\gamma_m = 0(\sqrt{\alpha_m})$, then $x_m^{\alpha_m} \to x$ as $m \to \infty$.

Proof. This follows directly from (2.7), (2.4) and Lemma 3. #

THEOREM 2. If $x \in R(K^*)$ and $\gamma_m = 0(\sqrt{\alpha_m})$, then $||x_m^{\alpha_m} - x|| = 0(\sqrt{\alpha_m})$.

Proof. If $x = K^*w$, then by (1.4)

$$x^\alpha = (K^*K + \alpha I)^{-1}K^*KK^*w$$
$$= K^*(KK^* + \alpha I)^{-1}KK^*w.$$

However, by the spectral radius formula it follows immediately that

$$||(K^*K + \alpha I)^{-1}KK^*|| \leq 1, \text{ and hence we have}$$

$$||(I - P_m)x^\alpha|| = ||(I - P_m)K^*(KK^* + \alpha I)^{-1}KK^*w|| \leq \gamma_m||w||.$$

Therefore, by Lemma 3,

$$||x^\alpha - x_m^\alpha|| \leq \sqrt{1 + \gamma_m^2/\alpha}\ \gamma_m||w||.$$

Setting $\alpha = \alpha_m$ and using (2.5) and (2.7) gives the result. #

In the same way, using (2.6), we can establish:

THEOREM 3. If $x \in R(K^*K)$ and $\gamma_m = 0(\alpha_m)$, then
$$||x_m^{\alpha_m} - x|| = 0(\alpha_m).$$

Consider now the case in which g is not precisely known but an approximation $\tilde{g}$ satisfying

$$||g - \tilde{g}|| \leq \delta$$

is in hand, where $\delta > 0$ is a known bound for the error in g. The approximation to x using the inexact right hand side will be denoted by $\tilde{x}_m^\alpha$ and is defined to be the minimizer in V_m of F_α with g in (1.3) replaced by $\tilde{g}$. As in (2.1) this is equivalent to

$$(K\tilde{x}_m^\alpha - \tilde{g}, K\xi) + \alpha(\tilde{x}_m^\alpha, \xi) = 0$$

for all $\xi \in V_m$, that is

$$[\tilde{x}_m^\alpha - x, \xi] = -\alpha(x,\xi) - (g - \tilde{g}, K\xi) \qquad (2.8)$$

for all $\xi \in V_m$.

LEMMA 4. $||x_m^\alpha - \tilde{x}_m^\alpha|| \leq ||K||\delta/\alpha$.

Proof. Subtracting (2.8) from (2.5) we have

$$[x_m^\alpha - \tilde{x}_m^\alpha, \xi] = (g - \tilde{g}, K\xi) \quad \text{for all } \xi \in V_m.$$

Setting $\xi = x_m^\alpha - \tilde{x}_m^\alpha$ and using $\alpha||\cdot||^2 \leq |\cdot|^2$, this gives

$$\alpha||x_m^\alpha - \tilde{x}_m^\alpha||^2 \leq |x_m^\alpha - \tilde{x}_m^\alpha|^2$$

$$= (g - \tilde{g}, K(x_m^\alpha - \tilde{x}_m^\alpha)) \leq \delta||K||\ ||x_m^\alpha - \tilde{x}_m^\alpha||. \quad \#$$

Combining Lemma 4 with the inequality

$$||x - \tilde{x}_m^\alpha|| \leq ||x - x_m^\alpha|| + ||x_m^\alpha - \tilde{x}_m^\alpha||$$

and using Theorems (1-3), we obtain the following results. In each case we assume that $m = m_\delta \to \infty$ as $\delta \to 0$.

THEOREM 1.1. If $\alpha = \alpha(m_\delta)$ and $\gamma = \gamma(m_\delta)$ satisfy $\gamma = O(\sqrt{\alpha})$ and $\delta = o(\alpha)$, then $\tilde{x}_m^\alpha \to x$ as $\delta \to 0$.

THEOREM 2.1. If $x \in R(K^*)$, $\alpha = C\,\delta^{2/3}$ and $\gamma = \gamma(m_\delta) = O(\delta^{1/3})$, then $||\tilde{x}_m^\alpha - x|| = O(\delta^{1/3})$.

THEOREM 3.1. If $x \in R(K^*K)$, $\alpha = C\,\delta^{1/2}$ and $\gamma = \gamma(m_\delta) = O(\delta^{1/2})$, then $||\tilde{x}_m^\alpha - x|| = O(\delta^{1/2})$.

3. EXAMPLES

For $r \geq 2$, denote by V_m^r the space of rth order splines on the uniform mesh of width $h = 1/m$, that is, V_m^r consists of functions in $C^{r-2}\,[0,1]$ which are piecewise polynomials of degree $r - 1$. Let H^p denote the Sobolev space of functions $u \in C^{p-1}\,[0,1]$ with $u^{(p-1)}$ absolutely continuous and the norm

$$||u||_p = \left\{ \sum_{i=0}^{p} ||u^{(i)}||^2_{L^2} \right\}^{1/2} .$$

Then V^r_m is a finite dimensional subspace of H^{r-1} which has the following (see e.g. [1]) well-known approximation property: for $u \in H^p$, $p \geq 1$, there is a constant C (independent of h) such that

$$\inf_{\phi \in V^r_m} ||u - \phi||_j \leq C\, h^{\min(p,r)-j}\, ||u||_p, \quad j = 0,1 \qquad (3.1)$$

Also, if $H^1_0 = \{u \in H^1:\ u(0) = u(1) = 0\}$, then (3.1) holds for $u \in H^p \cap H^1_0$ if V^r_m is replaced by $V^r_m \cap H^1_0$. In this section $\gamma^{(r)}_m$ will represent the number γ_m associated with V^r_m.

As a first example we consider zeroth order regularization. In this case we have $H_1 = H_2 = H^0$ $(= L^2)$ and

$$\gamma^{(r)}_m = \sup_{||u||_0 \leq 1} \{\inf_{\phi \in V^r_m} ||K^*u - \phi||_0\}$$

and hence by (3.1)

$$\gamma^{(r)}_m \leq C\, h^{\min(p,r)} \sup_{||u||_0 \leq 1} ||K^*u||_p .$$

However,

$$\frac{d^i}{dt^i}(K^*u)(t) = \int_0^1 \frac{\partial^i k}{\partial t^i}(s,t)u(s)ds$$

and hence

$$||K^*u||_p \leq ||u||_0\, ||k||_{0,p}$$

where

$$||k||^2_{0,p} = \int_0^1 \left\{ \sum_{i=0}^{p} \int_0^1 \left| \frac{\partial^i k}{\partial t^i}(s,t) \right|^2 ds \right\}^{1/2} dt .$$

Therefore $\gamma_m^{(r)} \leq C\, h^{\min(p,r)} ||k||_{0,p}$, where p indicates the smoothness of the kernel and r is the order of the spline approximation. In this case we therefore have, assuming the necessary smoothness on the kernel,

$$\gamma_m^{(r)} = O(h^r) = O(m^{-r}) \quad \text{for } p \geq r .$$

As an instance of higher order regularization, consider the case $H_1 = H_0^1$, $H_2 = L^2$. By H_0^1 we mean the completion of the space

$$D(L) = \{\phi \in C^2\,[0,1]:\ \phi(0) = \phi(1) = 0\}$$

with respect to the inner product

$$<\phi,\psi> = (L\phi,\psi) = (\phi,\psi) + (\phi',\psi')$$

where $L\phi = -\phi'' + \phi$ and $(\cdot,\cdot)$ is the usual L^2 inner product. In this case the Tikhonov functional is

$$F_\alpha(z) = ||Kz - g||_0^2 + \alpha||z||_1^2 .$$

Denote by $K^{\#}$ the adjoint of $K\colon\ H_0^1 \to L^2$ and by G the self-adjoint operator on L^2 generated by the Green's function for the operator L. Then $GLu = u$ for $u \in D(L)$ and

$$<u,GK^*v> = (Lu,GK^*v) = (GLu,K^*v) = (u,K^*v) = (Ku,v)$$

that is, $K^{\#} = GH^*$. Now,

$$\gamma_m^{(r)} = \sup_{||v||_0 \leq 1} \left\{ \inf_{\phi \in V_m^r \cap H_0^1} ||K^{\#}v - \phi||_1 \right\}$$

$$\leq C\, h^{\min(p,r)-1} \sup_{||v||_0 \leq 1} ||K^{\#}v||_p .$$

However, from standard estimates for boundary value problems,

$$||v||_p \leq C||Lv||_{p-2}, \quad v \in H^p,\ p \geq 2 ,$$

or equivalently

$$||Gv||_p \leq C||v||_{p-2}$$

Therefore

$$||K^{\#}v||_p = ||GK^{*}v||_p \leq C||K^{*}v||_{p-2}$$

and hence

$$\gamma_m^{(r)} \leq C\, h^{\min(p,r)-1}\, ||k||_{0,p-2}\ .$$

That is,

$$\gamma_m^{(r)} = O(h^{r-1}) = O(m^{1-r}) \quad \text{for } p \geq r\ .$$

To illustrate the asymptotic orders derived above consider the symmetric kernel given by $k(s,t) = s(1 - t)$, $0 \leq s \leq t \leq 1$. As example I we take $g(s) = (s^4 - 2s^3 + s)/24$ for which the minimal L^2-norm solution of (1.1) is $x(t) = (t - t^2)/2 = K1(t)$. For example II we take $g(s) = (-s^6 + 3s^5 - 5s^3 + 3s)/720$ and $x(t) = (t^4 - 2t^3 + t)/24 = K^2 1(t)$. As basis functions we use piecewise linear splines on a mesh of width $h = 1/m$ and hence $\gamma_m = O(h^2)$. Taking $\alpha_m = 10^{-3}h^4$, Theorem 2 predicts an error of order $O(h^2)$ for example I. If $\alpha_m = 10^{-3}h^2$, then Theorem 3 predicts an error of order $O(h^2)$ for example II. Below we display the computed L^2-errors rounded to three figures (the number in parentheses indicates multiplication by that power of 10) and the computed exponent in the error to two decimal places for each example.

Example I

m	16		32		64
error	.159 (-3)		.381 (-4)		.931 (-5)
		2.06		2.03	
			2.04		

Example II

m	16		32		64
error	.139 (-4)		.346 (-5)		.865 (-6)
		2.00		2.00	
			2.00		

Both examples we also tested with random error of amplitude δ in the right hand side to illustrate Theorems 2.1 and 3.1. Using $\delta = m^{-6}$ in example I and $\delta = m^{-4}$ in example II the theorems predict errors of the order $O(\delta^{1/3})$ and $O(\delta^{1/2})$, respectively. The computed results are as follows:

Example I

m	16		32		64
error	.169 (-3)		.381 (-4)		.921 (-5)
		.35		.34	
			.34		

Example II

m	16		32		64
error	.647 (-2)		.145 (-2)		.345 (-3)
		.54		.52	
			.53		

REFERENCES

1. BABUSKA, I. and AZIZ, A.K., Survey Lectures on the Mathematical Foundations of the Finite Element Method. pp. 3-345 of A.K. Aziz (Ed.), *The Mathematical Foundations of the Finite Element Method with Applications to Partial Differential Equations*. Academic Press, New York (1972).

2. GROETSCH, C.W., On a Regularization-Ritz Method for Fredholm Equations of the First Kind. Numerical Analysis Report 56, University of Manchester, 1980. To appear in *J. Integral Equations*.

3. MARTI, J.T., An Algorithm for Computing Minimum Norm Solutions of Fredholm Integral Equations of the First Kind. *SIAM J. Numer. Anal.* 15, 1071-1076 (1978).

4. MARTI, J.T., On the Convergence of an Algorithm for Computing Minimum Solutions of Ill-Posed Problems. *Maths. Comp.* 34, 521-527 (1980).

5. MOROZOV, V.A., Choice of Parameter for the Solution of Functional Equations by the Regularization Method. *Soviet Math. Doklady* 8, 1000-1003 (1967).

6. NASHED, M.Z. and SONG, M., Finite Element Method for Ill-Posed Linear Operator Equations. Preprint.

7. NATTERER, F., The Finite Element Method for Ill-Posed Problems. *RAIRO-Analyse Numerique* 11, 271-278 (1977).

8. TIKHONOV, A.N., The Regularization of Incorrectly Posed Problems. *Soviet Math. Doklady* 4, 1624-1627 (1963).

A COMPARISON OF VARIOUS INTEGRAL EQUATIONS FOR TREATING THE DIRICHLET PROBLEM

Søren Christiansen

The Technical University of Denmark

1. INTRODUCTION

We shall here investigate some integral equations used for treating the two-dimensional interior Dirichlet boundary value problem for Laplace's equation for a simply connected domain D with boundary denoted Γ^B

$$\Delta\Psi(x,y) = 0, \quad (x,y) \in D \tag{1-1a}$$

$$\Psi(x,y) = f(P), \ P = (x,y) \in \Gamma^B. \tag{1-1b}$$

On Γ^B the boundary value f is prescribed. The main part of the investigation is concentrated on four types of equations of the first kind (§ 1.1), while an equation of the second kind (§ 1.2) is stated for reason of comparison.

1.1 Equations of the First Kind

An auxiliary curve, denoted Γ^A, is either placed outside Γ^B, indicated by $\Gamma^A > \Gamma^B$, or placed coinciding with Γ^B, indicated by $\Gamma^A = \Gamma^B$. For the treatment of (1-1) four types of integral equations of the first kind are formulated in the following scheme. Here two further curves Γ and $\hat{\Gamma}$ are introduced; the position of the curves will be chosen below. (The use of the four curves Γ^A, Γ^B, Γ, and $\hat{\Gamma}$ was introduced in [4].)

$$\frac{1}{2\pi}\oint_{\Gamma} \ell n \frac{1}{\rho(\hat{Q},Q)} \Omega(Q) \, ds_Q + K\frac{\omega}{K} = T\{f;\hat{Q}\}; \quad \left\{\begin{matrix}\hat{Q} \in \hat{\Gamma} \\ \\ Q \in \Gamma\end{matrix}\right\}, \tag{1-2a}$$

$$W\oint_{\Gamma} \Omega(Q) \, ds_Q = W\alpha \quad ; \quad Q \in \Gamma . \tag{1-2b}$$

Here $\rho(\hat{Q},Q)$ is the distance between the two points $\hat{Q}$ and Q. The unknown quantities are the function Ω and the scalar ω, while the known quantities are the function f and the scalar α. The scalars K and W are to be chosen below. The system (1-2) with K = 0 and W = 0 was treated in [4].

The scheme (1-2) contains the two main procedures for deriving integral equations of the first kind, viz. (i) single layer potentials ([11] § 2.5, § 4.1), and (ii) Green's third identity ([11]§ 3.2, § 4.4). For each procedure two types are obtained depending upon whether $\Gamma^A > \Gamma^B$ or $\Gamma^A = \Gamma^B$.

1) Single layer potentials (cf. [4] Method I): Put $\Gamma := \Gamma^A$ and $\hat{\Gamma} := \Gamma^B$. The derivation shows that $T\{f;\hat{Q}\} = f(\hat{Q})$. When $\Gamma^A = \Gamma^B$ the system (1-2) has been treated by Hsiao & MacCamy [9], with K = 1 and W = 1, who show it is necessary to introduce the scalar ω and the equation (1-2b) in order to ensure existence and uniqueness. It is here conjectured that existence and uniqueness also hold when $\Gamma^A > \Gamma^B$ for $K \neq 0$ and $W \neq 0$. In case $\Gamma^A = \Gamma^B$ the choice of K and W is considered briefly in [6].

2) Green's third identity (cf. [4] Method II): Put $\Gamma := \Gamma^B$ and $\hat{\Gamma} := \Gamma^A$. The derivation shows that $T\{f;\hat{Q}\}$ is a certain integral transformation of f; that $\omega = 0$; that $\Omega = \partial\psi/\partial n$, where $\partial/\partial n$ indicates differentiation in the direction of the normal to Γ^B, such that $\alpha = 0$. Because $\omega = 0$ one may put K = 0. In this case the system (1-2) has a unique solution when $W \neq 0$ ([1] when $\Gamma^A = \Gamma^B$ and [2] when $\Gamma^A > \Gamma^B$). Introducing the term $K\omega$ leads to the system investigated by Hsiao & MacCamy [9]. We therefore use the system (1-2) with $K \neq 0$ and $W \neq 0$ also in the present case. When K = 0 the choice of W is considered in [5] for $\Gamma^A = \Gamma^B$, and in [3] for $\Gamma^A > \Gamma^B$.

1.2 Equation of the Second Kind

Equation (1-1) can also be formulated as an integral equation of the second kind ([11]§2.6)

$$\frac{1}{2}\Omega(P) + \frac{1}{2\pi}\oint_{\Gamma^B} \frac{\partial}{\partial n} \ln \frac{1}{\rho(P,Q)} \Omega(Q)\, ds_Q = f(P); \quad \left\{ \begin{array}{l} P \in \Gamma^B \\ Q \in \Gamma^B \end{array} \right\} \quad (1\text{-}3)$$

2. FORMULATION OF THE PROBLEM

The primary purpose of this paper is to investigate to what extent the various equations in the system (1-2) are suited to numerical solution.

For the numerical solution of (1-2) we can use a quadrature and collocation method, with the collocation points $\hat{Q}_i \in \hat{\Gamma}$

$i = 1,2,\cdots,N$, can be used. On Γ is chosen the points $Q_j \in \Gamma$, $j = 1,2,\cdots,N$.

The unknowns are $[\bar{\Omega}_1, \bar{\Omega}_2, \cdots, \bar{\Omega}_N, \frac{\bar{\omega}}{K}]^T$, where $\{\bar{\Omega}_j\}$ are approximations to Ω at the points $\{Q_j\}$, and $\bar{\omega}$ is an approximation to ω.

The left hand side of (1-2a) is thus replaced by the N sums

$$\sum_{j=i}^{N} \bar{M}_{ij}\bar{\Omega}_j + K(\frac{\bar{\omega}}{K}),$$

where the square matrix $\bar{\underset{\sim}{M}}$ may be constructed as in [4] § 4.

Furthermore the left hand side of (1-2b) is replaced by the sum

$$W\sum_{j=1}^{N} \bar{h}_j\Omega_j + 0(\frac{\bar{\omega}}{K}),$$

where $\{\bar{h}_j\}$ are (approximations to) the arc lengths between consecutive points on Γ.

Similarly the right hand side of (1-2) can be replaced by a known column vector.

We hereby obtain a system of linear algebraic equations with the matrix

$$\bar{\underset{\sim}{B}} = \left[\begin{array}{ccc|c} & & & K \\ & & & \cdot \\ & \{\bar{M}_{ij}\} & & \cdot \\ & & & \cdot \\ & & & K \\ \hline W\bar{h}_1 & \cdots & W\bar{h}_N & 0 \end{array}\right] \tag{2-1}$$

The stability of the equations with the matrix $\bar{\underset{\sim}{B}}$ may be characterized by the condition number ([7] Eq. 5.5.10)

$$\bar{\kappa}_p = \| \bar{\underset{\sim}{B}} \|_p \; \| \bar{\underset{\sim}{B}}^{-1} \|_p \tag{2-2}$$

where $\| \cdot \|_p$ denotes the p-norm of the matrix. It turns out to be convenient to apply the 2-norm condition number, in which case the definition (2-2) can also be formulated as ([8] p. 205)

$$\bar{\kappa}_2 = \frac{\max\{\bar{\sigma}_k\}}{\min\{\bar{\sigma}_k\}} \; ; \quad k = 0,1,2,\cdots,N \tag{2-3}$$

where $\{\bar{\sigma}_k\}$ are the singular values of the matrix $\underset{\sim}{\bar{B}}$ ([8] p. 203).

The purpose of the paper is to investigate how κ_2 is influenced by (i) the mutual position of the two curves Γ and $\hat{\Gamma}$, (ii) a geometrical scaling of the two curves, whereby all lengths are multiplied by the same positive factor μ, (iii) the value of K, and (iv) the value of W.

For the sake of comparison the condition number for the matrix derived from (1-3) is found in a special case using an elementary method.

3. ANALYTICAL INVESTIGATIONS

In the special case where the curve(s) in question is/are (a) circle(s) it is possible to carry out some analytical investigations which in more general cases may be used as guidelines.

3.1 Equations of the First Kind

Let the curves Γ and $\hat{\Gamma}$ be two concentric circles C and $\hat{C}$ with radii c and $\hat{c}$, respectively, and let the points on C and $\hat{C}$ be equally spaced and such that Q_i and $\hat{Q}_i$, $i = 1,2,\cdots,N$, are placed on the same radius. For simplicity N is assumed to be even. Then the matrix $\underset{\sim}{\bar{M}}$ in (2-1) is symmetric and circulant, and has the eigenvalues, cf. [5] § 3.2

$$\bar{\lambda}_o, \bar{\lambda}_1, \cdots, \bar{\lambda}_{N/2} \qquad (3\text{-}1)$$

where $\bar{\lambda}_o$ and $\bar{\lambda}_{N/2}$ are simple while the remaining are double. These results hold irrespective of the quadrature rule used for deriving $\underset{\sim}{\bar{M}}$. Furthermore all the elements $\{\bar{h}_i\}$ in (2-1) are equal with the common value $\bar{h}$. We here choose $\bar{h} = 2\pi c/N$. These elements are multiplied by the weight W, and we put

$$H := W\bar{h} = W\,\frac{2\pi c}{N}\,. \qquad (3\text{-}2)$$

We now have a matrix of the form

$$\underset{\sim}{\bar{B}} = \left[\begin{array}{ccc|c} & & & K \\ & & & \cdot \\ & \underset{\sim}{\bar{M}} & & \cdot \\ & & & \cdot \\ & & & K \\ \hline H & \cdot\ \cdot\ \cdot & H & 0 \end{array}\right] \qquad (3\text{-}3)$$

with $\underset{\sim}{\overline{M}}$ being symmetric and circulant. For such a matrix the singular values can be expressed in closed form in terms of the eigenvalues (3-1) and H and K ([3] Eq. 5-2)

$$\overline{\sigma}_o^{\pm} = \left[\tfrac{1}{2}\left\{ \overline{\lambda}_o^2 + N(H^2+K^2) \pm \sqrt{(\overline{\lambda}_o^2 + N(H+K)^2)(\overline{\lambda}_o^2 + N(H-K)^2)}\right\}\right]^{\frac{1}{2}} \quad (3\text{-}4a)$$

$$\overline{\sigma}_k = |\overline{\lambda}_k|, \; k = 1,2,\cdots,N/2. \quad (3\text{-}4b)$$

The matrix $\underset{\sim}{\overline{M}}$ is derived from the integral operator in (1-2a), and in the present case of concentric circles with radii c and $\hat{c}$ the integral operator has the eigenvalues ([4] Eq. 5-7) (without a bar because they refer to (1-2))

$$\lambda_o = -c\,\ell nR \quad (3\text{-}5a)$$

$$\lambda_m = \frac{c}{2m}\gamma^{-m}; \quad m = 1,2\cdots \quad (3\text{-}5b)$$

where

$$R = \max(c,\hat{c}), \quad r = \min(c,\hat{c}), \quad (3\text{-}5c)$$

$$\gamma = R/r \geqq 1. \quad (3\text{-}5d)$$

(The two procedures of § 1.1 correspond to $\hat{c} \leqq c$ and $\hat{c} \geqq c$, respectively).

We now assume that $\{\overline{\lambda}_k\}$ can be approximated by $\{\lambda_m\}$ to a sufficiently high degree of accuracy. The validity of this assumption will depend upon the method used for the computation of the matrix elements $\{\overline{M}_{ij}\}$. In ([4] Appendix B) a comparison of the two sets of eigenvalues has been carried through for a specific method of computation. We therefore replace $\{\overline{\lambda}_k\}$ in (3-4) by the convenient expressions (3-5). Thus we obtain tractable closed form expressions for the singular values denoted $\{\sigma_m\}$, where σ without a bar refers to (1-2). These functions depend upon R with γ parameter; i.e. $\sigma_m(R;\gamma)$. Having the expressions $\{\sigma_m(R;\gamma)\}$ it is straightforward to compute

$$\kappa_2 = \kappa_2(R;\gamma) = \frac{\max\{\sigma_m(R;\gamma)\}}{\min\{\sigma_m(R;\gamma)\}}\; ; \; m = 0^+,0^-,1,2,\ldots,\frac{N}{2} \quad (3\text{-}6)$$

which is assumed to be an approximation to $\overline{\kappa}_2$.

As an illustration we consider the matrix (3 - 3) when W = 1, K = 1. From (3 - 2) we get H, and having H and K we here find after some manipulation, for $N \geqq 8$:

$$\kappa_2(1;\gamma) = N^{3/2} \cdot \begin{cases} \gamma^{N/2} & ; \quad \hat{c} \leq c , \qquad (3\text{-}7a) \\ \gamma^{N/2+1} & ; \quad c \leq \hat{c} . \qquad (3\text{-}7b) \end{cases}$$

This result indicates an unfavorable growth of κ_2 with increasing N, especially when $\gamma > 1$. It is of interest to investigate whether H and K can be chosen such that $\kappa_2(1;\gamma)$ increases less unfavorably with N. We now show that this is indeed possible.

By further manipulation of the expressions we find that the following choice is advantageous:

$$H = K, \qquad (3\text{-}8a)$$

$$H = c\eta N^{-\frac{1}{2}}, \qquad (3\text{-}8b)$$

$$\eta = \frac{1}{2\nu\gamma^{\nu}} \qquad (3\text{-}8c)$$

where ν is a positive constant to be determined suitably. The choice (3-8) results in

$$\kappa_2(1;\gamma) = \frac{N}{2}\,\gamma^{N/2-1}, \text{ when } 1 \leq \nu \leq \frac{N}{2}, \text{ and } \hat{c} \gtrless c, \qquad (3\text{-}9)$$

which shows a substantial reduction of the condition number as compared with (3-7). As seen from (3-9) $\kappa_2(1;\gamma)$ is independent of ν , for ν in a certain interval. In contrast $\kappa_2(R;\gamma)$, for $R \neq 1$, computed from (3-6), depends strongly on ν, in the same interval. An actual calculation of $\kappa_2(R;2)$, with N = 4, for different values of ν (also values outside the interval) shows that $\nu = 1$ gives a fairly large interval around R = 1 with a reasonably small value of κ_2. Therefore we choose $\nu = 1$ which we denote the optimal value. From this value of ν we obtain by combining (3-2) and (3-8) the following values, denoted optimal values, for K, H, and the weight W:

$$K^* = H^* = W^* \frac{2\pi c}{N} , \qquad (3\text{-}10a)$$

$$W^* = \frac{N^{\frac{1}{2}}}{2\pi \cdot 2\gamma} . \qquad (3\text{-}10b)$$

With these optimal values the optimal condition number (3-9) is obtained provided $R = \max(c,\hat{c}) = 1$. In order to obtain optimal results it is therefore necessary to scale the lengths such that the radius of the external circle equals one.

The optimal condition number (3-9) reveals the dependence upon the mutual position of the two curves C and $\hat{C}$. The smallest value of κ_2 is obtained when $\gamma = 1$, which corresponds to the limiting case when C and $\hat{C}$ coincide. In this case

$$\kappa_2(1;1) = \frac{N}{2} . \tag{3-11}$$

3.2 Equations of the Second Kind

The numerical solution of (1-3) can also be carried out using a quadrature and collocation method leading to a square $N \times N$ matrix for which the condition number can be determined from a formula analogous to (2-3). In particular, when the boundary curve Γ^B is a circle C, with radius c, and the N collocation points are taken to be equally spaced the square matrix is symmetric and circulant with the eigenvalues, cf. (3-1), $\overline{\Lambda}_o, \overline{\Lambda}_1, \cdots, \overline{\Lambda}_{N/2}$, where $\overline{\Lambda}_o$ and $\overline{\Lambda}_{N/2}$ are simple while the remaining are double. In this case the formula analogous to (2-3) becomes

$$\overline{\kappa}_2 = \frac{\max\{\overline{\Lambda}_k\}}{\min\{\overline{\Lambda}_k\}} , \ k = 0,1,\cdots,\frac{N}{2} . \tag{3-12}$$

The matrix is derived from the integral operator on the left hand side of (1-3). When polar coordinates are introduced in (1-3) with $Q = (c,\theta)$ and $P = (c,\theta_1)$ the left hand side of (1-3) leads to an eigenvalue problem

$$\frac{1}{2}\chi_m(\theta_1) + \frac{1}{2\pi}\int_o^{2\pi} \frac{1}{2c}\chi_m(\theta)\, c d\theta = \Lambda_m \chi_m(\theta_1)$$

with eigenvalues $\{\Lambda_m\}$ and eigenfunctions $\{\chi_m(\theta)\}$

$$\Lambda_o = 1 , \ \chi_o(\theta) = 1$$

$$\Lambda_m = \frac{1}{2} , \ \chi_m(\theta) = \begin{Bmatrix} \cos m\Theta \\ \sin n m\Theta \end{Bmatrix} , \ m = 1,2,\cdots .$$

When we assume that $\{\overline{\Lambda}_k\}$ can be approximated by $\{\Lambda_m\}$, we can insert the approximation in (3-12) and obtain an approximation to $\overline{\kappa}_2$, namely

$$\kappa_2 = 2, \tag{3-13}$$

which is independent of N and c.

4. NUMERICAL COMPUTATIONS

The results of § 3 are based upon circular curves with equally spaced points using the eigenvalues of the corresponding integral operator. Under less restrictive assumptions the condition number has to be computed numerically, which we here shall do for the first kind equation.

Following ([4] § 6) we consider two ellipses with coinciding axes, which we shall call concentric ellipses:

$$E : x = a\cos\theta,\ y = b\sin\theta;\ 0 \leqq \theta \leqq 2\pi, \tag{4-1a}$$

$$\hat{E} : x = \hat{a}\cos\hat{\theta},\ y = \hat{b}\sin\hat{\theta};\ 0 \leqq \hat{\theta} \leqq 2\pi. \tag{4-1b}$$

On E are chosen the N points Q_j described by $\theta_j := (j-1)2\pi/N$, $j = 1,2,\cdots,N$. On $\hat{E}$ are chosen the N collocation points $\hat{Q}_i$ described by $\hat{\theta}_i := (i-1)2\pi/N$, $i = 1,2,\cdots,N$. Each ellipse is characterized by the axis ratio and by the mean semi-axis:

$$E : \varepsilon = b/a,\quad d = (a+b)/2 \tag{4-2a}$$

$$\hat{E} : \hat{\varepsilon} = \hat{b}/\hat{a},\quad \hat{d} = (\hat{a}+\hat{b})/2\ . \tag{4-2b}$$

For the characterization of the mutual position of the two curves we introduce

$$\alpha = \hat{d}/d \tag{4-3}$$

whereby the two methods of § 1.1 are characterized by $\alpha \leqq 1$ and $\alpha \geqq 1$, respectively. The parameter γ, (3-5d), is generalized as $\max(\alpha,\alpha^{-1})$. The radius $R = \max(c,\hat{c})$, (3-5c), is generalized as the length

$$L = \max(d,\hat{d})\ . \tag{4-4}$$

When the points $\{Q_j\}$ and $\{\hat{Q}_i\}$ on the two curves are given, the matrix elements $\{\tilde{M}_{ij}\}$ can be computed (e.g. according to [4] Eq. 4-15). If the elements $\{\overline{h}_j\}$ were equal to $\overline{h}$, then (3-8a) would lead to the value $K = W\overline{h}$. In case the elements $\{\overline{h}_j\}$ are not equal, it seems natural to use their mean value. This choice combined with (3-10) leads to the scalar column constant

$$K^* = W^* \frac{1}{N}\sum_{j=1}^{N} \overline{h}_j\ . \tag{4-5}$$

When we have constructed a square (N+1) × (N+1) matrix we compute the singular values $\{\bar{\sigma}_k\}$ and the condition mumber κ_2, (2-3), by means of [10].

Actual numerical computations of $\bar{\kappa}_2$ have been carried out, giving the graph of $\bar{\kappa}_2$ as a function of L, (4-4). In case the values W^*, (3-10b), and K^*, (4 - 5), are used it turns out that the graph has the form of an upright V, with slightly curved "legs", situated such that the tip corresponds to L ≅ 1, which means that the minimum value of $\bar{\kappa}_2$ is obtained for L ≅ 1. (In case the values W^* and K^* are not used, the graph does not have such a simple V-shaped form, and an increase of the values of $\bar{\kappa}_2$ is also obtained.) It is of interest to solve the linear algebraic equations for such a scaling that it corresponds to a minimum value of $\bar{\kappa}_2$. Therefore we do not present the graphs of $\bar{\kappa}_2$, but only the value of $\bar{\kappa}_2$ for L = 1.

We consider cases with separated curves (i.e. $\gamma > 1$), and cases with coinciding curves (i.e. $\gamma = 1$). We choose $\gamma = 2$, and consider here both $\alpha = 0.5$ (i.e. single layer potentials) and $\alpha = 2.0$ (i.e. Green's third identity). When $\gamma = 1$ we naturally have $\varepsilon = \hat{\varepsilon}$, while when $\gamma > 1$ this need not be the case. Furthermore it turns out that the graph of $\bar{\kappa}_2$, using W^* and K^*, for concentric ellipses also depends upon ε and $\hat{\varepsilon}$. (The results of § 3.1 were derived for $\varepsilon = 1$ and $\hat{\varepsilon} = 1$.) When the ellipses are confocal (which case comprises the case $\varepsilon = 1$ and $\hat{\varepsilon} = 1$) relatively small values of κ_2 are obtained, while when using non-confocal ellipses larger values of κ_2 are obtained.

In the tables I, II, and III is given the condition number of the matrix derived from the integral equation of the first kind (1-2) when the curves Γ^A and Γ^B are concentric ellipses for three values of α, (4-3).

Table I (with $\alpha = 1$, i.e. coinciding ellipses, $\gamma = 1$) shows the effect of W and K; the unmodified equation (W = 1, K = 1), and the modified equation ($W = W^*$, $K = K^*$) are considered both for circles (b/a = 1.0) and ellipses (b/a = 0.5). For comparison the analytical results for a circle with c = 1 are also given. We notice here the importance of using $W = W_*$ and $K = K_*^*$. Consequently in Table II and Table III we use $W = W^*$ and $K = K^*$.

Table II (with $\alpha = 0.5$, i.e. $\gamma = 2.0$) shows the effect of the auxiliary curve; the boundary curve has $\hat{\varepsilon} = 1.0$ or $\hat{\varepsilon} = 0.5$, while the auxiliary curve has ε as shown. Here $\hat{\varepsilon} = 1.0$, $\varepsilon = 1.0$ and $\hat{\varepsilon} = 0.5$, $\varepsilon = 0.8462$ are confocal. For comparison the analytical results for concentric circles with c = 1 are also given. We notice here the importance of choosing a suitable curve Γ^A.

Table III (with $\alpha = 2.0$, i.e. $\gamma = 2.0$) shows the effect of the auxiliary curve; the boundary curve has $\varepsilon = 1.0$ or $\varepsilon = 0.5$ while the auxiliary curve has $\hat{\varepsilon}$ as shown. Here $\varepsilon = 1.0$, $\hat{\varepsilon} = 1.0$, and $\hat{\varepsilon} = 0.5$, $\varepsilon = 0.8462$ are confocal. For comparison the analytical results for concentric circles with c = 1 are also given. We notice here the importance of choosing a suitable curve Γ^A.

TABLE I

Condition number for concentric ellipses: $\alpha = 1.0$

N	Circle c = 1 Analytical results: (3-7)	(3-11)	Ellipses: d = (a+b)/2 = 1 Numerical results (2-3) [10]: W = 1, K = 1, $\frac{b}{a} = 1.0$	$\frac{b}{a} = 0.5$	W = W*, K = K*, $\frac{b}{a} = 1.0$	$\frac{b}{a} = 0.5$
4		2	10.9	11.0	1.8	2.0
8	22.6	4	19.6	21.2	3.5	4.3
12	41.6	6	35.8	42.1	5.1	7.1
16	64.0	8	55.0	68.1	6.9	10.0
20	89.4	10	76.8	98.5	8.6	13.0
24	117.6	12	100.9	132.5	10.3	15.9

TABLE II

Condition number for concentric ellipses: $\alpha = 0.5$

N	$\hat{\varepsilon} = 1.0$			$\hat{\varepsilon} = 0.5$	
	$\varepsilon = 1.0$ (3-9)	(2-3)	$\varepsilon = 0.5$ (2-3)	ε=0.8462 (2-3)	$\varepsilon = 0.5$ (2-3)
4	4	3.2	5.5	3.2	4.6
8	32	24.5	71.5	29.5	47.7
12	192	147	1465	186	695

TABLE III

Condition number for concentric ellipses: $\alpha = 2.0$

N	$\varepsilon = 1.0$			$\varepsilon = 0.5$	
	$\hat{\varepsilon} = 1.0$ (3-9)	(2-3)	$\hat{\varepsilon} = 0.5$ (2-3)	$\hat{\varepsilon}$=0.8462 (2-3)	$\hat{\varepsilon} = 0.5$ (2-3)
4	4	3.5	8.0	3.6	7.5
8	32	26.6	173.2	30.7	97.7
12	192	156.3	2886	188.3	1354

5. CONCLUDING REMARKS

Primarily we have investigated four types of integral equations of the first kind, written in a uniform way in (1-2), and especially the matrix (2-1) which results when (1-2) is to be solved numerically. This matrix will attain a relatively small condition number (CN) when the following procedures are carried out:

P1) In (1-2a) the unknown ω/K^* is used (instead of ω or ω/K) with K^* given in (4-5).

P2) Eq. (1-2b) is multiplied by the weight W^* given in (3-10b).

P3) When (1-2) is used with an auxiliary curve Γ^A, the choice of the position of this curve may influence strongly on the CN.

P4) Having chosen Γ^A suitably the two curves Γ^A and Γ^B have to be scaled such that the exterior mapping radius of Γ^A, denoted $d\{\Gamma^A\}$, is one, See below.

In order to get an approximation to $d\{\Gamma^A\}$, we approximate the curve Γ^A by an ellipse with semi-axes $\bar{a}$ and $\bar{b}$, and use $d\{\Gamma^A\} = (\bar{a}+\bar{b})/2$ as an approximation, cf. ([1] Eq. 3.2.b).

Having performed the procedures, P1 - P4, the matrix has a CN which can be approximated by (3-9), where for γ_B is used $d\{\Gamma^A\}/d\{\Gamma^B\} > 1$, or when $d\{\Gamma^A\} = 1$, we use $\gamma = (d\{\Gamma^B\})^{-1}$.

A much smaller CN is obtained when Γ^A coincides with Γ^B in which case $\gamma = 1$, and having performed the above procedures we get a CN equal to $\frac{N}{2}$ or O(N), in accordance with [6]. Therefore the scheme used by Hsiao & MacCamy [9], with coinciding curves, seems to be the preferable integral equation of the first kind. Here we shall not make a choice between the two procedures for deriving (1-2), cf. § 1.1; if the normal derivative $\partial\psi/\partial n$ is of special interest then the second procedure (Green's third identity) seems preferable because the unknown Ω is equal to $\partial\psi/\partial n$.

By the above procedures we can not obtain a CN smaller than O(N). If this is thought to be too large then the integral equation of the second kind (1-3) has to be used because it gives a matrix with a CN independent of N.

ACKNOWLEDGEMENTS

The present work was finished while the author held a research scholarship at the Technical University of Denmark.

Dr. phil. Niels Arley is thanked for reading and commenting on the manuscript.

The numerical calculations were carried out at the Northern Europe University Computing Center (NEUCC), The Technical University of Denmark.

REFERENCES

1. CHRISTIANSEN, S., Integral Equations without a Unique Solution can be made Useful for Solving some Plane Harmonic Problems. *Jour. Inst. Math. Applications* 16, 143-159 (1975).
2. CHRISTIANSEN, S., On Kupradze's Functional Equations for Plane Harmonic Problems. pp. 205-243 of R.P. Gilbert & R.J. Weinacht (Eds.), *Function theoretic methods in differential equations*. Research Notes in Mathematics 8, Pitman Pub. London (1976).
3. CHRISTIANSEN, S., Numerical Treatment of an Integral Equation Originating from a Two-Dimensional Dirichlet Boundary Value Problem. pp. 73-91 of J. Albrecht & L. Collatz (Eds.), *Numerical Treatment of Integral Equations*. International Series of Numerical Mathematics, ISNM 53, Birkhaüser Verlag, Basel (1980).
4. CHRISTIANSEN, S., Condition Number of Matrices Derived from Two Classes of Integral Equations. *Math. Meth. Appl. Sci.* 3, 364-392 (1981).
5. CHRISTIANSEN, S., On Two Methods for Elimination of Nonunique Solutions of an Integral Equation with Logarithmic Kernel. *Applicable Analysis* 13, 1-18 (1982).
6. CHRISTIANSEN, S., Numerical Investigation of an Integral Equation of Hsiao and MacCamy, *Zeit. Angew. Math. Mech.* 63, (1983).
7. DAHLQUIST, G. & Å BJÖRCK, *Numerical Methods*. Prentice-Hall, Inc.; Englewood Cliffs (1974).
8. FORSYTHE, G.E., M.A. MALCOLM & C.B. MOLER, *Computer Methods for Mathematical Computations*. Prentice-Hall, Inc.; Englewood Cliffs (1977).
9. HSIAO, G & R.C. MACCAMY, Solution of Boundary Value Problems by Integral Equations of the First Kind. *SIAM Review* 15, 687-705 (1973).
10. IMSL, *Singular Value Decomposition of a Real Matrix*. International Mathematical & Statistical Libraries, Inc., Houston, LSVDF. IMSL Library, Ed. 8, (1981).
11. JASWON, M.A. & G.T. SYMM. *Integral Equation Methods in Potential Theory and Elastostatics*, Academic Press, London (1977).

OPTIMAL GRID CONTACTS ON SOLAR CELLS

G. De Mey

Laboratory of Electronics
Ghent State University
Sint-Pietersnieuwstraat 41
9000 GHENT
BELGIUM.

1. INTRODUCTION

A solar cell is a thin semiconductor layer capable of converting the absorbed light into electric power. Physical details of this conversion process will be omitted here. It is only necessary to know that the solar cell requires two metallic contacts for the electric current transport. The back contact at the non illuminated side covers the whole surface. The front contact shows a finger pattern. If the front contact covers the whole surface, no light can be absorbed in the underlying semiconductor. On the other hand, if the front contact fingers are too far away from each other, the electric current generated by the absorbed light can hardly reach the front contact, decreasing the efficiency of the solar cell. Hence, an optimum between these two extreme cases may be found.

Looking at solar cells available on the market, one observes a wide variety of front contact patterns (fig. 1). The purpose of this contribution is to compare these various geometries. It requires the numerical solution of a partial differential equation, as will be outlined below. As several geometries are involved, the boundary integral equation method (BIEM) is the best suited for the numerical evaluation of the problem [1][2]. The original given PDE with the boundary conditions is then replaced by an equivalent integral equation along the boundary of the given geometry. The BIEM has been applied succesfully to analyse several problems in semiconductor components [3][4][5][6].

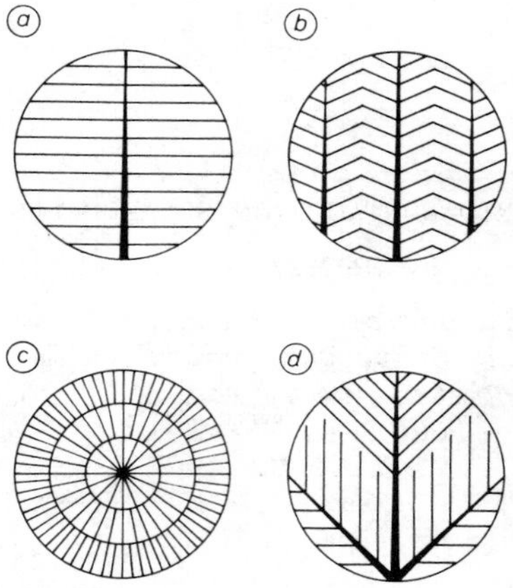

Fig.1. Various grid patterns used on solar cells.

2. BASIC EQUATIONS

Fig. 2 shows a cross sectional view of a solar cell. A potential difference V_o is applied between the back contact and the front contact. In each elementary part of the cell a current density J(V) is generated. When this current reaches the front surface, it has to be declined towards the grid contacts as shown on fig. 2. As the front surface has a non negligible electrical resistance (expressed by the so-called square resistance $R_\square$) the potential V(x,y) across the solar cell will be position dependent. Without going into the physical details, we just mention that the potential V satisfies the following equation :

$$\nabla^2 V = R_\square J(V) \qquad \text{in the (x,y) plane} \qquad (1)$$

with the boundary condition :

$$V = V_o \qquad \text{at a grid contact finger} \qquad (2)$$

J(V) is the current density delivered by a solar cell under influence of a voltage V [7]

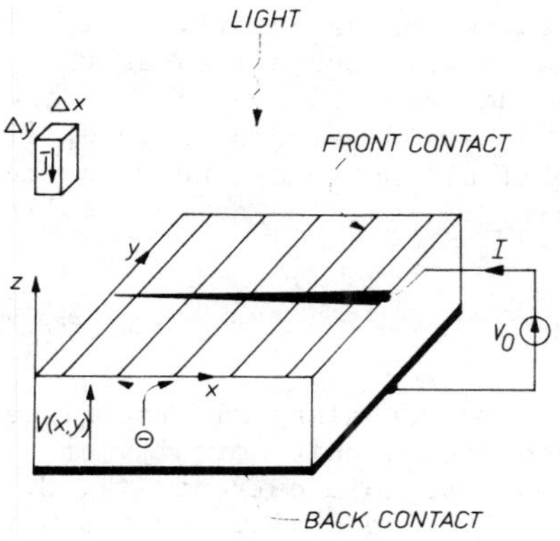

Fig.2. Cross sectional view of solar cell.

$$J(V) = -J_L + J_o(e^{qV/kT} - 1) \tag{3}$$

Besides V, all parameters in the right hand member of (3) can be considered as constants. Putting (3) in (1), one gets the nonlinear equation :

$$\nabla^2 V = R_\square(-J_L + J_o(e^{qV/kT} - 1)). \tag{4}$$

From physical considerations it can be stated that $V \geqslant V_o$. It should be noted that areas where V becomes much larger than V_o no longer contribute to the output current. This means that a part of the cell's surface is no longer active. In this case the grid contacts should be put closer together. As all actual solar cells are designed to use the major part of the current generated by the incident light, $V \approx V_o$ is a reasonable approximation. The exponential term in (4) can then be linearised. Putting :

$$V = V_o + \varphi \tag{5}$$

where $\varphi \ll V_o$, one gets :

$$\nabla^2\varphi - R_\square J_o \frac{q}{kT} e^{qV_o/kT} \varphi = R_\square(-J_L + J_o(e^{qV_o/kT} - 1)) \tag{6}$$

or :

$$\nabla^2\varphi - \frac{\varphi}{L^2} = R_\square J(V_o) \tag{7}$$

where the length L and $R_\square J(V_o)$ can be considered as constants.

In order to investigate the quality of the front grid contact, a quality factor Q will be defined as the ratio of the current delivered by the cell with respect to the current if the sheet resistance $R_\square$ would be zero. Indeed, if $R_\square = 0$, $\varphi = 0$ and hence $V = V_o$; all parts of the solar cell deliver the same current density $J(V_o)$. The current collected by the grid is found to be :

$$\frac{1}{R_\square} \oint_C \nabla\varphi . \bar{u}_n \, dc \tag{8}$$

where the integration extends along all the finger contacts, $\bar{u}_n$ being the outward directed unit normal vector. A perfect cell with $R_\square = 0$ gives rise to a current $J(V_o)S$, S being the surface of the solar cell. The quality factor Q is then :

$$Q = \frac{1}{S} \oint_C \nabla\varphi . \bar{u}_n \, dc \tag{9}$$

As the equation (7) for the function φ depends linearly on the right hand member $R_\square J(V_o)$, it is sufficient to solve (7) for $R_\square J(V_o) = 1$. The easiest way to compare different types of grid contacts is to calculate the current generated at a given voltage V_o. The optimal geometry will give the highest current or quality factor. As the length L depends upon V_o due to (7), the relation (9) has to be evaluated as a function of L for different grid geometries.

3. TRANSFORMATION TO AN INTEGRAL EQUATION

The technique will be outlined for the particular geometry shown on fig. 1.b. As the grid pattern is periodic only one unit cell has to be considered (fig. 3). B'A', A'A and AB are

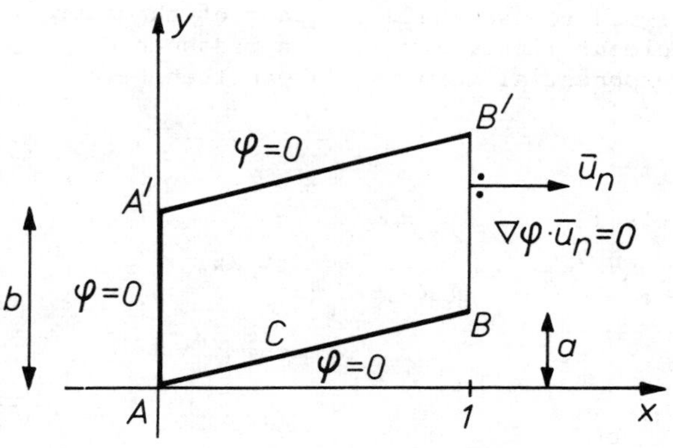

Fig.3. Unit cell of grid pattern.

metallic contact fingers. B'B is a (non metallic) axis of symmetry leading to the boundary condition $\nabla\varphi.\overline{u}_n = 0$. Inside C we have to solve the following equation :

$$\nabla^2\varphi - \frac{\varphi}{L^2} = 1 \qquad (10)$$

The Green's function $G(\overline{r}|\overline{r}')$ satisfies the equation :

$$\nabla^2 G - \frac{G}{L^2} = \delta(\overline{r}-\overline{r}') \qquad (11)$$

and is found to be :

$$G(\overline{r}|\overline{r}') = \frac{-1}{2\pi} K_o \left(\frac{|\overline{r}-\overline{r}'|}{L}\right) \qquad (12)$$

where K_o denotes the modified Bessel function of order zero. Multiplying (10) by $G(\overline{r}|\overline{r}')$ and (11) by φ one obtains after substracting and integrating over the whole surface inside C :

$$\oint_C (\varphi\nabla G.\overline{u}_n - G\nabla\varphi.\overline{u}_n)dc = \iint_S G(\overline{r}|\overline{r}')dS - \varphi(\overline{r}') \qquad (13)$$

where Green's theorem has been applied. Note that all integrations in (13) are performed with the coordinate $\overline{r}$ and not respect to $\overline{r}'$.

If $\overline{r}'$ in (13) is placed on the boundary C one obtains an integral equation. If φ is given on a part of C, the normal derivative $\nabla\varphi.\overline{u}_n$ becomes the unknown function and while on the remaining part of C, where $\nabla\varphi.\overline{u}_n = 0$, φ will be the unknown function.

4. NUMERICAL SOLUTION

A computer program has been written to solve the integral equation (13) for arbitrary polygonal geometries. Along each side another boundary condition can be imposed. The boundary is then divided into several intervals, wherein the unknown function (either φ or $\nabla\varphi.\overline{u}_n$) is replaced by an unknown constant. The solution of the integral equation is then reduced to the solution of an algebraic set. Finally the quality factor (9) can be easily calculated by a numerical integration.

In order to check the accuracy of the method, it has been applied to a square shaped form shown in fig. 4. The boundary conditions were chosen in such a way that an analytical solution could be found :

$$Q = 2L \text{ th} \frac{1}{2L} \qquad (14)$$

Fig. 4 compares (14) with the numerical data. For L = 0.4 the relative error is also plotted versus the number of unknowns n.

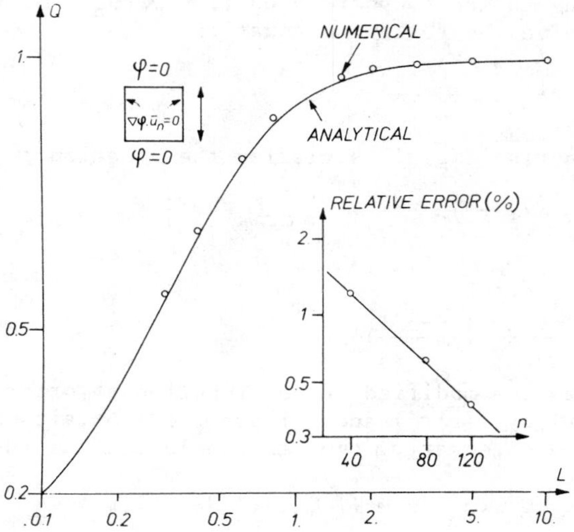

Fig.4. Quality factor and relative error calculated for a square geometry.

Note that again a 1/n law appears, which usually occurs when integral equations for potential problems are solved numerically [8].
Even when higher order approximations are used to represent the unknown function, still a 1/n law occurs [9].

5. RESULTS

The method has been applied to the structure shown in figs. 1a and 1b. For the numerical solution it is sufficient to consider only a single unit cell of the periodic pattern as shown on fig. 3. The dimension a has been varied in order to compare different kinds of grid patterns. If a = 0, then $b = b_o$; but for $a \neq 0$, b has been adjusted to obtain the same metallic contact length for a given cell area. Otherwise a fair comparison is not possible. Fig. 5 shows the quality factor Q as a function of a. Similar results were found for other b_o values. One observes that Q is almost independent of a. The conclusion is that there is no reason to introduce more complicated patterns as they do not influence the efficiency of the solar cell. These experiments also prove the usefulness of the BIEM when PDE have to be solved in various geometries such as those shown in fig. 1.

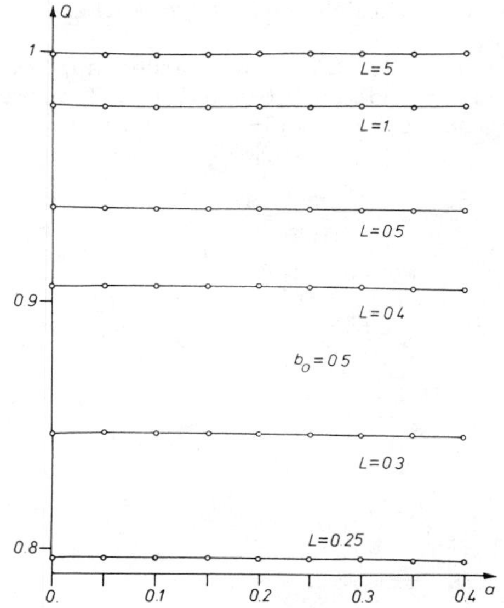

Fig.5. Quality factor as a function of a for various L-values.

REFERENCES

1. JASWON, M., and SYMM., *Integral equation methods in potential theory and elastostatics*, Academic Press, London (1977).
2. BREBBIA, C., *The boundary element method for engineers*, Pentech Press, London (1978).
3. JACOBS, B., and DE MEY, G., Theoretical analysis of Cu_2S-CdS solar cells with rough interfaces, *IEEEK Transactions on Electron Devices*, ED-28, 289-293 (1981).
4. DE MAY, G., Numerical application of integral equations in semi-conductor physics, *Proceedings of Colloquium Numerical Treatment of Integral Equations*, p. 99-114(ed. te Riele), Mathematisch Centrum, Amsterdam (1979).
5. DE MEY, G., Integral equation for the potential distribution in a Hall generator, *Electronics Letters*, 9, 264-266 (1973).
6. DE VISSCHERE, P., and DE MEY, G., Integral equation approach to the abrupt depletion approximation in semiconductor components, *Electronics Letters*, 13, 104-106 (1977).

7. SZE, S.M., *Physics of semiconductor devices*, Wiley, New York, p. 643 (1969).
8. DE MEY, G., Potential calculations in Hall plates, Invited review paper to be published in *Advances in Electronics and Electron Physics*.
9. STEVENS, K., and DE MEY, G., Higher order approximations for integral equations of potential problems, *International Journal of Electronics*, 45, 443-446 (1978).

ON THE CONSTRUCTION OF STABILITY POLYNOMIALS FOR MODIFIED R-K METHODS FOR VOLTERRA INTEGRO - DIFFERENTIAL EQUATIONS.

*Christopher T.H. Baker and †Joan C. Wilkinson

**University of Manchester,* †*Stanley Park C.H.S., Liverpool.*

1. INTRODUCTION

We commence by investigating, briefly, recurrence relations of the form

$$\sum_{\ell=0}^{m+1} P_\ell \hat{\underline{\rho}}_{n+1-\ell} + \sum_{\ell=0}^{m+1} Q_\ell \hat{\underline{\sigma}}_{n+1-\ell} = \underline{\Delta}'_{n+1} \tag{1.1}$$

$$\sum_{\ell=0}^{m+1} R_\ell \hat{\underline{\rho}}_{n+1-\ell} + \sum_{\ell=0}^{m+1} S_\ell \hat{\underline{\sigma}}_{n+1-\ell} = \underline{\Delta}''_{n+1} \;. \tag{1.2}$$

Here $\{P_\ell, Q_\ell, R_\ell, S_\ell\}$ are square matrices of fixed order, and $\{\hat{\underline{\rho}}_n, \hat{\underline{\sigma}}_n\}_{n\geq 0}$ is a sequence of vector pairs. Our results are of interest in the stability analysis of a wide class of numerical methods for the Volterra integro-differential equation

$$y'(x) = G(x,y(x), \int_0^x K(x,t,y(t))dt) \quad (x \geq 0) \tag{1.3}$$

given the initial condition $y(0) = y_0$, assuming suitable smoothness. The results are applicable due to the choice as a stability test equation of the "basic" equation

$$y'(x) = \xi y(x) + \eta \int_0^x y(t)dt + d(x) \tag{1.4}$$

and because we assume reducibility structure [1 ,5] in the numerical methods. The methods covered include classical methods treated in [3] and new γ - modified Runge-Kutta methods which are motivated by the corresponding methods, first introduced, for Volterra integral equations of the second kind, by Van der Houwen [4].

Our purpose is to gain insight into the construction of stability polynomials for *classical* and *modified Runge-Kutta methods*, which we describe below. This task is, in our view, a prerequisite to further study. Those of a perceptive humour may like to describe (1.4) as a second-order differential equation (it reduces to one on differentiation) but the approach to our methods is from the viewpoint of a system of integral equations and the modified methods are of particular interest. Incorporating the modified methods into a general analysis requires some rather special insight which we elucidate below.

2. STABILITY OF A RECURRENCE.

Consider the recurrence relations (1.1), (1.2). We have the following result (cf.[3]).

<u>THEOREM 2.1.</u> A stability polynomial for (1.1) and (1.2) is

$$\Sigma(\mu) \equiv \det\begin{bmatrix} P & Q \\ R & S \end{bmatrix} \tag{2.1}$$

where $P \equiv P(\mu) = \sum_{\ell=0}^{m+1} P_\ell \mu^{m+1-\ell}$, and likewise for $Q\equiv Q(\mu), R\equiv R(\mu)$ and $S \equiv S(\mu)$.

In general we wish to expand (2.1). We have a number of possibilities of which the following are illustrative (but not exhaustive).

Case 1 $SQ = QS$, and S^{-1} exists then $\Sigma(\mu) = \det[SP-QR]$;
Case 2 $PR = RP$, P^{-1} exists, then $\Sigma(\mu) = \det[PS-RQ]$;
Case 3 If S^{-1} exists then $\Sigma(\mu) = \det[P-QS^{-1}R]\det S$.

For remarks on *Case 3* see [2]. Observe that *Cases 1* and *2* (and, indeed, *Case 3*) correspond to elimination of either $\{\underline{\hat{\sigma}}_n\}$ or $\{\underline{\hat{\rho}}_n\}$ between (1.1) and (1.2). Thus, for example, if we denote by E the advancement operator $(E\underline{\hat{\rho}}_n = \underline{\hat{\rho}}_{n+1}, E\underline{\hat{\sigma}}_n = \underline{\hat{\sigma}}_{n+1})$ then (1.1) and (1.2) become

$$P(E)\underline{\hat{\rho}}_{n-m} + Q(E)\underline{\hat{\sigma}}_{n-m} = \underline{\Delta}'_{n+1};\quad R(E)\underline{\hat{\rho}}_{n-m} + S(E)\underline{\hat{\sigma}}_{n-m} = \underline{\Delta}''_{n+1},$$

where $P(E)$ denotes substitution of E for μ in $P(\mu)$, etc. We apply $S(E)$ to the first and $Q(E)$ to the second of these equations with some loss of information if either S or Q is singular. Subtracting now yields, if we make the assumption that $Q(E)S(E) = S(E)Q(E)$,

$$[S(E)P(E) - Q(E)R(E)]\underline{\hat{\rho}}_{n-m} = S(E)\underline{\Delta}'_{n+1} - Q(E)\underline{\Delta}''_{n+1} \tag{2.2}$$

for which a stability polynomial is *Case 1*, $\det[S(\mu)P(\mu)-Q(\mu)R(\mu)]$.

3. A GENERAL CLASS OF METHODS

Our techniques for discretizing (1.3) may be obtained from an integral equation approach. If we write (1.3) in integrated form, we obtain the pair of integral equations

$$y(x) = \int_0^x G(t,y(t),z(t))dt + y_0, \quad z(x) = \int_0^x K(x,t,y(t))dt, \tag{3.1}$$

and we seek approximations $\tilde{y}_j \simeq y(\tau_j)$ and $\tilde{z}_j \simeq z(\tau_j)$ at the points

$$\tau_j = ih + \theta_r h \quad (i=0,1,2,\ldots;\ r=0,1,\ldots,p;\ j=i(p+1)+r+1). \tag{3.2}$$

(Throughout, we reserve i,j,r for use with $r \equiv (j-1)\mathrm{mod}(p+1)$.) Here, the values θ_r are parameters defined by a Runge-Kutta (R-K) method (see below).

We require quadrature rules with abscissae $\{\tau_j\}$, of the form

$$\hat{Q}: \int_0^{\tau_j} \phi(t)dt \simeq h \sum_{k=0}^{(i+1)(p+1)} \Omega_{jk}\, \phi(\tau_k), \tag{3.3}$$

with weights $\hat{Q}' = \{\Omega'_{jk}\}$, $\hat{Q}'' = \{\Omega''_{jk}\}$. We define $\Omega'_{jk} = \Omega''_{jk} = 0$, for $k > (i+1)(p+1)$. A choice of vectors

$$\underline{\gamma}' = \{\gamma'_0,\ldots,\gamma'_p\}^T, \quad \underline{\gamma}'' = \{\gamma''_0,\ldots,\gamma''_p\}^T \qquad (3.4'),\ (3.4'')$$

with components in $[0,1]$ permits the replacement of (3.1) by

$$\begin{aligned}\tilde{y}_j &= h \sum_{k\geq 0} \Omega'_{jk} G(\tau_k,\tilde{y}_k,\tilde{z}_k) + y_0 \\ &\quad + \gamma'_r\{\tilde{y}_{i(p+1)} - (h \sum_{k=0}^{i(p+1)} \Omega'_{i(p+1)+1,k} G(\tau_k,\tilde{y}_k,\tilde{z}_k) + y_0)\}\end{aligned} \tag{3.5}$$

and

$$\begin{aligned}\tilde{z}_j &= h \sum_{k\geq 0} \Omega''_{jk}\, K(\tau_j,\tau_k,\tilde{y}_k) \\ &\quad + \gamma''_r\{\tilde{z}_{i(p+1)} - h \sum_{k=0}^{i(p+1)} \Omega''_{i(p+1)+1,k} K(ih,\tau_k,\tilde{y}_k)\}.\end{aligned} \tag{3.6}$$

The corresponding method will be denoted $M(\Omega',\underline{\gamma}';\Omega''\underline{\gamma}'')$.

These equations with $\gamma'_r = \gamma''_r = 0$ $(r=0,1,\ldots,p)$ define *classical* methods which are *modified* by the introduction of the parameters (3.4'), (3.4''). (The more natural class of methods requires only the choice of $\underline{\gamma}''$, with $\underline{\gamma}' = \underline{0}$.)

It remains to construct the rules $(\hat{Q}')$ and $(\hat{Q}'')$, of the form (3.3). We employ the quadrature rules

$$\int_0^{ih} \phi(t)dt \simeq \sum_{k=0}^{i} \omega_{ik}\, \phi(kh) \quad (i=1,2,\ldots),$$

associated with $Q = \{\omega_{ik}\}$ $(\omega_{ik} = 0,\ k > i)$, and the R-K parameters

$$\underline{\theta}|A = \begin{array}{c|cccc} \theta_0 & A_{00} & A_{01} & \cdots & A_{0p} \\ \theta_1 & A_{10} & A_{11} & \cdots & A_{1p} \\ \cdot & \cdot & \cdot & & \cdot \\ \cdot & \cdot & \cdot & & \cdot \\ \theta_{p-1} & A_{p-1,0} & A_{p-1,1} & \cdots & A_{p-1,p} \\ \hline \theta_p = 1 & A_{p0} & A_{p1} & & A_{pp} \end{array} .$$

Subsequently we shall employ rules Q', Q'' with weights ω'_{ik}, ω''_{ik} and R-K tableaux $[\underline{\theta}',A']$ and $[\underline{\theta}'',A'']$ with $\underline{\theta}'=\underline{\theta}''=\underline{\theta}$. Rules Q and tableaux $[\underline{\theta},A]$ permit the definition of $\hat{Q}$. "Extended R-K integration rules" [1] yield weights

$$\Omega_{jk}(A) = \begin{cases} A_{pt} & 0 < k \leq i(p+1), \\ A_{rt} & i(p+1) < k \leq (i+1)(p+1) \\ 0 & \text{otherwise}, \quad t \equiv (k-1)\bmod(p+1) \end{cases} \tag{3.7}$$

associated with $\underline{\gamma} = \underline{0}$ for definiteness, and we write $\hat{Q}=\Omega(A)$ for use in (3.3). The mixed quadrature - R-K rules [1] have weights

$$\Omega_{jk}[Q;A] = \begin{cases} \omega_{im}, & k = m(p+1),\ m \leq i \\ A_{rt} & i(p+1) < k \leq (i+1)(p+1) \\ 0 & \text{otherwise}, \quad t \equiv (k-1)\bmod(p+1) \end{cases} \tag{3.8}$$

and we write $\hat{Q} = \Omega[Q:A]$. Thus, the weights Ω'_{jk} and Ω''_{jk} may be defined as

$$\Omega'_{jk} = \Omega_{jk}(A') \tag{3.7'}$$

or

$$\Omega'_{jk} = \Omega_{jk}[Q',A'] \tag{3.8'}$$

and

$$\Omega''_{jk} = \Omega_{jk}(A'') \tag{3.7''}$$

or

$$\Omega''_{jk} = \Omega_{jk}[Q'',A'']. \tag{3.8''}$$

When $\Omega_{jk}(A)$ are taken as the weights defining Ω' or Ω'', the corresponding $\underline{\gamma}$ ($\underline{\gamma}'$ or $\underline{\gamma}''$) may be chosen arbitrarily. Computationally one would assume $\underline{\gamma} = \underline{0}$ but it is convenient, in §4, to suppose that $\underline{\gamma} = \{1,1,\ldots,1\}^T$. When $\Omega'_{jk} = \Omega_{jk}(A)$, the method may be regarded as an adaptation of an R-K method for an ordinary differential equation. This approach, and the resulting method, will be particularly attractive to those examining the topic from the viewpoint of ordinary differential equations.

We referred to the fact that structure will be required in the weights Ω_{jk}. Denoting by $\underline{\varepsilon}_0, \underline{\varepsilon}_1, \ldots \underline{\varepsilon}_{q-1}$ the columns of the identify matrix of order q and their sum by $\underline{\varepsilon}$ we have:

<u>DEFINITION 3.1</u> Quadrature rules with weights $\bar{Q} = \{\bar{\Omega}_{jk}\}$ corresponding to a prescribed ordering of the abscissae are called *block-reducible* when the weights $\bar{\Omega}_{jk}$ may be partitioned into matrices $\bar{V}_{nk}$ of order q with elements $\underline{\varepsilon}_\alpha^T \bar{V}_{nk} \underline{\varepsilon}_\beta = \bar{\Omega}_{nq+\alpha,kq+\beta}$ $(\alpha,\beta = 0,1,\ldots,q-1)$ satisfying, for fixed matrices $\left\{\bar{A}_\ell, \bar{B}_\ell\right\}_{\ell=0}^m$ where $\sum_{\ell=0}^m \bar{A}_\ell \underline{\varepsilon} = \underline{0}$, the relation

$$\sum_{\ell=0}^m \bar{A}_\ell \bar{V}_{n-\ell,k} = \bar{B}_{n-k} , \tag{3.9}$$

with the convention $\bar{B}_\ell = 0$ for $\ell \notin \{0,1,\ldots,m\}$. If $\bar{A}_0 = I$, $\bar{A}_1 = -I$ in (3.9), $\bar{A}_m = 0$ otherwise, $\bar{Q}$ will be termed *simply-block-reducible*. If $q=1$ in (3.9), $\bar{Q}$ will be called *reducible*.

'Reducibility' in one of the above senses will be assumed in the sequel. 'Reducibility' of Q yields 'reducibility' of the rules with weights $\Omega_{jk}[Q,A]$, under an associated partitioning.

4. BASIC STABILITY THEORY

We intend discussing the stability of a set of numerical methods, included in the description in §3, when they are applied to the basic test equation (1.4), viz. $y'(x) = \xi y(x) + \eta\int_0^x y(t)dt + d(x)$. We shall employ the notation

$$\underline{\rho}_{n+1} = \{\tilde{y}_{n(p+1)+1}, \ldots, \tilde{y}_{(n+1)(p+1)}\}^T,$$

$$\underline{\sigma}_{n+1} = \{\tilde{z}_{n(p+1)+1}, \ldots, \tilde{z}_{(n+1)(p+1)}\}^T$$

where $\{\tilde{y}_n\}$, $\{\tilde{z}_n\}$ are the values obtained for equation (1.4). We denote by $\underline{e}_0, \underline{e}_1, \ldots, \underline{e}_p, \underline{e}$ the successive columns of the identity matrix of order $p+1$ and their sum, whence we derive the matrix $E_p = \underline{e}\,\underline{e}_p^T$. We also write $\Gamma = \mathrm{DIAG}(\gamma_0, \gamma_1, \ldots, \gamma_p)$ Γ', Γ'' being defined in terms of γ', γ''. It is also convenient to write

$$\Gamma_p' = \Gamma' E_p, \Gamma_p'' = \Gamma'' E_p. \tag{4.1'),(4.1''}$$

Recall that we denote the columns of the identify matrix I_q of order q by $\underline{\varepsilon}_0, \underline{\varepsilon}_1 \ldots \underline{\varepsilon}_{q-1}$ (we shall employ a similar notation when $q = q'$, $q = q''$) and we write

$$J = [\underline{\varepsilon}_1|\underline{\varepsilon}_2|\ldots|\underline{\varepsilon}_{q-1}|\underline{0}],\ J^{\#} = [\underline{0}|\underline{0}\ \ldots\underline{0}|\underline{\varepsilon}_0].$$

The notation $\underline{\rho}_n$, $\underline{\sigma}_n$, will be used in a generic sense. Since we have to consider various methods applied to (1.4), we write

$\underline{\rho}_{n+1}^{(e)}$ for vectors generated using $\Omega_{jk}(A'), \underline{\gamma}' = \underline{0}$ in (3.5);

$\underline{\sigma}_{n+1}^{(e)}$ " " " " $\Omega_{jk}(A''), \underline{\gamma}'' = \underline{0}$ in (3.6),

(the "extended Runge-Kutta" formulae) and

$\underline{\rho}'_{n+1}$ for vectors generated using $\Omega_{jk}[Q',A'], \underline{\gamma}'$ in (3.5);

$\underline{\sigma}''_{n+1}$ for vectors generated using $\Omega_{jk}[Q'',A''], \underline{\gamma}''$ in (3.6);

(the "γ'- and γ'' - modified mixed formulae"). Thus, we find that (3.7') and (1.4) yield, from (3.5)

$$\underline{\rho}_{n+1}^{(e)} = \underline{E}_p\underline{\rho}_n^{(e)} + h\underline{A}'\{\xi\underline{\rho}_{n+1}^{(e)} + \eta\underline{\sigma}_{n+1}\} \tag{4.2}$$

and (3.8') and (1.4) yield, from (3.5),

$$\underline{\rho}'_{n+1} = \underline{\rho}^{\dagger}_{n+1} + \underline{\Gamma}'_p\{\underline{\rho}'_n - h\sum_{j=0}^{n}\omega'_{nj}(\xi\underline{\rho}'_j + \eta\underline{\sigma}_j)\} \tag{4.3a}$$

wherein

$$\underline{\rho}^{\dagger}_{n+1} = h\sum_{j=0}^{n}\omega'_{nj}\underline{E}_p\{\xi\underline{\rho}'_j + \eta\underline{\sigma}_j\} + hA'\{\xi\underline{\rho}'_{n+1} + \eta\underline{\sigma}_{n+1}\}. \tag{4.3b}$$

Similarly $\underline{\sigma}_{n+1}^{(e)} = h\sum_{j=0}^{n-1}E_pA''\underline{\rho}_{j+1} + hA''\underline{\rho}_{n+1}$ whence

$$\underline{\sigma}_{n+1}^{(e)} = E_p\underline{\sigma}_n^{(e)} + hA''\underline{\rho}_{n+1} \tag{4.4}$$

whilst

$$\underline{\sigma}''_{n+1} = \underline{\sigma}^{\dagger}_{n+1} + \Gamma''_p\{\underline{\sigma}''_n - h\sum_{j=0}^{n}\omega''_{nj}E_p\underline{\rho}_j\} \tag{4.5a}$$

wherein

$$\underline{\sigma}^{\dagger}_{n+1} = h\sum_{j=0}^{n}\omega''_{nj}E_p\underline{\rho}_j + hA''\underline{\rho}_{n+1}. \tag{4.5b}$$

Our task is to analyze the stability of (4.2) or (4.3a,b) where $\{\underline{\sigma}_n\}$ is replaced by $\{\underline{\sigma}_n^{(e)}\}$ or $\{\underline{\sigma}''_n\}$. In order to analyze the possible combinations of (4.2), or (4.3) with (4.4) or (4.5) we require some common structure which we shall develop. In particular we ask, when dealing with (4.3) or (4.5), that the rules $\{\omega'_{nj}\}$ or $\{\omega''_{nj}\}$ should be block-reducible. Then we require the matrices $\{V'_{n,j}\}$, $\{V''_{n,j}\}$ of order q', q'' respectively with the elements

$$\underline{\varepsilon}'^{T}_{\alpha} V'_{nj} \underline{\varepsilon}'_{\beta} = \omega'_{nq'+\alpha,jq'+\beta} (\alpha,\beta = 0,1,\ldots,q'-1) \tag{4.6'}$$

$$\underline{\varepsilon}''^{T}_{\alpha} V''_{nj} \underline{\varepsilon}''_{\beta} = \omega''_{nq''+\alpha,jq''+\beta} (\alpha,\beta = 0,1,\ldots,q''-1). \tag{4.6''}$$

in the (α,β) position, respectively. Associated with q',q'' (where appropriate) we define a parameter q which assumes, in our analysis of the indicated pair of equations, the value shown:

	Eq.(4.2) $(\underline{\rho}_n^{(e)})$	Eq.(4.3) $(\underline{\rho}'_n)$
Eq.(4.4) $(\underline{\sigma}_n^{(e)})$	$q = 1$	$q = q'$
Eq.(4.5) $(\underline{\sigma}'_n)$	$q = q''$	$q = q' = q''$

Table 4.1

Observe that in our analysis of (4.3) and (4.5) *we require* $q' = q''$ but if this is not the case we are able, by restructuring, to *take* $q = q'q''$. Recalling the notation J and $J^{\#}$ above, we derive from the matrices V'_{nj} and V''_{nj}, as appropriate, the matrices $\hat{V}'_{nj}$, $\hat{V}''_{nj}$ from the rule

$$\hat{V}_{nj} = V_{nj} J + V_{n,j+1} J^{\#}. \tag{4.7}$$

Having established our interpretation of q, we introduce the vectors $\hat{\underline{\rho}}_n^{(e)}$, $\hat{\rho}'_n$, $\hat{\underline{\sigma}}_n(e)$, $\hat{\underline{\sigma}}''_n$ obtained on replacing $\underline{\rho}_n$ and $\underline{\sigma}_n$ in an obvious manner in the notation

$$\hat{\underline{\rho}}_{n+1} = \{\underline{\rho}^T_{nq+1}, \underline{\rho}^T_{nq+2},\ldots,\underline{\rho}^T_{(n+1)q}\}^T \tag{4.8}$$

$$\hat{\underline{\sigma}}_{n+1} = \{\underline{\sigma}^T_{nq+1}, \underline{\sigma}^T_{nq+2},\ldots,\underline{\sigma}^T_{(n+1)q}\}^T \tag{4.9}$$

$(n=0,1,2,\ldots)$ with $\hat{\underline{\rho}}_0 = y(0)\,\{0,0,\ldots,0,1\}^T$, $\hat{\underline{\sigma}} = \underline{0} \in \mathbb{C}^{q(p+1)}$.

Finally, recalling that the Kronecker product $G \otimes H$ of the square matrices G and H is the partitioned matrix with elements $G_{\alpha\beta}H$, we can state the following results.

<u>LEMMA 4.1</u> If (a) $\hat{\underline{\rho}}_n = \hat{\underline{\rho}}_n^{(e)}$ in (4.8) and $\Gamma'_p = E_p$, V'_{nj} arbitrary, or (b) $\hat{\underline{\rho}}_n = \hat{\underline{\rho}}'_n$ in (4.8) and $\Gamma'_p = \Gamma' E_p, V'_{nj}$ satisfying (4.6'), then (a) (4.2) or (b) (4.3) yields

$$(I_q \otimes [I-\xi hA'])\hat{\underline{\rho}}_{n+1} = \xi h \sum_{j=0}^{n+1} \hat{V}'_{n+1,j} \otimes (E_p - \Gamma'_p)\hat{\underline{\rho}}_j + (I_q \otimes \Gamma'_p)\hat{\underline{\rho}}_n +$$

$$+ \eta h \sum_{j=0}^{n+1} \hat{V}'_{n+1,j} \otimes (E_p - \Gamma'_p)\underline{\hat{\sigma}}_j + \eta h (I_q \otimes A')\underline{\hat{\sigma}}_{n+1} + \underline{\delta}_{n+1}. \quad (4.10)$$

Proof By rearrangement.

LEMMA 4.2. If (a) $\underline{\hat{\sigma}}_n = \underline{\hat{\sigma}}_n^{(e)}$ in (4.9) and $\Gamma''_p = E_p$, V''_{nj} arbitrary, or (b) $\underline{\hat{\sigma}}_n = \underline{\hat{\sigma}}''_n$ in (4.9) and $\Gamma''_p = \Gamma'' E_p$, V''_{nj} satisfying (4.6"), then (a) (4.4) or (b) (4.5) yields

$$\underline{\hat{\sigma}}_{n+1} = h \sum_{j=0}^{n+1} \hat{V}''_{n+1,j} \otimes (E_p - \Gamma''_p)\underline{\hat{\rho}}_j + h(I_q \otimes A'')\underline{\hat{\rho}}_{n+1} + (I_q \otimes \Gamma''_p)\underline{\hat{\sigma}}_n . (4.11)$$

LEMMA 4.3 Suppose the quadrature rules $Q = \{\omega_{nj}\}$ are block-reducible with $\sum_{\ell=0}^{m} A_\ell V_{n-\ell,j} = B_{n-j}$ where $A_\ell = B_\ell = 0$ for $\ell \notin \{0,1,\ldots,m\}$. Then

$$\sum_{\ell=0}^{m+1} A_\ell \hat{V}_{n-\ell,j} = \hat{B}_{n-j}$$

where $\hat{B}_\ell = 0$ for $\ell \notin \{0,1,\ldots,m,m+1\}$ and

$$\hat{B}_\ell = B_{\ell-1} J^{\#} + B_\ell J. \quad (4.12)$$

Proof Immediate from (4.7).

The matrices $\hat{B}'_\ell$, $\hat{B}''_\ell$ are derived according to the rule (4.12), but when treating $\underline{\hat{\rho}}_n^{(e)}$ (respectively $\underline{\hat{\sigma}}_n^{(e)}$) the matrices $\hat{V}'_{nj}$ (respectively $\hat{V}''_{nj}$) are arbitrary and we then set $A'_0 = I, A'_\ell = 0$ otherwise (respectively $A''_0 = I$, $A''_\ell = 0$ otherwise) $\hat{B}'_\ell$ and $\hat{B}''_\ell$ being arbitrary.

LEMMA 4.4 (a) Let the rules $\{\omega'_{nj}\}$ be block-reducible. Then (4.10) yields (1.1) where $P_\ell = A'_\ell \otimes (I - \xi h A') - \xi h \hat{B}'_\ell \otimes (E_p - \Gamma'_\rho) - A'_{\ell-1} \otimes \Gamma'_p$ and $Q_\ell = -\eta h[\hat{B}'_\ell \otimes (E_p - \Gamma'_p) + A'_\ell \otimes A']$.

(b) Let the rules $\{\omega''_{nj}\}$ be block-reducible. Then (4.11) yields (1.2) where $R_\ell = -h[\hat{B}''_\ell \otimes (E_p - \Gamma''_p) + A''_\ell \otimes A'']$ and $S_\ell = (A''_\ell \otimes I) - (A''_{\ell-1} \otimes \Gamma''_p)$.

Observe that $P_\ell \equiv P_\ell(\xi h)$, $Q_\ell \equiv Q_\ell(\eta h)$, $R_\ell \equiv R_\ell(h)$ and S_ℓ is independent of ξ, η and h. As earlier, we write $P(\mu) = \sum_{\ell=0}^{m+1} P_\ell \mu^{m+1-\ell}$ etc. in what follows.

Our reference to equations (1.1) and (1.2) establishes the connection with Theorem 2.1. However, the result of *Case 1* requires $SQ = QS$, but in general the presence of Γ_p'' causes difficulties with this condition even under the reasonable assumption that $A_0'' = I, A_1'' = -I, A_\ell'' = 0$ otherwise (simple block-reducibility). For the combination (4.2) with (4.4) we can appeal instead to *Case 3* since S^{-1} is readily obtained, but to effect a general treatment we proceed as follows.

LEMMA 4.5 Under the assumptions of Lemma 4.4, and with the notation established above,

$$P(E)\ \hat{\underline{\rho}}_{n-m} + Q(E)\ \hat{\underline{\sigma}}_{n-m} = \underline{\Delta}'_{n+1} \tag{4.13}$$

and

$$R^*(E)\ \hat{\underline{\rho}}_{n-m} + S^*(E)\ \hat{\underline{\sigma}}_{n-m} = \underline{\Delta}'''_{n+1} \tag{4.14}$$

where

$$R^*(\mu) = [I_q \otimes \{(\mu-\gamma_p'')I + \Gamma_p''\}]R(\mu) \tag{4.15}$$

$$S^*(\mu) = [I_q \otimes \{(\mu-\gamma_p'')I+\Gamma_p''\}]S(\mu) = (\mu-\gamma_p'')\sum_{\ell=0}^{m+1}(A_\ell''\otimes I)\mu^{m-\ell} \tag{4.16}$$

and if $q = 1$ or the rules $Q'' = \{\omega_{nj}''\}$ are *simply* block-reducible, S^* commutes with Q.

Proof. Equation (1.2) reduces to (4.14). In equation (1.2) write n+1 in place of n and subtract γ_p'' times the original equation. Add to the result $(I_q \otimes \Gamma_p'')$ times the original (1.2) and (4.14) results with $\underline{\Delta}'''_{n+1} = \underline{\Delta}''_{n+2} - \gamma_p''\underline{\Delta}''_{n+1} + (I\otimes\Gamma_p'')\underline{\Delta}''_{n+1}$.

REMARK. Observe the simplification when considering the classical methods.

The analysis which led to (2.2) is now valid if R is replaced by R^* and S is replaced by S^*. Thus, we have the following general result.

THEOREM 4.1 Under the assumptions of Lemma 4.5

$$[S^*(E)P(E)-Q(E)R^*(E)]\hat{\underline{\rho}}_{n-m} = S^*(E)\underline{\Delta}'_{n+1} - Q(E)\underline{\Delta}'''_{n+1} \tag{4.17}$$

and the associated stability polynomial is

$$\det[S^*(\mu)\ \ P(\mu) - Q(\mu)R^*(\mu)] \tag{4.18}$$

Some specific results appear below. Theorem 4.2 is of special interest since (in view of remarks of Hairer) we might choose A'' with a sparse last row whilst A' might be conventional.

THEOREM 4.2 Consider the (extended) method $M(\Omega(A'), \underline{e}; \Omega(A''), \underline{e})$ applied to (1.4). The stability polynomial is

$$\det[\mu^2 Z - \mu(Z+E_p + \eta h^2 A'E_p A'') + E_p] \tag{4.19}$$

where

$$Z \equiv Z(\xi h, \eta h^2) = (I - \xi hA' - \eta h^2 A'A'') \tag{4.20}$$

REMARK Equation (4.19) may be expressed as

$$\det\frac{1}{\mu}[\lambda_0\{\mu(I-\lambda_0 hA')-E_p\}\{(\mu-1)I+E_p\}\{\mu(I-\lambda_1 hA'')-E_p\}$$
$$-\lambda_1\{\mu(I-\lambda_1 hA')-E_p\}\{(\mu-1)I+E_p\}\{\mu(I-\lambda_0 hA'')-E_p\}]. \qquad (4.19')$$

If $A = A' = A''$ then (4.19') reduces to a result of [3].

The results of Theorem 4.3 and 4.4 require the definitions of

$$\Psi^*(\lambda h) = I\otimes[\mu(I - \lambda hA') - E_p]\mu^m \qquad (4.21)$$

and of $\Psi'(\lambda h)$ and $\Psi''(\lambda h)$ obtained by inserting primes on A_ℓ, $\hat{B}_\ell$, A and Γ_p in the definition

$$\Psi(\lambda h) = \sum_{\ell=0}^{m+1}\{A_\ell\otimes\{\mu(I-\lambda hA)-\Gamma_p\}-\lambda h\mu\hat{B}_\ell\otimes(E_p-\Gamma_p)\}\mu^{m-\ell} \qquad (4.22)$$

THEOREM 4.3 Consider the (extended/modified) method $M(\Omega(A'),\underline{e};\Omega[Q'' A''],\underline{\gamma}'')$ applied to (1.4), and assume the rules Q'' are simply block-reducible (or reducible). Then if $\underline{\gamma}'' = \underline{e}$, the stability polynomial is (4.19).

For the general $\underline{\gamma}''$ the stability polynomial is

$$\det\frac{1}{\mu}[\lambda_0\Psi^*(\lambda_0 h)[(\mu-\gamma_p'')(I\otimes I)+I\otimes\Gamma_p'']\Psi''(\lambda_1 h)$$
$$-\lambda_1\Psi^*(\lambda_1 h)[(\mu-\gamma_p'')(I\otimes I)+I\otimes\Gamma_p'']\Psi''(\lambda_0 h)] \qquad (4.23)$$

The methods covered by the preceding results are of particular interest. The following is the general result.

THEOREM 4.4 The determinant (4.18) reduces, for the method $M(\Omega[Q',A'],\underline{\gamma}';\Omega[Q'',A''],\underline{\gamma}'')$ where Q', Q'' satisfy the assumptions of Theorem 4.3 to (4.23) with $\Psi^*(\lambda h)$ replaced by $\Psi'(\lambda h)$.

REFERENCES

1. BAKER, C.T.H. Families of structured quadrature rules and some ramifications of reducibility *Int. Ser. Num. Math.* 57 (ed. G. Hammerlin), 12-24 (1982).
2. BAKER, C.T.H., MAKROGLOU, A., and SHORT, E. Regions of stability in the numerical treatment of Volterra integro-differential equations. *SIAM J. Numer. Anal.* 16 809-910 (1979).
3. BAKER, C.T.H. and WILKINSON J.C. Stability of Runge-Kutta methods applied to a basic integro-differential equation. Manuscript October 1981 (A revised version of *Num.Anal.Tech. Rept.* No. 50, University of Manchester, June 1980.)
4. HOUWEN, P.J. VAN DER. Convergence and stability results in R-K type methods for Volterra integral equations *BIT* 20 375-377 (1980)
5. WOLKENFELT, P.H.M. The numerical analysis of reducible quadrature methods for Volterra integral and integro-differential equations. Acad. Proefschrift., Amsterdam (1981).

ON THE STABILITY OF NUMERICAL METHODS FOR VOLTERRA INTEGRAL EQUATIONS OF THE SECOND KIND

S. Amini

Plymouth Polytechnic

1. INTRODUCTION

Recent results of [1,3] show that it is possible to obtain stability polynomials for a large but structured class of numerical methods as applied to the convolution test problem.

$$f(x) = 1 + \int_0^x \left(\sum_{s=0}^{S} \lambda_s (x-y)^s\right) f(y)dy. \tag{1.1}$$

It is clear that (S+1) times differentiation of (1.1) yields an (S+1)-st order differential equation with constant coefficients. Results of [1,3] in effect demonstrate that it is possible to obtain, by placing sufficient restrictions on the numerical methods, differencing procedures (discrete analogue of differentiation), which on sufficient application to the discrete system would yield finite term recurrence relations with constant coefficients between the approximate values $\{f_j\}$.

We shall discuss briefly the qualitative behaviour of the solutions of (1.1) in order to define the concept of (A_0,S)-stability; see [1,2]. We shall also indicate why most conventional step-by-step methods would fail to satisfy this criterion for $S\geq 1$.

2. DEFINITIONS

Uniform stability of the solutions of Volterra integral equations with respect to certain allowable perturbations in their forcing functions is a desirable property for theoretical and numerical stability investigations. Such a property ensures that bounded perturbations in the (input) forcing function can only induce bounded perturbations in the (output) solution function.

The solution of the linear system

$$\underline{f}(x) = \underline{g}(x) + \int_{x_0}^{x} K(x,y)\underline{f}(y)dy \qquad x \geq x_0 \qquad (2.1)$$

assuming sufficient smoothness of g and K, may be represented in the form

$$\underline{f}(x) = \underline{g}(x) + \int_{x_0}^{x} R(x,y)\underline{g}(y)dy \qquad x \geq x_0 \qquad (2.2)$$

where R is the resolvent kernel. Integration by parts in (2.2) assuming differentiability of g yields the representation

$$\underline{f}(x) = \Phi(x,x_0)\underline{g}(x_0) + \int_{x_0}^{x} \Phi(x,y)\underline{g}'(y)dy \qquad x \geq x_0 \qquad (2.3)$$

where $\Phi(x,y) = I + \int_{y}^{x} R(x,t)dt$ and is known as the differential resolvent. The uniform boundedness of Φ over $x \geq y \geq x_0$ is necessary and sufficient for uniform stability of (2.1) with respect to constant changes in $\underline{g}(x)$, while the stronger condition

$$\int_{x_0}^{x} \| R(x,y) \| dy \leq M < +\infty \qquad \text{all } x \geq x_0 \qquad (2.4)$$

is necessary and sufficient for the uniform stability of (2.1) with respect to continuous perturbations in $\underline{g}(x)$; see [1] and references therein.

We may now define RT, the region of theoretical stability of (1.1).

Definition 2.1 $\underline{\lambda} = [\lambda_0, \lambda_1, \ldots, \lambda_S]^T \varepsilon RT \subset R^{S+1}$ if and only if (1.1) is uniformly stable with respect to constant (real) changes in its forcing function.

Definition 2.2 A numerical method is said to be (A_0,S)-stable if and only if on application to (1.1) the discrete solutions $\{f_j(h)\}$ remain uniformly bounded over positive integer j and h>0 whenever $\underline{\lambda} \varepsilon RT$.

The above definitions may be trivially modified to allow $\lambda_s \varepsilon C$ for s = 0,1,...,S. In this case a method preserving the appropriate qualitative behaviour of (1.1) for all h>0, will be referred to as (A,S)-stable.

The proof of the following theorem and other interesting results concerning the stability behaviour of (1.1) may be found in [1].

Theorem 2.1

$S = 0;\quad RT = \{\lambda_0 \mid \lambda_0 \le 0\}$

$S = 1;\quad RT = \{(\lambda_0,\lambda_1) \mid \lambda_0 \le 0, \lambda_1 < 0\}$

$S = 2;\quad RT = \{(\lambda_0,\lambda_1,\lambda_2) \mid \lambda_0 \le 0, \lambda_1 \le 0, \lambda_2 < 0, \lambda_0\lambda_1 + 2\lambda_2 \ge 0\}.$

3. RESULTS

Application of a multi-lag method (defined in [6]) to a non-linear Volterra integral equation of the second kind yields

$$f_n = -\sum_{i=1}^{k} \alpha_i^* \tilde{I}_{n-i}(x_n) + h\sum_{i=0}^{k} \beta_i^* K(x_n, x_{n-i}, f_{n-i}) + g(x_n) \quad (3.1)(a)$$

with

$$\tilde{I}_\nu(x) = h\sum_{j\ge 0} w_{\nu j} K(x, x_j, f_j) \quad (3.1)(b)$$

where $\alpha_0^* = 1$ and $\{\alpha_i^*, \beta_i^*\}$ are the coefficients of a (ρ^*,σ^*) linear multi-step method. A modification of the above method was also defined, with the aim of improving the stability properties of (3.1) by replacing $\tilde{I}_{n-i}(x_n)$ in (3.1)(a) with $\tilde{I}_{n-i}(x_n)+r_{n-i}$ where $r_\nu = f_\nu - g(x_\nu) - \tilde{I}_\nu(x_\nu)$ (residual).

If we assume that the rules $\{\omega_{nj}\}$ are (ρ,σ)-reducible then it is a simple exercise to show that the multi-lag method in this case reduces to a $(\tilde{\rho},\tilde{\sigma})$-reducible method, where $\tilde{\rho}(\mu) = \mu^k\rho(\mu)$ and $\tilde{\sigma}(\mu) = \rho(\mu)\sigma^*(\mu) - \rho^*(\mu)\sigma(\mu) + \mu^k\sigma(\mu)$. The stability polynomial for the modified multi-lag method is obtained and analysed in [2] and shall not be stated here.

It is shown in [5] that there are no $(A_0,1)$-stable reducible quadrature methods while [2] shows that neither the modified multi-lag nor the semi-implicit mixed Runge-Kutta methods, employing reducible quadrature rules may be $(A_0,1)$-stable.

4. REMARKS AND CONCLUSIONS

A large class of numerical methods yield unbounded solutions when applied to (1.1) with $S = 1$ and $(\lambda_0,\lambda_1)\varepsilon RT$, for $h \ge h_0(\lambda_0,\lambda_1)$. Consider the case $(\lambda_0 = 0,\ \lambda_1 < 0)\varepsilon RT$. Most conventional methods in this case reduce to an explicit rule for discretizing the integral operator (since $K(x,x)=0$). The polynomial convolution kernels (with $S \ge 1$) are unbounded and it seems unlikely that such methods could yield bounded solutions for all $h > 0$. It is natural to search for an $(A_0,1)$-stable method, among those, yielding an implicit discretization of the integral operator in this case; (fully implicit Runge-Kutta type methods say).

It is shown in [4] that the following one stage Beltyukov Runge-Kutta method

$$f_{n+1} = g(x_{n+1})+hK(x_n+ch,x_{n+1},f_{n+1})+h\sum_{j=0}^{n} K(x_{n+1},x_j,f_j)$$

is $(A_0,1)$-stable provided $c>\frac{5}{4}$. Let us notice that for $c>1$ this method yields an implicit discretization of the integral operator. It is shown in [1] that the above method is not however $(A_0,2)$-stable for any value of c.

Finally, let us observe that if the scalar function $R(x,y)$ is the resolvent kernel corresponding to $K(x,y)$, then $\tilde{R}(x,y)=\psi(x,y)R(x,y)$ and $\hat{R}(x,y)=\psi(y,x)R(x,y)$ are resolvent kernels corresponding to $\tilde{K}(x,y)=\psi(x,y)K(x,y)$ and $\hat{K}(x,y)=\psi(y,x)K(x,y)$ respectively where $\psi(x,y)=\phi(x)/\phi(y)$ with $\phi(x)\neq 0$ for $x\geq x_0$; ([1]). Therefore it may be possible to extend the results of stability analysis for a test problem to a larger class of equations.

REFERENCES

1. AMINI, S., *Analysis of Stability Behaviour in the Treatment of Certain Volterra Integral Equations*. PhD Thesis, University of Manchester (1980).
2. AMINI,S., Stability Analysis of Methods Employing Reducible Quadrature Rules for Solutions of Volterra Integral Equations. Report No: CS-81-02, *University of Bristol* (1981).
3. AMINI, S., BAKER, C.T.H., VAN DER HOUWEN, P.J. and WOLKENFELT, P.H.M., Stability Analysis of Numerical Methods for Volterra Integral Equations with Polynomial Convolution Kernels. Report NW 109/81, *Mathematisch Centrum, Amsterdam* (to appear in *J. of Integral Equations*)(1981).
4. BRUNNER, H., NORSETT, S.P. and WOLKENFELT, P.H.M., On V_0-stability of Numerical Methods for Volterra Integral Equations of the Second Kind. Report NW 84/80,*Mathematisch Centrum, Amsterdam* (1980).
5. WOLKENFELT, P.H.M., Stability Analysis of Reducible Quadrature Methods for Volterra Integral Equations of the Second Kind. Report NW 79/80, *Mathematisch Centrum, Amsterdam* (1980).
6. WOLKENFELT, P.H.M., Modified Multi-lag Methods for Volterra Functional Equations. Report NW 108/81, *Mathematisch Centrum, Amsterdam* (1981).

SPLINE-BLENDED SUBSTITUTION KERNELS
OF OPTIMAL CONVERGENCE

Lothar Bamberger and Günther Hämmerlin

University of Munich (Germany)

1. INTRODUCTION

In this paper we are concerned with the method of degenerate substitution kernels, in particular with substitution kernels generated by spline-blending. We consider the question of convergence properties of approximate solutions using different spline approximation schemes.

It can be shown that substitution kernels of this kind lead to approximate eigenvalues which converge to the exact values like $O(h^{\ell})$, where ℓ depends not on the approximation quality but rather on the quality of the numerical integration formula induced by the splines. We consider polynomial splines; the above mentioned fact becomes essential for polynomials of even degree: As is well known, the order of accuracy of the adjoint quadrature formula is greater by one than the accuracy of approximation. The term optimal is to be understood in this sense. For degenerate kernels constructed as tensor-products of one-dimensional splines, this convergence behaviour has already been proved by E. Schäfer ([5]); we utilize the deduction in [5] having blending-splines in view.

In particular, certain local spline approximations and orthogonal projection splines are discussed. Whereas the former splines reflect the optimal convergence behaviour described above also for spline-blending, the latter splines prove to produce even higher convergence order taking as a criterion the error relative to $\|\cdot\|_2$.

2. APPROXIMATING SPLINES

To begin with the splines under discussion, let us consider the interval $I := [0,1]$, the step-length $h := 1/(p+1)$, $(p \in \mathbb{N})$, and a partition π given by the equidistant knots $x_i = (i-m)h$, $(i=1,2,\ldots,2m+p; m \in \mathbb{N})$. The set

$$S(D^m,\pi) := \{s \in C^{m-2}(I) \mid D^m s(x) = 0 \quad \text{for } x \in (x_i, x_{i+1}),\ (i=m,m+1,\ldots,m+p)\}$$

defines a space of polynomial splines of order m (or degree (m-1), $m \geq 1$), subordinate to the partition π. The dimension of S is (m+p).

A local basis of $S(D^m,\pi)$ is provided by the B-Splines, which can be introduced, e.g., by

$$B_i^m(x) = (-1)^m mh[x_i,\ldots,x_{i+m}]G(.,x),$$

$$G(y,x) := (x-y)_+^{m-1},\ (i=1,2,\ldots,m+p),$$

(see [6], § 4.2). Using B-Splines, we are able to develop several linear approximation schemes in order to approximate a continuous function f defined on a sufficiently large neighbourhood of I.
Examples:

(2.1) ([6], S. 224):

Let $\lambda_i f := \sum_{j=1}^{m} \alpha_j^{(m)} f(t_{ij}), \quad (i=1,2,\ldots,m+p),$

$$t_{ij} := x_i + jh - \frac{h}{2}, \quad (j=1,2,\ldots,m),$$

be linear functionals. Then

$$(P_h f)(x) = \sum_{i=1}^{m+p} (\lambda_i f) B_i^m(x)$$

defines a local spline approximation of f, if the $\alpha_j^{(m)}$ are chosen to imply $P_h g = g$ for every polynomial $g \in \mathbb{P}_{m-1}$, where $\mathbb{P}_r$ denotes the space of all polynomials having degree less than or equal to r.

(2.2) Orthogonal projection:

Let $s = P_h^{\perp} f \in S(D^m, \pi)$ be the spline uniquely determined by the orthogonality conditions

$$\int_0^1 [s(x) - f(x)] B_i^m(x)\,dx = 0, \quad (i=1,2,\ldots,m+p).$$

$P_h^{\perp} f$ is the orthogonal projection of f onto the space of splines of order m.

3. DEGENERATE SUBSTITUTION KERNELS

In order to solve the eigenvalue problem

$$(3.1) \qquad \kappa\varphi = K\varphi$$

$$K\varphi(x) = \int_0^1 k(x,y)\varphi(y)\,dy, \quad k \in L_2(I \times I),$$

approximately, we apply the method of degenerate substitution kernels. Given an approximation $\tilde{k}(x,y)$ in degenerate representation

(3.2) $\tilde{k}(x,y) = \sum_{i,j=1}^{n} c_{ij}\psi_i(x)\eta_j(y)$,

the equation

(3.3) $\tilde{\kappa}\tilde{\varphi} = \tilde{K}\tilde{\varphi}$

$$\tilde{\kappa}\tilde{\varphi}(x) = \int_0^1 \tilde{k}(x,y)\tilde{\varphi}(y)\,dy$$

can be solved exactly. (3.3) is equivalent to the eigenvalue problem of the matrix E = CP,

$C := (c_{ij})_{i,j=1}^{n}$ and $P := (\int_0^1 \eta_i(x)\psi_j(x)\,dx)_{i,j=1}^{n}$

([2],[4]) .

In order to derive estimates for the quality of the approximation to the eigenvalues κ_i of (3.1) by the eigenvalues $\tilde{\kappa}_i$ of (3.3), we refer to results due to E. Schäfer ([5]); in [5] approximating kernels $\tilde{k}(x,y) := k_h(x,y)$ are considered which appear as tensor products of one-dimensional splines generated by certain approximation schemes and related to the mesh-length h. The results in [5] utilized here are summarized in the following

(3.4) Theorem: Let H be one of the Banach spaces $L_q(I)$, $1 \le q < \infty$, or $(C(I), \|\cdot\|)$. Let $K, K_h : H \to H$ be compact integral operators and $\kappa \neq 0$ an eigenvalue of K of algebraic multiplicity μ . We further assume $\lim_{h\to 0} \|K - K_h\| = 0$. Then the following estimate holds: For h sufficiently small there exist exactly μ eigenvalues $\tilde{\kappa}_1, \tilde{\kappa}_2, \ldots, \tilde{\kappa}_\mu$ of K_h converging to κ if $h \to 0$. Moreover, there exists a constant c, depending on κ and K, but independent of h, such that

$$|\kappa - \frac{1}{\mu} \sum_{i=1}^{\mu} \tilde{\kappa}_i | \leq c(\|K(K-K_h)K\| + \|K - K_h\|^2) .$$

In [5] conditions (V1)-(V3) are given, allowing us to determine the exact order of convergence of this estimate. These assumptions concern the uniform boundedness, the degree of approximation and the degree of integration associated with polynomial spline operators P_h. It is established, too, that the local spline approximation (2.1) belongs to this class.

Instead of (V1)-(V3) we specify modified conditions which can be deduced from (V1)-(V3) and can be utilized for our purposes directly later on. These assumptions, which, therefore, are valid for the local spline approximation (2.1), are:

(3.5.1) Degree of approximation: There exist $d_2 > 0$ and $n_2 \in \mathbb{N}$ such that

$$\| f - P_h f\|_\infty \leq d_2 h^{n_2} \| D^{n_2} f\|_\infty$$

for every $f \in C^{n_2}(I)$.

(3.5.2) Degree of integration rule: There exist $d_3 > 0$, which only depends on $\sum_{i=0}^{n_3} \|D^i g\|_\infty$, and $n_3 \in \mathbb{N}$ such that

$$|\int_0^1 g(x)(f - P_h f)(x)dx| \leq d_3 h^{n_3} (\sum_{i=n_2}^{n_3} \|D^i f\|_\infty)$$

for every $f, g \in C^{n_3}(I)$.

(3.5) For (2.1), the assumptions hold with $n_2 = m$, $n_3 = m$ for m even and with $n_2 = m$, $n_3 = m+1$ for m odd. In the latter case, we rediscover the improvement of the accuracy of approximation by polynomials by integration.

4. SUBSTITUTION KERNELS USING BLENDING-SPLINES

Let P_h be a spline approximation operator as described in section 2. The blending-spline approximation of the kernel k(x,y) by means of P_h is given by

(4.1) $\tilde{k}(x,y) = P_h^x\, k(x,y) + P_h^y\, k(x,y) - P_h^x\, P_h^y\, k(x,y)$.

We can establish the following error bound:

(4.2) Let $k \in C^{2n_2}(I \times I)$, $P_h : C(I) \to S(D^m, \pi)$, satisfying (3.5.1) and each of the operator pairs P_h^x and P_h^y, D_x and P_h^y, D_y and P_h^x commute.
Then the error of the approximation (4.1) is bounded by

$$\| k - \tilde{k} \|_\infty \leq d_2^2\, h^{2n_2} \| D_x^{n_2} D_y^{n_2} k \|_\infty \quad .$$

Proof:

$$\| k - \tilde{k} \|_\infty = \| (k - P_h^x k) - P_h^y (k - P_h^x k) \|_\infty \leq$$

$$\leq d_2 h^{n_2} \| D_y^{n_2} (k - P_h^x k) \|_\infty =$$

$$= d_2 h^{n_2} \| (D_y^{n_2} k) - P_h^x (D_y^{n_2} k) \|_\infty \leq$$

$$\leq d_2^2 h^{2n_2} \| D_x^{n_2} D_y^{n_2} k \|_\infty \quad . \quad *$$

Since $\| K - K_h \| \leq \| k - \tilde{k} \|_\infty$, we receive from (4.2) immediately the estimate

(4.3) $\| K - K_h \| \leq c_2 h^{2n_2}$.

The local spline approximation (2.1), e.g., satisfies the assumptions (4.2); therefore the estimate (4.3) is valid.

Now let us use blending-splines to construct degenerate substitution kernels. We adjust (4.1), represented by B-splines,

$$\tilde{K}(x,y) = \sum_{i=1}^{m+p} B_i^m(x) g_i(y) + \sum_{j=1}^{m+p} h_j(x) B_j^m(x) -$$

$$- \sum_{i,j=1}^{m+p} a_{ij} B_i^m(x) B_j^m(y)$$

to a degenerate kernel (3.2) by the definition ([3]):

(4.4) $$n := 2(m+p)$$

$$\psi_i(x) := B_i^m(x) \; , \quad (i=1,2,\ldots,m+p) \; ,$$

$$\eta_j(y) := B_j^m(y) \; , \quad (j=1,2,\ldots,m+p) \; ,$$

$$\psi_{m+p+j}(x) := h_j(x) \; , \quad (j=1,2,\ldots,m+p) \; ,$$

$$\eta_{m+p+i}(y) := g_i(y) \; , \quad (i=1,2,\ldots,m+p) \; .$$

This definition implies

$$C = \left(\begin{array}{c|c} -a_{ij} & I_{m+p} \\ \hline I_{m+p} & 0 \end{array} \right) .$$

In order to apply theorem (3.4), we need an estimation of the operator-norm $\| K(K - K_h)K \|$. For this purpose, let us define the following notation: If $u, v : I \times I \to \mathbb{R}$, then $u * v : I \times I \to \mathbb{R}$ is defined by

$$(u * v)(x,y) := \int_0^1 u(x,\xi)v(\xi,y)d\xi \ .$$

Our sequence of bounds is

$$\|K(K-K_h)K\| \le \|k*[(k-P_h^x k) - P_h^y(k-P_h^x k)]*k\|_\infty \le$$

$$\le d_3 h^{n_3}\{\sum_{i=n_2}^{n_3} \|D_y^i(k*[k-P_h^x k])\|_\infty\} \le$$

$$\le d_3 h^{n_3}\{\sum_{i=n_2}^{n_3} \|k*(D_y^i k - P_h^x D_y^i k)\|_\infty\} \le$$

$$\le d_3 h^{n_3}\{\sum_{i=n_2}^{n_3} d_3' h^{n_3} \sum_{j=n_2}^{n_3} \|D_x^j D_y^i k\|_\infty\} \ ,$$

or, essentially,

$$\|K(K-K_h)K\| \le c_3 h^{2n_3} \ . \tag{4.5}$$

The two bounds (4.3) and (4.5) allow us to establish ($n_2 := m$) the result:

(4.6) <u>Theorem:</u> Let m be the order of a polynomial spline, $k \in C^{2n_3}(I \times I)$, $n_3 \ge m$ as in (3.5), and P_h be the corresponding spline operator. Let K_h be defined as the integral operator of the spline-blended substitution kernel generated by the operator P_h. Then the estimate holds

$$|\kappa - \frac{1}{\mu}\sum_{i=1}^{\mu} \tilde{\kappa}_i| \le c(c_3 h^{2n_3} + c_2 h^{4m}) =$$

$$= O(h^{\min(4m,2n_3)}) \ .$$

We are particularly interested in those splines for which n_3 is optimally high. As mentioned above, the local approximation (2.1) of even degree (m odd) warrants $n_3 = m+1$, so that $\min(4m, 2n_3) = 2m+2$.

For odd degree polynomials (2.1), we receive $n_3 = m$ and consequently $\min(4m, 2n_3) = 2m$. Therefore, we perceive that piecewise constant and linear spline-blending assures convergence $O(h^4)$, quadratic and cubic spline-blending $O(h^8)$.

(4.7) Let us now cast a glance at the orthogonal projection (2.2). For this approximation, $\|\cdot\|_2$ is the most suitable norm. Since the orthogonal projection $P_h^{\perp}$ gives $\|g - P_h^{\perp}g\|_2 \leq \|g - P_h g\|_2$ compared with any other spline operator, we have $\|g - P_h^{\perp}g\|_2 \leq b_1 h^m \|D^m g\|_2$ ([6], p.438). Further,

$$|\int_0^1 g(x)(f-P_h^{\perp}f)(x)dx| = |\int_0^1 (g-P_h^{\perp}g)(x)(f-P_h^{\perp}f)(x)dx| \leq$$

$$\leq \|g - P_h^{\perp}g\|_2 \|f - P_h^{\perp}f\|_2 \leq b_1^2 h^{2m} \|D^m g\|_2 \|D^m f\|_2 =$$

$$= b_2(g) h^{2m} \|D^m f\|_2 .$$

We recognise $n_3 = 2m$. Revising the previous estimations, now related to $\|\cdot\|_2$ and using theorem (3.4) ($q := 2$), we arrive again at theorem (4.6). $\min(4m, 2n_3) = 4m$ yields $O(h^{4m})$ convergence of the eigenvalues. That is to say $O(h^8)$ for linear blending, $O(h^{12})$ for quadratic blending and $O(h^{16})$ in the cubic orthogonal projection blending case.

5. CONCLUDING REMARKS

Numerical examples have been calculated ([1]) in order to examine the convergence orders predicted by the estimates given in this article. Among these were highly oscillating kernels like $\sin 9\pi(x+y)$ and very smooth ones like exp(xy). The examples cover the spline orders m = 1,2,3,4,5 when applying the local spline approximation (2.1) and m = 1,2,3 for (2.2). The numerical calculations uniformly corroborated the theoretical results.

In this paper, only polynomial splines were taken into consideration. Generalisations to different kinds of splines are possible. In particular, trigonometric splines have already been studied in [1] and led to corresponding results. But since polynomial splines seem to be the most transparent ones, we restrict ourselves here to these.

Acknowledgement: The authors appreciate helpful discussions with Dr. Schäfer on the presentation of the results.

REFERENCES

[1] L. Bamberger: Die Anwendung allgemeiner Splines zur numerischen Behandlung von Integralgleichungen. Diplomarbeit Universität München 1981.

[2] G. Hämmerlin: Zur numerischen Behandlung von homogenen Fredholmschen Integralgleichungen 2. Art mit Splines. *Spline Functions Karlsruhe 1975*, Lecture Notes in Mathematics vol. 501, ed. by K. Böhmer, G. Meinardus and W. Schempp, 92-98, Springer-Verlag 1976.

[3] G. Hämmerlin - W. Lückemann: The Numerical Treatment of Integral Equations by a Substitution Kernel Method Using Blending Splines. *Memorie della Accademia Nazionale de Scienze, Lettere et Arti di Modena - Serie VI*, Vol. XXI - 1979, 1-15 (1981).

[4] G. Hämmerlin - L.L. Schumaker: Procedures for Kernel Approximation and Solution of Fredholm Integral Equations of the Second Kind. *Numer. Math.* 34, 125-141 (1980).

[5] E. Schäfer: Spectral Approximation for Compact Integral Operators by Degenerate Kernel Methods. *Numer. Funct. Anal. and Oprtimiz.*, 2, 43-63 (1980).

[6] L.L. Schumaker: *Spline Functions: Basic Theory*. John Wiley & Sons 1981.

ON A REGULARIZATION METHOD FOR FREDHOLM EQUATIONS OF THE FIRST KIND USING SOBOLEV SPACES

J. T. Marti

ETH Zürich

1. INTRODUCTION

Let $Kf = g$ be a Fredholm integral equation of the first kind, with given K and g explicitly written out as

$$\int_0^1 k(s,t)f(t)dt = g(s) , \quad s \in [0,1] \tag{1}$$

where k is a real function on the square $[0,1]^2$. The ill-posedness [9] of most equations of the form (1) is the origin of frequently occurring difficulties when dealing with methods for solving (1) numerically. The trouble with collocation methods is that a discretization process transforms (1) into another problem (often a system of linear equations which may be solved by the well known singular value decomposition) which can lead to solutions which may or may not deviate strongly from the (minimum norm) solution of (1). On the other hand, regularization methods (e.g. Tikhonov's method [8]) may suffer from the fact that approximate solutions obtained by these methods are dependent on the chosen regularization parameter λ . Therefore, it is important to define algorithms giving λ in a way which does not crucially depend on the intuition of the user of a regularization method. However, some a priori knowledge of the shape or smoothness of the minimum norm solution of (1) will probably always be needed if one has to solve ill-posed problems.

In this paper, it is shown that Sobolev spaces may serve as a tool for a partial regularization of equations of the type (1), where a Sobolev space $H^m(0,1)$ for nonnegative integers m and on the real open interval $(0,1)$ (for the sake of simplicity) is defined as the completion of the set $C^m(0,1)$ of bounded continuous and m-times bounded continuously differentiable real functions on $(0,1)$ with respect to the Sobolev norm given by

$$\| f \|_m = \left(\sum_{i=0}^m \int_0^1 f^{(i)}(t)^2 \, dt\right)^{1/2} , \qquad f \in C^m(0,1) .$$

In the usual way, the norm $\| \; \|_m$ of these Hilbert spaces is generated by the scalar product $(\; , \;)_m$ defined by

$$(f,g)_m = \sum_{i=0}^m \int_0^1 f^{(i)}(t) g^{(i)}(t) dt , \qquad f,g \in H^m(0,1) .$$

Numerical results, as well as a comparison with the collocation method, are presented for a severely ill-posed example of (1), where

$$k(s,t) := \exp(-st) , \quad g := Kf , \quad f(t) := 1 - \exp(-t) ,$$
$$s,t \in (0,1) .$$

2. AN ALGORITHM FOR THE CHOICE OF A REGULARIZATION PARAMETER

In order to be more specific on the assumptions made for (1), let K be a bounded linear operator from the real Sobolev space $H^0(0,1)$ $(= L_2(0,1))$ into $H^m(0,1)$ $(m = 0,1)$. For $m = 0$ this is the case if k is in $L_2((0,1)^2)$ or if the functions $\int_0^1 |k(s,.)| ds$ and $\int_0^1 |k(.,t)| dt$ are in $L_\infty(0,1)$. For $m = 1$

a sufficient condition for the boundedness of k is that k and its first partial derivative k_s are both in $L_2((0,1)^2)$ [5]. If $R(K)$ denotes the range of K, let g be an element of $R(K)\oplus R(K)^{\perp}$, where by definition (1) is called an ill-posed problem if this last set is distinct from $H^m(0,1)$ which is the case if and only if the generalized inverse of K (having $R(K)\oplus R(K)^{\perp}$ as domain of definition and its range in $H^o(0,1)$) is not bounded [3]. For later use, let P be the orthogonal projection from $H^m(0,1)$ onto the closure $\overline{R(K)}$ of $R(K)$.

The following algorithm [6] can serve to find a good choice of a regularization parameter which requires from the user just a minimum of knowledge about the minimum norm solution of f_o of (1):

Let $\{V_n\}$ $(n = 2^j,\ n \in \mathbb{N})$ be an increasing sequence of finite dimensional subspaces of $H^o(0,1)$ with union dense in $H^o(0,1)$. This sequence uniquely defines a sequence $\{a_n\}$ of nonnegative numbers given by

$$a_n := \inf\{ \| Kv - g \|_m : v \in V_n \} .$$

It is easy to see that each a_n may be computed by an application of the singular value decomposition (e.g. using the procedure MINFIT [2]) which allows the determination of the minimum a_n^2 of the quadratic form

$$\underline{x}^T B_n \underline{x} - 2\underline{w}_n^T \underline{x} + (g,g)_m , \qquad \underline{x} \in \mathbb{R}^{m(n)} ,$$

for a basis $\{v_1,\ldots,v_{m(n)}\}$ of V_n, where the $m(n)$ by $m(n)$ matrix B_n and the $m(n)$-vector $\underline{w}_n$ have entries $(Kv_i,Kv_j)_m$ and $(Kv_i,g)_m$ respectively. For later use let us also introduce the Gramian M_n of the basis with entries $(v_i,v_j)_o$. The only information on the minimum norm solution f_o of (1) is required by the knowledge of a sequence $\{b_n\}$ of positive numbers such that

$$\lim_n \| P_n f_o - f_o \|_o / b_n = 0 ,$$

where P_n is the orthogonal projection from $\overset{o}{H}(0,1)$ onto V_n. The following geometric idea then leads to an explicit construction of a sequence of regularization parameters λ_n : Let f_n be the (unique) nearest point to 0 in the closed convex set

$$V_n \cap [f_o + (a_n^2 - \| g - Pg \|_o^2 + b_n^2)^{1/2} K^{-1}(U_o)] ,$$

where $K^{-1}(U_o)$ is the inverse image by K of the unit ball U_o of $\overset{o}{H}(0,1)$. The idea for the modification of the original algorithm [6] by taking g in $R(K) \oplus R(K)^{\perp}$ and $(a_n^2 - \|g-Pg\|_o^2 + b_n^2)^{1/2}$ instead of $a_n + b_n$ is due to M. Trummer [10] who himself has been inspired by the recent paper of Ch. Groetsch [4] on this algorithm. Algebraically, the above definition of

$$f_n = \sum_{i=1}^{m(n)} x_i v_i$$

in V_n means that one has to solve the following nonlinear system

$$-\lambda \underline{x}^T M_n \underline{x} - \underline{w}_n^T \underline{x} + (g,g)_m - a_n^2 - b_n^2 = 0$$

$$(B_n + \lambda M_n) \underline{x} = \underline{w}_n$$

for the pair $(\lambda, \underline{x})$ (depending on n) in $\mathbb{R} \times \mathbb{R}^{m(n)}$ by the methods of Newton and Cholesky. According to numerical experience just a few Newton steps are sufficient for the computation of the regularization parameter λ , where the corresponding $m(n)$-vector $\underline{x}$ yields the approximation f_n in V_n for f_o .

3. THE ALGORITHM FOR LINEAR AND CUBIC SPLINE SPACES

Let $h := 2^{-n}$ and let $\{0,h,2h,\dots,1\}$ be an equidistant partition of $[0,1]$. Two sorts of bases for V_n have been used, both consisting of B-splines of degree r, where r is 1 or 3 respectively and $m(n) = n+r$. Based on a recurrence relation of C. de Boor [1] one obtains for B-splines of degree r with support $[0,(r+1)h]$ for its value at the point t the number c_1, where

$$c_p := 2\delta_{pq}, \quad 1 \leqq p \leqq r+1, \quad \text{where} \quad q := \sup\{j \leqq t/h+1 : j \in \mathbb{N}\}$$

$$\text{and} \quad [(s := (t/h-q+1)/p, \quad c_q := sc_q + (1-s)c_{q+1}),$$

$$1 \leqq q \leqq r-p+2], \qquad 2 \leqq p \leqq r+1.$$

The corresponding (symmetric) Gramians M_n can be determined analytically. For $r = 1$, $6nM_n$ is triangular with main diagonal 2,4,4,...,4,2 and off-diagonal elements 1. For $r=3$, $7!4nM_n$ is a quintic band matrix with main diagonal 20,1208,2396, 2416,2416,..., first off-diagonal 129,1062,1191,1191,..., second off-diagonal 60,120,120,... and 1 in the third off-diagonal. The entries of B_n and $\underline{w}_n$ have to be determined by numerical integration using the first m derivatives of K and g. The scalar products $(\ ,\ )_m$ involved are computed by the trapezoidal rule on the mesh points $0,h,2h,\dots,1$, where the evaluation of $Kv_i(jh)$ $(0 \leqq j \leqq n)$ is done by applying $r+3$-point Gaussian quadrature on each of the subintervals of length h in $\operatorname{supp}(v_i)$.

If f_o lies in $H^{r+1}(0,1)$ then due to well known results on spline interpolation [7] one has

$$\| P_n f_o - f_o \|_o = O(h^{r+1}) .$$

Therefore, a reasonable choice of b_n is, e.g., $b_n = 10^{-3} h^{r+1} (-\ln h)$. The convergence proofs of [6,4,10] then yield an upper bound for the convergence rate of about $O(b_n^{1/2}) = O(h^{(r+1)/2})$.

4. NUMERICAL EXAMPLES

Let k and g in (1) have the special values

$$k(s,t) = e^{-st}$$

and

$$g(s) = \int_0^s k(s,t) f(t) dt$$

$$= [1 - \exp(-s)((1 - 1/e)s + 1)]/s(s+1) ,$$

where $s,t \in [0,1]$ and

$$f(t) = 1 - \exp(-t) .$$

Since the linear integral operator of (1) defined this way is injective (from $L_2(0,1)$ into itself), it is clear that $f_o := f$ is the minimum norm solution of (1). In the following table 1 one can compare the number of subintervals of equal length and the computing time needed to obtain L_2-error norms of 0.002 for the regularization method using the Sobolev spaces $H^o(0,1)$ and $H^1(0,1)$ with an algorithmic choice of

the regularization parameters described in the last sections. The regularization based on $H^1(0,1)$ requires the first derivatives of k and g. Obviously, $k_s(s,t) = -tk(s,t)$, and g_s is given by

$$g_s(s) = [s-(s-1)/e-(2s+1)g(s)]/s(s+1) .$$

TABLE 1

Represented are the pairs n, t_c in the form $n(t_c)$, where

n = number of equidistant intervals needed for L_2-error norms of 0.002,

t_c = computing time in seconds on a CDC 6400/6500,

r = degree of the spline functions

$H^m(0,1)$ = Sobolev space used in the method to measure the residual functions

r =	1	3
$H^o(0,1)$	128 (30)	4 (0.2)
$H^1(0,1)$	16 (1)	4 (0.2)

A comparison with the collocation method with equidistant collocation points $s_i = t_i = (2i-1)/2n$, $1 \leqq i \leqq n$, shows (table 2) that the last method never attains the above level of an L_2-error norm of 0.002. Moreover, the collocation method shows an unstable behavior of the error norm when n becomes large. Here, the systems of linear equations have been solved again with the procedure MINFIT mentioned in Section 2.

TABLE 2

$H^o(0,1)$-error norm e for collocation with n equidistant mesh points

n =	2	4	8	16	32	64	128
e =	.04	.09	.17	.04	.02	.03	.03

REFERENCES

1. DE BOOR, C., On Calculation with B-splines. *J. Approx. Theory* 6, 50-62 (1972).
2. GOLUB, G., and REINSCH, C., Singular Value Decomposition and Least Squares Solutions. *Numer. Math.* 14, 403-420 (1970).
3. GROETSCH, C.W., *Generalised Inverses of Linear Operators*. M. Dekker Inc., New York (1977).
4. GROETSCH, C.W., A Regularization-Ritz Method for Fredholm Equations of the First Kind. *J. Integral Equations* 4, 173-182 (1982).
5. HOIDN, H.-P., *Die Kollokationsmethode angewandt auf die Symmsche Integralgleichung*. ETH Zurich, in preparation.
6. MARTI, J.T., On the Convergence of an Algorithm Computing Minimum-Norm Solutions of Ill-Posed Problems. *Math. Comp.* 34, 521-527 (1980).
7. SCHULTZ, M.H., *Spline Analysis*. Prentice-Hall, Englewood Cliffs (1973).
8. TIKHONOV, A.N., Regularization of Incorrectly Posed Problems. *Soviet Math. Dokl.* 4, 1035-1038 (1963).
9. TIKHONOV, A.N., and ARSENIN, V.Y., *Ill-Posed Problems*. Winston, Washington (1977).
10. TRUMMER, M., thesis in preparation.

REDUCIBLE QUADRATURE METHODS
FOR VOLTERRA INTEGRAL EQUATIONS

Paul H.M. Wolkenfelt

Technische Hogeschool Eindhoven, the Netherlands

ABSTRACT

A formal relationship between quadrature rules and linear multistep methods (ρ,σ) for ordinary differential equations is exploited for the generation of quadrature weights. Employing the quadrature rules constructed in this way, step-by-step methods for Volterra integral and integro-differential equations are defined and convergence and stability results are presented.

Also, for this class of quadrature rules, a simple characterization of both the repetition factor and numerical stability is given in terms of the location of the zeros of the polynomial ρ from which results with respect to a conjecture of Linz can be derived in a rather straightforward fashion.

Finally, a class of so-called modified multilag methods is presented which display the same stability behaviour as (ρ,σ)-reducible quadrature methods but have the additional advantage of being more efficient.

The material presented in this lecture comprises the main results from the papers [1,2,3,4] and the author's dissertation [5].

REFERENCES

1. WOLKENFELT,P.H.M., The construction of reducible quadrature rules for Volterra integral and integro-differential equations *IMA Journal of Numerical Analysis* (1982), to appear.
2. WOLKENFELT, P.H.M., Reducible quadrature methods for Volterra integral equations of the first kind, *BIT* 21, 232-241 (1981).
3. WOLKENFELT, P.H.M., On the relation between the repetition factor and numerical stability of direct quadrature methods for second kind Volterra integral equations, submitted for publication.
4. WOLKENFELT, P.H.M., Modified multilag methods for Volterra functional equations, *Math. Comp.* (1983), to appear.
5. WOLKENFELT, P.H.M., *The numerical analysis of reducible quadrature methods for Volterra integral and integro-differential equations*, dissertation, Mathematisch Centrum, Amsterdam,1981.

AN AUTOMATIC ITERATION SCHEME AND ITS APPLICATION TO NONLINEAR OPERATOR EQUATIONS

Wolfgang Volk

Hahn-Meitner-Institut Berlin

1. INTRODUCTION

The aim of this paper is to present an iteration process which acts with a sequence of contractive operators. The investigation of such algorithms began with the works of Ehrmann [3], Schmidt [7] and others in the late fifties. They treat the iteration scheme

$$x_n = T_n x_{n-1} \tag{1.1}$$

for given $x_0 \in X$ and $(T_n : X \to X)_{n \in N}$, where X is a generalized metric space. In the present paper it is assumed for simplicity that (X,d) is a (complete) metric space.

Since it is not meaningful to change the iteration operator in each step it is desirable to get an a posteriori criterion for varying the operators. An heuristic approach to this problem has been given by Atkinson [2] p. 168ff, where in addition estimations for the distance to the true solution are developed, yielding a criterion for the termination of the process. Both problems, the control of the variation of the operators and the development of a criterion for the termination of the iterations, are treated in the second section of this paper, where practicable exactness is achieved in the first case.

The third part of the present paper is concerned with the applicability of the iteration process to nonlinear operator equations. First a theory for approximate solutions is developed; and under slightly stronger assumptions, the convergence of the iterations is proved. The theorem of Kantorovich [1], [4], [5] and [6] is of fundamental importance for the investigations.

2. AN AUTOMATIC CONTROL FOR ITERATIONS WITH A SEQUENCE OF CONTRACTIONS

First the assumptions have to be presented such that the iteration scheme can be applied.

Let (X,d) be a metric space, $x^* \in X$ be the desired (isolated) solution of a given problem and let an approximate method be given, which yields a sequence of approximate solutions $(x_n^*)_{n\in N}$ with the property that there exist $C\geq 0$ and $\rho\in(0,1)$ such that for any $n\in N$ the inequality

$$d(x^*,x_n^*) \leq C\cdot\rho^n \tag{2.1}$$

holds. This situation for example is given for most of the approximate methods for solving linear operator equations of the second kind (cf. [2], [8], [9] and [11]). The quantity ρ^n is known to be h_n^p, where h_n is the discretization parameter and p is the order of convergence. For the application of the algorithm stated in this section the quantities h_n must be proportional to k^n, where k is a natural number exceeding 1.

In addition, let a sequence of subspaces of X named $(X_n)_{n\in N}$ with $X_n \subset X_{n+1}$ for $n\in N$ and a sequence of contractions $(T_n:X_n \to X_n)_{n\in N}$ be given in such a way that for any $n\in N$ $x_n^*\in X_n$ is the uniquely determined fixed point of T_n. For the following investigations let q_n denote the contraction constant of the mapping T_n.

Subsequently the a posteriori criterion for the variation of the iteration operators $(T_n)_{n\in N}$ is derived under the assumption that neither C and x^* nor x_n^* are known for large n. The aim of this development is to construct a sequence $(y_n)_{n\in N}$, which converges to the true solution x^* as fast as the sequence $(x_n^*)_{n\in N}$, which means that there exists $\tilde{C}\geq 0$ such that for any $n\in N$

$$d(x^*,y_n) \leq \tilde{C}\cdot\rho^n \tag{2.2}$$

holds.

In the following description of the process the sequence $(a_n)_{n\in N}$ is constructed additionally, which is needed for the verification of the desired property of $(y_n)_{n\in N}$.

1. Compute $y_1:=x_1^*$ and $y_2:=x_2^*$ by the given approximate method and define $d(x_1^*,x_2^*)=:b_1=:a_1$.

2. Set $y_3:=y_2$ and iterate with T_3 ($\tilde{y}_3:=y_3$; $y_3:=T_3\tilde{y}_3$) until

   $$[\ d(y_3,x_3^*) \leq]\ \frac{q_3}{1-q_3}\cdot d(y_3,\tilde{y}_3) \leq \alpha\cdot\rho^2\cdot b_1$$

   holds. Then

$$\begin{aligned}
d(x_2^*,x_3^*) &\leq d(x_2^*,y_3) + d(y_3,x_3^*) \leq d(y_2,y_3) + \frac{q_3}{1-q_3}\cdot d(y_3,\tilde{y}_3)=:b_2 \\
&\leq d(x_2^*,x^*) + d(x^*,x_3^*) + d(x_3^*,y_3) + \frac{q_3}{1-q_3}\cdot d(y_3,\tilde{y}_3) \\
&\leq C(\rho^2+\rho^3) + 2\cdot\frac{q_3}{1-q_3}\cdot d(y_3,\tilde{y}_3) \leq C(\rho^2+\rho^3) + 2\alpha\cdot\rho^2\cdot a_1 =:a_2 .
\end{aligned}$$

3. Set $y_n := y_{n-1}$ and iterate with T_n ($\tilde{y}_n := y_n$; $y_n := T_n\tilde{y}_n$) until

$$[\ d(y_n,x_n^*) \le]\ \frac{q_n}{1-q_n}\cdot d(y_n,\tilde{y}_n) \le \alpha\cdot\rho^2\cdot b_{n-2}$$

holds. Then

$$\begin{aligned}
d(x_{n-1}^*,x_n^*) &\le d(x_{n-1}^*,y_{n-1}) + d(y_{n-1},y_n) + d(y_n,x_n^*)\\
&\le \frac{q_{n-1}}{1-q_{n-1}}\cdot d(y_{n-1},\tilde{y}_{n-1}) + d(y_{n-1},y_n) + \frac{q_n}{1-q_n}\cdot d(y_n,\tilde{y}_n) =: b_{n-1}\\
&\le \frac{q_{n-1}}{1-q_{n-1}}\cdot d(y_{n-1},\tilde{y}_{n-1}) + d(y_{n-1},x_{n-1}^*) + d(x_{n-1}^*,x^*) +\\
&\quad + d(x^*,x_n^*) + d(x_n^*,y_n) + \frac{q_n}{1-q_n}\cdot d(y_n,\tilde{y}_n)\\
&\le 2\alpha\cdot\rho^2\cdot b_{n-3} + C(\rho^{n-1}+\rho^n) + 2\alpha\cdot\rho^2\cdot b_{n-2}\\
&\le 2\alpha\cdot\rho^2\cdot a_{n-3} + C(1+\rho)\cdot\rho^{n-1} + 2\alpha\cdot\rho^2\cdot a_{n-2} =: a_{n-1}\ .
\end{aligned}$$

4. Raise n and continue at step 3.

The constant α is assumed to be given, its relevance should become clear in theorem 2.2.

The idea for this algorithm arises from the observation that, if $d(x^*,x_n^*) = O(\rho^n)$, also $d(x_n^*,x_{n-1}^*) = O(\rho^n)$. Under the assumption of $d(x_{n+1}^*,x_n^*) = \rho\cdot d(x_n^*,x_{n-1}^*)$ for $n\in N$ it is meaningful to terminate the iterations with the operator T_n if

$$\begin{aligned}
[\ d(y_n,x_n^*) \le]\ &\frac{q_n}{1-q_n}\cdot d(y_n,\tilde{y}_n) \le \alpha\cdot d(x_{n+1}^*,x_n^*) =\\
&= \alpha\cdot\rho^2\cdot d(x_{n-1}^*,x_{n-2}^*) \le \alpha\cdot\rho^2\cdot b_{n-2}\ .
\end{aligned} \tag{2.3}$$

As it will be shown in the following this assumption is not necessary for the correctness of the algorithm.

2.1 Lemma [11]: For any $n\in N$

$$a_n \le \frac{C}{2\alpha}\cdot\rho^n\cdot\sum_{i=1}^{n}(2\alpha)^i\cdot(1+\rho)^i\ . \tag{2.4}$$

This lemma can be proved by induction, but the proof will be omitted here.

2.2 Theorem [11]: If α is chosen such that $0 < 2\alpha\cdot(1+\rho) < 1$, then

$$d(x^*,y_n) = O(\rho^n)\ . \tag{2.5}$$

Proof: Since $d(x^*,y_n) \le d(x^*,x_n^*) + d(x_n^*,y_n)$ and $d(x^*,x_n^*) = O(\rho^n)$ it remains to show that $d(x_n^*,y_n) = O(\rho^n)$. But by the assumption the estimate

$$d(x_n^*,y_n) \le \alpha\cdot\rho^2\cdot b_{n-2} \le \alpha\cdot\rho^2\cdot a_{n-2} < \frac{C}{2}\cdot\rho^n\cdot\frac{1}{1-2\alpha\cdot(1+\rho)}$$

holds. □

Obviously the presupposition of the theorem holds for any $\rho\in(0,1)$ if α is chosen to be less than or equal to 1/4. The choice $\alpha = 1/8$ yields $d(x_n^*,y_n) \le C\cdot\rho^n$ for $n\in N$ and arbitrary ρ.

In the following second part of this section the problem of termination of the iterations will be considered. The application of this discussion is not restricted to the iteration process stated above; it is still applicable to any sequence $(y_n)_{n\in N}$ with $d(x^*,y_n) \le \tilde{C}\cdot\rho^n$, where $\tilde{C}$ is unknown and an estimation of the quantity $d(x^*,y_n)$ is wanted for some n.

So the iterations may be terminated if the estimation for $d(x^*,y_n)$ is less than the precision wanted and therefore the consideration may be reduced to find good estimations.

By the assumptions it is evident that

$$x^*\in\bigcap_{n\in N} B(y_n,\tilde{C}\cdot\rho^n)\ , \tag{2.6}$$

where $B(x,r)$ denotes the closed ball with the radius $r>0$ centered at $x\in X$.

Let

$$D_2 := \frac{d(y_1,y_2)}{1-\rho}\cdot\rho^{-1} \tag{2.7}$$

and for $n\ge 3$

$$D_n := \max\{D_{n-1},\ \frac{d(y_{n-1},y_n)}{1-\rho}\cdot\rho^{1-n}\}\ . \tag{2.8}$$

It turns out that the sequence $(D_n)_{n\ge 2}$ is monotonely increasing and bounded by $\tilde{C}(1+\rho)/(1-\rho)$. For $D := \lim D_n$ the following relation holds: For any $n\in N$ $B(y_{n+1},D\cdot\rho^{n+1}) \subset B(y_n,D\cdot\rho^n)$. Hence for any $n\in N$ $d(x^*,y_n) \le D\cdot\rho^n$. This approach does not guarantee the estimate $d(x^*,y_n) \le D_n\cdot\rho^n$, which is only available in current computation. By numerical results [11] it is suggested that the sequence $(D_n^*)_{n\ge 3}$ defined by

$$D_n^* := D_n + (D_n - D_{n-1})\cdot\rho\cdot(1-\rho)^{-1} \tag{2.9}$$

for $n\ge 3$ works really better than the sequence $(D_n)_{n\ge 3}$. An extensive presentation of the theoretical background for the definition of the sequence $(D_n^*)_{n\ge 3}$ is given in [11].

3. THE APPROXIMATE SOLUTION OF NONLINEAR OPERATOR EQUATIONS

The theory developed in this section is based on the theorem of Kantorovich ([1], [4], [5] and [6]), which gives an existence and uniqueness result and moreover provides a constructive method for the computation of the solution of nonlinear equations.

3.1 Theorem of Kantorovich: Let $A:X \to Y$ be an operator acting between the two Banach spaces X and Y and let $x_o \in X$. Assume that $A'(x_o)^{-1}:Y \to X$ is continuous and define $\eta := \|A'(x_o)^{-1} A(x_o)\|$. If there exists $r>0$ such that

(i) there exists $\bar{\kappa} \geq 0$ such that for any $x,y \in B(x_o,r)$ $\|A'(x)-A'(y)\| \leq \bar{\kappa} \cdot \|x-y\|$,

(ii) for $\kappa := \bar{\kappa} \cdot \|A'(x_o)^{-1}\|$ the inequality $\eta \cdot \kappa < 1/2$ holds, and

(iii) $\underline{r} := \dfrac{1-\sqrt{1-2\cdot\eta\cdot\kappa}}{\kappa} \leq r < \bar{r} := \dfrac{1+\sqrt{1-2\cdot\eta\cdot\kappa}}{\kappa}$

then the equation $A(x) = 0$ possesses a unique solution $x^* \in B(x_o,r)$ and the estimate $\|x^*-x_o\| \leq \underline{r}$ holds. If $\eta \cdot \kappa = 1/2$ the assertions are true if the hypothesises are satisfied for $r = \underline{r} = \bar{r} = \kappa^{-1}$. The modified Newton's method, working by the iteration rule

$$x_\nu := x_{\nu-1} - A'(x_o)^{-1} A(x_{\nu-1}) , \tag{3.1}$$

yields the sequence $(x_\nu)_{\nu \in N}$ converging to x^*.

The next theorem provides a theory of the existence, uniqueness, and convergence of the solutions of approximate equations.

3.2 Theorem: Let $A:X \to Y$ be an operator acting between the two Banach spaces X and Y and let $x^* \in X$ be an isolated solution of the equation $A(x) = 0$. Let $(A_n:X_n \to Y_n)_{n\in N}$ be a sequence of differentiable operators, where for $n \in N$ X_n and Y_n are subspaces of X and Y respectively. Let $(Q_n:X \to X)_{n \in N}$ be a sequence of projections (not necessarily linear or continuous) converging pointwise to the identity and with $X_n = Q_n[X]$ for any $n \in N$. Further requirements are:

(i) there exists $\bar{\kappa} \geq 0$ such that for any $n \in N$, $x,y \in X_n$ $\|A_n'(x)-A_n'(y)\|_{X_n \to Y_n} \leq \bar{\kappa} \cdot \|x-y\|$,

(ii) $(A_nQ_n:X \to Y)_{n \in N}$ converges pointwise to A, and

(iii) for any $x \in X$ the set $\{\|A_n'(Q_nx)^{-1}\|_{Y_n \to X_n} \mid n \in N\}$ is bounded.

Then there exists $n_o \in N$ such that for $n \geq n_o$ there exists a solution $x_n^* \in X_n$ of the equation $A_n(x) = 0$ and this solution is the only one in $B(Q_nx^*, r_n)$ for

$$\underline{r}_n := \frac{1-\sqrt{1-2\cdot\eta_n\cdot\kappa}}{\kappa} \leq r_n < \frac{1+\sqrt{1-2\cdot\eta_n\cdot\kappa}}{\kappa} =: \bar{r}_n , \tag{3.2}$$

where

$$\eta_n := \|A_n'(Q_n x^*)^{-1} A_n(Q_n x^*)\| \tag{3.3}$$

and

$$\kappa := \bar{\kappa} \cdot \sup\{\|A_m'(Q_m x^*)^{-1}\| \mid m \in N\} . \tag{3.4}$$

Furthermore

$$\|x^* - x_n^*\| \le \underline{r}_n \to 0 \text{ with } n \to \infty . \tag{3.5}$$

Proof: For $n \in N$ $\eta_n \le \sup\{\|A_m'(Q_m x^*)^{-1}\| \mid m \in N\} \cdot \|A_n(Q_n x^*) - A(x^*)\|$. According to the assumptions $(\eta_n)_{n \in N}$ converges to 0; which yields the existence of $n_o \in N$ such that for $n \ge n_o$

$$\eta_n \cdot \bar{\kappa} \cdot \sup\{\|A_m'(Q_m x^*)^{-1}\| \mid m \in N\} < \frac{1}{2} .$$

Hence the hypothesises of theorem 3.1 are satisfied for sufficiently large n. From theorem 3.1

$$\|x^* - x_n^*\| \le \|x^* - Q_n x^*\| + \|Q_n x^* - x_n^*\| \le \|x^* - Q_n x^*\| + \underline{r}_n \le$$

$$\le \|x^* - Q_n x^*\| + 2 \cdot \eta_n \to 0$$

with $n \to \infty$. □

It is clear that the hypothesises of theorem 3.2 are needed only in a neighbourhood of x* rather than in the whole space X. For the investigation of the applicability of the iteration scheme of section 2 it is necessary to suppose the uniform Lipschitz continuity of the operators $\{A_n \mid n \in N\}$.

3.3 Theorem: Let A and $(A_n)_{n \in N}$ be given as in theorem 3.2 with the additional assumptions that there exists $\kappa^* \ge 0$ such that for any $n \in N$ and $x, y \in X_n$ $\|A_n(x) - A_n(y)\| \le \kappa^* \cdot \|x - y\|$ and that the subspaces X_n $(n \in N)$ are nested. Then there exists $n_1 \in N$ such that the iteration process of section 2 works with the operators

$$T_n := (x \mapsto x - A_n'(y_{n-1})^{-1} A_n(x)) \tag{3.6}$$

for $n \ge n_1$ provided there exist $C \ge 0$ and $\rho \in (0,1)$ such that for $n \ge n_1$ $\|x^* - x_n^*\| \le C \cdot \rho^n$. The algorithm has to be started with the index n_1 instead of 1.
Proof: Let $M := \sup\{\|A_m'(Q_m x^*)^{-1}\|_{Y_m \to X_m} \mid m \in N\}$. Then there exists $n_1 \ge n_o$ such that for any $n \ge n_1$

$$\left(1 + \frac{1}{2[1 - 2\alpha(1+\rho)]}\right) \cdot C \cdot \rho^{n-1} + \|x^* - Q_n x^*\| \le \frac{1}{2 \cdot M \cdot \bar{\kappa}}$$

and

$$2 \cdot M \cdot \kappa^* \cdot C \cdot \left(\frac{1}{2[1 - 2\alpha(1+\rho)]} + 1 + \rho\right) \cdot \bar{\kappa} \cdot M \cdot \rho^{n-1} < \frac{1}{2} .$$

Thus for $n \geq n_1$ (remember that $y_{n_1} = x^*_{n_1}$) it can be shown inductively that the operator

$$A'_n(y_{n-1})^{-1} : Y_n \to X_n$$

exists by proposition 1.3 in [1] and is bounded by

$$\frac{\|A'_n(Q_n x^*)^{-1}\|}{1-\|A'_n(Q_n x^*)^{-1}\|\cdot\|A'_n(y_{n-1})-A'_n(Q_n x^*)\|} \leq \frac{M}{1-M\cdot\bar{\kappa}\cdot\|y_{n-1}-Q_n(x^*)\|} \leq$$

$$\frac{M}{1-M\cdot\bar{\kappa}\cdot[\|y_{n-1}-x^*_{n-1}\|+\|x^*_{n-1}-x^*\|+\|x^*-Q_n x^*\|]} \leq \frac{M}{1-M\bar{\kappa}\cdot(2M\bar{\kappa})^{-1}} \leq 2M$$

since $\|y_{n-1}-x^*_{n-1}\| \leq \frac{C}{2}\cdot\rho^{n-1}\cdot\frac{1}{1-2\alpha(1+\rho)}$ and $\|x^*_{n-1}-x^*\| \leq C\cdot\rho^{n-1}$.
Moreover

$$\eta_n\bar{\kappa}M \leq \|A'_n(y_{n-1})^{-1}\|\cdot\|A_n(y_{n-1})\|\cdot\bar{\kappa}M \leq 2M\cdot\|A_n(y_{n-1})-A_n(x^*_n)\|\cdot\bar{\kappa}M$$

$$\leq 2M\kappa^*\cdot[\|y_{n-1}-x^*_{n-1}\| + \|x^*_{n-1}-x^*_n\|]\cdot\bar{\kappa}M$$

$$\leq 2M\kappa^*\cdot[\frac{C}{2}\cdot\rho^{n-1}\cdot\frac{1}{1-2\alpha(1+\rho)} + C(1+\rho)\cdot\rho^{n-1}]\cdot\bar{\kappa}M$$

$$\leq 2M\kappa^*\cdot C\cdot[\frac{1}{2[1-2\alpha(1+\rho)]} + 1 + \rho]\cdot\bar{\kappa}M\cdot\rho^{n-1} < \frac{1}{2}. \;\square$$

4. THE APPLICATION OF THE ALGORITHM TO URYSOHN'S INTEGRAL EQUATION

This section is devoted to Urysohn's integral equation

$$x - \int_0^1 k(\cdot,s,x(s))\,ds = f, \tag{4.1}$$

where $f\in C[0,1]$ and $k, D_3k \in C([0,1]^2\times R)$. In addition it is assumed that k and D_3k are Lipschitz continuous with respect to the third variable, which means that there exists $L\geq 0$ such that for any $t,s\in[0,1]$ and $y,z\in R$ $|k(t,s,y)-k(t,s,z)| \leq L\cdot|y-z|$ and $|D_3k(t,s,y)-D_3k(t,s,z)| \leq L\cdot|y-z|$. For the sake of clarity, differentiation is written in operator context, so that for multivariate functions D_i denotes the partial differentiation with respect to the i-th argument.

The Urysohn integral equation can be written as

$$A(x) = x - K(x) - f = 0 \tag{4.2}$$

where A and K are nonlinear operators acting from and into $X=C[0,1]$. Throughout this section let C[0,1] be furnished with the supremum norm. An approximate operator sequence $(A_n)_{n\in N}$ can be constructed by the modified collocation method [8], [9], [11] in the following manner

$$\begin{aligned} A\ &: X_n \to X_n \\ g &\mapsto P_n(t\mapsto g(t)-\int_0^1 P_n k(t,\cdot,g(\cdot))(s)ds-f(t)) =: P_n(g-K_n(g)-f) \end{aligned} \tag{4.3}$$

where $X_n=P_n[X]=S^2_{\Delta_n}$ denotes the space of continuous piecewise linear functions related to the grid $\Delta_n:=\{i\cdot 2^{-n}|i=0(1)2^n\}$ [11]. Let P_n denote the linear Hermite interpolation operator [10] and let $Q_n=P_n$ for any $n\in N$. It is well known [10] that for any $g\in C[0,1]$

$$\|g-P_ng\| \le \omega(g;2^{-n}) \to 0 \tag{4.4}$$

with $n\to\infty$, where $\omega(g;\cdot):=(h\mapsto \sup\{|g(t)-g(s)|\ |t,s\in[0,1] \text{ and } |t-s|\le h\})$ denotes the modulus of continuity of g. This means that $(P_n)_{n\in N}$ is pointwise convergent to the identity.

Furthermore, the uniform Lipschitz continuity of the operator families $\{A_n'|n\in N\}$ and $\{A_n|n\in N\}$ depends essentially on the Lipschitz continuity of D_3k and k and can easily be shown.

For arbitrary $g\in C[0,1]$ the inequality

$$\begin{aligned}\|A_n(P_ng)-A(g)\| &= \|P_n^2g-P_nK_n(P_ng)-P_nf-g+K(g)+f\| \\ &\le \|(I-P_n)g\| + \|(I-P_n)K(g)\| + \|P_n\|\cdot\|K(g)-K_n(g)\| + \\ &\quad + \|P_n\|\cdot\|K_n(g)-K_n(P_ng)\| \\ &\le \omega(g;2^{-n})+\omega(K(g);2^{-n})+1\cdot\sup\{\omega(k(t,\cdot,g(\cdot));2^{-n})|t\in[0,1]\} \\ &\quad + 1\cdot L\cdot\omega(g;2^{-n})\end{aligned}$$

holds and since the last expression tends to 0 with $n\to\infty$ according to the uniform continuity of the relevant functions, the pointwise convergence is assured.

The hypothesis (iii) of theorem 3.2 can only be shown in a somewhat relaxed form, but it is sufficient for the proof of theorem 3.2 since only the boundedness of the set $\{A_n'(Q_nx^*)^{-1}\mid n\ge n_o\}$ has to be required: For any x in some neighbourhood of x^* there exists n_x such that for any $n\ge n_x$ the set $\{\|A_n'(Q_nx)^{-1}\|\,|n\ge n_x\}$ is bounded. Provided $A'(x^*)^{-1}$ exists, the proof of this condition can be reduced to show the presuppositions of theorem 2.1 in [11], which is stated in the following in a reduced form for the given situation of this section.

4.1 Theorem: Let X be a Banach space, $L:X\to X$ and $M:X\to X$ be linear continuous operators such that L and L-M are invertible, $(M_n)_{n\in N}$ a sequence of linear continuous operators with the properties that $\{M_n|n\in N\}$ is collectively compact and $(M_n)_{n\in N}$ is pointwise convergent to M, $(P_n)_{n\in N}$ be a sequence of linear continuous projections which converges pointwise to the identity and for each $n\in N$ P_n has a finite dimensional range. Further requirements are: For each $n\in N$ the operator $P_nL|_{P_n[X]}:P_n[X]\to P_n[X]$ is continuously invertible and

$$\sup\{\|(P_nL|_{P_n[X]})^{-1}\|\,|n\in N\} < \infty . \tag{4.5}$$

Then there exists $n_o \in N$, such that for $n \geq n_o$ the operator $P_n(L-M_n)|_{P_n[X]}: P_n[X] \to P_n[X]$ is continuously invertible and

$$\sup\{\|(P_n(L-M_n)|_{P_n[X]})^{-1}\| \,|\, n \geq n_o\} < \infty \,. \tag{4.6}$$

Clearly $X=C[0,1]$ is a Banach space, $L=I$ is invertible and L and $M=K'(x)$ are linear and continuous for fixed $x \in C[0,1]$. Now for any $n \in N$ $P_n I|_{P_n[X]}$ equals the identity onto the subspace $P_n[X]$ and is therefore invertible, linear, continuous, and bounded by 1. So it remains only to show the collective compactness of the set $\{K_n'(x)|n \in N\}$ and the pointwise convergence of $(K_n'(x))_{n \in N}$ to $K'(x)$, which can easily be done.

The crucial point is the determination of $\rho \in (0,1)$ for the application of the algorithm of section 2, but by theorem 3.2 the inequality

$$\|x^* - x_n^*\| \leq \underline{r}_n = \frac{1-\sqrt{1-2\cdot\eta_n\cdot\kappa}}{\kappa} \leq 2\cdot\eta_n$$

$$\leq 2\cdot\sup\{\|A_m'(P_m x^*)^{-1}\| \,|\, m \geq n_o\}\cdot\|A_n(P_n x^*) - A(x^*)\| \tag{4.7}$$

holds. If the kernel function k and the right hand side of Urysohn's integral equation (4.1) are sufficiently smooth,

$$\|A_n(P_n x^*) - A(x^*)\| = O(2^{-np}) \tag{4.8}$$

with $p \in (0,2]$, which means that

$$\|x^* - x_n^*\| = O(\rho^n) \tag{4.9}$$

with $\rho = 2^{-p} < 1$.

Finally a numerical example is given. Consider the equation

$$x(t) - \int_0^1 2\cdot\ln(t+s+x(s))\cdot(1+x(s))\,ds = \tag{4.10}$$

$$= \exp(t) - 2(1+t+e)[\ln(1+t+e)-1] + 2(1+t)[\ln(1+t)-1]$$

for $t \in [0,1]$, where the exponential function is a solution.

Since all appearing functions are infinitely differentiable it turns out that $\rho = 2^{-2} = 0.25$.

The following table contains the norm of the error functions $x^* - y_n$ and the estimations $D_n^* \cdot \rho^n$ which are determined during the elaboration of the iteration process.

TABLE 1

n	$\|x^* - y_n\|_\infty$	$D_n^* \cdot \rho^n$
3	0.600e-2	0.660e-2
4	0.150e-2	0.165e-2
5	0.375e-3	0.412e-3
6	0.936e-4	0.103e-3

The norm of the error functions are computed approximately by taking the maximum of the absolute values on a subset of 1001 uniformly distributed points of [0,1]. Another modification appears in the computation of the quantities D_n^* where the formula

$$D_n^* = 2 \cdot D_n - D_{n-1} \tag{4.11}$$

is used, which takes into account that possibly the quantities D_n^* are independent of the derived order of convergence p (cf. [11] p. 50ff). Thus in formula (2.9) for the determination of D_n^* the quantity $\rho = 2^{-p}$ is replaced by 0.5 .

REFERENCES

1. ANSELONE, P.M., *Collectively Compact Operator Approximation Theory*. Prentice Hall, Englewood Cliffs, N.J. (1971).

2. ATKINSON, K.E., *A Survey of Numerical Methods for the Solution of Fredholm Integral Equations of the Second Kind*. SIAM (1976).

3. EHRMANN, H., Iterationsverfahren mit veränderlichen Operatoren. *Arch. Rational Mech. Anal.* 4, 45-64 (1959).

4. KANTOROWITSCH, L.W. and AKILOW, G.P., *Funktionalanalysis in normierten Räumen*. Akademie Verlag, Berlin (1964).

5. POTRA, F.A. and PTAK, V., Sharp error bounds for Newton's process. *Numer. Math.* 34, 63-72 (1980).

6. RALL, L.B., *Computational Solution of Nonlinear Operator Equations*. Wiley, New York-London-Sydney-Toronto (1969).

7. SCHMIDT, J.W., Konvergenzuntersuchungen und Fehlerabschätzungen für ein verallgemeinertes Iterationsverfahren. *Arch. Rational Mech. Anal.* 6, 261-276 (1960).

8. TÖPFER, H.-J. and VOLK, W., Die numerische Behandlung von Integralgleichungen zweiter Art mittels Splinefunktionen. pp. 228-243 of J. Albrecht and L. Collatz (Eds.), *Numerische Behandlung von Integralgleichungen*. Birkhäuser, Basel-Boston-Stuttgart (1980).

9. VOLK, W., *Die numerische Behandlung Fredholm'scher Integralgleichungen zweiter Art mittels Splinefunktionen*. Bericht des Hahn-Meitner-Instituts HMI-B 286, Berlin (1979).

10. VOLK, W., *Fehlerabschätzungen für einige Spline-Interpolationsaufgaben*. Bericht des Hahn-Meitner-Instituts HMI-B 328, Berlin (1980).

11. VOLK, W., *Die numerische Behandlung von linearen Integrodifferentialgleichungen*. Dissertation, Freie Universität Berlin (1982).

LINEAR MULTISTEP METHODS FOR VOLTERRA INTEGRAL EQUATIONS OF THE SECOND KIND

P.J. van der Houwen* and H.J.J. te Riele*

* *Mathematical Centre, Amsterdam*

1. INTRODUCTION

The simplest linear multistep (LM) method for solving the Volterra integral equation of the second kind

$$y(t) = g(t) + \int_{t_0}^{t} K(t,\tau,y(\tau))d\tau, \quad t \in I:=[t_0,T], \tag{1.1}$$

is obtained by writing down this equation in a sequence of equidistant points

$$t_n := t_0 + nh, \; n = 0(1)N \; (h \text{ fixed and } t_N = T) \tag{1.2}$$

and by approximating the integral term by some suitably chosen quadrature formula. Such a method is called a *direct quadrature* (DQ) method for (1.1). Recently, several other LM methods for solving (1.1) have been proposed (cf. the *indirect backward differentiation* method in [5] and the *multilag* and *modified multilag* methods in [9](see also[12]).

In this paper a general class of linear multistep methods is presented which includes all these methods, and many others (Section 2). This enables us to give a uniform treatment of the problems of consistency (Section 3), of convergence (Section 4) and of stability (Section 6). Since the ordinary differential equation $dy/dt = f(t,y)$, $y(t_0)=y_0$, is a special case of the differentiated version of (1.1), the relation with linear multistep methods for ODEs is analyzed and fixed terms recurrence relations are derived for a class of convolution kernels(Section 5). Finally, two numerical experiments are reported (Section 7).

Space prevents us including the detailed proofs of the theorems presented here. These may be found in [6]. A number of additional numerical experiments which support and confirm the theory, may also be found in [6].

The work presented here can easily be extended to Volterra integral equations of the *first* kind, and to Volterra integro-differential equations (cf.[6]).

2. A GENERAL CLASS OF LM METHODS FOR SOLVING (1.1)

Let us associate with (1.1) the so-called *lag* term

$$Y(t,s) := g(t) + \int_{t_0}^{s} K(t,\tau,y(\tau))d\tau \tag{2.1}$$

for $(t,s) \in S := \{(t,s): t_0 \le s \le t \le T\}$. Note that $Y(t,t) \equiv y(t)$. Let y_n and $Y_n(t)$ denote numerical approximations to $y(t_n)$ and to $Y(t,t_n)$, respectively, and let

$$K_n(t) := K(t,t_n,y_n),\ n \ge 0. \tag{2.2}$$

Usually, $Y_n(t)$ will be computed by a quadrature formula of the form

$$Y_n(t) = g(t) + h \sum_{j=0}^{n} w_{n,j} K_j(t),\ n \ge n_0 , \tag{2.3}$$

where the $w_{n,j}$ are given weights and n_0 is sufficiently large to ensure the required order of accuracy. We assume that this quadrature formula is of order r, i.e.,

$$E_n(h;t) := \int_{t_0}^{t_n} K(t,\tau,y(\tau))d\tau - h \sum_{j=0}^{n} w_{n,j} K(t,t_j,y(t_j)) = \mathcal{O}(h^r) \tag{2.4}$$

as $h \to 0$, $n \to \infty$, with $t_n = t_0 + nh$ fixed. Our general LM *method* for (1.1) consists of the quadrature *formula* (2.3) and the LM *formula*

$$\sum_{i=0}^{k} \alpha_i y_{n-i} + \sum_{i=0}^{k} \sum_{j=-k}^{k} \beta_{ij} Y_{n-i}(t_{n+j}) = h \sum_{i=0}^{k} \sum_{j=-k}^{k} \gamma_{ij} K_{n-i}(t_{n+j}),\ n = k(1)N,\ k \text{ fixed}, \tag{2.5}$$

where the parameters α_i, β_{ij} and γ_{ij}, $i=0(1)k$, $j=-k(1)k$, are to be prescribed. From this scheme the quantities $y_k, y_{k+1}, \ldots, y_N$ can be computed successively. The quantities $y_1, \ldots, y_{k-1}$ and $Y_1(t), \ldots, Y_{k-1}(t)$ are assumed to be precomputed by some starting method. Since the kernel $K(t,\tau,y)$ is not necessarily defined outside S, we usually require (cf. Figure 1) that $\beta_{ij} = \gamma_{ij} = 0$, for $j < -i$.

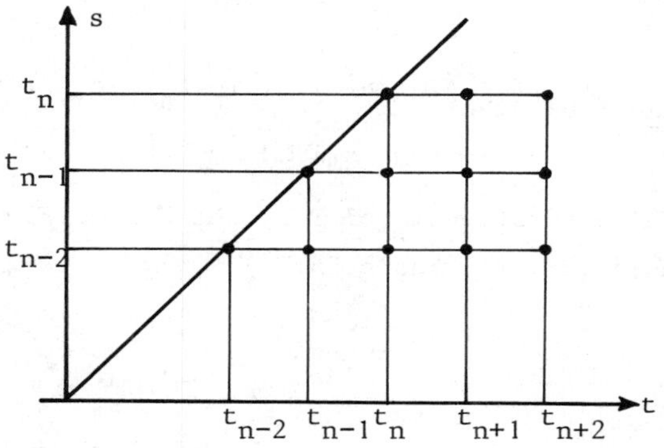

FIG. 1. Points in the (t,s)-plane needed in (2.5) for k = 2

Furthermore, it will be assumed that the points t_j are equally spaced (cf. (1.2)) although most of the analysis can be carried through for non-uniform spacing (compare a similar situation in the analysis of LM methods for ODEs). It is convenient to characterize the formula (2.5) by the matrices

$$A = (\alpha_i),\ B = (\beta_{ij}),\ C = (\gamma_{ij}) \tag{2.6}$$

where the row index i assumes the values 0(1)k and the column index j the values -k(1)k. We now describe four subclasses of (2.5) from which we will borrow several illustrating examples in this paper.

2.1 *Direct quadrature methods*

Consider the LM formula defined by the (1x1) matrices

$$A = 1,\ B = -1,\ C = 0,\ \text{for which (2.5) reduces to} \qquad y_n = Y_n(t_n). \tag{2.7}$$

Evidently, this is the *direct quadrature* (DQ) method, described in the introduction.

2.2 *Indirect linear multistep methods*

We formally derive this subclass by applying a linear multistep method for ODEs with coefficients α_i and γ_i, i=0(1)k, to the differentiated version of (1.1) (cf. [5]) :

$$y'(t) = K(t,t,y(t)) + Y_t(t,t), \tag{2.8}$$

where $Y_t(t,t)$ denotes the partial derivative of $Y(t,s)$ with respect to its first variable t, in the point (t,t). This yields the scheme

$$\sum_{i=0}^{k} \alpha_i y_{n-i} = h \sum_{i=0}^{k} \gamma_i K_{n-i}(t_{n-i}) + h \sum_{i=0}^{k} \gamma_i Y_t(t_{n-i},t_{n-i}) , \quad n \geq k. \tag{2.9}$$

Now we approximate the derivative Y_t of Y by the k-step *forward differentiation* formula (cf. [1, Table 25.2])

$$Y_t(t_{n-i},t_{n-i}) \approx -\frac{1}{h} \sum_{\ell=0}^{k} \delta_\ell Y(t_{n+\ell-i},t_{n-i}). \tag{2.10}$$

Using (2.3) we obtain

$$\sum_{i=0}^{k} \alpha_i y_{n-i} + \sum_{i=0}^{k} \sum_{j=-i}^{k-i} \gamma_i \delta_{i+j} Y_{n-i}(t_{n+j}) = h \sum_{i=0}^{k} \gamma_i K_{n-i}(t_{n-i}), \quad n \geq k, \tag{2.11a}$$

or, equivalently, the generating matrices

$$A=\begin{pmatrix} \alpha_0 \\ \alpha_1 \\ \vdots \\ \alpha_k \end{pmatrix}, \quad B=\begin{pmatrix} 0 & \gamma_0\delta_0 & \gamma_0\delta_1 & \cdots & \gamma_0\delta_k \\ & \gamma_1\delta_0 & \gamma_1\delta_1 & \cdots & \gamma_1\delta_k \\ & \cdot & \cdot & & \cdot \\ \cdot & \cdot & & \cdot & 0 \\ \gamma_k\delta_0 & \gamma_k\delta_1 & \cdots & \gamma_k\delta_k & \end{pmatrix}, \quad C=\left(\begin{array}{ccc|c} 0 & & \gamma_0 & \\ & \gamma_1 & & \\ & \cdot & & 0 \\ \cdot & & 0 & \\ \gamma_k & & & \end{array}\right). \tag{2.11b}$$

These matrices generate an *indirect linear multistep* (ILM) method. When the α_i and γ_i are the coefficients of a backward differentiation method, (2.11) represents the IBD (indirect backward differentiation) method, analyzed in [5]. We notice that for this IBD method we have $\gamma_i=0$, $i=1(1)k$, and $\gamma_0\delta_j=\alpha_j$, $j=0(1)k$.

2.3 Multilag methods

In Wolkenfelt et.al. [12] we find methods which can be characterized by the matrices

$$A=\begin{pmatrix} \alpha_0 \\ 0 \\ \vdots \\ 0 \end{pmatrix}, \quad B=\begin{pmatrix} & 0 & \\ & \alpha_1 & \\ 0 & \vdots & 0 \\ & \alpha_k & \end{pmatrix}, \quad C=\begin{pmatrix} & \gamma_0 & \\ & \gamma_1 & \\ 0 & \vdots & 0 \\ & \gamma_k & \end{pmatrix}. \tag{2.12}$$

Here, the α_i and γ_i, $i=0(1)k$, may be the coefficients of any LM method for ODEs. If the lag term $Y_n(t)$ is computed by using a quadrature rule which is (α,γ)-reducible (see Section 2.5), then the resulting method turns out to be equivalent to a DQ method based on the same (α,γ)-reducible quadrature rule (provided, of course, that the starting values are identical). Thus, a different implementation of the same method was used for the stability analysis of DQ methods. However, as was pointed out by Wolkenfelt [9], this implementation requires a lot of additional arithmetic operations and, although suitable for theorerical analysis, it is not recommendable in actual computations. In order to avoid this disadvantage, he proposed to compute the lag term simply by a quadrature rule of the form (2.3) to obtain the *multilag* (ML) methods.

2.4 Modified multilag methods

In [9] Wolkenfelt also introduced a modification of the ML methods, viz., the so-called *modified multilag* (MML) methods, characterized by the matrices

$$A=\begin{pmatrix}\alpha_0\\ \alpha_1\\ \vdots\\ \alpha_k\end{pmatrix},\quad B=\begin{pmatrix} 0 & & -\alpha_1 & \alpha_1 & \\ & \cdot^{\cdot^{\cdot}} & & \vdots & \cdot\;\; 0\\ & & 0 & & \\ -\alpha_k & & & \alpha_k & \end{pmatrix},\quad C=\begin{pmatrix} & \gamma_0 & \\ & \gamma_1 & \\ 0 & \vdots & 0\\ & \gamma_k & \end{pmatrix}. \tag{2.13}$$

The α_i and γ_i are, again, the coefficients of any LM method for ODEs.

2.5 The quadrature weights of the lag term

In order to define a specific LM method for (1.1) we have not only to specify the generating matrices A, B and C, but also the quadrature weights $w_{n,j}$ in (2.3). An important family of quadrature formulas, including the well-known Gregory quadrature formulas, are the so-called *reducible quadrature formulas* [8]. The weights $w_{n,j}$ in such formulas are recursively defined by the equations

$$\sum_{i=0}^{k^*} a_i w_{n-i,j} = \begin{cases} 0 & \text{if } j = 0(1)n-k^*-1 \\ b_{n-j} & \text{if } j = n-k^*(1)n \end{cases},\quad n=k^*,k^*+1,\ldots, \tag{2.14}$$

where the a_i and b_i, $i=0(1)k^*$, are the coefficients of some given LM method for ODEs. Here, we define $w_{n,j}=0$ for $j>\max(n,k^*-1)$, and the "starting weights" $w_{n,j}$, $0\le n,j \le k^*-1$ are assumed to be prescribed.

Defining the characteristic polynomials

$$\rho(z) := \sum_{i=0}^{k^*} a_i z^{k^*-i}, \quad \sigma(z) := \sum_{i=0}^{k^*} b_i z^{k^*-i}, \tag{2.15}$$

the quadrature formulas generated by (2.14) are said to be (ρ,σ)-*reducible*. We note that the characteristic polynomials of the Adams-Moulton methods generate the weights of the Gregory formulas. The backward differentiation methods generate rather unconventional quadrature rules, which were analysed in [11].

3. CONSISTENCY OF THE LM FORMULA (2.5)

Let us associate with the LM formula (2.5) the difference-differential operator L_n defined by

$$L_n(Y) := \sum_{i=0}^{k} \Big\{ \alpha_i Y(t_{n-i}, t_{n-i}) + \\ + \sum_{j=-k}^{k} [\beta_{ij} - \gamma_{ij} h \tfrac{\partial}{\partial s}] \, Y(t_{n+j}, t_{n-i}) \Big\}, \tag{3.1}$$

where $Y(t,s)$ is an arbitrary function, differentiable with respect to s on $t_0 \le s \le T$. As in the case of LM methods for ODEs, the operator L_n is introduced in order to operate on test functions Y of sufficient differentiability (cf. e.g. Lambert [7,p.23]). Unlike the ODE case, the relation of the operator L_n with the LM formula (2.5) is not immediate, and needs some explanation. Suppose that $Y(t,s)$ is defined by (2.1) with $y(t)$ the exact solution of (1.1). Observing that $Y(t,t)=y(t)$ and $\frac{\partial Y}{\partial s}(t,s)=K(t,s,y(s))$, and using (2.4) we find, on substitution of $y(t)$ into (2.5), the equation

$$\sum_{i=0}^{k} \Big\{ \alpha_i y(t_{n-i}) + \sum_{j=-k}^{k} [\beta_{ij} Y_{n-i}(t_{n+j}) - h\gamma_{ij} K_{n-i}(t_{n+j})] \Big\} = \\ = L_n(Y) - \sum_{i=0}^{k} \sum_{j=-k}^{k} \beta_{ij} E_{n-i}(h; t_{n+j}). \tag{3.2}$$

Thus, the exact solution of (1.1) satisfies the method {(2.3)-(2.5)} apart from the residual terms in the right-hand side of (3.2). In this section, we concentrate on the first residual term.

<u>Definition 3.1</u> The operator (3.1) and the associated LM formula (2.5) are said to be *consistent of order* p if for all $Y \in C^{p+1}[S]$, $L_n(Y) = O(h^{p+1})$ as $h \to 0$ with nonvanishing error constant. If Y corresponds to the theoretical solution of (1.1), then $L_n(Y)$ will be called the *local truncation error* of (2.5). ⊠

The following theorem provides the consistency conditions in terms of the parameters α_i, β_{ij} and γ_{ij}.

Theorem 3.1 The operator (3.1) and the associated LM formula (2.5) are consistent of order p if

$$\sum_{i=0}^{k} [(-i)^q \alpha_i - \sum_{j=-k}^{k} j^{q-\ell}(-i)^{\ell-1}(i\beta_{ij} + \ell\gamma_{ij})] =: C_{q\ell} = 0 \quad (3.3)$$

for $q = 0(1)p$ and $\ell = 0(1)q$ (with $(-i)^{\ell-1}\ell := 0$ if $\ell=i=0$). ⊠

Corollary 3.1 Let $\tilde{p}$ be the order of consistency of the LM method for ODEs defined by the coefficients $\{\alpha_i, \gamma_i\}$ employed in the sections 2.2, 2.3 and 2.4. Then the order of consistency p of the LM formula (2.5) for (1.1) is given by $p = \infty$ for the DQ method, $p = \min\{k, \tilde{p}\}$ for the ILM method, and $p = \tilde{p}$ for both the ML method and the MML method. ⊠

If the LM formula (2.5) is consistent of order p, then the local truncation error $L_n(Y)$ can be expressed in terms of the constants defined in (3.3) as follows:

$$L_n(Y) = h^{p+1} \sum_{\ell=0}^{p+1} C_{p+1,\ell} \binom{p+1}{\ell} \left(\frac{\partial}{\partial t}\right)^{p+1-\ell} \left(\frac{\partial}{\partial s}\right)^{\ell} Y(t,s)\Big|_{t=s=t_n} + \; + \mathcal{O}(h^{p+2}) \text{ as } h \to 0. \quad (3.4)$$

It is of some interest now to compare the values of the error constants $C_{p+1,\ell}$, $\ell=0(1)p+1$, for the various subclasses given in Section 2. We have evaluated and simplified the expressions for these constants as much as possible:

For the ILM method Corollary 3.1 gives p=k, under the (reasonable) assumption that $\tilde{p} \geq k$. We then find

$$C_{p+1,\ell} = (-1)^{p-1} \sum_{i=0}^{k} [i^p\{i\alpha_i + (p+1)\gamma_i\} - R] , \quad (3.5)$$

where $R = k!\gamma_i$ if $\ell = 0$ and $R = 0$ if $\ell = 1(1)p+1$.

For both the ML and MML method we find

$$C_{p+1,\ell} = \begin{cases} 0, & \text{if } \ell = 0(1)p, \\ (-1)^{p-1} \sum_{i=0}^{k} i^p\{i\alpha_i + (p+1)\gamma_i\}, & \text{if } \ell = p+1. \end{cases} \quad (3.6)$$

We have computed the numerical values of the error constants for two usual choices of the coefficients $\{\alpha_i, \gamma_i\}_{i=0}^{k}$, viz., the *backward differentiation* (BD) method, for which $\tilde{p}$=k, and the *Adams-Moulton* (AM) method for which $\tilde{p}$=k+1. Table 1 gives the values of the relevant constants $C_{p+1,\ell}$ where p is prescribed by Corollary 3.1. Note that the (M)ML-AM methods have p=k+1, whereas the other methods have p=k.

TABLE 1

Error constants in (3.4) for various choices of $\{\alpha_i,\gamma_i\}$ *in (2.5)*

method	$\{\alpha_i,\gamma_i\}$			k=1	k=2	k=3	k=4	k=5
ILM	BD	$C_{k+1,0}$	:	-2	0	-72/11	0	-14400/137
		$C_{k+1,>0}$	:	-1	-4/3	-36/11	-288/25	-7200/137
	AM	$C_{k+1,0}$	:	-1	2	-6	24	-120
		$C_{k+1,>0}$	:	0	0	0	0	0
ML & MML	BD	$C_{k+1,<k+1}$	:	0	0	0	0	0
		$C_{k+1,k+1}$	:	-1	-4/3	-36/11	-288/25	-7200/137
	AM	$C_{k+2,<k+2}$	:	0	0	0	0	0
		$C_{k+2,k+2}$	:	-1/2	-1	-19/6	-27/2	-863/12

The *order of convergence* of the LM method is dictated not only by its order of consistency, but also, of course, by the quadrature error (2.4) and by the errors in the starting values $y_1,\ldots,y_{k-1}$. In the next Section we shall analyze the convergence of the LM method $\{(2.3)-(2.5)\}$.

4. CONVERGENCE

Similarly as with LM methods for ODEs, a *necessary condition* for convergence of the LM method $\{(2.3)-(2.5)\}$ is that the characteristic polynomial

$$\alpha(z) := \sum_{i=0}^{k} \alpha_i z^{k-i} \tag{4.1a}$$

satisfies the *root condition,* i.e., its roots are on the unit disk, those on the unit *circle* being simple.

In the *sufficient conditions* for convergence the parameters β_{ij} and γ_{ij} are also involved. We define

$$\beta(z) := \sum_{i=0}^{k} \beta_i z^{k-i}, \quad \beta_i := \sum_{j=-k}^{k} \beta_{ij}\,; \tag{4.1b}$$

$$\gamma(z) := \sum_{i=0}^{k} \gamma_i z^{k-i}, \quad \gamma_i := \sum_{j=-k}^{k} \gamma_{ij}\,. \tag{4.1c}$$

Furthermore, we will use the notation

$$\Delta K_n(t) = K(t,t_n,y(t_n)) - K(t,t_n,y_n),$$

$$\Delta E_n(h) = \max_{\substack{i\le j\le n\\ \ell\le k}} \left|E_i(h;t_{j+\ell}) - E_i(h;t_j)\right|,$$

$$E_n(h) = \max_{i \le j \le n} |E_i(h;t_j)|,$$

$$T_n(h) = \max_{i \le n} |L_i(Y)| \quad \text{and} \quad \delta(h) = \max_{j \le k-1} |y(t_j) - y_j|.$$

$E_n(h)$ is the maximal error arising in the approximation of the lag terms $Y(t,t_n)$ by $Y_n(t)$ (cf. (2.4)). $T_n(h)$ may be considered as the maximal local truncation error of the LM formula, and $\delta(h)$ is the maximal starting error. We now formulate a general convergence theorem which provides an estimate for the *global error*

$$\varepsilon_n = y(t_n) - y_n. \tag{4.2}$$

We assume that K satisfies the Lipschitz conditions

$$|\Delta K_\ell(t)| \le L_1|\varepsilon_\ell| \text{ and } |\Delta K_\ell(t) - \Delta K_\ell(t^*)| \le L_2|t-t^*||\varepsilon_\ell|,$$

where L_1 and L_2 are the Lipschitz constants.

Theorem 4.1 Let $\alpha(z)$ satisfy the root condition.

(i) If $\alpha(z) = \alpha_0 z^k$ then there exists a constant $C > 0$ such that

$$|\varepsilon_n| \le C\,[h\delta(h) + E_N(h) + \Delta E_N(h) + T_N(h)],\ n=k(1)N. \tag{4.3a}$$

(ii) If $\beta(z) \equiv 0$ then there exists a constant $C > 0$ such that

$$|\varepsilon_n| \le C\,[\delta(h) + h^{-1}\{\Delta E_N(h) + T_N(h)\}],\ n=k(1)N. \tag{4.3b}$$

⊠

Now it is easy to derive the following

Corollary 4.1 Let $\delta(h) = \mathcal{O}(h^q)$, $E_N(h) = \mathcal{O}(h^r)$ (as in (2.4)), $\Delta E_N(h) = \mathcal{O}(h^{r+1})$ as $h \to 0$ and let $\{\alpha_i,\gamma_i\}$ in (2.5) be the coefficients of a $\tilde{p}$ - th order consistent LM method for ODEs. Then the order of convergence p^* of the LM method for (1.1) is given by: $p^* = \min(q+1,r)$ for the DQ method, $p^* = \min(q,r,p)$ for the ILM method, $p^* = \min(q+1,r,p+1)$ for the ML method and $p^* = \min(q,r,p)$ for the MML method, where p is the order of consistency of (2.5), given by Corollary 3.1. ⊠

5. RELATION WITH LM METHODS FOR ODEs

The Volterra equation (1.1) contains the classes of ordinary differential equations as special cases. For example, if in (1.1) $g(t) \equiv$ constant then

$$K(t,\tau,y) = f(\tau,y) \quad \to \quad \frac{dy}{dt} = f(t,y) \tag{5.1a}$$

$$K(t,\tau,y) = (t-\tau)f(\tau,y) \to \frac{d^2y}{dt^2} = f(t,y) \tag{5.1b}$$

, etc.

Therefore, it is natural to ask to what method the LM formula (2.5) reduces when it is applied to the special cases (5.1). Furthermore, one may ask for the relationship with LM methods for ODEs of the form(5.1). In order to formulate this relationship we introduce, in addition to the polynomials $\alpha(z)$, $\beta(z)$ and $\gamma(z)$ (cf. (4.1)), the polynomials

$$\bar{\beta}(z) := \sum_{i=0}^{k} \bar{\beta}_i z^{k-i}, \quad \bar{\beta}_i := \sum_{j=-k}^{k} j\beta_{ij}; \tag{5.2a}$$

$$\bar{\gamma}(z) := \sum_{i=0}^{k} \bar{\gamma}_i z^{k-i}, \quad \bar{\gamma}_i := \sum_{j=-k}^{k} j\gamma_{ij}. \tag{5.2b}$$

We shall also employ the shift operator E defined by $Ey_n = y_{n+1}$.

<u>Theorem 5.1</u> Let $g(t) \equiv$ constant.
(i) If $K(t,\tau,y) \equiv f(\tau,y)$ then the formula (2.5) reduces to

$$\alpha(E)y_n + \beta(E)Y_n(t_n) = h\gamma(E)f(t_n,y_n), \quad n \geq 0. \tag{5.3a}$$

(ii) If $K(t,\tau,y) \equiv (t-\tau)f(\tau,y)$, $\beta(z) \equiv 0$, and if the weights $w_{n,j}$ in(2.3) are (ρ,σ)-reducible, then the formula (2.5) reduces to

$$\alpha(E)\rho(E)y_n + h^2[\sigma(E)\bar{\beta}(E) - \rho(E)\bar{\gamma}(E) - k\rho(E)\gamma(E) + \rho(E)\gamma'(E)E]f(t_n,y_n) = 0, \tag{5.3b}$$

where γ' denotes the derivative of γ. ⊠

From part (i) of this theorem it follows that the LM formula (2.5), when applied to the integrated form of the first order equation $dy/dt=f(t,y)$, reduces to a linear multistep method $\{\alpha,\gamma\}$ for this equation, provided that $\beta(z)\equiv 0$. This statement holds, *irrespective of the weights* $w_{n,j}$ used in the definition of the lag term $Y_n(t)$. In other words, if the matrix B is chosen such that the *row sums vanish* ($\beta_i=0$) then our linear method is in fact an LM method for ODEs whenever the Volterra equation (1.1) is a (first order) ODE. Such linear methods will be called (α,γ)-*reducible*. The recurrence relation (5.3a) plays an important rôle in the stability analysis of Volterra equations with (5.1a) as test kernel. In particular, for (α,γ)-reducible methods, the ODE-stability theory directly applies and may suggest suitable polynomials α and γ for the construction of stable numerical methods for solving equation (1.1).

<u>Example 5.1</u> The ILM and the MML methods are (α,γ)-reducible, whereas the DQ and the ML methods are not. ⊠

Part (ii) of Theorem 5.1 provides us with further information about how we should choose the weights $w_{n,j}$ and the matrices A, B and C in order to construct a suitable integration method. Observe, that here the structure of the matrices B and C is such, that the same set of polynomials $\{\rho,\sigma,\alpha,\gamma\}$ may lead to *different* recurrence relations.

Example 5.2 Let both $\{\alpha,\gamma\}$ and $\{\rho,\sigma\}$ be the trapezoidal rule, i.e., $\alpha(z)=\rho(z)=z-1$ and $\gamma(z)=\sigma(z)=\frac{1}{2}(z+1)$. Now it is a simple calculation to find that (5.3b) reduces to a LM method $\{\rho^*,\sigma^*\}$ for second order ODEs with $\rho^*(z)=(z-1)^2$ both for the ILM and the MML method, but with $\sigma^*(z)=\frac{1}{4}(z+1)^2$ for the ILM and $\sigma^*(z)=z$ for the MML method, respectively. (Note that both methods have order 2 (cf. [7, p.253]), with error constant -1/6 for the ILM and 1/12 for the MML method, respectively.) ☒

For an extension of Theorem 5.1 to the case of a general convolution kernel $K(t,\tau,y) = \Sigma_{\ell=0}^{m} (t-\tau)^{\ell} f_{\ell}(\tau,y)$, the reader is referred to [6].

6. V_0-STABILITY

Definition 6.1 A discretization method for (1.1) is said to be *V_0-stable* if $y_n \to 0$ as $n \to \infty$ whenever it is applied, with fixed stepsize $h > 0$, to the test equation

$$y(t) = y_0 + \int_0^t \{\lambda + \mu(t-\tau)\}y(\tau)d\tau, \tag{6.1}$$

with arbitrary $(\lambda,\mu) \in Q_{\lambda,\mu} := \{(\lambda,\mu): \lambda < 0, \mu \le 0\}$. ☒

Wolkenfelt[10] has shown that the DQ method (2.7) can *not* be V_0-stable when the quadrature weights in (2.3) are (ρ,σ)-reducible. This negative result raised the question of whether V_0-stable methods for (1.1) do exist at all. Brunner, Nørsett and Wolkenfelt[4] answered this question affirmatively for a certain class of so-called one-stage implicit *Runge-Kutta* methods. In the class of *LM* methods analysed in the present paper, V_0-stable methods do also exist. In particular, they occur in the subclass of ILM methods. To see this, we observe that for the ILM methods we have

$$\bar{\beta}(z) = -\gamma(z), \quad \bar{\gamma}(z) = -k\gamma(z) + z\gamma'(z), \tag{6.2}$$

and from Theorem 5.1 we derive the following result.

Theorem 6.1 Let the conditions of Theorem 5.1 part (ii) be satisfied, then the LM method, when applied to the test equation (6.1), assumes the form

$$\{\rho(E)[\alpha(E) - h\lambda\gamma(E)] + $$
$$+ h^2\mu[\sigma(E)\bar{\beta}(E) - \rho(E)(\bar{\gamma}(E)+k\gamma(E)-E\gamma'(E))]\}y_n = 0. \tag{6.3}$$

In the ILM case this equation reduces to

$$\{\rho(E)[\alpha(E) - h\lambda\gamma(E)] - h^2\mu\sigma(E)\gamma(E)\}y_n = 0. \tag{6.3'}$$

☒

Since equation (6.3') is identical to the one obtained by Brunner and Lambert ([3]) in their stability analysis of numerical methods for the test integro-differential equation

$$\frac{dy}{dt}(t) = \lambda y(t) + \mu \int_0^t y(\tau)d\tau, \tag{6.4}$$

we may find examples of V_0-stable ILM methods just by inspecting the stability regions given by Brunner and Lambert. In this way, we immediately conclude from [3] that the four combinations, with $\{\alpha,\gamma\}$ and $\{\rho,\sigma\}$ defining either the trapezoidal rule or the backward Euler rule, are V_0-stable methods. It turns out that the MML versions of these methods are *not* V_0-stable. (In fact, as communicated to us by S. Amini, no MML methods can be V_0-stable.) In Figure 2 the stability regions of both the IML and the MML methods are given. Evidently, the ILM methods have considerably larger regions of stability.

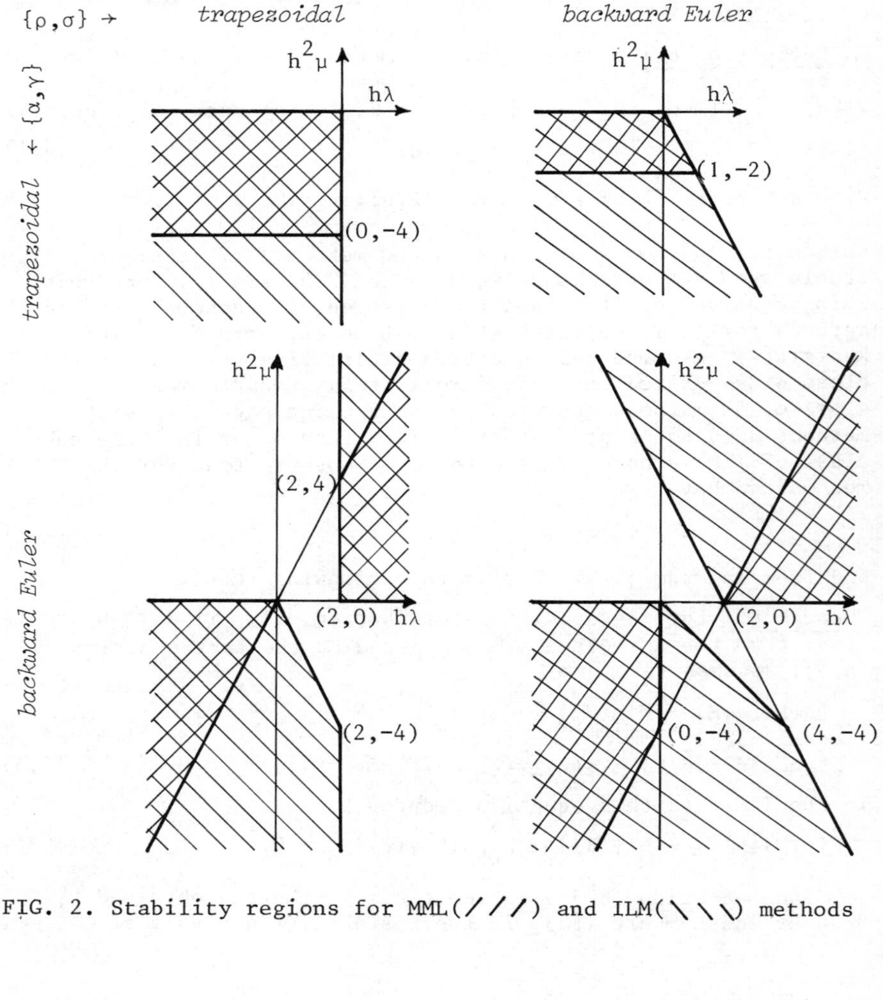

FIG. 2. Stability regions for MML(///) and ILM(\\\) methods

7. NUMERICAL EXPERIMENTS

In this Section we illustrate by a few numerical experiments the convergence theorem 4.1 and the improved stability behaviour of the ILM and the MML methods. In the tables of results we list the accuracy obtained, by

$$A(h) := -\log_{10}(|\text{relative error at the end point}|), \qquad (7.1)$$

i.e., the number of correct digits in the numerical solution. The pair $\{\rho,\sigma\}$ used for computing the lag term will always define a Gregory formula of order r; the pair $\{\alpha,\gamma\}$ defines either an Adams-Moulton or a backward differentiation formula of order $\tilde{p}$. Methods are denoted by, e.g., ILM(G_r-$BD_{\tilde{p}}$).

7.1 Order of convergence

In the first experiment we integrated the equation

$$y(t) = 1 + \sin(t) - \cos(t) - \int_0^t y(\tau)d\tau, \quad 0 \le t \le 2. \qquad (7.2)$$

The starting values were taken from the exact solution y(t) = sin(t). The generating characteristic polynomials $\{\rho,\sigma,\alpha,\gamma\}$ were chosen such that, according to Corollary 4.1, all methods listed in Table 2 are just of order $p^*= 5$. In this Table the values of A(h) and the corresponding *effective order* p^*_{eff} are presented, where $p^*_{eff} = [A(h) - A(2h)]/\log_{10}2$

TABLE 2

Tests of order of convergence

h^{-1}	DQ(G_5)	ILM(G_5-AM_6)	ML(G_5-AM_4)	MML(G_5-AM_5)
4	5.0	3.8	4.7	5.3
	)*5.7*	)*5.0*	)*5.5*	)*5.1*
8	6.7	5.3	6.3	6.8
	)*5.4*	)*5.2*	)*5.3*	)*5.1*
16	8.3	6.9	7.9	8.4
	)*5.2*	)*5.1*	)*5.2*	)*5.1*
32	9.9	8.5	9.5	9.9

From the results we see that the effective order tends to the asymptotic order as h decreases. We also see that the ILM method is less accurate than the other methods, which may be explained by its larger error constants (cf. Table 1).

7.2 Stability

In the second experiment we chose an example in which the

kernel has a large Lipschitz constant (obtained by modifying an example given by Bownds[2]):

$$y(t) = 50(1-t^2)\ln(1+t) + 75t^2-51t+1 - 100\int_0^t \ln(1+t-\tau)y(\tau)d\tau, \quad 0 \le t \le 4. \tag{7.3}$$

Again, the starting values were taken from the exact solution $y(t) = 1-t$. The results listed in Table 3 clearly show the better stability properties of the ILM method (a negative A(h)-value may be interpreted as an unstable behaviour). In particular, we observe the only marginally better performance of the MML methods when compared with the ML methods.

TABLE 3

Stability tests

h^{-1}	$DQ(G_5)$	ILM		ML		MML	
		G_5-BD_5	G_5-AM_6	G_5-BD_4	G_5-AM_4	G_5-BD_5	G_5-AM_5
4	-4.3	+2.5	+1.2	-2.8	-4.3	-2.6	-2.8
8	-6.5	+2.2	+2.2	-6.2	-5.6	-2.7	-4.9
16	+2.3	+2.6	+4.4	-6.9	+3.7	-2.4	+5.6

REFERENCES

1. ABRAMOWITZ, M. & I.A. Stegun (Eds.), *Handbook of mathematical functions*, Nat. Bureau of Standards, Washington D.C., (1964).
2. BOWNDS, J.M., Theory and performance of a subroutine for solving Volterra integral equations, to appear in *Computing*.
3. BRUNNER, H. & J.D. LAMBERT, Stability of numerical methods for Volterra integro-differential equations, *Computing* 12, 75-89 (1974).
4. BRUNNER, H., S.P. NØRSETT & P.H.M. WOLKENFELT, On V_0-stability of numerical methods for Volterra integral equations of the second kind, *Report NW 84/80*, Mathematical Centre, Amsterdam (1980).
5. HOUWEN, P.J. VAN DER & H.J.J. TE RIELE, Backward differentiation type formulas for Volterra integral equations of the second kind, *Numer. Math.* 37, 205-217 (1981).
6. HOUWEN, P.J. VAN DER & H.J.J. TE RIELE, Linear multistep methods for Volterra integral and integro-differential equations, *Report NW /82*, Mathematical Centre, Amsterdam (1982).
7. LAMBERT, J.D., *Computational methods in ordinary differential equations*, John Wiley & Sons, London etc. (1973).
8. MATTHYS, J., A-stable linear multistep methods for Volterra integro-differential equations, *Numer. Math.* 27, 85-94 (1976).

9. WOLKENFELT, P.H.M., Modified multilag methods for Volterra functional equations, to appear in *Math. Comp.*.
10. WOLKENFELT, P.H.M., Stability analysis of reducible quadrature methods for Volterra integral equations of the second kind, *ZAMM* 61, 399-401(1981).
11. WOLKENFELT, P.H.M., The construction of reducible quadrature rules for Volterra integral and integro-differential equations, to appear in *IMA J. Numer. Anal.*.
12. WOLKENFELT, P.H.M., P.J. VAN DER HOUWEN & C.T.H. BAKER, Analysis of numerical methods for second kind Volterra equations by imbedding techniques, *J. Integral Eq.* 3, 61-82 (1981).

ON THE MAXIMUM LIKELIHOOD REGULARIZATION OF FREDHOLM CONVOLUTION EQUATIONS OF THE FIRST KIND

A.R. Davies

University College of Wales, Aberystwyth

1. INTRODUCTION

Consider the Fredholm integral equation of the first kind of convolution type

$$(Kf)(x) \equiv \int_{-\infty}^{\infty} k(x-y)f(y)dy = g(x), \quad -\infty < y < \infty, \tag{1.1}$$

where k and g are known functions in $\mathcal{L}_2(\mathbb{R})$, and $f \in H^p(\mathbb{R})$ is to be found. If ^ denotes Fourier transformation, then from the convolution theorem we have

$$\hat{k}(\omega)\hat{f}(\omega) = \hat{g}(\omega), \tag{1.2}$$

whence

$$f(y) = \frac{1}{2\pi} \int_{-\infty}^{\infty} \frac{\hat{g}(\omega)}{\hat{k}(\omega)} \exp(i\omega y)d\omega. \tag{1.3}$$

The ill-posedness of (1.1) is reflected by the fact that any small perturbation ε in g, whose transform $\hat{\varepsilon}(\omega)$ does not decay faster than $\hat{k}(\omega)$ as $|\omega| \to \infty$, will result in a perturbation in $\hat{g}(\omega)/\hat{k}(\omega)$ which will grow without bound. When g is inexact, therefore, we may seek a stable or *filtered* approximation to f given by

$$f_\lambda(y) = \frac{1}{2\pi} \int_{-\infty}^{\infty} Z(\omega;\lambda) \frac{\hat{g}(\omega)}{\hat{k}(\omega)} \exp(i\omega y)d\omega, \tag{1.4}$$

where $Z(\omega;\lambda)$ is a filter function dependent on a parameter λ.

Filters may be constructed in several ways, either directly for the convolution kernel [1], or as a special case of general Fredholm I equations [2], provided in the latter case it is realized that in (1.1) the operator K is not compact and the Fourier transform (FT) here plays the rôle of singular function expansions in the context of compact operators.

In this paper we restrict attention to filters generated from regularization theory. The smoothing functional

$$C(f;\lambda) = \|Kf - g\|_2^2 + \lambda\Omega[f] \tag{1.5}$$

is minimized in an appropriate subspace of $\mathcal{L}_2$, where $\Omega[f]$ is a stabilizing functional in the form of a smoothing norm

$$\Omega[f] = \|Lf\|^2 \tag{1.6}$$

and L is a linear operator. The regularization parameter λ controls the trade-off between smoothness, as imposed by Ω, and the extent to which (1.1) is satisfied.

Two essential decisions are needed in this approach:
(i) the choice of smoothing norm Ω for a given problem;
(ii) given Ω, the choice of λ in some optimal sense.
Apart from the important work in [1], [3-7], relatively little effort has been expended on (i), which reflects the difficulty of the problem. In contrast, several methods have been proposed for choosing λ (see [8], for example, for a list). Most of these methods depend on *a priori* knowledge of the noise level in the data, usually the variance σ^2. In the comparative study in [8], such methods were found to perform badly when σ was not known accurately, and in many cases even when σ was known exactly. The cross-validation method of Wahba [9], which does not require prior knowledge of σ, proved dramatically superior.

In this paper we construct a maximum likelihood (ML) method which determines λ optimally, requiring no prior knowledge of σ. In the case of numerical deconvolution, at least, the method compares extremely well with Wahba's method. Our construction of the method is a simple extension of the ideas of Anderssen and Bloomfield [10,11], who consider the problem of numerically differentiating noisy data.

We restrict attention to regularization of order p, where L in (1.6) is the pth order differential operator, $Lf = f^{(p)}$, and the norm in (1.6) is $\mathcal{L}_2$. The minimizer of (1.5) in H^p is then given by (1.4) where

$$Z(\omega;\lambda) = \frac{|\hat{k}(\omega)|^2}{|\hat{k}(\omega)|^2 + \lambda\omega^{2p}} . \tag{1.7}$$

2. APPROXIMATION AND SOLUTION METHOD

We assume that the support of each function f, g and k is essentially finite and contained within the interval [0,1), possibly by a change of variable. Let T_{N-1} denote the space of trigonometric polynomials of degree at most N-1 and period 1. We shall look for a filtered solution of (1.1) within the space T_{N-1} for the following reasons:

(i) The discretization error in the convolution may be made precisely zero at the grid points.
(ii) Fast Fourier Transform (FFT) routines are easily employed in the solution procedure, reducing the number of arithmetical operations to O(N log N).
(iii) The adoption of T_{N-1} as the approximating function space is itself a regularizing feature.

Let g and k be given at N equally spaced points, $x_n = nh$, $n=0,\ldots,N-1$, with spacing $h = 1/N$. Then g and k are interpolated by g_N and $k_N \in T_{N-1}$, where

$$g_N(x) = \frac{1}{N}\sum_{q=0}^{N-1} \hat{g}_{N,q} \exp(i\omega_q x), \tag{2.1}$$

$$\hat{g}_{N,q} = \sum_{n=0}^{N-1} g_n \exp(-i\omega_q x_n), \tag{2.2}$$

and $g(x_n) = g_n = g_N(x_n), \quad \omega_q = 2\pi q,$ (2.3)

with similar expressions for k_N.

If (1.1) is now replaced by

$$(K_N f_N)(x) \equiv \int_0^1 k_N(x-y) f_N(y)\,dy = g_N(x), \tag{2.4}$$

where k_N is periodically continued outside $[0,1)$, then we may prove (i) above.

Lemma 2.1 Let $\phi \in T_{N-1}$ and $\underline{\phi} = (\phi(x_o),\ldots,\phi(x_{N-1}))^T \in \mathbb{R}^N$. Then the N×N matrix

$$K = \Psi \,\mathrm{diag}(h\hat{k}_{N,q})\Psi^H \tag{2.5}$$

where Ψ is the unitary matrix with elements

$$\Psi_{rs} = \frac{1}{\sqrt{N}} \exp\left(\frac{2\pi}{N} irs\right), \quad r,s=0,\ldots,N-1, \tag{2.6}$$

has the property

$$(K\underline{\phi})_n = (K_N\phi)(x_n). \tag{2.7}$$

Thus, from the finite support hypothesis and (2.3), it follows that at $\{x_n\}$, (1.1) is exactly equivalent to the discrete system

$$(K\underline{f})_n = g_n, \tag{2.8}$$

where K is given in (2.5) and $\underline{f} = (f_N(x_o),\ldots,f_N(x_{N-1}))^T$.

In T_{N-1} it is easily shown that f_λ in (1.4) is approximated by

$$f_{N;\lambda}(x) = \sum_{q=0}^{N-1} Z_{q;\lambda} \frac{\hat{g}_{N,q}}{\hat{k}_{N,q}} \exp(i\omega_q x) \tag{2.9}$$

where the discrete pth order filter is

$$Z_{q;\lambda} = \frac{|\hat{k}_{N,q}|^2}{|\hat{k}_{N,q}|^2 + N^2\lambda\tilde{\omega}_q^{2p}}, \tag{2.10}$$

$$\text{and } \tilde{\omega}_q = \begin{cases} \omega_q, & 0 \le q < \tfrac{1}{2}N, \\ \omega_{N-q}, & \tfrac{1}{2}N \le q < N-1. \end{cases} \tag{2.11}$$

To show (ii) above, we note that $\sqrt{N}\Psi^H$ is the discrete FT matrix representing (2.2), and so (2.8) is equivalent to the diagonal system

$$h\hat{k}_{N,q}\hat{f}_{N,q} = \hat{g}_{N,q}. \tag{2.12}$$

After regularization, (2.12) is replaced by

$$h\hat{k}_{N,q}\hat{f}_{N,q;\lambda} = Z_{q;\lambda}\hat{g}_{N,q}, \tag{2.13}$$

so that $f_{N;\lambda}(x)$ may be found by multiplying the FFT of $\{g_n\}$ by the filter, dividing by the FFT of $\{k_n\}$, and then taking the inverse FFT.

3. THE FILTER IN A STOCHASTIC SETTING

Several authors [9-15] have treated filters in various statistical frameworks. In this section we relate the pth order convolution filter (2.10) to certain spectral densities which play a rôle in the ML optimization of λ in the next section.

Assume that the data $\{g_n\}$ are noisy, and that there is an underlying function $u_N \in T_{N-1}$ such that

$$g_n = u_N(x_n) + \varepsilon_n \equiv u_n + \varepsilon_n. \tag{3.1}$$

We identify both $\{u_n\}$ and $\{\varepsilon_n\}$ with independent stationary stochastic processes. Since in general the expectation $E(u_n)$ is not zero, it is suggested in [10,11] that the data $\{g_n\}$ be *detrended* so that u_n becomes weakly stationary. This would involve subtracting from the data the values of a smooth function of roughly the same shape as u_N. In [8] it is argued that this precaution is not always necessary in the context of deconvolution, and the reader may refer to this paper for details.

In the limit $N \to \infty$, $h \to 0$, for any discrete process X_n we may write (see, for example, [16])

$$X_n = \int_0^1 \exp(2\pi i\omega n)\,dS_X(\omega) \tag{3.2}$$

where $S_X(\omega)$ is a stochastic process defined on $[0,1)$. The essential property of S_X we require is

Lemma 3.1 The variance of any integral

$$\int \theta(\omega)\,dS_X(\omega)$$

is given by

$$\int |\theta(\omega)|^2\,dG_X(\omega),$$

where

$$dG_X(\omega) = E(|dS_X(\omega)|^2).$$

$G_X(\omega)$ may be interpreted as a spectral distribution function, and accordingly we shall write $dG_X(\omega) = P_X(\omega)d\omega$ where $P_X(\omega)$ is a spectral density.

Now consider $f_N \in \mathcal{T}_{N-1}$, with $\underset{\sim}{f} = (f_n) \equiv (f_N(x_n))$ defined by

$$(K\underline{f})_n = u_n, \qquad n=0,\ldots,N-1,$$

with K given by (2.5). From (3.2) we have

$$\begin{aligned} f_n &= \sum_{m=0}^{N-1} \{(K^{-1})_{mn} \int_0^1 \exp(2\pi i\omega m)\,dS_u(\omega)\} \\ &= \int_0^1 [\hat{k}_N(\omega)]^{-1} \exp(2\pi i\omega n)\,dS_u(\omega) \end{aligned} \tag{3.3}$$

where

$$\hat{k}_N(\omega) = \frac{1}{N} \sum_{n=0}^{N-1} k_n \exp(-2\pi i\omega n). \tag{3.4}$$

Assume that f_n is estimated by $\sum_{m=0}^{N-1} \ell_m g_{n-m}$, where $\{\ell_m\}$ is a filter which we shall relate to $Z_{q;\lambda}$, and $\{g_n\}$ is periodically continued for $n \notin [0,N)$. Then the error

$$f_n - \sum_{m=0}^{N-1} \ell_m g_{n-m} \tag{3.5}$$

is given by

$$\int_0^1 \exp(2\pi i\omega n)([\hat{k}_N(\omega)]^{-1} - \hat{\ell}_N(\omega))dS_u(\omega) - \int_0^1 \exp(2\pi i\omega n)\hat{\ell}_N(\omega)dS_\varepsilon(\omega) \tag{3.6}$$

where $\hat{\ell}_N(\omega)$ is defined as in (3.4). From Lemma 3.1 the variance of this error is clearly

$$\int_0^1 |[\hat{k}_N(\omega)]^{-1} - \hat{\ell}_N(\omega)|^2 P_u(\omega)d\omega + \int_0^1 |\hat{\ell}_N(\omega)|^2 P_\varepsilon(\omega)d\omega \tag{3.7}$$

which is minimized when

$$\hat{\ell}_N(\omega)\hat{k}_N(\omega) = \frac{P_u(\omega)}{P_u(\omega) + P_\varepsilon(\omega)} . \tag{3.8}$$

Since the Fourier coefficients of the filtered solution (2.5) must satisfy

$$\hat{f}_{N,q;\lambda} = h\hat{\ell}_{N,q}\hat{g}_{N,q} = Z_{q;\lambda}\hat{g}_{N,q}[h\hat{k}_{N,q}]^{-1}$$

we find

$$Z_{q;\lambda} = h^2\hat{\ell}_{N,q}\hat{k}_{N,q} .$$

Thus from the observation $h\hat{\ell}_{N,q} = \hat{\ell}_N(qh)$, $h\hat{k}_{N,q} = \hat{k}_N(qh)$, we have from (3.8):

<u>Theorem 3.1</u> In the limit $N \to \infty$, $h \to 0$, the variance of the error $f_N(x_n) - f_{N;\lambda}(x_n)$ is minimized at x_n by the choice of filter

$$Z_{q;\lambda} = \frac{P_u(qh)}{P_u(qh) + P_\varepsilon(qh)} . \tag{3.9}$$

4. OPTIMIZATION BY ML

We now simply relate the filter (3.9) to the pth order filter (2.10). Assuming that the errors are uncorrelated, $P_\varepsilon(\omega)$ has the form

$$P_\varepsilon(\omega) = \sigma^2 = \text{constant} \tag{4.1}$$

where σ^2 is the unknown variance of the noise in the data. Choosing

$$P_u(\omega) = \frac{\sigma^2 |\hat{k}_N(\omega)|^2}{\lambda\tilde{\omega}^{2p}} \tag{4.2}$$

where

$$\tilde{\omega} = \begin{cases} 2\pi N\omega & , \quad 0 \leq \omega < \tfrac{1}{2}, \\ 2\pi N(1-\omega), & \tfrac{1}{2} \leq \omega < 1 \end{cases}$$

then yields (2.10) from (3.9). Moreover, the spectral density for $\{g_n\}$ is then

$$P_g(\omega) = P_u(\omega) + P_\varepsilon(\omega) = \sigma^2 \left[1 + \frac{|\hat{k}_N(\omega)|^2}{\lambda\tilde{\omega}^{2p}} \right]$$

whence

$$P_g(qh) = \sigma^2(1 - Z_{q;\lambda})^{-1}. \tag{4.3}$$

The statistical likelihood of any suggested values of σ^2 and λ may now be estimated from the data. Following Whittle [17], the logarithm of the likelihood function of P_g is given approximately by

$$\text{const.} - \tfrac{1}{2} \sum_{q=0}^{N-1} [\log P_g(qh) + I(qh)/P_g(qh)] \tag{4.4}$$

where

$$I(\omega) = \left| \sum_{n=0}^{N-1} g_n \exp(-2\pi i\omega n) \right|^2$$

is the periodogram of the data, with

$$I(qh) = |\hat{g}_{N,q}|^2.$$

We now maximize (4.4) with respect to σ^2 and λ. The partial maximum with respect to σ^2 may be found exactly (in terms of λ) with the maximizing value of σ^2 given by

$$\sigma^2 = \frac{1}{N} \sum_{q=1}^{N-1} |\hat{g}_{N,q}|^2 (1-Z_{q;\lambda}). \tag{4.5}$$

The maximum with respect to λ may then be found by *minimizing*

$$V_{ML}(\lambda) = \tfrac{1}{2}N \log\left[\sum_{q=1}^{N-1} |\hat{g}_{N,q}|^2 (1-Z_{q;\lambda}) \right] - \tfrac{1}{2} \sum_{q=1}^{N-1} \log(1-Z_{q;\lambda}). \tag{4.6}$$

Thus the optimal regularization parameter is given by the minimizer of a simple function of λ, depending on the known Fourier coefficients $\hat{g}_{N,q}$ and $\hat{k}_{N,q}$. No prior knowledge of σ^2 is assumed but an *a posteriori* estimate is given by (4.5).

An estimate of the order of convergence is given in the following:

<u>Theorem 4.1</u> If $f \in C^P[0,1)$ then

$$E(\|f-f_{N;\lambda}\|^2) = O(h^{2p}) + \lambda^2 \sum_{q=0}^{N-1} \frac{\tilde{\omega}_q^{4p}|\hat{f}_{N,q}|^2}{(|\hat{k}_{N,q}|^2 + N^2\lambda\tilde{\omega}_q^{2p})^2}$$

$$+ N\sigma^2 \sum_{q=0}^{N-1} \frac{|\hat{k}_{N,q}|^2}{(|\hat{k}_{N,q}|^2 + N^2\lambda\tilde{\omega}_q^{2p})^2}. \quad (4.7)$$

<u>Proof</u> Let $f^o_{N;\lambda}$ denote the filtered solution for exact data with a given value of λ. Then the total error may be written

$$f - f_{N;\lambda} = (f-f_N) + (f_N - f^o_{N;\lambda}) + (f^o_{N;\lambda} - f_{N;\lambda})$$

where the three terms represent the aliasing error (projection onto T_{N-1}), the regularization error, and the error resulting from the noise in the data, respectively. Assuming that these three sources of error are mutually independent we then have

$$E(\|f - f_{N;\lambda}\|^2) = E(\|f-f_N\|^2) + E(\|f_N - f^o_{N;\lambda}\|^2) + E(\|f^o_{N;\lambda} - f_{N;\lambda}\|^2). \quad (4.8)$$

Let $\hat{f}_q$ denote the exact Fourier coefficients of f on [0,1). Then

$$f(x) \sim \frac{1}{N} \sum_{q=-\infty}^{\infty} \hat{f}_q \exp(i\omega_q x),$$

where

$$|\hat{f}_q| = O(Nq^{-P})$$

and

$$\hat{f}_{N,q} = \hat{f}_q + \sum_{r=1}^{\infty} (\bar{\hat{f}}_{Nr-q} + \hat{f}_{Nr+q}).$$

Thus for the first term in (4.8) we have

$$\|f-f_N\|^2 = \frac{1}{N^2}[\sum_{q=0}^{N-1} |\hat{f}_q-\hat{f}_{N,q}|^2 + 2\sum_{q=N}^{\infty} |\hat{f}_q|^2] = O(N^{-2p}) = O(h^{2p}).$$

For the second term, from (2.9) and (2.10) we have

$$\|f_N - f^o_{N;\lambda}\|^2 = \frac{1}{N^2} \sum_{q=0}^{N-1} (1-Z_{q;\lambda})^2 |\hat{f}_{N,q}|^2 = \lambda^2 \sum_{q=0}^{N-1} \frac{\tilde{\omega}_q^{4p}|\hat{f}_{N,q}|^2}{(|\hat{k}_{N,q}|^2 + N^2\lambda\tilde{\omega}_q^{2p})^2}.$$

For the third term we have

$$E(\|f^{o}_{N;\lambda}-f_{N;\lambda}\|^2) = E\left[\sum_{q=0}^{N-1} Z^2_{q;\lambda}\frac{|\hat{\varepsilon}_{N,q}|^2}{|\hat{k}_{N,q}|^2}\right] = N\sigma^2\sum_{q=0}^{N-1}\frac{Z^2_{q;\lambda}}{|\hat{k}_{N,q}|^2},$$

since

$$\sigma^2 = E(\varepsilon_n^2) = \frac{1}{N}E(|\hat{\varepsilon}_{N,q}|^2).$$

The result (4.7) follows immediately.

A similar result to Theorem 4.1 has been given by Lukas [5].

5. COMMENTS

It is of interest to compare the performance of regularization by ML with that of regularization by Wahba's cross-validation method [9], wherein the optimal λ is found by minimizing

$$V_{CV}(\lambda) = \frac{\frac{1}{N}\sum_{q=0}^{N-1}(1-Z_{q;\lambda})^2|\hat{g}_{N,q}|^2}{\left[1-\frac{1}{N}\sum_{q=0}^{N-1}Z_{q;\lambda}\right]^2}$$

In the comparative study in [8] it was found that, in general, there is little to choose between the two methods in the case of deconvolution using pth order regularization in T_{N-1}. In the limit of exact data ($\sigma \to 0$) it was found in a few cases that $V_{CV}(\lambda)$ ceased to possess a well-defined minimum in the sense that $dV_{CV}/d\lambda$ was essentially zero in an interval $[0,\lambda_1]$, $\lambda_1 > 0$, and positive for $\lambda > \lambda_1$. In these cases a minimum of $V_{ML}(\lambda)$ was clearly defined. In the opposite limit of high noise level ($\sigma >> 0$), both V_{CV} and V_{ML} were liable to possess more than one minimum, although oscillations in V_{ML} tended to set in at a higher σ than for V_{CV}. Further details are to be found in [8].

Whereas both CV and ML appear quite powerful in the regularization of problems which are not too highly ill-posed, in contrast, when the decay of $|\hat{k}_{N,q}|$ with q is rapid, reflecting high ill-posedness, neither method is particularly successful. In such problems it is advantageous to impose *a priori* constraints, such as non-negativity, on the solution [18-20].

ACKNOWLEDGEMENT

I wish to thank Professor G. Wahba for her comments on the first draft of this paper and for bringing to my attention Dr. Lukas' thesis [5] and references [6,7].

REFERENCES

1. TIKHONOV, A.N. and ARSENIN, V.Y., *Solution of ill-posed problems*. Wiley/Winston, London (1977).
2. MILLER, G.F., Fredholm integral equations of the first kind. pp.175-188 of L.M. Delves and J. Walsh (Eds.), *Numerical Solution of Integral Equations*, Oxford University Press, Oxford (1974).
3. ARE'FEVA, M.V., Asymptotic estimates for the accuracy of optimal solutions of equations of convolution type. *USSR Comp. Math. Math. Phys.* 14, 19-33 (1974).
4. CULLUM, J., The effective choice of smoothing norm in regularization. *Math. Comp.* 33, 149-170 (1979).
5. LUKAS, M., Ph.D. Thesis, Australian National University, Canberra (1980).
6. GAMBER, H.A., Choice of an optimal shape parameter when smoothing noisy data. *Comm. Statist. A - Theory Methods* 8, 1425-1435 (1979).
7. WAHBA, G. and WENDELBERGER, J., Some new mathematical methods for variational objective analysis using splines and cross-validation. *Monthly Weather Review* 108, 1122-1143 (1980).
8. DAVIES, A.R. and MALEKNEJAD, K., On the statistical regularization of Fredholm convolution equations of the first kind. *To be published.*
9. WAHBA, G., Practical approximate solutions to linear operator equations when the data are noisy. *SIAM J. Numer. Anal.* 14, 651-667 (1977).
10. ANDERSSEN, R.S. and BLOOMFIELD, P., Numerical differentiation procedures for non-exact data. *Numer. Math.* 22, 157-182 (1974).
11. ANDERSSEN, R.S. and BLOOMFIELD, P., A time series approach to numerical differentiation. *Technometrics.* 16, 69-75 (1974).
12. STRAND, O.N. and WESTWATER, E.R., Statistical estimation of the numerical solution of a Fredholm integral equation of the first kind. *J. Ass. Comput. Mach.* 15, 100-114 (1968).
13. FRANKLIN, J.N., Well-posed stochastic extensions of ill-posed linear problems. *J. Math. Anal. Applic.* 31, 682-716 (1970).
14. TURCHIN, V.F., KOSLOV, V.P. and MALKEVICH, M.S., The use of mathematical-statistics methods in the solution of incorrectly posed problems. *Soviet Phys. Usp.* 13, 681-702 (1971).
15. KLEIN, G., On spline functions and statistical regularization of ill-posed problems. *J. Comp. Appl. Math.* 5, 259-263 (1979).

16. COX, D.R. and MILLER, H.D., *The Theory of Stochastic Processes*. Methuen, London (1965).
17. WHITTLE, P., Some results in time series analysis. *Skand. Actuarietidskr.* 35, 48-60 (1952).
18. WAHBA, G., Constrained regularization for ill-posed linear operator equations, with applications in meteorology and medicine. Technical Report 646, Dept. of Statistics, Univ. of Wisconsin, Madison (1981).
19. AL-FAOUR, O.M., Ph.D. Thesis, Univ. of Wales, Aberystwyth (1981).
20. REDSHAW, T.C., Ph.D. Thesis, Univ. of Wales, Aberystwyth (1982).

STABILITY AND STRUCTURE IN NUMERICAL METHODS FOR VOLTERRA EQUATIONS

Christopher T.H. Baker

University of Manchester

1. INTRODUCTION

1.1. 'Of course, the first thing to do was to make a grand survey'.

We concentrate on numerical methods for Volterra equations of the second kind:

$$f(x) - \int_0^x K(x,t,f(t))\,dt = g(x) \qquad (x \geq 0) \tag{1.1}$$

where g and K are prescribed and satisfy convenient smoothness assumptions; we seek f. Also of interest are Abel equations where H is smooth and

$$K(x,t,v) = H(x,t,v)/(x-t)^{\sigma} \qquad (0<\sigma<1). \tag{1.2}$$

If one regards f,g,K as vector-valued then (1.1) represents a system of integral equations; such a system arises on integrating the first of the equations

$$y'(x) = G(x,y(x),z(x)), \quad z(x) = \int_0^x K(x,t,y(t))\,dt \quad (x \geq 0) \tag{1.3}$$

where $y(x)$ is sought given $y(0)$. Thus methods for (1.1) yield, after simplification, methods for integro-differential equations. In particular, methods for (1.1) yield, on simplifying, associated methods for the initial-value problem

$$y'(x) = F(x,y(x)), \quad y(0) = y_0, \qquad x \geq 0; \tag{1.4}$$

set $K(x,t,v) = F(t,v)$ and $g(x) = y_0$ in (1.1).

1.2 '"Do you mean that you think you can find out the answer...?" said the March Hare.'

Various methods proposed for the numerical solution of (1.1) upon a "mesh" $\{\tau_k\}$ may be derived by setting

$x=\tau_k$ in (1.1) and discretizing. *Quadrature methods* result with $\tau_k = kh$ and the rules

$$Q: \int_0^{ih} \phi(t)dt \simeq h \sum_{k=0}^{i} \omega_{ik}\phi(kh) \quad (h>0;\ i=1,2,3,\ldots). \quad (1.5)$$

We obtain $\tilde{f}_0 = g(0)$, $\tilde{f}_i \simeq f(ih)$ and

$$\tilde{f}_i - h\sum_{k=0}^{i} \omega_{ik} K(ih,kh,\tilde{f}_k) = g(ih) \quad (i=1,2,\ldots). \quad (1.6)$$

Rules Q have been obtained from Gregory rules, or a composite rule together with a 'starting' rule or 'end' rule; we call the latter Kobayasi rules because of the systematic study [16].

EXAMPLE 1.1 Consider (a) $\omega_{ii}=\theta$, $\omega_{i0}=1-\theta$; $\omega_{ij}=1$, $0<j<i$ (Nevanlinna [20]). Of interest are the repeated *trapezium, Euler,* and *backward Euler,* rules ($\theta=1/2$, $\theta=0$, $\theta=1$) (b) *Gregory rules* [1], e.g. $\omega_{i0}=\omega_{ii}=5/12$, $\omega_{i1}=\omega_{i,i-1}=13/12$, $\omega_{ij}=1$ $(1<j<i-1)$ (c) *Simpson's & trapezium rule*: for $s\geq 1$, $\omega_{2s,0}=\omega_{2s,2s}=1/3$, $\omega_{2s,2r+1}=4/3$ $(r=0,1,\ldots,s-1)$, $\omega_{2s,2r}=2/3$ $(r=1,2,\ldots,s-1)$, $\omega_{2s+1,r}=\omega_{2s,r}$ $(r=0,1,\ldots,2s-1)$, $\omega_{2s+1,2s}=1/3+1/2$, $\omega_{2s+1,2s+1}=1/2$ (d) *trapezium & Simpson's rule*: denote the weights in (c) by ω'_{ij} and define $\omega_{ij} = \omega'_{i,i-j}$. In (c), (d) the basic rule is Simpson's rule, repeated with the trapezium rule as an 'end' correction (c) or a 'starting' rule (d).

Many *Runge-Kutta methods* can be derived (cf.[13]) as starting formulae with auxiliary discretizations of a lag term. However, classical extended and mixed R-K methods can be viewed as extensions of the quadrature methods. These R-K methods use abscissae $\theta_0,\theta_1,\ldots,\theta_p=1$ and values A_{rs} $(r,s=0,1,\ldots,p)$ of a tableau $[\underline{\theta}|A]$ associated with formulae

$$\int_0^{\theta_r h} \phi(t)dt \simeq h\sum A_{rs}\phi(\theta_s h) \quad (r=0,1,\ldots,p). \quad (1.7)$$

If we set $x = ih + \theta_r h$ in (1.1) and discretize we can consider the integral on $[0,ih]$ and that on $[ih,ih+\theta_r h]$ separately. For the first interval we have the choice of the rules Q in (1.4) or the i-times repeated version of (1.6) with $r=p$ $(\theta_p=1)$. Writing $\tau_j = ih + \theta_r h$ $(j=i(p+1)+r+1,\ \tau_0 = 0)$ the discretized equations are $\tilde{f}_0 = g(0)$, $\tilde{f}_j \simeq f(\tau_j)$ where

$$\tilde{f}_j - h\sum_{k=0}^{(i+1)(p+1)} \Omega_{jk} K(\tau_j,\tau_k,\tilde{f}_k) = g(\tau_j). \quad (1.8)$$

According to the choice above, $\Omega_{jk} = \Omega_{jk}[Q,A]$ for mixed quadrature-R-K methods or $\Omega_{jk} = \Omega_{jk}(A)$ for the extended R-K (cf. page 36 for the formulae). Equations (1.6) fit into the framework (1.8); set

$$\Omega_{jk} = \Omega_{jk}(Q) \equiv \omega_{jk}, \quad \tau_j = jh. \tag{1.9}$$

These methods are best described by illustrations:

EXAMPLE 1.2 Conventional R-K tableau may be taken from the study of (1.4). Examples of $[\underline{\theta}|A]$ are:

$$\text{(a)} \quad \left[\begin{array}{c|cc} \frac{1}{2} & \frac{1}{2} & 0 \\ 1 & 1 & 0 \end{array}\right]; \qquad \text{(b)} \quad \left[\begin{array}{c|ccc} \tau_0 & \tau_0 & 0 & 0 \\ \tau_1 & \tau_1-\tau_0 & \tau_0 & 0 \\ 1 & \frac{1}{2} & \frac{1}{2} & 0 \end{array}\right]$$

where $\tau_0 = (3\pm\sqrt{3})/6$. For (a), the array of weights $\Omega_{jk}(A)$ has the appearance shown in (c) whilst those in (d) correspond

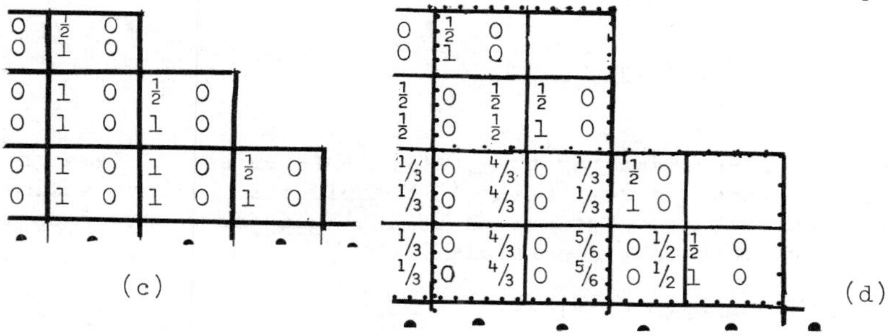

to the choice $\Omega_{jk}[Q,A]$ using the rules Q of Example 1.1(c).

The methods above are inadequate for Abel equations. For these (and for certain Volterra problems) product integration may be employed; the techniques can be amalgamated where the kernel decomposes into sums of weakly - singular and well-behaved kernels. Suppose for example $K(x,y,v) = \kappa(x,y)v$.

If $\kappa(x,y)$ is smooth or weakly-singular, "generalized quadrature" rules

$$P: \int_0^{ih} \kappa(x,t)\ \phi(t)dt \simeq \sum_k v_{ik}\ \phi(kh) \tag{1.10}$$

can be constructed. Approximating $\phi(t)$ by a piecewise polynomial $\tilde{\phi}_i(t)$ agreeing with $\phi(t)$ for $t \in \{kh\}_{k\geq 0}$ and integrating $\kappa(x,t)\tilde{\phi}(t)$ yields sums of the type required in (1.10).

EXAMPLE 1.3. To describe generalizations of rules in Example 1.2, let $P_m[\alpha,\beta]$, $P_m(\alpha,\beta]$ and $P_m[\alpha,\beta)$ be those functions coinciding with some polynomial of degree m on $[\alpha,\beta]$, $(\alpha,\beta]$ or $[\alpha,\beta)$ respectively. We then take (a)(i) to generalize the Euler rule $\tilde{\phi}_i(x) \in P_0\ [kh,(k+1)h)\ k=0,1,\ldots,i-1$; (ii) for the generalized backward Euler rule, $\tilde{\phi}_i(x) \in P_0(kh,(k+1)h]$; for

(iii) the generalized trapezium rule, $\tilde{\phi}_i(x) \varepsilon C[0,ih] \cap P_1[kh,(k+1)h]$ $k=0,1,\ldots,i-1$; (b) for the generalized Simpson's & trapezium rule, $\phi_{2r}(x) \varepsilon C[0,2rh] \cap P_2[2sh,2(s+1)h]$, $\tilde{\phi}_{2r+1}(x) \varepsilon C[0,(2r+1)h] \cap P_2[2sh,2(s+1)h] \cap P_1[2rh,(2r+1)h]$ $(s=0,1,\ldots,r-1)$, etc.

In the case where

$$\kappa(x,y) = k(x-y) \tag{1.11}$$

the weights $v_{k,i}$ assume a convenient pattern. Thus, those of the generalized trapezium rule form the array

$$\begin{array}{lllll} 0 & & & & \\ \nu_1'' & \nu_0' & & & \\ \nu_2'' & \nu_1 & \nu_0' & & \\ \vdots & \vdots & \vdots & \ddots & \\ \nu_i'' & \nu_{i-1} & \nu_{i-2} & \cdots & \nu_0' \\ \vdots & & & & \ddots \end{array} \tag{1.12}$$

where the expressions for the coefficients arise by elementary calculus; those corresponding to Example 1.3 (b) have a similar block-structure. Extensions to analogues of a set of collocation - RK methods are possible.

2. CONVERGENCE AND STABILITY

2.1. *'... she waited ... to see if she was going to shrink any further.'*

It may be hoped that as h shrinks to zero our approximate solutions *converge*, for $\{\tau_j\}$ on compact intervals, to the true values. Results exist for second-kind equations under reasonable conditions; see [2, 15] for example. The concept of *zero-stability*, valuable in discussing (1.4), does not preoccupy us when considering (1.1): it amounts to boundedness of the weights, and convergence then follows from consistency. Convergence as $h \to 0$ is here taken as *sine qua non* but modelling properties when computing with realistic h is of concern and traditionally discussed under 'stability' and 'qualitative behaviour'.

2.2. *'"When I use a word", Humpty Dumpty said..."it means just what I choose it to mean... ."'*

'Stability' is a much used term with varying interpretations. Stability involves the response to perturbation; the stability of (1.1) and that of the discretized equation should be related.

Consider the linear version of (1.1):

$$f(x) - \lambda\int_0^x \kappa(x,t)\, f(t)dt = g(x). \qquad (2.1)$$

The solution, expressible in the form

$$f(x) = g(x) + \lambda\int_0^x \rho_\lambda(x,t)\, g(t)dt, \qquad (2.2)$$

depends upon the '*resolvent kernel*' $\rho_\lambda(x,t)$; if $g(x)$ is perturbed by an amount $\delta g(x)$ then $\delta f(x)$ is obtained by substitution of δf for f and δg for g. Some early stability analysis [16, 17, 23] involved study of the dominant term (as $h\to 0$) of the error to determine whether it mirrored the response of δf to δg. Where there were unwanted parasitic components of the dominant error term (corresponding to replacement of λ in (2.2) by *unwanted* values $\lambda\upsilon_i$) the method was regarded as "unstable". This analysis is asymptotic as $h\to 0$ but extends to non-linear equations.

2.3. *'"I'm glad I've seen that done," thought Alice.'*

In studying the discretized equations the intimate connection with methods for (1.4) can be exploited. It is recognised that the use of a Gregory formulae (1.5) with $K(x,t,v) = F(t,v)$, $g(x) \equiv y_0$, is equivalent to the use of an Adams-Moulton rule for (1.4), with appropriate starting values. Euler formulae are similarly related to the formulae of the same name in Example 1.1(a), whilst the extended R-K method using a conventional RK tableau is equivalent to the normal RK method. The respective formulae for (1.1) and (1.4) are associated when that for (1.1) reduces to the method for (1.4), if $K(x,t,v)=F(t,v)$, $g(x)=1$.

EXAMPLE 2.1. The quadrature methods using the rules of Example 1.1 are associated with linear multistep or *cyclic linear multistep* methods, in a non-unique manner.

DEFINITION 2.1. The rules

$$\int_0^{\tau_j} \phi(t)dt \simeq \sum_{k=0}^{(i+1)(p+1)} \Omega_{jk}\, \phi(\tau_j) \qquad (i\equiv j \bmod (p+1))$$

in (1.8) are (block-) reducible when the corresponding block-lower triangular array of weights Ω_{jk} can be partitioned into submatrices $V_{n\ell}$ of order q such that

$$\sum_{\ell=0}^{m} A_\ell\, V_{n-\ell,j} = B_{n-j} \qquad \text{(m finite and fixed)} \qquad (2.3)$$

for a set of fixed matrices $\{A_\ell, B_\ell\}$ which vanish for $\ell \notin \{0,1,\ldots,m\}$. We usually ask $\sum_\ell A_\ell\{1,1,\ldots,1\}^T = \underline{0}$. If $A_0 = I$, $A_1 = -I$, $A_\ell = 0$ otherwise, the rules are called simply-block-reducible. When $\Omega_{jk} = \Omega_{jk}(Q)$, $q=1$, and $A_\ell = \alpha_\ell, B_\ell=\beta_\ell$ with $V_{n\ell} = \omega_{n\ell}$, the rules Q are called $\{\rho^*,\sigma^*\}$-reducible

$$\rho^*(\mu) = \sum_{\ell} \alpha_{\ell}\mu^{m-\ell}, \sigma^*(\mu) = \sum_{\ell} \beta_{\ell}\mu^{m-\ell}. \tag{2.4}$$

EXAMPLE 2.2. The weights in the extended R-K methods and in the quadrature methods of Example 1.1, as well as the derived mixed quadrature - R-K methods, can be partitioned ([5], eq.) as

$$\begin{matrix} | & | & & \diagdown & & \diagdown & \diagdown & \\ W & W_0 & W_0 \cdots\cdots\cdots W_0 & & W_1 & \cdots\cdots W_{p-1} & W_p \end{matrix} \quad \text{(etc.)}. \tag{2.5}$$

Such rules are simply-block-reducible. For extended R-K methods take $p = 1$, $W_1 = A$ and $W_0 = W$ equal to the matrix each of whose rows is the last row of A. The Gregory rules are simply reducible ($W_0, W_1, \ldots, W_p$ are scalar and W is a row-vector).

Suppose $K(x,t,f(t)) = \sum X_s(x)Y_s(t,f(t))$, in (1.1), the sum being finite. Then (1.1) is equivalent to a system of ordinary differential equations (o.d.e.s). If the weights Ω_{jk} reduce to an associated method for an o.d.e., then (1.8) is equivalent to applying the associated method to the differential system (apart from terms involving g). This equivalence indicates the relevance of the stability analysis for methods for o.d.e.s (including recent concepts such as G-stability and algebraic stability [12]). Further insight can be looked for in the case of non-separable kernels by *imbedding*. In the case of *convolution kernels* $(K(x,y,v)=k(x-y)\phi(v))$, and in particular for polynomial convolution kernels, one is involved with the analysis of constant-coefficient finite term recurrence relations. (Observe that *simple* - block-reducibility is frequently required to facilitate such analysis if $q>1$ in Definition 2.1.)

2.4. *'This sounded very hopeful,...'*

For the basic integral equation

$$f(x) - \lambda\int_0^x f(t)\,dt = g(x) \tag{2.6}$$

the analysis of numerical methods using reducible integration rules can be brought into sharp focus. Assuming (2.3), the approximate values $\tilde{f}_j$ define vectors satisfying a recurrence relation whose auxiliary or 'stability' polynomial is

$$\Sigma(\mu) \equiv \det \sum_{\ell} \{A_{\ell} - \lambda h\, B_{\ell}\}\mu^{m-\ell} \tag{2.7}$$

and the recurrence is called *strictly stable* if this polynomial is *Schur* (has its zeros in $|\mu|<1$), and *stable* if it is (*simple*) *von Neumann* i.e. has zeros in $|\mu|\leq 1$ with those of modulus unity being semi-simple. Reflecting the study of (2.6), with $g(x)=1$, a method is *weakly A-stable* (resp. *A-stable*) if it is stable (resp. strictly stable) for $\text{Re}(\lambda h)\leq 0$ (resp. <0). Definitions of $A(\alpha)$-, A_0-, *L-stability* also mirror those for the o.d.e. $y'(x) = \lambda y(x)$.

EXAMPLE 2.3. For extended R-K methods, stability is governed by the condition $|\hat{\mu}(\lambda h)|\leq 1$ where [5] $\hat{\mu}(\lambda h)$ is a rational approximation to $\exp(\lambda h)$ reducing to the usual form if $[\underline{\theta}|A]$ is a conventional tableau. A $\{\rho^*,\sigma^*\}$-reducible quadrature method for (2.6) is strictly stable when $\rho^*(\mu) - \lambda h\sigma^*(\mu)$ is Schur, or equivalently $r^*(\mu) - \lambda h\, s^*(\mu)$ is *Hurwitz* where $r^*(\mu) = \rho^*(\zeta)$, $s^*(\mu) = \sigma^*(\zeta)$, $\zeta = (\mu+1)/(\mu-1)$. The method is $A(\alpha)$-stable when $\{r^*(\mu)/s^*(\mu)\}^{\beta}$ is, for $\beta = (2-2\alpha/\pi)^{-1}$, *positive* i.e. is a regular map of $\mathrm{Re}(\mu)>0$ onto itself. The A-stability condition can be written $\mathrm{Re}\{\rho^*(\mu)/\sigma^*(\mu)\}>0$ for $|\mu|>1$. See [8]. With weak A-stability, equalities can obtain in the latter inequalities.

Mixed quadrature - RK methods can be analysed [5] if the rules Q are (block-) reducible. Mixed methods using A-stable Q and A-stable $[\underline{\theta}|A]$ need not be A-stable; such results are of concern and have led to *modified methods* [14].

Whilst stability is a clear-cut concept it is frequently appropriate to consider whether $\Sigma(\mu/\nu)$ is Schur or (simple-) von Neumann. Thus with $\nu = \exp(\lambda h)$ one achieves criteria for "relative stability", and additional insight is available.

Study of the basic equation gives insight into more general results, suggests new methods and yields information concerning parasatic components (§2.2), and into methods for first kind equations [4]. It extends also to methods for problems of the form (1.3).

3. EXPLOITING STRUCTURE

3.1. "... you should say what you mean"

The discussion of §2 has indicated a number of results but we now propose to remedy an omission by giving a formal definition of *stability*. To simplify we consider convolution equations; though the concepts are more general, these have useful structure.

Consider the equation $f - \lambda\, k*f = g$ where $*$ denotes convolution:

$$f(x) - \lambda\int_0^x k(x-y)f(y)dy = g(x). \tag{3.1}$$

A perturbation $\delta(x)$ in $g(x)$ results in a change $\varepsilon(x)$ in $f(x)$

$$\varepsilon(x) = \delta(x) + \lambda\int_0^x r_\lambda(x-y)\,\delta(y)dy \tag{3.2}$$

where $r_\lambda(x)$ is the "resolvent kernel" for $k(x)$: r_λ satisfies (3.1) when $g(x)=k(x)$. The study of *'analytical'* stability of (1.1) (as opposed to stability of a *numerical* scheme) is concerned with the behaviour of the response $\varepsilon(x)$ to various classes of input perturbations $\delta(x)$. Various definitions are quoted by Tsalyuk [25] in his review. We offer the following simplified definition.

DEFINITION 3.1. Equation (3.1) is stable in (F,G) where F,G are normed function spaces if $\varepsilon \in F$ whenever $\delta \in G$ and

$\|\varepsilon\|_F \leqslant M\|\delta\|_G$, where $M<\infty$ is independent of δ. Equation (3.1) is asymptotically stable in (F,G) when it is stable in (F,G) and further $\varepsilon(x) \to 0$ as $x \to 0$ for all δ of finite norm $\|\delta\|_G$.

Clearly, different properties result upon different choices of F,G and norms. Let (i) $C_\infty = \{\phi \mid \phi \in C[0,\infty), \|\phi\| = \sup_{x\geqslant 0} |\phi(x)|\}$ (ii) $C_\infty^0 = \{\phi \in C_\infty, \lim_{x\to 0} \phi(x) = 0\}$; (iii) $\mathbb{R} = \{\phi \mid \phi(x) = \gamma, x\geqslant 0, \|\phi\| = |\gamma|\}$, and let $L_p[0,\infty)$ have the usual meaning; these are possible choices for F,G. The choice $\mathbb{R}$ for G is given plausibility by the form of the equation for the *differential resolvent*, to which the resolvent is related.

EXAMPLE 3.1 (a). Equation (2.6) is stable in $(C_\infty,\mathbb{R})$ if $\lambda\leqslant 0$ and asymptotically stable (i.e. stable in $(C_\infty^0,\mathbb{R})$) if $\lambda<0$.
(b) If $k(x-y)=(x-y)^{-\frac{1}{2}}$ and $g(x)=1$ then $f(x)= \exp(\lambda^2\pi x)[1+\mathrm{erf}(\lambda\sqrt{\pi x})]$ where $\mathrm{erf}(z) = (2/\sqrt{\pi})\int_0^z \exp(-t^2)dt$ and the stability results of (a) obtain again. (c) If $F=L_2[0,\infty)$ and $\int_0^x\int_0^t k(s-t)\phi(t)\phi(s)ds\,dt>0$ for all $x>0$, all $\phi\in F$, then (3.1) is stable in (F,F) for $\lambda\leqslant 0$. [Show that $\|\varepsilon\|\leqslant\|\delta\|$ using $\|\delta\|^2 = \|\varepsilon-\lambda k*\varepsilon\|^2 = \|\varepsilon\|^2-2\lambda(\varepsilon,k*\varepsilon)+\lambda^2\|k*\varepsilon\|^2$.]

We observe that F,G are meaningful for our present purposes when they consist of functions defined upon $[0,\infty)$.

Tsalyuk [25] summarizes certain stability conditions in terms of r_λ, in particular when $r_\lambda \in L_1[0,\infty)$. The convolution structure in (3.1) can be exploited because of the rôle to be played by the *Laplace transform* [9] $\mathcal{L}\{\phi\}(s)=\int_0^\infty \exp(-st)\phi(t)dt$, $s\in\mathbb{C}$. Under modest conditions, $\mathcal{L}\{f\} = \mathcal{L}\{g\}/[1-\lambda\mathcal{L}\{k\}]$; with $g\equiv k$ we obtain $\mathcal{L}\{r_\lambda\}$. In principle, the Laplace transform is a useful tool (the result of Example 3.1(b) emerges from its use) but requires a technical ability in computing and inverting the transforms.

The connection between the *Laplace transform* and the *Fourier transform* is well-known. It permits, in particular, the derivation of the *Paley-Wiener* theorem which can be viewed as a criterion for stability [18; 19, p.407; 25]. *Let* $k\in L_1[0,\infty)$ $(\int_0^\infty |k(t)|dt<\infty)$. *Then* $r_\lambda\in L_1[0,\infty)$ *if and only if* $1-\lambda\mathcal{L}\{k\}(s) \neq 0$ *for all* $\mathrm{Re}(s)\geqslant 0$. Specific stability deductions are given in [25, §2] for example.

Now consider the discretized equations which provide approximate solutions to (3.1). The quadrature, R-K, and product-integration techniques, described above, yield a lower-triangular or block-lower triangular system of equations. The pertinent feature, mimicking the convolution property in (3.1) is (see [3]) the bordered *isoclinal* or bordered *block-isoclinal* nature of the matrix of coefficients: the equations assume the form in (3.3), namely:

$$\left[\begin{array}{c|cccc} I-\lambda V' & & 0 & & \\ \hline V_0 & I-\lambda V^{(0)} & & & \\ V_1 & -\lambda V^{(1)} & I-\lambda V^{(0)} & & \\ V_2 & -\lambda V^{(2)} & -\lambda V^{(1)} & I-\lambda V^{(0)} & \\ \cdot & \ddots & \ddots & \ddots & \\ \cdot & & & & \\ \cdot & & & & \end{array}\right] \begin{bmatrix} \tilde{\underline{f}}_0 \\ \tilde{\underline{f}}_1 \\ \tilde{\underline{f}}_2 \\ \tilde{\underline{f}}_3 \\ \cdot \\ \cdot \\ \cdot \end{bmatrix} = \begin{bmatrix} \underline{g}_0 \\ \underline{g}_1 \\ \underline{g}_2 \\ \underline{g}_3 \\ \cdot \\ \cdot \\ \cdot \end{bmatrix} . \tag{3.3}$$

EXAMPLE 3.2. (a) Assume Ω_{jk} can be partitioned as shown in (2.5). Let the corresponding array of values $k(\tau_j-\tau_k)$ be partitioned similarly to assume the form

$$\left[\begin{array}{c|ccccc} K_i & K^{(i)} & \cdots & K^{(2)} & K^{(1)} & K^{(0)} \end{array}\right]$$

Then $V^{(i)} = K^{(i)} * W_{p-i}$ where $W_{-r} = W_0$ for $r>0$, $V_i=K_i*W$, and $*$ denotes the Schur (or element-wise) product of matrices. (b) Suppose $k(x-y)=(x-y)^{-\sigma}H(x-y)$ and the weights displayed in (1.12) arise from the choice $K(x,y) = (x-y)^{-\sigma}$. Then $V^{(i)}$ is scalar, $V^{(i)} = \nu_i\, H(ih)$.

The system (3.3) of equations is solvable in a block-by-block fashion if $I-V'$ and $I-V^{(0)}$ are invertible, which we presume. The coefficient matrix is of such a form that when $g(x)$ is perturbed by an amount $\delta(x)$ we can write

$$\left[\begin{array}{c|c} I-\lambda V_0 & 0 \\ \hline -\lambda V & I-\lambda W \end{array}\right] \left[\begin{array}{c} \underline{\varepsilon}' \\ \hline \underline{\varepsilon}'' \end{array}\right] = \left[\begin{array}{c} \underline{\delta}' \\ \hline \underline{\delta}'' \end{array}\right] \tag{3.4}$$

for the relation between the perturbations in g and $\tilde{f}$. The infinite matrix $I-\lambda W$ is *"block-isoclinal"* [3] , the (i,j)-th block depending upon $i-j$, and block-lower triangular. Thus, (3.3) presents an analogue of the convolution structure in (3.1). In discussing the numerical stability of (3.3) we require concepts similar to those of Definition 3.1.

DEFINITION 3.2. Equation (3.3) is stable in (F,G) where F,G are normed *sequence* spaces if $\underline{\varepsilon} \in F$ whenever $\underline{\delta} \in G$ and $\|\varepsilon\|_F \le M\|\delta\|_G$ for some $M < \infty$ independent of $\underline{\delta}$. It is asymptotically stable if, further, $\varepsilon_i \to 0$ as $i \to \infty$.

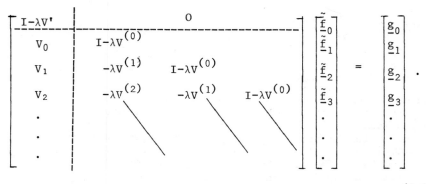

If we aim to preserve certain qualitative behaviour, the choice of F, G should mimic appropriate choices of F, G in Definition 3.1). The choices

$$\ell_p = \{\underline{\phi} \mid \|\underline{\phi}\|_p = (\Sigma|\phi_i|^p)^{1/p} < \infty\} \quad (\|\phi\|_\infty = \sup|\phi_i|),$$
$$\ell_\infty^0 = \{\underline{\phi} \mid \underline{\phi} \in \ell_\infty, \phi_i \to 0 \text{ as } i \to \infty\} \text{ and } \mathbb{R} = \{\underline{\phi} \mid \phi_i = \gamma \ \forall i\}$$

suggest themselves *inter alia*.

With appropriate conditions on V (boundedness, or decay, of successive rows) stability in (3.4) is generally resolved by attention to the block-isoclinal matrix $I-\lambda W$. By assumption, λ and h assume those values for which (3.3) can be solved in block-by-block fashion, but solvability in (F,G) is not assured. We therefore consider the *formal* solution of equations

$$[I-\lambda W]\ \underline{\varepsilon} = \underline{\delta} \tag{3.5}$$

for appropriate infinite-dimensional vectors (or, sequences) $\underline{\varepsilon},\underline{\delta}$. Now generating functions (or, equivalently, z-transforms [9]) have been employed in solving finite-term recurrence relations since the times of De Moivre and Laplace, and they assume a rôle here where the relations are rather different.

DEFINITION 3.3. The symbol of $I-\lambda W$, partitioned as indicated by (3.4) and (3.3), is the formal power series $I-\lambda W(\xi)$ where

$$W(\xi) = W^{(0)} + W^{(1)}\xi + W^{(2)}\xi^2 + \ldots \tag{3.6}$$

(where $W(\xi)$ is the *symbol* of W). The *generating function* $\underline{\phi}(\xi)$ of the infinite-dimensional vector (or, sequence) $\underline{\phi}$ partitioned as shown here is the formal power series $\underline{\phi}(\xi)$:

$$\underline{\phi} = \begin{bmatrix} \underline{\phi}_0 \\ \underline{\phi}_1 \\ \underline{\phi}_2 \\ \cdot \\ \cdot \\ \cdot \end{bmatrix}, \quad \underline{\phi}(\xi) = \underline{\phi}_0 + \underline{\phi}_1\xi + \underline{\phi}_2\xi^2 + \ldots \tag{3.7}$$

THEOREM 3.1. Given (3.5), and conformable partitioning, $[I-\lambda W(\xi)]\underline{\varepsilon}(\xi) = \underline{\delta}(\xi)$. Thus

$$\underline{\varepsilon}(\xi) = [I-\lambda W(\xi)]^{-1}\ \underline{\delta}(\xi) \tag{3.8}$$

where

$$[I-\lambda W(\xi)]^{-1} := \text{ADJ } [I-\lambda W(\xi)]/\det[I-\lambda W(\xi)], \tag{3.9}$$

ADJ denoting the *adjugate* (or "classical-adjoint").

The first statement follows from (3.5) and the rules for

multiplying formal power series; the second then provides (for any n) successive components $\underline{\varepsilon}_0, \underline{\varepsilon}_1, \ldots \underline{\varepsilon}_n$ of $\underline{\varepsilon}$ in terms of $\underline{\delta}_0, \underline{\delta}_1, \ldots, \underline{\delta}_n$; the arbitrariness of n yields the formal power series $\underline{\varepsilon}(\xi)$. (Equation (3.9) determines the inverse of successive partioned segments of $I-\lambda W$.)

Theorem 3.1. provides in principle a tool for analysing the stability of (3.5) and thence (3.4); in the scalar case W is isoclinal, $I-\lambda W(\xi)$ becomes

$$1-\lambda w(\xi) = 1-\lambda w_0 - \lambda w_1 \xi - \lambda w_2 \xi^2 \ldots, \tag{3.10}$$

and the generating function of $\underline{\phi}$ is $\phi(\xi) = \phi_0 + \phi_1 \xi + \phi_2 \xi^2 + \ldots$. Setting $\xi = z^{-1}$ the result is a relation between z-transforms. (z-transforms are intimately related [9] to the Laplace transforms which prove valuable for (3.1); known results concerning z-transforms are at our disposal.)

Observe that to analyze stability by the approach suggested the spaces F, G must be characterized by properties of the associated generating functions.

EXAMPLE 3.1. (a) $\phi \in \ell_1$ if and only if $\phi(\xi)$ is absolutely convergent on $|\xi| = 1$. Suppose in particular $p_n(z) = \Sigma a_\ell z^{n-\ell}$ is a Schur polynomial in z, then we readily see that $[\xi^n p_n(1/\xi)]^{-1}$ is the generating function of some $\underline{\phi} \in \ell_1$. More generally if $\psi(\xi)$ is absolutely convergent for $|\xi| = 1$ and has no zeros satisfying $|\xi| \leq 1$, $\phi(\xi) = [\psi(\xi)]^{-1}$ is absolutely convergent Zygmund [26 , p147] and generates $\underline{\phi} \in \ell_1$. (The result of Zygmund is the analogue of a result of Wiener, upon the convergence of reciprocals of Fourier Series, which plays a rôle in the proof of the Paley-Wiener theorem.) (b) If $\phi(\xi) = (1-\xi)^{-1}$ then $\underline{\phi} \in \ell_\infty$ although $\underline{\phi} \notin \ell_1$. More generally, if $p_n(z)$ is now *simple* von Neumann in the discussion of (a), $\underline{\phi} \in \ell_\infty^n$.

The foregoing remarks indicate an approach and a technique which we now illustrate by re-deriving earlier results. We consider the quadrature or R-K methods (1.8) applied to the basic equation (2.6), subject to constant perturbations δ_j in $g(\tau_j)$ ($\underline{\delta} \in \mathbb{R}$). With (block-)reducibility we obtain (3.4) where $I - \lambda W$ is block-isoclinal and the submatrices of V are constant. Then (3.5) results with $\underline{\varepsilon} = \underline{\varepsilon}''$; $(I-\lambda W)\underline{\varepsilon}'' = \underline{\delta} := \underline{\delta}' + \lambda V(I-\lambda V_0)^{-1}\underline{\delta}'$ (except for certain exceptional values of h where $I-\lambda V_0$ is singular) and $\underline{\delta}$ partitions so that

$$\underline{\delta}(\xi) = \underline{\delta}_0(1+\xi+\xi^2+\ldots) = \underline{\delta}_0(1-\xi)^{-1}.$$

Theorem 3.1 now yields $\underline{\varepsilon}(\xi)$ and we exploit reducibility structure: the relation (2.3) yields

$$A(\xi)\ W(\xi) = h\,B(\xi) \tag{3.11}$$

whence $A(\xi)\,(I-\lambda W(\xi)) = A(\xi) - \lambda h\ B(\xi)$ where $A(\xi)$, $B(\xi)$ are polynomials: $A(\xi)-\lambda h\ B(\xi) = \sum_\ell \{A_\ell - \lambda h\ B_\ell\}\xi^\ell$.

By Theorem 3.1,

$$\underline{\varepsilon}(\xi) = \frac{\mathrm{ADJ}[\ A(\xi) - \lambda B(\xi)\ ]\ A(\xi)\ \underline{\delta}_0}{(1-\xi)\ \det[\ A(\xi) - \lambda h B(\xi)\ ]}. \tag{3.12}$$

Commonly (as in $\{\rho^*,\sigma^*\}$-reducibility, where $\rho^*(1) = 0$, or simple-reducibility) $A(\xi)$ has the factor $(1-\xi)$ and this cancels the corresponding factor in the denominator of (3.12). It follows from Examples 3.1 that $\sqrt{\underline{\varepsilon}} \in \ell_1$ if the stability polynomial of (2.7), $\bar{\Sigma}(\mu) = \det(\mu^m\{A(\mu^{-1}) - \lambda h B(\mu^{-1})\})$ is Schur, and $\underline{\varepsilon} \in \ell_\infty$ if the roots of the stability polynomial lie on the unit disk, those on the boundary being simple. We then recover the recognised stability criteria as those for stability in $(\ell_\infty, \mathbb{R})$ and $(\ell_1, \mathbb{R})$; in particular the stability polynomial $\rho^*(\mu) - \lambda h\sigma^*(\mu)$ for $\{\rho^*,\sigma^*\}$-reducible Q is derived from

$$\varepsilon(\xi) = \frac{\delta_0}{1-\xi} \times \frac{1}{1-\lambda w(\xi)} \tag{3.13}$$

with $w(\xi) = h\sigma^*(z)/\rho^*(z)$, $z = \xi^{-1}$.

For the case considered, $w(\xi) = \sum w_\ell \xi^\ell$ and (1.5) is a convolution quadrature with $h\,\omega_{i,j} = w_{i-j}$ for $j \geq j_0$, $i \geq i_0$. The criterion (Example 2.3) for A-stability can be written in terms of w (cf. [22]); for weak A-stability, the condition

$$\mathrm{Re}\{w(\xi)\} \geq 0 \quad \text{for} \quad |\xi| \leq 1 \tag{3.14}$$

is necessary. We shall blur some of the finer differences.

For general equations of the form (3.5) we seek an analogue of the Paley-Wiener theorem. The latter is closely related to the study of equations of the form $f(x) - \int_{-\infty}^{\infty} k(x-y)f(y)dy=g(x)$ [19] and it is not surprising to find the required results arising in the study [11] of the discrete analogue of Wiener-Hopf equations.

Considering (3.5) where $W = (w_{i-j})$ is isoclinal and lower-triangular, $w(\xi)$ being continuous on the unit circle, stability in (ℓ_2,ℓ_2) is *"determined by"* [11] the symbol $1-\lambda w(\xi)$. In particular, if $1-\lambda w(\xi) \neq 0$ for $|\xi|=1$ and $2\pi\,\mathrm{ind}[\ 1-\lambda w(\xi)] \equiv \arg[\ 1-\lambda w(\theta)]_0^{2\pi} = 0$ (equivalently w being pole-free in the unit circle, $1-\lambda w(\xi) \neq 0$ for $|\xi| \leq 1$) the equations are stable in (ℓ_2,ℓ_2). Further, conditions for stability in (ℓ_p,ℓ_p) are indicated [11 , p38] and the preceding conditions suffice if $w(\xi)$ is absolutely convergent for $|\xi| = 1$. In particular, the result quoted by Lubich [18] as an analogue of the Paley-Wiener theorem (though derivable by appealing to Zygmund's analogue, given above, of Wiener's theorem) is seen as a special case. Finally, the extension to the case where W in (3.5) is block-isoclinal is indicated [11, Chapter 8] in terms of $\det[I-\lambda W(\xi)]$. and appropriate indices.

REMARK Observe that there are instances where block-isoclinal structure can apparently be replaced by scalar-isoclinal struct-

structure (as in the treatment of mixed quadrature-RK methods where Q is $\{\rho^*,\sigma^*\}$-reducible, on employing $[I-\lambda W^{(0)}]^{-1} = [I-\lambda h\ A^*K^{(0)}]^{-1}$ to produce the scalar equations). Observe further that certain methods for convolution equations of the form (1.3) yield equations of the type discussed on simplifying.

Note that product integration methods for weakly-singular problems [15,22] are of particular interest and although at pains to show that their structure makes the technique described applicable, we have not addressed the associated technical difficulties.

3.2.' "Sentence first-verdict afterwards" '.

Frequently, A-stable RK tableau employed in mixed quadrature-RK methods are not A-stable. van der Houwen was motivated by this to derive modified mixed methods in which A-stability is restored (for the basic equation).

An approach with a differing emphasis involves limiting consideration to methods which are "naturally" A-stable and considering the class of problems for which the property-stability for $Re(\lambda) \leqslant 0$ - is maintained. The works [20,21] have something of this flavour. In particular we shall place our own perspective on the results of Lubich [18], which relate to A-stable quadrature methods, omitting details.

Consider a weakly A-stable convolution quadrature, having a symbol $w(\xi)$ satisfying (3.14). The relation (3.14) is precisely the characterization for the sequence $\{w_n\}$, $(n=0,1,2,\ldots)$ with $w_{-n} = \bar{w}_n$ to be a positive definite sequence [6,p.113].

DEFINITION 3.4. $\{w_n\}$ is positive definite if and only if $w_{-n} = \bar{w}_n$ and $\sum_0^n\sum_0^n w_{k-j}\, z_k\bar{z}_j \geqslant 0$ for all sets of complex z_k.

Observe that the segments of $W+W^*$ are then positive semi-definite. If the quadrature method is applied to (3.1), $\underline{w} \in \ell_\infty$ and $\underline{k} \in \ell_1$ then stability in $(\ell_\infty,\ell_\infty)$ obtains for all λ with $Re(\lambda) \leqslant 0$ if

$$(w\circ k)(\xi) = w_0k_0 + w_1k_1\xi + w_2k_2\xi^2 + \ldots$$

is absolutely convergent for $|\xi| = 1$ and $Re\{w\circ k(\xi)\} \geq 0$ for $|\xi| \leq 1$. We therefore ask for conditions which ensure that $\{w_nk_n\}$ is a positive definite sequence.

The following and related definitions are prominent in harmonic analysis.

DEFINITION 3.5. A L_1-function $k(x)$ is of positive type if and only if $k(-x) = \overline{k(x)}$ and

$$\int_{-\infty}^{\infty}\int_{-\infty}^{\infty} k(x-y)\phi(x)\overline{\phi(y)}\,dy \geqslant 0 \tag{3.15}$$

for all continuous integrable ϕ.

This characterization permits extensions; see [7].

For our present purposes, $k(x)$ is continuous. If $k(x)$ is of positive type and continuous it is called positive-definite, in the sense of Mathias and Bochner. Then (3.16) reduces to

$$\sum_{i=0}^{n} \sum_{j=0}^{n} k(\tau_i - \tau_j)\, z_i\, \bar{z}_j \geq 0 \qquad \text{for all } \{z_j\} \qquad (3.16')$$

for every sequence $\{\tau_i\}$. In particular, the sequence $k_n = k(nh)$ is then, for every $h>0$, positive definite. By a theorem of Schur [10] it follows that $\underline{wok}$ is a positive sequence and the recurrence is stable in $(\ell_\infty, \ell_\infty)$ for all λ with $\mathrm{Re}(\lambda) \leq 0$, reflecting the corresponding behaviour of the integral equation. Note that $k(x) \equiv 0$ is positive-definite!

Though the Hadamard product $(wok)(\xi)$ of $w(\xi)$ and $k(\xi)$ can sometimes be expressed in closed form [9] the use of positive definiteness eliminates the need for intricate analysis; the results of Lubich [18] are correspondingly appealing and elegant. Observe that use of (3.16) or (3.16'), the use of Bochner's characterization theorem [10] , or the construction of $L\{k\}$ (see [18]) to verify positive definiteness present varying problems. However, the literature does contain examples of necessary conditions and of sufficient conditions. The works of Mathias and Bochner can be traced through [6].

4. ACKNOWLEDGEMENTS

Quotations are from the works of Lewis Caroll. The author has, over a period, enjoyed a number of conversations with H. Brunner, P.J. van der Houwen, D. Kershaw, and P.H.M. Wolkenfelt and is indebted to Ch. Lubich for the opportunity to see in advance the statement of his results published in these proceedings.

We await with interest the further details promised by Lubich. Extensions and further developments are a matter for continuing interest. Use of the symbol and generating function was presaged in [3].

REFERENCES

1. BAKER, C.T.H. *The numerical treatment of integral equations* Clarendon Press, Oxford (1977).
2. BAKER, C.T.H. Runge-Kutta methods for Volterra integral equations of the second kind. pp.1-13 of G.A. Watson (Ed.) *Numerical analysis*. Springer-Verlag Lect. Notes. Math 630, Berlin (1978).
3. BAKER, C.T.H. Families of structured quadrature rules and some ramifications of reducibility. pp.12-24 of G. Hämmerlin (Ed.) *Numerical integration*. Birkhauser-Verlag, ISNM 57, Basel (1982).

4. BAKER, C.T.H. and KEECH, M.S. Stability regions in the numerical treatment of Volterra integral equations. *SIAM J. Numer. Anal.* 15, 394-417 (1978).
5. BAKER, C.T.H. and WILKINSON, J.C. Stability analysis of Runge-Kutta methods applied to a basic Volterra integral equation. *J. Austral. Math. Soc. (Ser.B)* 22, 515-538 (1981).
6. BECKENBACH, E.F. and BELLMAN, R. *Inequalities.* Springer-Verlag, Berlin (1965).
7. COOPER, J.L.B. Positive definite functions of a real variable. *Proc. London Math. Soc.* (3) 10, 54-66 (1966).
8. DAHLQUIST, G. Positive functions and some applications to stability questions for numerical methods. pp.1-29 of C. de Boor and G. Golub (Eds.) *Recent advances in numerical analysis.*
9. DOETSCH, G. *Guide to the applications of the Laplace and z-transforms.* Van Nostrand Reinhold, New York (1971).
10. DONOGHUE, W.F. (Jr.) *Distributions and Fourier transforms.* Academic Press, New York (1969).
11. GOHBERG, I.C. and FELDMAN, I.A. *Convolution equations and projection methods for their solution.* Transl. Math. Mon.41, American Math. Soc., Providence (1974).
12. HALL, G. Initial-value problems for ordinary differential equations. pp. 95-111 of C.T.H. Baker and C. Phillips (Eds.) *The numerical solution of nonlinear problems.* Clarendon Press, Oxford, (1981).
13. HAIRER, E. *These proceedings.*
14. HOUWEN, P.J. VAN DER. Convergence and stability results in Runge-Kutta methods for Volterra integral equations. *BIT* 20, 375-377 (1980).
15. KERSHAW, D. *These proceedings.*
16. KOBAYASI, M. On numerical solution of the Volterra integral equations of the second kind by linear multistep methods. *Rep. statis. Applic. Res. (JUSE)* 13, 1-21 (1966).
17. LINZ, P. The numerical solution of Volterra integral equations by finite difference methods. *MRC Tech. Rept.* 825, Madison (1967).
18. LUBICH, C. *These proceedings.*
19. MILLER,R.K. *Nonlinear Volterra integral equations.* Benjamin, Menlo Park (1971).
20. NEVANLINNA,O. Positive quadratures for Volterra equations. *Computing* 16 349-357 (1976).
21. NEVANLINNA,O. On the numerical solution of some Volterra equations on infinite intervals. *Tech. Rept.*2, Inst. Mitt.-Leffler (1976).
22. NEVANLINNA, O. *These proceedings.*
23. NOBLE, B. Instability when solving Volterra integral equations of the second kind by multistep methods pp. 23-29 of J.Ll. Morris (Ed.) *Conf. num. sol. diff. eq.* Springer-Verlag Lect. Notes. Math. 109 (1969).

24. PALEY, R.A.C. and WIENER, N. *Fourier transforms in the complex plane*. Amer. Math. Soc., Providence, (1934).
25. TSALYUK, Z.B. Volterra integral equations. *J. Soviet Math.* 12, 715-758 (1979).
26. WOLKENFELT, P.H.M. *The numerical analysis of reducible quadrature methods for Volterra integral and integro-differential equations*. Acad. Proefschrift, Math. Centrum, Amsterdam (1981).
27. ZYGMUND, A. *Trigonometric series, Volume 1*. University Press, Cambridge (1976).

THE SOLUTION OF VOLTERRA EQUATIONS OF THE FIRST KIND IN THE PRESENCE OF LARGE UNCERTAINTIES

*Peter Linz

* *Department of Mathematics, University of California at Davis*

1. INTRODUCTION

The problem of numerically solving the Volterra equation of the first kind

$$\int_0^t k(t,s)f(s)ds = g(t) \quad , \quad 0 \leq t \leq T, \qquad (1.1)$$

has been studied extensively. A number of methods have been thoroughly analyzed with respect to their accuracy, convergence, and stability properties. Much of this work was done under the assumption that the kernel k(t,s) and the right side g(t) are known without error, and that the approximating equations can be solved exactly. When, as is often the case, equation (1.1) arises in the analysis of experimental data, these assumptions may not be realistic. As was pointed out in [2], appreciable perturbations in g(t) can make the standard numerical methods useless. It is therefore surprising that, with the exception of the work on the simple Abel equation (see, for example, Anderssen [1]), the problem of solving (1.1) in the presence of data errors has received so little attention.

That the numerical solution of (1.1) can cause difficulties has been recognized. Schmaedeke [4] suggests that (1.1) be treated essentially like a Fredholm integral equation of the first kind, and consequently uses standard Tikhonov-type regularization. This seems a little extreme. Methods for Volterra equations of the first kind are not nearly as ill-conditioned as methods for Fredholm equations of the first kind. A complete regularization (with its accompanying difficulties and unresolved questions) is too complicated an approach. A simpler way, much used in practice, is to use some *data smoothing* techniques to reduce the errors in the given information before solving the equation by a standard numerical method. Radziuk [3] advocates the application of *least-squares* methods directly to (1.1). On the basis of some numerical examples Radziuk suggests that the least-squares method gives better results than data smoothing. At the present, there is virtually no theory to allow us to quantify such claims. It is the aim of this paper to make a preliminary analysis for judging the merits of the least-squares method versus smoothing, and to suggest some topics for further study.

2. SENSITIVITY OF THE SOLUTION

Our first task is to establish a plausible criterion by which different methods can be compared. Order of convergence and numerical stability, which are used in the analysis of standard methods, do not work well here since they do not address the effect of perturbations in the data. These errors will affect the solution. The best one can hope for is that their influence will be small. To keep the influence of an error on the solution small, the solution must be relatively insensitive to changes in the data. Such intuitive considerations lead us to introduce the following definition.

DEFINITION 1. Assume that we have points $0 \leq t_0 < t_1 < \ldots < t_N \leq T$, and values $g_0, g_1, \ldots, g_N$, such that g_i is the value given for $g(t_i)$. Let $\{f_j, j=0,1,\ldots,n\}$ denote a discrete approximation to $f(t)$, computed by some algorithm, using values $g_0, g_1, \ldots, g_N$. Let $\{f_j^k\}$ denote another approximation, computed by the same algorithm, using right hand sides $g_0, g_1, \ldots, g_{k-1}, g_k + \varepsilon, g_{k+1}, \ldots, g_N$. The *sensitivity* $S(\varepsilon)$ of the algorithm is then defined as

$$S(\varepsilon) = \max_{0 \leq k \leq N} \max_{0 \leq j \leq n} |f_j - f_j^k| . \tag{2.1}$$

In words, the sensitivity measures the maximum effect on the approximate solution of a single perturbation ε in the right side. The sensitivity is closely related to, but not identical with the *condition number* of an algorithm.

This definition considers only errors in g(t) and not in k(t,s). It can be shown that, in general, errors in the kernel affect the solution less than errors in the right side. Therefore, for simplicity, we consider only perturbations in g(t).

For example, consider the *midpoint* method for equally spaced points $t_i - t_{i-1} = h$. Let

$f_j = F_{j+1/2}$ be an approximation to $f(t_j + h/2)$ defined by

$$h \sum_{j=0}^{r-1} k(t_r, t_j + h/2) F_{j+1/2} = g_r, \qquad (2.2)$$

$$r = 1, 2, \ldots, N.$$

Then, as shown in [2], the sensitivity of the algorithm is

$$S(\varepsilon) = O(\varepsilon/h). \qquad (2.3)$$

In the midpoint method, an error in g is magnified by a factor 1/h. For typical values of h, even an error of a few percent may make the results meaningless. There is strong evidence that the sensitivity of other methods, such as higher order block-by-block methods is also of order ε/h. Therefore, when data for g is subject to large uncertainties, the common methods for solving Volterra integral equations of the first kind will fail.

3. DATA SMOOTHING

One of the most popular methods of smoothing experimental data is the least-squares method. Given a set of experimental data $\{t_i, g_i, i=0,1,\ldots,N\}$

and m continuous, linearly independent basis functions $\varphi_j(t)$, $j=1,2,\ldots,m$, we represent the data by

the linear combination

$$\sum_{j=1}^{m} c_j \varphi_j(t) \tag{3.1}$$

The coefficients c_j are chosen so as to minimize

$$\sum_{i=0}^{N} \left\{ \sum_{j=1}^{m} c_j \varphi_j(t_i) - g_i \right\}^2, \tag{3.2}$$

and can be determined by solving the *normal equation*

$$A_N \underline{c} = \underline{b} . \tag{3.3}$$

Here $\underline{c}$ is a vector with components $c_1, c_2, \ldots, c_m$, A_N is an $m \times m$ matrix with elements

$$(a_N)_{ij} = \sum_{\ell=0}^{N} \varphi_i(t_\ell)\varphi_j(t_\ell), \tag{3.4}$$

and $\underline{b}$ has components

$$b_i = \sum_{\ell=0}^{N} \varphi_i(t_\ell) g_\ell . \tag{3.5}$$

By choosing an appropriate set of basis functions (e.g. cubic B-splines) we can produce a smooth approximation to the data. In addition to smoothing, the method also makes the result relatively insensitive to changes in the data values.

THEOREM 1. Assume that $t_i - t_{i-1} = h$, $t_0 = 0$, $t_N = T$. Let $\{c_j\}$ be the solution computed by equations (3.3) - (3.5). Let $\{\hat{c}_j\}$ be the solution to these equations using $\hat{g}_\ell$ instead of g_ℓ, where

$$\hat{g} = g_\ell, \quad \ell = 0, 1, \ldots, k-1, k+1, \ldots, N,$$

$$\hat{g}_k = g_k + \varepsilon .$$

Then, for sufficiently large N, and fixed m

$$(c_j - c_j) = O(\varepsilon/N). \tag{3.6}$$

Proof: Let A be the matrix with elements

$$a_{ij} = \int_0^T \varphi_i(s)\varphi_j(s)ds.$$

Then

$$(a_N)_{ij} = \frac{N}{T} a_{ij} + O(1).$$

Since the φ_j are assumed to be linearly independent, A is nonsingular, so that for large N, A_N is also nonsingular and

$$A_N^{-1} = \frac{1}{N} B + O(1/N^2)$$

where B is a bounded matrix. Then

$$\begin{aligned} \underline{c} - \hat{\underline{c}} &= A_N^{-1}(\underline{b} - \hat{\underline{b}}) \\ &= \frac{1}{N} B(\underline{b} - \hat{\underline{b}}) + O(1/N^2) \end{aligned}$$

where

$$(\underline{b}-\hat{\underline{b}})^T = (\varphi_1(t_k)\varepsilon, \varphi_2(t_k)\varepsilon, \ldots, \varphi_m(t_k)\varepsilon).$$

Thus

$$c_j - \hat{c}_j = O(\varepsilon/N).$$

Consider now the method in which the data is first smoothed and the g_r are recomputed by

$$\tilde{g}_r = \sum_{j=1}^{m} c_j \varphi_j(t_r) \quad , \quad r=0,1,\ldots,N. \tag{3.7}$$

These values are then used in the midpoint method (2.2). Then the effect of a perturbation in g on $\tilde{g}_r$ is $O(\varepsilon/N)$. The midpoint method magnifies the error by a factor $1/h = N/T$ and we see that the sensitivity of the algorithm is $O(\varepsilon)$. We conclude that this algorithm magnifies inherent errors by at most a small amount, and is considerably better than an immediate application of the midpoint method.

4. LEAST-SQUARES APPROXIMATION TO EQUATION (1.1)

A more direct way of using the least-squares method is to approximate the unknown f(t) in (1.1) by

$$F(t) = \sum_{j=1}^{m} c_j \varphi_j(t), \tag{4.1}$$

then choose the coefficients c_j such that

$$\sum_{i=0}^{N} \left\{ \int_0^{t_i} k(t_i, s) \sum_{j=1}^{m} c_j \varphi_j(s) ds - g_i \right\}^2 \tag{4.2}$$

is minimized. The normal equations for this case have the form (3.3) with

$$(a_N)_{ij} = \sum_{\ell=0}^{N} \chi_i(t_\ell) \chi_j(t_\ell), \tag{4.3}$$

$$b_i = \sum_{\ell=0}^{N} \chi_i(t_\ell) g_\ell, \tag{4.4}$$

where

$$\chi_j(t) = \int_0^t k(t,s) \varphi_j(s) ds. \tag{4.5}$$

THEOREM 2. Assume that $t_i - t_{i-1} = h$, $t_0 = 0$, $t_N = T$. Also assume that $\{\varphi_j\}$ is a set of continuous and linearly independent basis functions. Then the algoritnm described by equations (4.1) - (4.5) has sensitivity of order ε/N.

Proof: The set $\{\chi_j\}$ is linearly independent. This follows directly from tne facts tnat $\{\varphi_j\}$ is linearly independent and that the Volterra operator has no eigenvalues. Therefore, the χ_j satisfy the same conditions as the φ_j in Theorem 1, and, equation (3.6) holds for the coefficients c_j in (4.1). But F(t) is a sum of a fixed number of terms. Thus

$$F(t_i) - \hat{F}(t_i) = O(\varepsilon/N).$$

Since usually N >> 1, the least-squares method not only does not magnify errors, but smoothes the data and reduces the sensitivity of the method. This conclusion is in agreement with the empirical observations of Radziuk [3].

5. CONCLUSIONS

The qualitative arguments presented in this paper, together with empirical observations, indicate that the least-squares method described in Section 4, is an effective way of solving Volterra integral equations of the first kind in the presence of significant data errors. However, the analysis presented here must be considered only a first step. A number of important questions remain to be answered. One is the influence on the accuracy of the choice of basis in (4.1). As with all least-squares algorithms, a poor choice can lead to significant ill-conditioning in the solution of the normal equations. In our numerical computations we used local bases for spline approximations (e.g. hat functions for linear splines, B-splines for the cubic case). For small m, no unusual ill-conditioning was detected, but the situation for larger m remains unclear. Another important practical question is how to subdivide the interval [0,T]. So far, we have used only equally spaced points. However, to increase the approximating capability while keeping m small, unequally spaced points are sometimes necessary. This points to the need for an adaptive algorithm, with stepsize chosen by the algorithm, using known statistical information on the data. Both these questions are currently under investigation.

REFERENCES

1. ANDERSSEN, R.S., Stable Procedures for the Inversion of Abel's Equation. *J. Inst. Math. Appl.* *17*, 329-342 (1976).

2. LINZ, P., A Survey of Methods for the Solution of Volterra Integral Equations of the First Kind. pp. 183-194 of R.S. Anderssen, F. R. deHoog, and M.A. Lukas (Eds), *The Application and Numerical Solution of Integral Equations*, Sijthoff and Noordhoff, Alphen aan den Rijn (1980).

3. RADZIUK, J., The Numerical Solution from Measurement Data of Linear Integral Equations of the First Kind. *Int. J. Numer. Methods Eng.* *11*, 729-740 (1977).

4. SCHMAEDEKE, W.W., Approximate Solutions for Volterra Integral Equations of the First Kind. *J. Math. Anal. Appl.* *23*, 604-613 (1968).

COLLOCATION AS A PROJECTION METHOD AND SUPERCONVERGENCE FOR VOLTERRA INTEGRAL EQUATIONS OF THE FIRST KIND

P. P. B. Eggermont

Department of Mathematical Sciences
University of Delaware
Newark, Delaware 19711/USA

1. INTRODUCTION

We study the superconvergence properties of the numerical solution of Volterra integral equations of the first kind

$$\mathcal{V}f(x) = \int_0^x k(x,y)\, f(y)\, dy = g(x), \quad x \in (0,1) \tag{1.1}$$

using collocation and piecewise polynomials. Given an integer p, and *collocation parameters* $0 < \eta_0 < \eta_1 < \dots < \eta_p \le 1$, problem (1.1) is approximated by

$$\mathcal{V}f_n(x_{iq}) = g(x_{iq}), \quad 0 \le i \le n-1, \quad 0 \le q \le p, \tag{1.2}$$

with $x_{iq} = (i+\eta_q)/n$, and where we require f_n to be in

$$S(p,n) = \{f : f|_{\sigma_j} \in \pi_p,\ \sigma_j = ((j-1)/n, j/n], 1 \le j \le n\}. \tag{1.3}$$

Under suitable conditions on k and g, it can be shown, BRUNNER [2], that $\|f_n - f\|_\infty = O(n^{-p-1})$, $n \to \infty$. If there are identifiable points y_{jr} for which $\max|f_n(y_{jr}) - f(y_{jr})| = O(n^{-p-2})$, then we have superconvergence. BRUNNER [2,3] has shown that superconvergence can occur only if $\eta_p = 1$, and exhibited some choices of η_q's for which there is indeed superconvergence. It is shown in this paper that $\eta_p = 1$ is in fact sufficient, and that superconvergence occurs at the points $y_{jr} = (j+u_r)/n$, $0 \le j \le n-1$, $0 \le r \le p$, where the u_r's are the zeroes of

$$\frac{d}{dx}\{x \prod_{q=0}^{p} (x-\eta_q)\}. \tag{1.4}$$

We prove this result using the projection method approach, rather than BRUNNER'S *differencing* technique [2]. The utility of the projection method approach is apparent for other types of Volterra equations as well, e.g. those studied by ANDRADE *et al.* [1], see EGGERMONT [5].

The paper contains the following sections. 2. Preliminaries. 3. Collocation as a projection method and superconvergence. 4. Boundedness of the projectors. 5. Superconvergence. 6. Numerical experiments. 7. Concluding remarks.

2. PRELIMINARIES

Let L^∞ be the space of (all equivalence classes of) bounded, measurable functions on (0,1), with norm denoted by $\|\cdot\|$, and let Lip_0 be the space of Lipschitz continuous functions on (0,1) that vanish at 0. On Lip_0 we define the norm $|\cdot|_{Lip}$ by

$$|g|_{Lip} = \sup_{x \neq y} |g(x) - g(y)|/|x-y|. \tag{2.1}$$

Define the operator $\mathcal{W} : L^\infty \to Lip_0$ by

$$\mathcal{W}f(x) = \int_0^x f(y)\, dy. \tag{2.2}$$

It is easily shown that $\mathcal{W}$ is an isometric isomorphism. Also, under suitable assumptions on k, see below, $\mathcal{W}^{-1}\mathcal{V}$ is an isomorphism of L^∞, see e.g., KRASNOSEL'SKII *et al.* [6].

Notations, definitions. For $f \in S(p,n)$ or $f \in C(0,1)$, $r_n f$ denotes the $(p+1)n$ vector of components $f(y_{jr})$, $0 \le j \le n-1$, $0 \le r \le p$. Similarly, $\rho_n g$ has the components $g(x_{iq})$. The interpolation operator p_n is given by

$$p_n r_n f(y) = \sum_r f(y_{jr})\, \ell_r(ny-j),\ y \in \sigma_j, \tag{2.3}$$

where the ℓ_r's are the Lagrangean fundamental polynomials for the points $u_0, u_1, \ldots, u_p$. To describe the interpolation error we define

$$\Psi(y) = \{\prod_r (x-u_r)\}/(p+1)! \tag{2.4}$$

Assumptions. We assume that $g \in C^{p+3}$, $k \in C^{p+3}$ and $k(x,x) \neq 0$ on [0,1]. Then f, the solution of (1.1) is in C^{p+2}. For convenience we also assume that $k(x,x) = 1$ on [0,1].

3. COLLOCATION AS A PROJECTION METHOD AND SUPERCONVERGENCE

Representing f_n in (1.2) as $p_n r_n f_n$, (1.2) reduces to a block-triangular system of equations, see BRUNNER[2, equation (2.5b)],

$$V_n r_n f_n = \rho_n g. \tag{3.1}$$

For n large enough, the diagonal blocks of V_n are invertible, so then V_n has an inverse. Then let $P_n(V) : \mathrm{Lip}_0 \to \mathrm{Lip}_0$, and $Q_n(V) : L^\infty \to L^\infty$ be defined by

$$P_n(V) = V p_n V_n^{-1} \rho_n \tag{3.2}$$

$$Q_n(V) = V^{-1} P_n(V) V. \tag{3.3}$$

It is easily seen that $P_n(V)$ and $Q_n(V)$ are idempotent, i.e., they are *projectors*, and that (1.2) is equivalent to

$$V f_n = P_n(V) g. \tag{3.4}$$

Equation (3.4) is the standard formulation of a *projection method*, see, e.g., PHILLIPS [7]. Since $V p_n r_n f = P_n(V) V p_n r_n f$, we then get (note the subscripts)

$$f_n - p_n r_n f = Q_n(V)(f - p_n r_n f). \tag{3.5}$$

From the theory of Lagrangean interpolation we know that for $f \in C^{p+2}$,

$$f - p_n r_n f = n^{-p-1} \Psi_n(y) + O(n^{-p-2}) \tag{3.6}$$

where

$$\Psi_n(y) = f^{(p+1)}(j/n) \Psi(ny-j), \; y \in \sigma_j. \tag{3.7}$$

From (3.5) we then obtain that

$$n^{p+2} \| r_n(f_n - f) \| \le n \, \| Q_n(V) \Psi_n \| + O(\| Q_n(V) \|). \tag{3.8}$$

It is now seen that the boundedness of $Q_n(V)$ as $n \to \infty$ plays an important role. In the next section we show that $\| Q_n(V) \|$ is bounded in n, and that $n \| Q_n(V) \Psi_n \| = O(1)$, thus establishing the superconvergence. The special choice of the u_r's mentioned before only plays a role in the latter part; the projectors are independent of the u_r's.

4. BOUNDEDNESS OF THE PROJECTORS

The proof of the boundedness of the $Q_n(V)$ goes in two steps. Note that the norms $\| Q_n(V) \|$, $|P_n(V)|_{\mathrm{Lip}}$ are equivalent, uniform in n, since $|g|_{\mathrm{Lip}} = \| W^{-1} g \|$ and $W^{-1} V$ is an isomorphism. First we show that $P_n(W) : Lip_0 \to Lip_0$ is bounded in n, and then we show under the assumptions on k that this implies that $P_n(V)$ is bounded in n, for n large enough.

4.1 *Boundedness of* $P_n(W)$. It is essentially well-known that $P_n(W)$ performs continuous, piecewise polynomial interpolation. To be precise, for $g \in Lip_0$, $x \in \sigma_i$,

$$[P_n(W)g](x) = \sum_q g(x_{iq})L_q(nx-i) + g(x_{i-1,q})L_{-1}(nx-i), \tag{4.1}$$

where L_s, $-1 \le s \le p$ are the Lagrangean fundamental polynomials for the points $0, \eta_0, \eta_1, \ldots, \eta_p(=1)$. It follows that

$$|P_n(W)g|_{Lip} \le c|g|_{Lip} \tag{4.2}$$

for some c independent of n and g. This says that $|P_n(W)|_{Lip}$, and hence $\| Q_n(W) \|$, is bounded in n.

We include a proof of (4.2). On σ_i we have for all constant functions γ,

$$\begin{aligned} \| (P_n(W)g)' \|_{L^\infty(\sigma_i)} &= \|(P_n(W)(g - \gamma))' \|_{L^\infty(\sigma_i)} \\ &\le c\, n \| P_n(W)(g - \gamma) \|_{L^\infty(\sigma_i)} . \end{aligned}$$

By taking the infimum over γ we obtain that

$$\| (P_n(W)g)' \|_{L^\infty(\sigma_i)} \le c|g|_{Lip(\sigma_i)} .$$

Since both $P_n(W)g$ and g are continuous on $[0,1]$, (4.2) follows.

4.2 *Boundedness of* $|P_n(V)|_{Lip}$. It is notationally more convenient to prove that $\| Q_n(V) \|$ is bounded in n. Let $f \in L^\infty$. From (3.3) we obtain

$$\| Q_n(V)f \| \le \| p_n \| \, \| V_n^{-1} W_n \| \, \| W_n^{-1} \rho_n V f \| . \tag{4.3}$$

Clearly, $\| p_n \| < \infty$, uniformly in n. In Section 4.3 we show that $\| V_n^{-1} W_n \|$ is bounded in n for n large enough. To estimate the remaining factor in (4.3) we have

$$\| W_n^{-1} \rho_n V f \| \le \| W_n^{-1} \rho_n W \| \, \| W^{-1} V \| \, \| f \| . \tag{4.4}$$

Since $W^{-1}V$ is an isomorphism, and since $W_n^{-1}\rho_n W = r_n Q_n(W)$, it now follows from (4.3) and Section 4.1 that

$$\| Q_n(\mathcal{V})f \| \le c\| f \| \tag{4.5}$$

for some constant c, for all f and all n large enough.

4.3 *Boundedness of* $\| \mathcal{V}_n^{-1}\mathcal{W}_n \|$. Let $\psi \in S(p,n)$. If

$$\| \mathcal{W}_n^{-1}\mathcal{V}_n r_n\psi \| \ge c\| \psi \| \ge c\| r_n\psi \| \tag{4.6}$$

for some c independent of ψ we are done. The last inequality of (4.6) is obvious. To prove the first inequality, note that

$$\| \mathcal{W}_n^{-1}\mathcal{V}_n r_n\psi \| = \| r_n Q_n(\mathcal{W})\mathcal{W}^{-1}\mathcal{V}\psi \| . \tag{4.7}$$

Since $Q_n(\mathcal{W})f \in S(p,n)$ for all f, there exists a positive constant c independent of n and ψ such that

$$\| r_n Q_n(\mathcal{W})\mathcal{W}^{-1}\mathcal{V}\psi \| \ge c\| Q_n(\mathcal{W})\mathcal{W}^{-1}\mathcal{V}\psi \| . \tag{4.8}$$

Now, $\mathcal{W}^{-1}\mathcal{V} = \mathcal{I} + \mathcal{U}$, where $\mathcal{U}$ is given by

$$\mathcal{U}f(x) = \int_0^1 v(x,y)\, f(y)\, dy, \tag{4.9}$$

with

$$v(x,y) = k_x(x,y) \quad \text{for} \quad 0 \le y \le x \le 1 \tag{4.10}$$

and zero otherwise, so for all $\psi \in S(p,n)$,

$$Q_n(\mathcal{W})(\mathcal{I} + \mathcal{U})\psi = \psi + \mathcal{U}_n\psi, \tag{4.11}$$

with

$$\mathcal{U}_n\psi(x) = \int_0^1 [Q_n(\mathcal{W})v(\cdot,y)](x)\, \psi(y)\, dy. \tag{4.12}$$

From (3.3) and the block-structure of $\mathcal{W}_n$ it follows that $[Q_n(\mathcal{W})v(\cdot,y)](x)$ vanishes for $0 < \lceil nx\rceil/n < y < 1$. Here $\lceil z\rceil$ is the least integer not less than z. So $\mathcal{U}_n$ may be written as $\mathcal{U}_n = \mathcal{S}_n + \mathcal{T}_n$, where $\mathcal{S}_n$ is the "Volterra part" of $\mathcal{U}_n$ and $\mathcal{T}_n$ is the "remainder",

$$\mathcal{S}_n\psi(x) = \int_0^x [Q_n(\mathcal{W})v(\cdot,y)](x)\psi(y)\, dy. \tag{4.13}$$

From the boundedness of $Q_n(\mathcal{W})$, uniformly in n, it follows that $I + S_n$ has a bounded inverse, uniformly in n, and $\| T_n \| = O(n^{-1})$, $n \to \infty$. By a standard perturbation argument it follows for n large enough that $\| (I + U_n) f \| \geq c \| f \|$, for some positive c, for all n large enough, and for all $f \in L^\infty$. From (4.7) and (4.11) we then obtain the first inequality of (4.6).

5. SUPERCONVERGENCE

Now we estimate the term $n \| Q_n(V) \Psi_n \|$ in (3.8). We show that

$$\rho_n V \Psi_n = n^{-1} \rho_n A^{(n)} \Psi_n \tag{5.1}$$

where $A^{(n)}$ is a Volterra integral operator with kernel $a_n(x,y)$ which is uniformly bounded in n and has a uniformly bounded derivative with respect to x. Assuming (5.1) has been proven, we obtain that

$$\begin{aligned} n \| Q_n(V) \Psi_n \| &= \| Q_n(V) V^{-1} A^{(n)} \Psi_n \| \\ &\leq \| Q_n(V) \| \, \| V^{-1} W \| \, \| W^{-1} A^{(n)} \| \, \| \Psi_n \| \\ &\leq c \| \Psi_n \| \end{aligned} \tag{5.2}$$

for some constant c for all n large enough. In other words, $n \| Q_n(V) \Psi_n \|$ is bounded in n.

Now we prove (5.1). By (2.4) and the property (1.4) of the u_r's, we have that

$$\int_0^{\eta_q} \Psi(y) \, dy = 0, \quad q = 0, 1, \ldots, p. \tag{5.3}$$

Since, by a change of variable,

$$\begin{aligned} V\Psi_n(x_{iq}) &= n^{-1} f^{(p+1)}(i/n) \int_0^{\eta_q} k(x_{iq}, (y+i)/n) \Psi(y) \, dy \\ &+ n^{-1} \sum_{j=0}^{i-1} f^{(p+1)}(j/n) \int_0^1 k(x_{iq}, (y+j)/n) \Psi(y) \, dy \end{aligned} \tag{5.4}$$

we may, by (5.3), subtract $k(x_{iq}, j/n)$ from the kernels in each of the integrals in (5.4), so that

$$V\Psi_n(x_{iq}) = n^{-1} A^{(n)} \Psi_n(x_{iq}), \tag{5.5}$$

where $A^{(n)}$ is the Volterra integral operator with kernel

$a_n(x,y)$ defined by

$$a_n(x,y) = n[k(x,y) - k(x,(j-1)/n)], \quad y \in \sigma_j \,. \tag{5.6}$$

It follows that a_n is bounded by

$$\sup \left\{ \left| \frac{\partial}{\partial y} k(x,y) \right| \; : \; 0 \le y \le x \le 1 \right\}$$

and $\partial a_n/\partial x$ is bounded by the sup of $|\partial^2 k/\partial x \partial y|$. This proves (5.1).

6. NUMERICAL EXPERIMENTS

We present some numerical results for the integral equation

$$\int_0^x (1 + x - y)\, f(y)\, dy = -1 + x + e^{-x}, \quad x \in [0,T] \tag{6.1}$$

for $T = 1$ and $T = 10$. (For $T = 10$ we transform it by a change of variable back to an integral equation on $[0,1]$). The results were computed not by the collocation method but by the related quadrature method, DE HOOG and WEISS [4]. For both equations we used $p = 2$ and $\underline{\eta} = (1/3, 2/3, 1)$. The zeroes of $\frac{d}{dx}\{x \prod_q (x-\eta_q)\}$ are

$$0.1273220038 \qquad 0.5 \qquad 0.8726779962 \tag{6.2}$$

In Table 1 we show the results for $\underline{\underline{u}} = (0, 1/2, 1)$. It should be remarked that for the case $T = 10$ the errors near $x = 0$ are about a factor 1000 larger than those for $x > 1/2$. In Table 2 we show maximum errors for the u_r's given by (6.2). The orders of convergence 3 and 4 (superconvergence) are nicely supported by these data for $T = 1$, and more or less so for $T = 10$. However, in this case too it is clear that superconvergence occurs.

T	n = 10	n = 20	n = 40
1	$.77_{10}{-4}$	$.10_{10}{-4}$	$.13_{10}{-5}$
10	$.40_{10}{-1}$	$.72_{10}{-2}$	$.11_{10}{-2}$

Table 1

Convergence of $\|r_n(f_n - f)\|$

T	n = 10	n = 20	n = 40
1	$.15_{10}{-6}$	$.99_{10}{-8}$	$.63_{10}{-9}$
10	$.83_{10}{-3}$	$.73_{10}{-4}$	$.54_{10}{-5}$

Table 2

Superconvergence of $\|r_n(f_n - f)\|$

7. CONCLUDING REMARKS

We have analyzed a well-known collocation method for Volterra integral equations of the first kind, using the projection method approach. We have shown that superconvergence can also be brought within the scope of projection methods. Essentially the same methods apply to the iterated Volterra integral equation of the first kind

$$\int_0^x (x-y)\, h(x-y)\, f(y)\, dy = g(x), \quad x \in (0,1).$$

Superconvergence occurs, only if the collocation parameters satisfy $0 < \eta_0 < \eta_1 < \ldots < \eta_{p-1} = \eta_p = 1$, at the zeroes of the *second* derivative of

$$x^2 \prod_{q=0}^{p} (x - \eta_q)$$

(so some collocation points *coalesce*), cf. [5].

ACKNOWLEDGEMENTS

Tables 1 and 2 were prepared with the assistance of Mr. Paul Townsend.

REFERENCES

1. ANDRADE, C., N. B. FRANCO, and S. McKEE, Convergence of linear multistep methods for Volterra first kind equations with $k(t,t) \equiv 0$. *Computing* 27, 189-204 (1981).
2. BRUNNER, H., Discretization of Volterra integral equations of the first kind (II). *Numer. Math.* 30, 117-136 (1978).
3. BRUNNER, H., Superconvergence of collocation methods for Volterra integral equations of the first kind. *Computing* 21, 151-157 (1979).
4. DE HOOG, F., and R. WEISS, On the solution of Volterra integral equations of the first kind. *Numer. Math.* 21, 22-32 (1973).
5. EGGERMONT, P. P. B., Collocation for Volterra integral equations of the first kind with iterated kernel. Submitted for publication.
6. KRASNOSEL'SKII, M. A., et al., *Approximate solution of operator equations*. Wolters-Noordhoff, Groningen (1972).
7. PHILLIPS, J. L., The use of collocation as a projection method for solving linear operator equations. *SIAM J. Numer. Anal.* 9, 14-28 (1972).

ON THE STABILITY OF DISCRETE VOLTERRA EQUATIONS

Olavi Nevanlinna

Helsinki University of Technology

1. INTRODUCTION

We discuss some stability concepts for quadratures for Volterra equations which are useful in preserving qualitative properties of the equations.

In chapter 2 we present the concepts and demonstrate that if the quadrature can be traced back to a linear multistep method, then the concepts agree with those familiar ones for ordinary differential equations.

In chapter 3 we show how the concepts can be used in treating the error behavior of discrete nonlinear Volterra equations. In chapter 4 we include a simple "order barrier" for nonlinear integro-differential equations.

Definitions and notations, if not indicated, are essentially as in [1] and [3].

2. STABILITY CONCEPTS FOR QUADRATURES

We shall mainly discuss quadratures W which are of the form

$$\int_0^{nh} \phi(s)\,ds \simeq h \sum_{j=0}^{n} w_{nj}\phi(jh).$$

Definition 2.1. A quadrature W is called *stable* if

$$\sup_{j,n\geq 0} |w_{nj}| < \infty . \quad \square \tag{2.1}$$

W will also denote the following operator on sequences

$$W\xi_n = h \sum_{j=0}^{n} w_{nj}\xi_j.$$

Definition 2.2. A quadrature W is called *strongly stable*, if it is stable and there are an integer k and a constant c such that, after a possible redefinition of w_{nj} for $0 \leq j < k$, we have for all real sequences ξ

$$\sum_{n=0}^{N} \xi_n W\xi_n \geq -ch \sum_{n=0}^{N} \xi_n^2, \text{ for all } N. \quad \square \tag{2.2}$$

Definition 2.3. A quadrature W is called *A-positive*, if it is strongly stable with constant $c = 0$. □

In [5], [6] A-positive quadratures were called positive. The following stronger version of A-positivity is easily seen to be equivalent to the strict positivity considered in [6].

Definition 2.4. A quadrature W is called *strictly A-positive*, if it is strongly stable with a negative c. □

Example 2.1. The repeated Simpson's rule has weights $w_{nj} = \omega_{n-j}$, where $\omega = \{\frac{1}{3}, \frac{4}{3}, \frac{2}{3}, \frac{4}{3}, \ldots\}$ (we omit discussing w_{nj} for $j = 0,1,2,3$ as long as they are bounded). Hence it is stable. However, it is not strongly stable, since choosing $\xi_n = (-1)^n$ we have $\sum_0^N \xi_n W\xi_n \simeq -\text{const.}\ N^2$, while $\sum_0^N \xi_n^2 = N+1$. □

Observe, however, that every stable quadrature satisfies

$$\left|\sum_0^N \xi_n W\xi_n\right| \leq C(N+1)\, h \sum_0^N \xi_n^2. \tag{2.3}$$

Example 2.2. Let $w_{nn} = 0$, $w_{nj} = 1$ for $j < n$. Then W is strongly stable: Set $\omega = \{\frac{1}{2}, 1, 1, \ldots\}$, then

$$\sum_0^N \xi_n\ \omega * \xi_n \geq 0, \tag{2.4}$$

and therefore

$$\sum_0^N \xi_n W\xi_n \geq -\frac{h}{2} \sum_0^N \xi_n^2. \quad \square$$

Hence, there exist explicit ($w_{nn} = 0$) strongly stable quadratures. However (compare with Theorem 3.1 in [4]):

Theorem 2.1. For a consistent and explicit strongly stable quadrature we have $c \geq \frac{1}{2}$.

Proof. Let W_c denote the "quadrature" obtained from W by adding c to the diagonal weights w_{nn}. Hence W_c is A-positive if W is strongly stable. By Lemma 2.2 in [5] A-positive quadratures satisfy $|w_{nj}| \leq w_{jj} + w_{nn}$ which here implies $|w_{nj}| \leq 2c$. For $j < k$ we only know that the weights are bounded, $|w_{nj}| \leq C$, say. By consistency,

$$n = \sum_{j=0}^{n-1} w_{nj} = \sum_{j=0}^{k-1} w_{nj} + \sum_{j=k}^{n-1} w_{nj} \leq kC + (n-k)2c.$$

Letting n tend to infinity we conclude that $1 \leq 2c$. □

Example 2.3. Inequality (2.4) says that the repeated trapezoidal rule is A-positive. Since A-positivity implies $|w_{nj}| \leq w_{nn} + w_{jj}$, it is not strictly A-positive. □

Example 2.4. The implicit version of the quadrature of Example 2.2 is strictly A-positive. Clearly we can take $c = -\frac{1}{2}$. □

All of the quadratures above can be obtained from linear multistep methods (ρ,σ). Expanding $\frac{\sigma}{\rho}(\zeta) = \sum_0^\infty \omega_n \zeta^{-n}$ we obtain with $w_{nj} = \omega_{n-j}$ the (ρ,σ)-reducible quadratures, except that for $0 \leq j < k$ the weights w_{nj} have to be computed separately.

Definition 2.5. Let S denote the stability region of (ρ,σ), i.e. $\mu \in S$ if $\rho - \mu\sigma$ satisfies the root condition. The method is called *stable* if $0 \in S$, and *strongly stable* if there exists $R > 0$, such that $D_R = \{\mu \in \mathbb{C} \mid |\mu+R| \leq R\} \subset S$. □

Theorem 2.2. The (ρ,σ)-reducible quadrature is stable if and only if the linear multistep method (ρ,σ) is stable. □

Proof. See [3], especially Lemma 5.5. □

Theorem 2.3. The (ρ,σ)-reducible quadrature is strongly stable if and only if the linear multistep method (ρ,σ) is strongly stable. □

Proof. The condition $D_R \subset S$ means $\frac{\rho}{\sigma}(\zeta) \notin D_R$ for $|\zeta| > 1$, which in turn can be written as

$$\text{Re } \phi(\zeta) > 0 \quad \text{for } |\zeta| > 1, \tag{2.5}$$

where $\phi(\zeta) = \dfrac{2R\sigma(\zeta) + \rho(\zeta)}{\rho(\zeta)}$.

Now we use the following

Lemma 2.1. [6]. A convolution quadrature Ω with weights $w_{nj} = \omega_{n-j}$ is A-positive if and only if ω is bounded and

$$\mathrm{Re} \sum_{n=0}^{\infty} \omega_n r^n e^{-in\tau} \geqq 0 \quad \text{for} \quad \tau \in [-\pi,\pi],\ r \in (0,1). \quad \square \tag{2.6}$$

If the (ρ,σ)-reducible quadrature Ω is strongly stable, then Ω_c is A-positive, where Ω_c is the convolution "quadrature" given by the sequence $\omega_c = \{\omega_0+c,\omega_1,\omega_2,\ldots\}$. Hence

$$\mathrm{Re} \sum_{n=0}^{\infty} \omega_n r^n e^{-in\tau} \geqq -c, \tag{2.7}$$

which is essentially the same condition as (2.5) with $c = 1/2R$. □

Definition 2.6. A linear multistep method is called *A-stable* if its stability region contains the left half plane. □

The following result is Theorem 4.1 in [6]:

Theorem 2.4. The (ρ,σ)-reducible quadrature is A-positive if and only if the linear multistep method (ρ,σ) is A-stable. □

Proof. The result follows here as a corollary to the previous theorem. □

Theorem 2.5. The (ρ,σ)-reducible quadrature is strictly A-positive if and only if the linear multistep method (ρ,σ) is A-stable and additionally

$$\mathrm{Re} \frac{\sigma}{\rho}(\zeta) \geqq c > 0 \quad \text{for} \quad |\zeta| > 1. \quad \square \tag{2.8}$$

Proof is obvious using Lemma 2.1. □

Theorem 2.6. If (2.8) holds, then the method (ρ,σ) is at most of first order. □

Proof. Condition (2.8) corresponds to the fact that S contains the complement of the disc $\{\mu \in \mathbb{C} \mid |\mu - \frac{1}{2c}| < 1/2c\}$. This implies that the method (ρ,σ) is only of first order, which is seen by expanding $\frac{\rho}{\sigma}(e^{i\tau})$ around $\tau = 0$. □

We add here a remark concerning weakly singular kernels. Consider operators

$$(K\phi)(t) = \int_0^t a(t-s)b(t,s)\phi(s)ds,$$

where $a \in L^1_{loc} \cap C(0,\infty)$, and b is continuous on $0 \leqq s \leqq t < \infty$.

These operators are natural to discretize in the form $h \sum_{j=0}^{n} \omega_{n-j} b(nh,jh)\phi(jh)$, where ω is associated with the function a. Define two sequences α and β as follows:

$$\alpha_k = h^{-1} \int_0^h a((k+1)h - s)ds, \quad k = 0,1,...$$

$$\beta_0 = h^{-2} \int_0^h sa(h-s)ds$$

$$\beta_k = h^{-2} \int_0^h \{sa((k+1)h-s) + (h-s)a(kh-s)\}ds, \ k = 1,2,...$$

and associate with them the convolution quadratures P_i as follows:

$$P_1\xi_n = h \sum_{j=0}^{n} \alpha_{n-j}\xi_j$$

$$P_2\xi_n = h \sum_{j=0}^{n} \beta_{n-j}\xi_j .$$

Then the following holds:

Theorem 2.7 [6]. If $a(t)$ is nonnegative, nonincreasing and convex for $t > 0$, then P_1 is A-positive. If $a(t) \not\equiv 0$, then it is strictly A-positive. □

Theorem 2.8 [6]. If $a \in C^{\infty}(0,\infty)$ and $(-1)^j a^{(j)}(t) \geqq 0$ for $j = 0,1,...$ and $t > 0$, then P_2 is A-positive. It is strictly A-positive if additionally $a(t) \not\equiv a(0)$. □

3. STABILITY RESULTS FOR DISCRETIZED INTEGRAL EQUATIONS

In this chapter we derive one typical stability result for the discretization of

$$x(t) + \int_0^t a(t,s)g(x(s))ds = f(t), \tag{3.1}$$

which illustrates the concepts of the previous section.

We make the following assumptions:

(Ha): a is real valued, continuous and bounded on $0 \leqq s \leqq t < \infty$, $|a(t,s)| \leqq \alpha$, and for all T and ϕ we have

$$\int_0^T \phi(t) \int_0^t a(t,s)\phi(s)ds\, dt \geqq 0.$$

$$\left.\begin{array}{l} g \text{ maps } R^d \text{ into } R^d \text{ and satisfies for some } L>0 \\ \text{the "circle condition":} \\ (x-y,g(x)-g(y)) \geqq \frac{1}{L}\,|g(x)-g(y)|^2. \end{array}\right\} \quad \text{(Hg)}$$

For shortness we write A for the operator $A\phi(t) = \int_0^t a(t,s)\phi(s)ds$. By (Ha) A is a densely defined positive operator in L^2. Furthermore, we write WA for the operator in ℓ^2: $WA\xi_n = h \sum_{j=0}^{n} w_{nj}\, a(nh,jh)\xi_j$.

Consider now

$$x_n - y_n + WA(g(x_n)-g(y_n)) = p_n. \tag{3.2}$$

If we only know that our quadrature W is stable, the the best we can do is to use the fact that W maps bounded kernels into bounded kernels. Since g is Lipschitz-continuous by (Hg), we obtain a standard bound

$$\sup_{0\leqq n\leqq T/h} |x_n - y_n| \leqq \frac{\exp\{LC\alpha T/(1-hLC\alpha)\}}{1 - hLC\alpha} \sup_{0\leqq n\leqq T/h} |p_n|. \tag{3.3}$$

If the quadrature is strongly stable, then we can obtain, for small enough h, a bound which does not grow exponentially with $L\alpha T$. Here our main tool is provided by the following result.

Theorem 3.1. WA is a positive operator in ℓ^2 whenever A is a positive operator in L^2 if and only if W is an A-positive quadrature. □

This is Theorem 2.1 in [5]. In [6] we presented a generalization to operators with matrix valued kernels. Thus the restriction to scalar valued kernels in (3.1) is not essential.

If W is strongly stable, then W_c is A-positive, where W_c is obtained from W by adding c to the diagonal elements w_{nn}^c. Hence $WA = W_cA - hca(nh,nh)$, and we can rewrite (3.2) in the form

$$x_n - y_n + W_cA(g(x_n)-g(y_n)) - hca(nh,nh)\,(g(x_n)-g(y_n)) = p_n. \tag{3.4}$$

Multiplying (3.4) by $g(x_n) - g(y_n)$, summing and using Theorem 3.1 yields

$$\sum_{n=0}^{N} (x_n-y_n, g(x_n)-g(y_n)) - hc \sum_{n=0}^{N} a(nh,nh)\,|g(x_n)-g(y_n)|^2$$

$$\leq \sum_{n=0}^{N} (p_n, g(x_n) - g(y_n)). \tag{3.5}$$

The left hand side of (3.5) can be bounded from below by

$$(\frac{1}{L} - hc\alpha) \sum_{n=0}^{N} |g(x_n)-g(y_n)|^2.$$

Hence we have the following:

Theorem 3.2. If W is strongly stable and (Ha), (Hg) hold and the step h is small enough: $h < 1/c\alpha L$, then we have a stability bound in the form

$$(\sum_{n=0}^{N} |g(x_n)-g(y_n)|^2)^{1/2} \leq \frac{L}{1-hc\alpha L} (\sum_{n=0}^{N} |p_n|^2)^{1/2}. \quad \square \tag{3.6}$$

Setting $c = 0$ we see that the estimate holds for A-positive quadratures without step size restriction:

Theorem 3.3. If W is A-positive and (Ha), (Hg) hold, then (3.6) holds with $c = 0$. □

In order to formulate a result for strictly A-positive quadratures, we change slightly our hypotheses: let (H'a) be as (Ha) except that instead of assuming a to be bounded assume that $\inf_{t>0} a(t,t) = \delta > 0$; in (Hg) we let L tend to infinity, so that (H'g) assumes that g is monotone in R^d.

Theorem 3.4. If W is strictly A-positive and (H'a) and (H'g) hold, then the inequality (3.6) holds with constant $L/(1-hc\alpha L)$ replaced by $1/h|c|\delta$. □

Remark 3.1. Observe that by setting $g(x) = Lx$ and letting L tend to infinity we arrive at a Volterra equation of first kind. Combining Theorem 2.8 and Theorem 3.4 we conclude e.g. that the product trapezoidal is stable for the Abel equation.

Remark 3.2. In the definition of strong stability we allowed some freedom in choosing w_{nj} for $j = 0,1,\ldots,k-1$. If these values were redefined, then we shall obtain in the right hand side of (3.5) a term of the form

$$\sum_{n=0}^{N} (v_n, h \sum_{j=0}^{k-1} \Delta_{nj} a(nh,jh) v_j), \tag{3.7}$$

where Δ_{nj} denotes the difference in the weights, and $v_n = g(x_n)-g(y_n)$. This yields an additional term

$h(N+1)k\alpha D \max_{0 \leq j < k} |g(x_j)-g(y_j)|$ into the ℓ^2 sum of p_n in (3.6). Here $D = \sup |\Delta_{nj}|$.

4. A REMARK ON INTEGRO-DIFFERENTIAL EQUATIONS

In [5] and [6] we discussed some integro-differential equations, where the history term was linear. Nonlinearity in the history term poses a technical difficulty which we now demonstrate.

Consider the equation

$$x'(t) + Ag(x(t)) = f(t), \tag{4.1}$$

with operator A as in (Ha). Suppose that $g = \text{grad } G$. A standard trick is to multiply (4.1) by $g(x(t))$ and to integrate:

$$G(x(T)) - G(x(0)) \leq \int_0^T (g(x(s)),f(s))ds. \tag{4.2}$$

Assuming that g can be bounded using G one can derive bounds for $G(x(T))$.

The crucial point to be carried over in discretization, in addition of preserving the positivity of A, is the formula

$$\int_0^T (x'(t),g(x(t)))dt = G(x(T)) - G(x(0)). \tag{4.3}$$

If g is the identity then a formula replacing (4.3) is obtained using G-stability [2], but for nonlinear g the situation is different.

Suppose we use a linear multistep method (ρ,σ) and an A-positive quadrature W to bring (4.1) into the form

$$h^{-1} \rho x + \sigma WAg(x) = \sigma f \tag{4.4}$$

where e.g. ρx denotes the sequence $\{\sum_{i=0}^{k} \alpha_i x_{n+i}\}$. Suppose $g(0) = 0$, set $x = 0$ for negative indices and redefine σf for indices around 0 so that (4.4) holds for all n. Operating by σ^{-1} we obtain

$$h^{-1} \gamma * x + WAg(x) = f \tag{4.5}$$

where γ is given by $\frac{\rho}{\sigma}(\zeta) = \sum_0^\infty \gamma_j \zeta^{-j}$. Multiplying by $g(x)$ we are led to bound

$$\sum_{0}^{N} (g(x), \gamma * x) \tag{4.6}$$

from below.

Suppose that G is additionally convex. Then (4.6) is non-negative if γ satisfies: $\gamma_0 > 0$ and $\gamma_j \leqq 0$ for $j \geqq 1$, see e.g. Section 3 in [7].

Theorem 4.1. Suppose that the linear multistep method (ρ,σ) has the sign structure: $\gamma_0 > 0$, $\gamma_j \leqq 0$ for $j \geqq 1$, then $p \leqq 1$ and among the first order methods the implicit Euler method has the smallest error constant. □

Proof. Set $f(z) = \sum_{0}^{\infty} \gamma_j z^j + \log z$. If $p \geqq 2$, then $f^{(2)}(1) = 0$ but here $f^{(2)}(1) = \sum_{2}^{\infty} j(j-1)\gamma_j - 1 < 0$. Clearly $f^{(2)}(1)$ is maximized when $\gamma_j = 0$ for $j \geqq 2$, which is implicit Euler. □

Observe that for the implicit Euler method it is easy to simulate the steps of the proof for the continuous equation.

REFERENCES

1. H. BRUNNER: A survey of recent advances in the numerical treatment of Volterra integral and integro-differential equations, to appear in *J. Comput. Appl. Math.*

2. G. DAHLQUIST: G-stability is equivalent to A-stability, *BIT* 18 (1978), 384-401.

3. P. HENRICI: *Discrete variable methods in ordinary differential equations*, J. Wiley, New York, 1962.

4. R. JELTSCH and O. NEVANLINNA: Stability of explicit time discretizations for solving initial value problems, *Numer. Math.* 37, 61-91, 1981.

5. O. NEVANLINNA: Positive quadratures for Volterra equations, *Computing* 16 (1976), 349-357.

6. O. NEVANLINNA: On the numerical solution of some Volterra equations on infinite intervals, *Rev. Anal. Num. Théor. Appr.* 5 (1976), 31-57 (also: Report 2/1976, Institut Mittag-Leffler).

7. O. NEVANLINNA and F. ODEH: Multiplier techniques for linear multistep methods, *Numer. Funct. Anal. and Optimiz.*, 3 (1981), 377-423.

WATER-WAVE PROBLEMS AND THE NULL-FIELD METHOD

P.A. Martin

Department of Mathematics, University of Manchester.

1. INTRODUCTION.

A time-harmonic wave is incident upon a fixed, horizontal cylinder which is partially immersed in the surface of deep water, under gravity; the wave is modified by the presence of the cylinder and induces hydrodynamic forces upon it. Under well-known assumptions [2], we can formulate a two-dimensional, linear boundary-value problem (S, say) for the *total potential* ϕ: ϕ is harmonic in the fluid (D), $\partial\phi/\partial n = 0$ on the wetted surface of the cylinder (∂D), and $K\phi + \partial\phi/\partial y = 0$ on the mean free surface (F); if ϕ_I is the (given) potential corresponding to the incident wave, then $\phi - \phi_I \equiv \phi_D$ must satisfy a radiation condition (the presence of the cylinder can only generate outgoing waves at ∞); and the fluid motion must vanish as $y \to \infty$ (see Fig. 1). Also, $K = \omega^2/g$ where g is the acceleration due to gravity and a time-dependence of $\exp(-i\omega t)$ has been suppressed throughout.

Under certain geometrical conditions on ∂D, it is known that S has a unique solution for *all* real values of the parameter K [2].

2. BOUNDARY INTEGRAL EQUATIONS

The usual method for solving S is to derive an integral equation over ∂D. Let $G_0(P,Q)$ denote the potential at $P \in D$ due to a simple source at Q, i.e. G_0 is a fundamental solu-

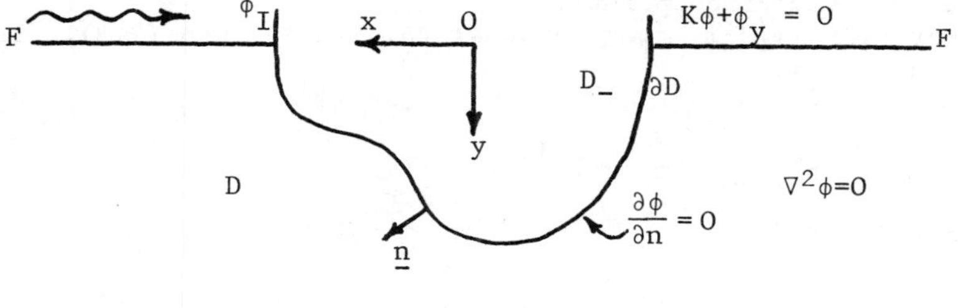

FIG. 1. The boundary-value problem, S, for ϕ.

tion [3,5]. Using Green's theorem, we derive three equations depending on the location of the field point P:

$$2\pi\phi_D(P) = -\int \phi(q)M(P;q)ds_q, \qquad P \in D, \tag{2.1}$$

$$\pi\phi(p) + \int \phi(q)M(p;q)ds_q = 2\pi\phi_I(p), \qquad p \in \partial D, \tag{2.2}$$

$$\int \phi(q)M(P\_;q)ds_q = 2\pi\phi_I(P\_), \qquad P\_ \in D\_, \tag{2.3}$$

where $D\_$ denotes the interior of the cylinder, all integrations are over ∂D and $M(P;q) = (\partial/\partial n_q)G_0(P,q)$. (2.2) is an integral equation for ϕ on ∂D; it is uniquely solvable except when K is an *irregular value* (IV), i.e. when K coincides with an eigenvalue of a certain interior problem [3,5]. If K is not an IV, we can substitute the unique solution of (2.2) into (2.1), and then $\phi_I + \phi_D$ solves S [4]. Note that IV's are unphysical, because we know that S is uniquely solvable for all K, i.e. IV's are a consequence of the method of solution. We shall now describe a method which does not suffer from this artificial difficulty; it is based on solving (2.3), which holds for all $P\_$ in $D\_$. An alternative method has been described by Ursell [5]; he replaces G_0 by a different fundamental solution, and obtains an integral equation which is uniquely solvable at any given value of K.

3. THE NULL-FIELD METHOD

Ursell [5] has obtained a bilinear expansion for G_0:

$$G_0(P,Q) = \sum_{m=0}^{\infty} \alpha_m(P)\,\Phi_m(Q) \tag{3.1}$$

for $|OP|<|OQ|$. Here, α_m and Φ_m are harmonic and satisfy the free-surface condition. In addition, α_m are regular whilst Φ_m are singular at O and satisfy the radiation condition. Φ_m are called *multipole potentials;* the set consists of a source at O, a horizontal dipole at O, and an infinite number of wave-free potentials.

For $|OP\_|<|Oq|$, we can substitute (3.1) into (2.3). If we expand Φ_I as

$$2\pi\phi_I(P\_) = \sum_{m=0}^{\infty} d_m\,\alpha_m(P\_), \tag{3.2}$$

we obtain

$$\int \phi\Psi_m ds = d_m \qquad (m = 0,1,\ldots) \tag{n-f}$$

where $\Psi_m(q) = (\partial/\partial n_q)\Phi_m(q)$ are *known* functions and d_m are *known constants*. (n-f) is our infinite system of *null-field equations*, from which the function $\phi(q)$, $q \in \partial D$ is to be determined. For many applications of the null-field (or T-matrix) method in other branches of mathematical physics, see [6]. We have [3,4]

Theorem 1 The null-field equations, (n-f) have a unique solution $\phi(q)$ for all real values of K. Moreover, the solution of S is given by (2.1).

We can use theorem 1 to prove

Theorem 2. The set $Y \equiv \{\Psi_m(q)\}$ (m=0,1,...) is a *basis* for $L_2(\partial D)$.

Proof. We must show that Y is (i) complete and (ii) linearly independent; note that the functions in Y are not orthogonal over ∂D.

(i) With the usual inner product for $L_2(\partial D)$, we have Y is complete iff, given $g \in L_2(\partial D)$, $(g,\Psi_m) = 0$ for all $m \geq 0 \Rightarrow g=0$. But (n-f) may be written as $(\phi^*,\Psi_m) = d_m^*$, m=0,1,..., and these are uniquely solvable; in particular, $\phi_I \equiv 0 \Rightarrow d_m = 0 \Rightarrow \phi=0$. (The asterisk denotes complex conjugate.)

(ii) Consider the set $X \equiv \{X_m(q)\}$ (m=0,1,...) where $(X_n,\Psi_m) = \delta_{nm}$; X is *biorthonormal* to Y. If X exists, then it is trivial to show that Y is linearly independent. Now, choose $\phi_I = \alpha_n/2\pi \Rightarrow d_m = \delta_{nm}$; since (n-f) are always solvable, we can assert that X_n exists; hence X exists.

Results corresponding to theorems 1 and 2 can also be proved in three dimensions, and for water of constant finite depth.

For deep water and the particular incident wave

$$\phi_I = \exp(-Ky-iKx)$$

(corresponding to a train of regular surface waves from $x=+\infty$), we can use theorem 2 to obtain a convergent numerical scheme for solving (n-f).

We use the 'method of projection', as devised by Gregory and Gladwell [1] for treating various boundary-value problems for the semi-infinite elastic strip. Further details and a numerical example may be found in [4].

REFERENCES

1. GREGORY, R.D. and GLADWELL, I., The cantilever beam under tension, bending or flexure at infinity. *J. Elast.* (1982) To appear.
2. JOHN, F., On the motion of floating bodies. II. *Comm. Pure Appl. Math.* 3, 45-101 (1950).
3. MARTIN, P.A., On the null-field equations for water-wave radiation problems. *J. Fluid Mech.* 113, 315-332 (1981).

4. MARTIN, P.A., On the null-field equations for water-wave scattering problems. *J. Fluid Mech.* Submitted for publication.
5. URSELL, F., Irregular frequencies and the motion of floating bodies. *J. Fluid Mech.* 105, 143-156 (1981).
6. VARADAN, V.K. and VARADAN, V.V. (Eds.), *Acoustic, electromagnetic and elastic wave scattering - Focus on the T-matrix approach,* Pergamon Press, New York (1980).

A REVIEW OF LINEAR MULTISTEP METHODS AND PRODUCT INTEGRATION METHODS AND THEIR CONVERGENCE ANALYSIS FOR FIRST KIND VOLTERRA INTEGRAL EQUATIONS

Sean McKee

Hertford College, Oxford.

1. INTRODUCTION

Lax's equivalence theorem [23] for parabolic partial differential equations relating stability and consistency to convergence is well known to all numerical analysts. Equally in ordinary differential equation theory we are aware of Dahlquist's pioneering work [9] on the convergence of linear multistep methods and his introduction of the concept of stability (sometimes called zero- or D-stability). It is however not commonly known that a similar theory exists relating a concept of stability and consistency to convergence of linear multistep methods for solving Volterra integral equations of the first kind.

The purpose of this article is to present a succinct historical survey of the development of linear multistep methods [LMM's] (or product integration methods [PIM's]) and their analysis of convergence for solving nonsingular (singular) Volterra integral equations [VIE's] of the first kind.

2. LMM'S AND PIM'S FOR FIRST KIND VOLTERRA INTEGRAL EQUATIONS

Consider the linear VIE of the first kind

$$\int_0^t \frac{k(t,s)}{(t-s)^\alpha} y(s)ds = f(t) \ , \ 0 \leq \alpha < 1 \qquad (2.1)$$

with $f(0) = 0$. Let $y(t)$ and $k(t,s)$ be continuous functions defined on $[0,T]$ and $\{0 \leq s \leq t \leq T\}$ respectively with sufficiently many continuous derivatives.

We shall be interested in two cases: (i) $\alpha = 0$ in which case (2.1) is a nonsingular VIE and (ii) $0<\alpha<1$ in which case (2.1) is the singular Abel's equation. The first case shall be denoted by (2.1'). To demonstrate that (2.1') has a unique solution it is usual to assume that $k(t,t) \neq 0$ (to prevent degeneracy) so that on differentiation (2.1') becomes a second kind VIE; the usual techniques for second kind equations then apply (see e.g. Yosida [37]). Interesting results on existence and smoothness of the solution of (2.1) can be found in Atkinson [5].

By a LMM we mean the replacement of the integral in (2.1') by a quadrature rule or sums of (possibly different) quadrature rules. Similarly for (2.1) by a PIM we usually mean the replacement of $k(t,s)y(s)$ by some polynomial approximation or by several different approximations over small intervals of $0 \leq t \leq T$.

More formally the resulting method can be written as

$$\Phi_h y = 0 \text{ where } \Phi_h : R^{N+1} \to R^{N+1}$$

where

$$\left[\Phi_h y\right]_i = \begin{cases} y_i - \tilde{y}_i, & i=0,1,\ldots,r-1 \\ h^{1-\alpha} \sum_{j=o}^{i} w_{ij} k(t_i,t_j) y_j - f(t_i), & i=r,r+1,\ldots,N \end{cases} \qquad (2.2)$$

where $\tilde{y}_i$, $i=0,1,\ldots,r-1$ are given starting values and w_{ij} are quadrature weights (or sums of quadrature weights). Note h is the mesh spacing and N is such that $Nh = T$. Further $t_i = t_{i-1}+h$ with $t_0 = 0$.

Example: Product Simpson and $\frac{3}{8}$th rule ($k(t,s) \equiv 1$ for clarity)

$$h^{1-\alpha} A_N y = g$$

where

$$A_N = \begin{pmatrix} 1 \\ 0 & 1 \\ \alpha_{02} & \alpha_{12} & \alpha_{22} \\ \beta_{03} & \beta_{13} & \beta_{23} & \beta_{33} \\ \alpha_{04} & \alpha_{14} & (\alpha_{24}+\alpha_{02}) & \alpha_{12} & \alpha_{22} \\ \alpha_{05} & \alpha_{15} & (\alpha_{25}+\beta_{03}) & \beta_{13} & \beta_{23} & \beta_{33} \\ \vdots & \vdots & (\alpha_{26}+\alpha_{04}) & \alpha_{14} & (\alpha_{24}+\alpha_{02}) & \alpha_{12} & \alpha_{22} \\ & & \vdots & & & & & \ddots \end{pmatrix}$$

and where

$y = (y_0, y_1, .., y_N)^T$, $g = (h^{1-\alpha}\tilde{y}_0, h^{1-\alpha}\tilde{y}_1, f_2, .., f_N)^T$,

$\tilde{y}_0$, $\tilde{y}_1$ are given starting values and $f_i = f(t_i)$.
The elements of the matrix A_N are as follows:

$$\alpha_{0k}=\tfrac{1}{2}\int_0^2\frac{(p-1)(p-2)}{(k-p)^\alpha}dp;\alpha_{1k}=-\int_0^2\frac{p(p-2)}{(k-p)^\alpha}dp;\alpha_{2k}=\tfrac{1}{2}\int_0^2\frac{p(p-1)}{(k-p)^\alpha}dp;$$

$$\beta_{03}=-\frac{1}{6}\int_0^3\frac{(p-1)(p-2)(p-3)}{(3-p)^\alpha}dp;\beta_{13}=\tfrac{1}{2}\int_0^3\frac{p(p-2)(p-3)}{(3-p)^\alpha}dp;$$

$$\beta_{23}=-\tfrac{1}{2}\int_0^3\frac{p(p-1)(p-3)}{(3-p)^\alpha}dp;\beta_{33}=\frac{1}{6}\int_0^3\frac{p(p-1)(p-2)}{(3-p)^\alpha}dp.$$

Note that when $\alpha = 0$ these reduce to the familiar weights $\{\alpha_{0k},\alpha_{1k},\alpha_{2k}\} \equiv \{1/3,\ 4/3,\ 1/3\}$ and $\{\beta_{03},\beta_{13},\beta_{23},\beta_{33}\} \equiv \{3/8,\ 9/8,\ 9/8,\ 3/8\}$. This method is of course well known not to converge (e.g. Linz [24]) but the familiarity of the two rules make them useful for illustrative purposes.

3. HISTORICAL REVIEW

This section will attempt to give a brief review of LMM's and PIM's for the nonsingular and the singular VIE's of the first kind.

3.1 *Nonsingular VIE's*

Historically the idea of solving VIE's by finite difference methods probably dates back to Fox and Goodwin [13] although they do not treat first kind equations explicitly. In 1961 Jones [20] studied the trapezoidal rule for solving first kind VIE's with convolution kernels and noted that the solutions generated oscillated about the exact solution. In 1967 Kobayasi [22] provided a careful convergence argument for the trapezoidal rule. At about the same time Linz [24] and [25] considered several finite difference methods proving that the mid-point rule was convergent and that high order Newton-Gregory formulae were not. It would appear that it was he who first introduced a stability polynomial but was only able to demonstrate non-convergence by considering the case $k(t,s) \equiv 1$ (which of course was sufficient). Unable to find high accuracy convergent methods he proffered the remark, "We have not found any convergent methods based on higher order quadrature methods. While it is possible that such methods do exist, it is clear that their convergence properties have to be examined carefully". Linz's work which was clearly influenced by the earlier substantial paper of his supervisor Noble [28] was important not perhaps so much for the work itself but for the catalytic effect it had on other researchers. A certain simple method was suggested by Anderson and Whyte [1] (see also Squire [29]) but it was Weiss [31] who in his Ph.D. thesis produced high accuracy block-by-block methods which he proved were convergent (see also de Hoog & Weiss [10] and [11]). Other authors then followed, notably Brunner's student Gladwin [14] (see also Gladwin and Jeltsch [16] and Gladwin [15]) who studied classes of quadrature rules and associated polynomial root conditions and further deduced the existence of high accuracy methods. However it was not until 1975 that Holyhead et al. [18] (see also Holyhead [17]) produced a general concept of stability with an associated root condition; in this paper a general theorem demonstrates that consistency plus stability imply convergence. These results although aimed at cyclic interpolatory-type methods are really quite general and essentially subsume the results of de Hoog and Weiss [10] as Holyhead [17] showed. A further paper introducing the concept of weak stability followed (Holyhead and McKee [19]) tackling the convergence problem using generating functions as the essential tools. Taylor [30] in an interesting paper derived stable methods by 'inverting' backward differentiation formulae. In 1977 Keech [21] derived a semi-explicit third order method (which may be viewed as a Runge-Kutta method) while Andrade and McKee [3] derived a fourth and a sixth order method, the latter having all its zeros of its associated polynomial at the origin (a local differentiation method (see [17])). McKee [27] unified the two papers [18] and [19] under the assumption that consistency could be expressed as an asymptotic expansion and in a further paper Andrade et al. [2] considered the problem of solving 1st kind VIE's directly when $k(t,t) \equiv 0$. Also in 1981 Wolkenfelt [34] in a substantial Ph.D. thesis (see also Wolkenfelt [35]) considered reducible quadrature methods.

3.2 Singular equations. The idea of using PIM's for

solving VIE's dates back to Young [36] although he, like Fox and Goodwin, did not explicitly advocate them for first kind equations. Linz [26] was probably the first man to suggest the use of such methods. In 1971 he presented some high order methods and outlined the convergence arguments. The product mid-point and the trapezoidal rule were first theoretically justified by Weiss and Anderssen [33] and Weiss [32]. However Weiss was only able to prove convergence for $\alpha\varepsilon[0.1292,1)$ and indeed it was not until 1981 that Eggermont [12] showed that the product trapezoidal was convergent for all $\alpha\varepsilon(0,1)$. In 1973 Benson [6] in his Ph.D. thesis proved the convergence of the product mid-point and trapezoidal rule for the equation with the kernel $k(t,s)(t-s)^{-\alpha}(t+s)^{-\beta}, 0<\alpha,\beta<1$. Atkinson [4] considered and proved the convergence of the product trapezoidal rule for the equation with the kernel $k(t,s)(t^2-s^2)^{-\frac{1}{2}}$. Holyhead [17] in his Ph.D. thesis showed that his general analysis for the non-singular equation could be extended and that his concepts of stability and weak stability were relevant. His general convergence proofs however depended upon a certain root condition being satisfied. In 1978 Branca [7] constructed order 2 and 3 methods for the nonlinear equation with $\alpha = \frac{1}{2}$. More recently Cameron [8] has considered two families of methods, the so-called implicit and explicit backward difference product integration methods (IBDPIM's and EBDPIM's). Using a simple sufficient condition he has been able to determine theoretically the precise range of α for which the IBDPIM's are convergent. Although the EBDPIM are clearly convergent for all α (from substantial numerical verification) a convergence analysis has proved illusory despite an elegant characterisation of stability through the ideas of fractional and discrete fractional integration.

4. A CONVERGENCE THEOREM

As the example illustrates all LMM's and PIM's can be characterised by an A_N matrix with $k(t,s) \equiv 1$. We shall denote such a matrix by A_N^1.

The matrix $D_N^{1-\alpha}$ is a lower triangular semi-circulant matrix whose columns are generated by the coefficients of $(1-z)^{1-\alpha}$.

Definition 1: A method will be said to be consistent of order p if there exists C, independent of h, such that

$$\left|\left[D_N^{1-\alpha}A_N\Delta_N y(t) - h^{-1+\alpha}D_N^{1-\alpha}\Delta_N f(t)\right]_i\right| \leq Ch^p,$$

$$i=r,r+1,\ldots,N.$$

(Δ_N is the usual restriction operator)

Definition 2: A method will be said to be zero-stable if there exists a constant M, independent of h, such that

$$||(D_N^{1-\alpha}A_N^1)^{-1}||_\infty \leq M.$$

For simplicity we assume that we are given r exact starting values.

Theorem (Holyhead [17] and Cameron [8]))
If the method is consistent of order p and it is zero-stable then it is convergent of order p.

From the previous section it is clear that there are many results of this type. We have chosen to include this one as it would appear to be the most general.

Finally definitions 1 and 2 while admitting a straightforward convergence argument do nevertheless disguise a wealth of technical difficulties. In practice zero-stability would be determined through an equivalent root-condition which we now illustrate for the example.

Let $g_j(z) = (1-z)^{1-\alpha}h_j(z)$ where

$$h_1(z) = \alpha_{22}+\beta_{23}z+(\alpha_{24}+\alpha_{02})z^2+(\alpha_{25}+\beta_{03})z^3+\ldots$$

and

$$h_2(z) = \beta_{33}+\alpha_{12}z+\beta_{13}z^2+\alpha_{14}z^3+\ldots$$

To demonstrate the method is zero-unstable we must show that at least one of the roots of

$$\begin{vmatrix} g_1^1(z) & g_1^2(z) \\ g_2^2(z) & g_2^1(z) \end{vmatrix} = 0$$

where $g_j^n(z) = \frac{1}{2}\sum_{\ell=1}^{2}(-1)^{2-(\ell-1)(n-1)}g_j((-1)^{\ell-1}z)$

lies inside the unit circle. For $\alpha = 0$ this is easily demonstrable.

The general root condition can be found in Holyhead and McKee [19].

ACKNOWLEDGEMENTS

The author would like to acknowledge financial support from the Science and Engineering Research Council.

REFERENCES

1. ANDERSON, A.S. and WHYTE, E.T., Improved numerical methods for Volterra integral equations of the first kind. *Comput. J.* 14, 442-443 (1971).
2. ANDRADE, C., FRANCO, N.B. and McKEE, S., Convergence of linear multistep methods for Volterra first kind equations with $k(t,t) \equiv 0$. *Computing* 27, 189-204 (1981).
3. ANDRADE, C. and McKEE, S., On optimal high accuracy linear multistep methods for first kind Volterra integral equations. *BIT* 19, 1-11 (1979).
4. ATKINSON, K.E., The numerical solution of an Abel integral equation by a product trapezoidal method. *SIAM J. Numer. Anal.* 11, 97-101 (1974).
5. ATKINSON, K.E., An existence theorem for Abel integral equations. *SIAM J. Math. Anal.* 5, 729-736 (1974).
6. BENSON, M.P. *Errors in Numerical Quadrature for Certain Singular Integrands, and the Numerical Solution of Abel Integral Equations.* Ph.D. thesis, University of Wisconsin, Madison (1973).
7. BRANCA. H.W., The nonlinear Volterra equation of Abel's kind and its numerical treatment. *Computing* 20, 307-324 (1978).
8. CAMERON, R.F. *Direct Solution of Applicable Volterra Integral Equations.* D.Phil. thesis, University of Oxford (1981).
9. DAHLQUIST, G., Convergence and stability in the numerical integration of ordinary differential equations. *Math. Scand.* 4, 33-53 (1956).
10. de HOOG, F. and WEISS, R., High order methods for Volterra integral equations of the first kind. *SIAM J. Numer. Anal.* 10, 647-664 (1973).
11. de HOOG, F. and WEISS, R., On the solution of Volterra integral equations of the first kind. *Numer. Math.* 21, 22-32 (1973).

12. EGGERMONT, P.P.B., A new analysis of the trapezoidal discretization method for the numerical solution of Abel-type integral equations. *J. of Integral Equs.* 3, 317-332 (1981).
13. FOX, L. and GOODWIN, E.T. *The numerical solution of nonsingular linear integral equations.* Phil. Trans. R. Soc. 245, 501-534 (1953).
14. GLADWIN, C.J. *Numerical Solution of Volterra Integral Equations of the First Kind.* Ph.D. thesis, Dalhousie University, Nova Scotia (1975).
15. GLADWIN, C.J., Quadrature rule methods for Volterra integral equations of the first kind. *Math. Comp.* 33, 705-716 (1979).
16. GLADWIN, C.J. and JELTSCH, R., Stability of quadrature rules for first kind Volterra integral equations. *BIT* 14, 144-151 (1974).
17. HOLYHEAD, P.A.W. *Direct Methods for the Numerical Solution of Volterra Integral Equations of the First Kind.* Ph.D. thesis, University of Southampton (1976).
18. HOLYHEAD, P.A.W., McKEE, S. and TAYLOR, P.J., Multistep methods for solving linear Volterra integral equations of the first kind. *SIAM J. Numer. Anal.* 12, 698-711 (1975).
19. HOLYHEAD, P.A.W. and McKEE, S., Stability and convergence of multistep methods for linear Volterra integral equations of the first kind. *SIAM J. Numer. Anal.* 13, 269-292 (1976).
20. JONES, J.G., On the numerical solution of convolution integral equations and systems of such equations. *Math. Comp.* 15, 131-142 (1961).
21. KEECH, M.S., A third order, semi-explicit method in the numerical solution of first kind Volterra integral equations. *BIT* 17, 312-320 (1977).
22. KOBAYASI, M., On numerical solution of the Volterra integral equations of the first kind by trapezoidal rule. *Rep. Statist. Applic. Res. Tokyo* 14, 1-14 (1967).
23. LAX, P.D. and RICHTMYER, R.D., Survey of the stability of linear finite difference equations. *Comm. Pure Appl. Math.* 9, 267-293 (1956).
24. LINZ, P. *Numerical Methods of Volterra Integral Equations with Applications to Certain Boundary Value Problems.* Ph.D. thesis, University of Wisconsin, Madison (1968).
25. LINZ, P., Numerical methods for Volterra integral equations of the first kind. *Comput. J.* 12, 393-397 (1969).

26. LINZ, P., Product integration methods for Volterra integral equations of the first kind. *BIT* 11, 413-421 (1971).
27. McKEE, S., Best convergence rates of linear multistep methods for Volterra first kind equations. *Computing* 21, 343-358 (1979).
28. NOBLE, B. The numerical solution of non-linear integral equations and related topics. pp. 215-318 of P.M. Anselone (Ed.), *Nonlinear Integral Equations*. University of Wisconsin Press, Madison (1964).
29. SQUIRE, W., Numerical solution of linear Volterra equations of the first kind. *Aerospace Engineering Rep. TR-15, West Virginia University* (1969).
30. TAYLOR, P.J., The solution of Volterra integral equations of the first kind using inverted differentiation formulae. *BIT* 16, 416-425 (1976).
31. WEISS, R. *Numerical Procedures for Volterra Integral Equations*. Ph.D. thesis, Australian National University, Canberra (1972).
32. WEISS, R., Product integration for the generalised Abel equation. *Math. Comp.* 26, 177-190 (1972).
33. WEISS, R. and ANDERSSEN, R.S., A product integration method for a class of first kind Volterra equations. *Numer. Math.* 18, 442-456 (1972).
34. WOLKENFELT, P.H.M. *The Numerical Analysis of Reducible Quadrature Methods for Volterra Integral and Integro-Differential Equations*. Ph.D. thesis, Mathematisch Centrum, Amsterdam (1981).
35. WOLKENFELT, P.H.M., Reducible quadrature methods for Volterra integral equations of the first kind. *BIT* 21, 232-241 (1981).
36. YOUNG, A., The application of approximate product-integration to the numerical solution of integral equations. *Proc. R. Soc. A* 224, 561-573 (1954).
37. YOSIDA, K. *Lectures on Differential and Integral Equations*. Wiley Interscience, New York (1960).

COMMENTS ON THE PERFORMANCE OF A FORTRAN SUBROUTINE FOR CERTAIN VOLTERA EQUATIONS

John M. Bownds*

University of Arizona

1. INTRODUCTION

In a previous summary [2], the author considered a basic technique for solving Volterra Integral Equations of the Second Kind which amounts to a conversion to a system of ordinary differential equations by using either the exact decomposability of the kernel or by approximating the kernel by one which decomposes. At that time, it appeared that the approach may not produce highly accurate results and therefore was restricted to use when a quick, low accuracy numerical solution was needed, since numerical results at that time seemed to show a rather consistent absolute error of about 3% on problems commonly occurring in the literature. The point of this article is to indicate that after a certain amount of experimentation and analysis, the author has seen that improved approaches to approximating kernels, more accurate and adaptive ordinary differential equations solvers, and simply better programming techniques have made it possible to produce an algorithm which, at this point, seems competitive as far as accuracy is concerned and in many cases is more economical in the number of function evaluations needed. This algorithm currently is in the form of a FORTRAN subroutine which has been designed to be "user-oriented"; this code is referred to here as VE1. Based on techniques detailed in [3], the programming development was aided in large part by collaboration with L. Appelbaum [1]. The current version of VE1 appears to be portable to the extent that it has been checked by portability software [6] and, otherwise, has run successcully on a variety of computers. Pending the elimination of any further "bugs" which may be uncovered, it is certainly the author's desire to offer this code by publication in the open literature; it is currently still under review.

2. FEATURES OF THE CODE VE1

The following will summarize what is felt to be the salient

*Work supported by NSF Grant No. MCS-7902038.

features of VE1.

The current subroutine applies to scalar integral equations of the form

$$u(x) = f(x) + \int_a^x k(x - t)g(u(t))dt, \tag{1}$$

or

$$u(x) = f(x) + \int_a^x \sum_{j=1}^m c_j(x)g_j(t,u(t))dt, \quad a \le x \le b, \tag{2}$$

where all functions concerned are at least continuous and a given solution is unique. (One notes that this, then, does not cover the surely important case where $k(x - t) = (x - t)^{-\alpha}$.)

In the case of (1), the code automatically constructs a Chebyshev Polynomial approximation to k. It has been the author's experience that since this is a one-time procedure. and since these polynomials enjoy such convenient orthogonality, recurrence, and rapid convergence properties, this does not account for very much computational effort at all; it is however counted in the overall function evaluations. It must also be noted here that continuous functions with singular derivatives, as the approximation theory suggests, are not necessarily approximated well this way; on the other hand, this approximation is very near the best possible polynomial approximation and, for smooth kernels, it is frequently extremely accurate.

When a Chebyshev approximation is made for the kernel in (1), the user of VE1 may opt to compute a simultaneous approximation to the error in the solution due to the truncation of the Chebyshev Series. This can be of use in general, and may eventually provide for an adaptive version of the code wherein the user need not set the degree of the approximation.

In the case of (2), if an original kernel $K(x,t,u)$ is decomposable then VE1 is usually as accurate as the o.d.e. solver being used since there are no kernel errors. Otherwise, the user may supply an approximation of the form in (2), such as interpolating polynomials, splines, truncated series, etc. The author has seen that the current version of VE1, using the basic Lagrange Polynomials in several dimensions produce much more accurate results than originally reported in [4]; the error analysis for these choices was given in this reference.

For the purposes of showing some of the features of VE1, several test equations are considered here; most of these results are reported in [3], and a very large collection is available directly from the author.

The test equations to be considered here are

$$\text{(I)} \qquad u(x) = 1 + x - \cos x - \int_0^x \cos(x - t)u(t)dt, \quad 0 \le x \le 2,$$

with exact solution $u = x$, and

$$\text{(II)} \quad u(x) = f(x) - 4\int_0^x \ln(1 + x - t)u(t)dt, \quad 0 \leq x \leq 4,$$

where

$$f(x) = 2 - 5x + (x + 1)(3x - 1)$$
$$+ 2(x + 1)(1 - x) \ln(x + 1),$$

with exact solution $u = 1 - x$.

Equation (I) is useful because the kernel is at the same time: of convolution type, exactly decomposable, and amenable to two-dimensional interpolation. The results in using VE1 with the various options are summarized in Table 1. In all cases, the input error tolerance for the o.d.e. solver is 10^{-8}; also if ε is the actual error then Accuracy = $\log \varepsilon^{-1}$ and if N is the number of times the right-hand side of the system of o.d.e.'s is evaluated, K is the number of times the kernel is evaluated, and n is the number of terms in the approximation (either m or $m + 1$), then Effort = $\log(nN + K)$.

TABLE 1

Comparison of Options for Equation (I) in VE1

Option	m	Endpoint Accuracy	Effort
Exact decomposition Chebyshev expansion	2	11.1	2.2
	5	6.2	2.8
	10	13.2	3.4
Two-dimensional Lagrange interpolation	10	10.4[(1)]	3.0
	20	8.6[(1)]	3.4

Note:
[(1)] Since the error is measured only at $x = 2$, error oscillation can produce smaller apparent error than the actual global error observed at grid points.

Table 2 is a tabulation of the computed approximate solution error due to series truncation for Equation (I)

Regarding Equation (II), it is shown in [3] that VE1, with the Chebyshev option, is capable of producing much higher accuracy (3 to 4 orders of magnitude) than that reported using other methods [5] at an expense of less than one order of magnitude in function evaluations.

TABLE 2

Solution Error Due to Chebyshev Series Truncation in VE1; Tabulation points are 0:0.1:2.0

m	Actual Global Error	Computed (Predicted) Error
5	1.1 (- 6)	4.0 (- 5)
10	2.7 (-13)	6.5 (-12)
20	2.7 (-15)	Machine Epsilon 2.2 (-16)

TABLE 3

Endpoint Accuracy and Function Evaluations for Equation (II)

m	Endpoint Accuracy	Effort
10	5.3	3.4
20	10.1	4.1

In summary, the code VE1 seems to be capable of efficiently solving Volterra equations of the second kind, subject to the restrictions mentioned above. On the other hand, in situations where the Chebyshev Series approximation is slowly converging, this code may not be cost effective.

Finally, it is noted that VE1 is moduled in such a way that users may change o.d.e. solvers to suit their needs; the o.d.e. subroutine currently used by the author is the Adams Method code found in [7].

REFERENCES

1. APPELBAUM, L. and BOWNDS, J., A FORTRAN Subroutine for Solving Volterra Integral Equations, *Univ. of Arizona Tech. Report* (1981).

2. BOWNDS, J., On an Initial-Value Method for Quickly Solving Volterra Integral Equations: A Review, *J. Opt. Thy. Appl.*, 24, 133-151.

3. BOWNDS, J., Theory and Performance of a Subroutine for Solving Volterra Integral Equations, *Computing*, to appear.

4. BOWNDS, J. and WOOD, B., On Numerically Solving Nonlinear Volterra Integral Equations with Fewer Computations, *SIAM J. Num. Anal.* 13, 705-719.

5. GAREY, L., Solving Nonlinear Second Kind Volterra Equations by Modified Increment Methods, *SIAM J. Numer. Anal.* 12, 501-508.

6. RYDER, B. and HALL, a., The PFORT Verifier, *Computing Science Technical Report #12*, Bell Laboratories, Murray Hill, N. J., Apr. 1979, 32 pages.

7. SHAMPINE, L. and GORDON, M., *Computer Solution of Ordinary Differential Equations; the Initial Value Problem*, W. H. Freeman, San Francisco (1975).

THE NUMERICAL COMPUTATION OF TURNING POINTS OF NONLINEAR EQUATIONS

*A. Spence and §†A. Jepson

*University of Bath, †Stanford University
§Currently at University of Toronto.

1. INTRODUCTION

This paper is concerned with nonlinear problems which depend on one or more parameters. Such problems arise in many physical situations. For example, in the theory of thermal combustion in exothermic reactions problems of the form

$$\begin{aligned} Lx &= \lambda \exp[x/(1+\mu x)] , \\ Bx &= 0, \end{aligned} \tag{1.1}$$

occur, where L is a uniformly elliptic differential operator, B is a boundary operator, x is the (dimensionless) temperature, and λ and μ are real parameters related to various physical quantities. Of particular interest are the values λ^o, μ^o which correspond to the loss of criticality in the reaction [1],[2], [7]. (See Figure 1, section 5.) For theoretical and/or computational reasons [3], one can rewrite problems like (1.1) in the form

$$x + Kh(x,\lambda,\mu) = 0 \tag{1.2}$$

where h contains the nonlinearity and K is a linear compact operator. Thus (1.2) is of Hammerstein type with K representing an integral operator with Green's function kernel.

Parameter-dependent nonlinear integral equations of various types also arise naturally. In the theory of radiative transfer there are the H equation and its generalisation, the X-Y equations [17]; in the theory of surface waves there is Nekrasov's equation [8]; and there is the Ball-Zachariasen equation in diffractive scattering [21].

We tend to consider a general equation of the form

$$g(x,\lambda) = 0 \tag{1.3}$$

where g is a nonlinear mapping. The reason is that in such problems the form of the nonlinearity is usually more important than the integral or differential character of the equation. Also such problems will usually be solved by iteration using the linearised form of the nonlinear equation and so the properties of $g_x(x,\lambda)$ will be of crucial importance. In (1.3) it is normal to regard x as a function of λ and to consider $x(\lambda)$ as a solution curve or arc in $X \times \mathbb{R}$. The aim of bifurcation theory is to characterise $x(\lambda)$ for λ in some interval containing a special value of λ, say $\lambda = \lambda^o$. If

$$g_x^o \equiv g_x(x(\lambda^o),\lambda^o)$$

is nonsingular then $(x(\lambda^o),\lambda^o)$ is a *regular point* and the implicit function theorem provides existence and uniqueness results in a neighbourhood of $(x(\lambda^o),\lambda^o)$. Of special interest are the points where g_x^o is singular. Such points usually correspond to points of physical significance, for example, the loss of structural stability or spontaneous combustion. This paper is concerned with the calculation of a type of singular point called a turning point, which we define in section 2.

It is well known in the theory of nonlinear equations that if y^o is an *isolated* solution of $G(y) = 0$ (i.e. $G_y(y^o)$ is nonsingular) then two important results follow. First Newton's method for the calculation of y^o converges quadratically provided a good enough starting value is known. Second, if $G(y) = 0$ is approximated by a discretised equation, say, $\underline{G}(\underline{y}) = \underline{0}$, $\underline{y} \varepsilon \mathbb{R}^n$, $n \geq N_o$, then, under suitable conditions, [11],[22], a solution, $\underline{y}^o$ say, of $\underline{G}(\underline{y}) = \underline{0}$ exists and it can be found using Newton's method, which again converges quadratically. An error estimate can also be given. Hence the "isolatedness" of a solution is a desirable property. Therefore it is not surprising that for the problem of the calculation of $(x(\lambda^o),\lambda^o)$, where g_x^o is singular, some recent attention has been paid to the derivation of systems of equations (so called "extended systems") which have isolated solutions at points where g_x^o is singular. The main aim of this paper is to derive and analyse one such "extended" system.

The plan of the paper is as follows. In section 2 we summarise some standard bifurcation theory with special attention being paid to turning points. In section 3 we introduce an extended system and prove the key result on the isolatedness of a special type of turning point. Newton's method and some numerical details are discussed in section 4. Section 5 contains some discussion of an important degenerate case.

2. PRELIMINARY THEORY

In this section we derive some standard results on turning points of one parameter problem of the form

$$g(x,\lambda) = 0 , \tag{2.1}$$

where g is a C^3 mapping from $X\times\mathbb{R}$ to X, with X a Banach space. The treatment is based on the approach of Crandall and Rabinowitz [5].

Recall that we are interested in points where g_x^o is singular and we assume that g_x^o has an algebraically simple zero eigenvalue i.e.

$$N(g_x^o) = \text{span}\{\phi^o\}, \quad \phi^o \varepsilon X\setminus\{0\} , \tag{2.2a}$$

$$\text{Range } (g_x^o) = \{y\varepsilon X: \psi^o y = 0\}, \quad \psi^o \varepsilon X'\setminus\{0\}, \tag{2.2b}$$

$$\psi^o\phi^o = 1 . \tag{2.2c}$$

Hence g_x^o is a *Fredholm operator of index zero* on X and there is a natural decomposition of X into

$$X = \text{span}\{\phi^o\} + V$$

where V is a complement of $\{\phi^o\}$ in X, which we may take as

$$V = \{v\varepsilon X: \psi^o v = 0\} . \tag{2.3}$$

Following Crandall and Rabinowitz we introduce another variable to represent the solution curve of (2.1), say $(x,\lambda) = (x(s),\lambda(s))$, and we seek conditions which ensure that certain types of behaviour occur.

Differentiation of

$$g(x(s),\lambda(s)) = 0 \tag{2.4}$$

with subsequent evaluation at $s = s^o$, where $(x(s^o),\lambda(s^o))=(x^o,\lambda^o)$, gives

$$g_x^o \dot{x}_o + g_\lambda^o \dot{\lambda}_o = 0 \tag{2.5}$$

$$g_x^o \ddot{x}_o + g_\lambda^o \ddot{\lambda}_o = -(g_{xx}^o \dot{x}_o \dot{x}_o + 2g_{x\lambda}^o \dot{x}_o \dot{\lambda}_o + g_{\lambda\lambda}^o \dot{\lambda}_o^2) \tag{2.6}$$

where

$$\dot{\lambda}_o = \frac{d\lambda}{ds}\Big|_{s=s^o} \quad \text{etc.}$$

Assume

$$\psi^o g_\lambda^o \neq 0. \tag{2.7}$$

If (2.1),(2.2) and (2.7) hold then (x^o,λ^o) is called a *simple turning point*. Clearly (2.5) and (2.7) imply

$$\dot{\lambda}_o = 0, \quad \dot{x}_o = \beta\phi^o \quad \beta\varepsilon\mathbb{R} \tag{2.8}$$

and $\dot{x}_o$ is normalised by taking $\beta = 1$ i.e.

$$\dot{x}_o = \phi^o. \tag{2.9}$$

The implicit function theorem can be used to prove the existence of a smooth curve $\{(x(s),\lambda(s)) : |s-s^o|<\delta\}$ where $x(\cdot),\lambda(\cdot)$ are C^3 mappings. (see figure 1, section 5). From (2.6) we derive

$$\ddot{\lambda}_o = -\psi^o g_{xx}^o \phi^o \phi^o / \psi^o g_\lambda^o \tag{2.10}$$

and this leads to the following definition. If (2.1),(2.2) and (2.7) hold and

$$\psi^o g_{xx}^o \phi^o \phi^o \neq 0 \tag{2.11}$$

then (x^o,λ^o) is a *simple quadratic turning point*. It is straightforward to show that if

$$\psi^o g_{xx}^o \phi^o \phi^o = 0 \tag{2.12}$$

then

$$\dddot{\lambda}_o = -(3\psi^o g_{xx}^o \phi^o v^o + \psi^o g_{xxx}^o \phi^o \phi^o \phi^o) / \psi^o g_\lambda^o \tag{2.13}$$

where v^o is the (unique) solution of

$$g_x^o v^o = -g_{xx}^o \phi^o \phi^o \qquad v^o \varepsilon V. \tag{2.14}$$

If (2.12) holds but

$$\dddot{\lambda}_o \neq 0 \qquad (2.15)$$

then (x^o,λ^o) is a *simple cubic turning point* (see figure 1, section 5).

We end by remarking that if $\psi^o g_\lambda^o = 0$ (i.e. (2.7) does not hold) then we may have a *bifurcation point*, where two distinct solution curves cross, but an extra non-degeneracy condition is needed for this. We do not consider this case here.

3. SIMPLE QUADRATIC TURNING POINTS

Assume that (x^o,λ^o) is a solution of (2.1) with (2.2), (2.7) and (2.11) holding i.e. (x^o,λ^o) is a simple quadratic turning point of (2.1). In this section we introduce an "extended" system which has an isolated solution at (x^o,λ^o).

Consider the extended system

$$G(y) \equiv \begin{bmatrix} g(x,\lambda) \\ \psi g_x(x,\lambda)\phi \end{bmatrix} = 0 \qquad y = (x,\lambda) \varepsilon X \times \mathbb{R} \qquad (3.1)$$

where ψ and ϕ satisfy

$$g_x(x,\lambda)\phi = \sigma\phi \quad , \qquad \phi = \phi(x,\lambda) \varepsilon X \qquad (3.2)$$

$$\psi g_x(x,\lambda) = \sigma\psi \quad , \qquad \psi = \psi(x,\lambda) \varepsilon X' \qquad (3.3)$$

with

$$\sigma = \sigma(x,\lambda) \varepsilon \mathbb{C}. \qquad (3.4)$$

Note that since g_x^o has an algebraically simple eigenvalue at (x^o,λ^o) there is a ball centred on (x^o,λ^o) in which $\sigma(x,\lambda)$ is an algebraically simple eigenvalue of $g_x(x,\lambda)$. We have the following theorem:

THEOREM 3.5 Assume (2.2) holds. Then $y^o = (x^o,\lambda^o)$ is a simple quadratic turning point of (2.1) if and only if y^o is an isolated solution of $G(y) = 0$.

PROOF We only prove the "only if" part here. We have

$$G_y(y) = \begin{bmatrix} g_x & g_\lambda \\ \psi_x g_x \phi + \psi g_x \phi_x & \psi_\lambda g_x \phi + \psi g_x \phi_\lambda \\ + \psi g_{xx} \phi & + \psi g_{x\lambda} \phi \end{bmatrix} \tag{3.6}$$

and we consider $G_y(y^o)\begin{bmatrix} u \\ \alpha \end{bmatrix} = 0$, $u \varepsilon X, \alpha \varepsilon \mathbb{R}$, i.e.

$$\begin{bmatrix} g_x^o & g_\lambda^o \\ \psi_o g_{xx}^o \phi_o & \psi_o g_{x\lambda}^o \phi_o \end{bmatrix} \begin{bmatrix} u \\ \alpha \end{bmatrix} = \begin{bmatrix} 0 \\ 0 \end{bmatrix}. \tag{3.7}$$

To simplify (3.6) we have used (3.2),(3.3) evaluated at (x^o, λ^o). Hence

$$g_x^o u + \alpha g_\lambda^o = 0, \tag{3.8}$$

$$\psi_o g_{xx}^o \phi_o u + \alpha \psi_o g_{x\lambda}^o \phi_o = 0, \tag{3.9}$$

and, using (2.7), (3.8) implies

$$\alpha = 0, \quad u = \beta \phi_o \qquad \beta \varepsilon \mathbb{R}.$$

Hence (3.9) and (2.11) give $\beta = 0$. Thus $N\{G_y(y^o)\} = \{0\}$.

It is straightforward to show that Range$\{G_y(y^o)\} = X \times \mathbb{R}$ and so the open mapping theorem gives that $G_y(y^o)$ is nonsingular.

We describe an efficient method of solving $G(y) = 0$ in the next section. However we conclude this section by mentioning that extended systems have been used by various authors [9], [10],[11],[14],[15],[19], both to prove theoretical results and to derive numerical methods for turning points and bifurcation points. The extended system used to compute simple quadratic turning points by previous authors [15],[19] has been

$$\left.\begin{aligned} & g(x,\lambda) = 0 \\ & g_x(x,\lambda)\phi = 0 \\ & \ell(\phi) - 1 = 0 \end{aligned}\right\} \quad H(z) = 0, \qquad z = (x, \phi, \lambda) \tag{3.10}$$

where ℓ is a suitable linear function in X'. A theorem similar to Theorem 3.5 is proved in [15] under the same assumptions. In the next section we compare the work done in solving the systems G(y) = 0 and H(z) = 0. (Note that in [15] the term *simple* turning point is used instead of *simple quadratic* turning point.)

4. NEWTON'S METHOD FOR SOLVING G(y) = 0.

In this section we consider the nonlinear problem

$$\underline{g}(\underline{x},\lambda) = 0 \qquad \underline{x}\in\mathbb{R}^n \ , \quad \lambda\in\mathbb{R} \tag{4.1}$$

since in most practical cases a continuous problem of the form (2.1) will be discretised using some appropriate numerical method (e.g. quadrature, finite differences, finite elements). We shall consider here only the problem of the calculation of a simple quadratic turning point $(\underline{x}^o,\lambda^o)$ of (4.1) given a good approximation $(\underline{x}_1,\lambda_1)$, obtained, say, by some continuation method. We shall not discuss any questions of the convergence of solutions of a discretised problem to solutions of a continuous problem. (However see [3],[16]).

Newton's method applied to

$$\underline{G}(\underline{y}) \equiv \begin{bmatrix} \underline{g}(\underline{x},\lambda) \\ \underline{\psi}^T \underline{g}_x(\underline{x},\lambda)\underline{\phi} \end{bmatrix} = \begin{bmatrix} 0 \\ 0 \end{bmatrix}, \qquad \underline{y} = (\underline{x},\lambda), \tag{4.2}$$

where

$$\underline{g}_x(\underline{x},\lambda)\underline{\phi} = \sigma\underline{\phi} \ , \ \underline{\psi}^T\underline{g}_x(\underline{x},\lambda) = \sigma\underline{\psi}^T \ , \tag{4.3}$$

gives rise to the sequence of linear problems

$$\underline{G}_y(\underline{y}_k)\underline{\Delta y}_k = -\underline{G}(\underline{y}_k) \qquad k=1,2,\ldots$$

where $\underline{G}_y(\underline{y}_k)$ has the form shown by (3.6). Now if the normalisation (2.2c) is used we have

$$\underline{\psi}_x\underline{\phi} + \underline{\psi}\underline{\phi}_x = 0 \tag{4.4}$$

and so

$$\underline{\psi}_x\underline{g}_x\underline{\phi} + \underline{\psi}^T\underline{g}_x\underline{\phi}_x = \sigma\left[\underline{\psi}_x\underline{\phi} + \underline{\psi}\underline{\phi}_x\right] = 0.$$

Similarly

$$\underline{\psi}_\lambda\underline{g}_x\underline{\phi} + \underline{\psi}^T\underline{g}_x\underline{\phi}_\lambda = \sigma\left[\underline{\psi}_\lambda\underline{\phi} + \underline{\psi}\underline{\phi}_\lambda\right] = 0.$$

Thus the Newton algorithm is as follows:

(o) Given $\underline{y}_k = (\underline{x}_k, \lambda_k)$, $\underline{G}(\underline{y}_k)$;

(i) Set up $\underline{g}_x(\underline{x}_k, \lambda_k) \equiv \underline{g}_x^k$, $\underline{g}_\lambda(\underline{x}_k, \lambda_k) \equiv \underline{g}_\lambda^k$;

(ii) Compute $LU = \underline{g}_x^k$ using a suitable factorisation algorithm and hence obtain $\underline{\phi}_k, \underline{\psi}_k^T$, where

$$\underline{g}_x^k \underline{\phi}_k = \sigma_k \underline{\phi}_k \; , \quad \underline{\psi}_k^T \underline{g}_x^k = \sigma_k \underline{\psi}_k^T \; ;$$

(iii) Set up $\underline{\psi}_k^T \underline{g}_{xx}^k \underline{\phi}_k \underline{\phi}_k$, $\underline{\psi}_k^T \underline{g}_{x\lambda} \underline{\phi}_k$

(iv) Set up the matrix

$$B_k \equiv \begin{bmatrix} \underline{g}_x^k & \underline{g}_\lambda^k \\ \underline{\psi}_k^T \underline{g}_{xx}^k \underline{\phi}_k & \underline{\psi}_k^T \underline{g}_{x\lambda}^k \underline{\phi}_k \end{bmatrix} \qquad (4.5)$$

(v) Solve

$$B_k \underline{\Delta y}_k = -\underline{G}(\underline{y}_k)$$

using a block-factorisation approach since the LU decomposition of $\underline{g}_x^k$ is known from (ii). (See the discussion below.)

(vi) Compute $\underline{y}_{k+1} = \underline{y}_k + \underline{\Delta y}_k$, evaluate $\underline{G}(\underline{y}_{k+1})$ and test for convergence.

For completeness we state

THEOREM 4.6 Assume that $\underline{g}$ is a C^2 mapping from $\mathbb{R}^n \times \mathbb{R}$ to $\mathbb{R}^n$ satisfying (2.2), (2.7) and (2.11) i.e. $(\underline{x}^o, \lambda^o)$ is a simple quadratic turning point of $\underline{g}(\underline{x}, \lambda) = \underline{0}$. Then, if $(\underline{x}_1, \lambda_1)$ is close enough to $(\underline{x}^o, \lambda^o)$ the Newton method (O)-(vi) produces a sequence $\{(\underline{x}_k, \lambda_k)\}$ k=1,2,... which converges quadratically to $(\underline{x}^o, \lambda^o)$.

(Obviously a rigorous condition on the closeness of $(\underline{x}_1, \lambda_1)$ to $(\underline{x}^o, \lambda^o)$ could be given as in the Newton-Kantorovich Theorem.)

It is worth expanding briefly the steps (ii) and (v). We obtain, after possibly reordering the equations using row pivoting,

$$\underline{g}_x^k = LU = \begin{bmatrix} L_{11} & \underline{0} \\ \underline{\ell}_{21}^T & 1 \end{bmatrix} \begin{bmatrix} U_{11} & \underline{u}_{12} \\ \underline{0}^T & u_{22} \end{bmatrix}$$

where u_{22} is "small" since g_x^k is "nearly" singular. The right and left eigenvectors are given approximately by, [13],

$$\underline{\phi}_k = \begin{bmatrix} -U_{11}^{-1}\,\underline{u}_{12} \\ 1 \end{bmatrix} \quad , \quad \underline{\psi}_k^T = \left[-\underline{\ell}_{21}^T L_{11}^{-1}, 1 \right] . \tag{4.7}$$

which can be normalised according to (2.2c). (But see remark 4.8 below.) The matrix B_k given by (4.5) can now be decomposed into the block LU form

$$B_k = \begin{matrix} n-1\left\{ \vphantom{L_{11}} \right. \\ \\ 2\left\{ \vphantom{\begin{matrix} \ell \\ I_2 \\ \ell \end{matrix}} \right. \end{matrix} \left[\begin{array}{c|cc} \overbrace{L_{11}}^{n-1} & \multicolumn{2}{c}{\overbrace{\underline{0} \quad \underline{0}}^{2}} \\ \hline \underline{\ell}_{21}^T & & \\ & \multicolumn{2}{c}{I_2} \\ \underline{\ell}_{31}^T & & \end{array} \right] \left[\begin{array}{c|cc} \overbrace{U_{11}}^{n-1} & \multicolumn{2}{c}{\overbrace{\underline{u}_{12} \quad \underline{u}_{13}}^{2}} \\ \hline \underline{0}^T & & \\ & \multicolumn{2}{c}{U_{22}} \\ \underline{0}^T & & \end{array} \right] \begin{matrix} \left. \vphantom{L_{11}} \right\}n-1 \\ \\ \left. \vphantom{\begin{matrix} \ell \\ I_2 \\ \ell \end{matrix}} \right\}2 \end{matrix}$$

where $\underline{\ell}_{31}, \underline{u}_{13}$ can be readily calculated and where the 2×2 matrix U_{22} can be shown to be nonsingular at $(\underline{x}^o, \lambda^o)$ provided (2.7),(2.11) hold. It is straightforward to show that the major part of the work involved in steps (ii)-(v) is

(a) 1 LU decomposition of $\underline{g}_x^k$,

(b) 6 forward or back substitutions,

(c) 1 (matrix)×(vector) multiplication.

For the extended system $\underline{H}(\underline{z}) = 0$ given by (3.10) it is shown in [9] that one Newton step involves

(a) 1 LU decomposition of $\underline{g}_x^k$,

(b) 10 forward or back substitutions,

(c) 3 (matrix)×(vector) multiplications,

and so the method presented here is an improvement.

REMARK 4.8 One may decide not to normalise the vectors $\underline{\phi}_k, \underline{\psi}_k$ given by (4.7). In this case the matrix B_k given by (4.5) is an approximation to $\underline{G}(\underline{y}_k)$ and the method (o)-(vi) is a quasi-Newton method where

$$\|B_k - \underline{G}(\underline{y}_k)\| = o\,(\|\underline{x}_k - \underline{x}^o\| + |\lambda_k - \lambda^o|).$$

Hence the conditions of Theorem 3.4 in [6] hold and the sequence $\{(\underline{x}_k, \lambda_k)\}$ k=1,2,... converges quadratically to $(\underline{x}^o, \lambda^o)$. This approach should be marginally more efficient than the full Newton's method.

5. SIMPLE CUBIC TURNING POINTS

We turn our attention in this section to nonlinear problems depending on two parameters which we write in the general form

$$f(x,\lambda,\mu) = 0 \tag{5.1}$$

where f is a C^3 mapping from $X\times\mathbb{R}^2$ to X, with X a Banach space. The material in this section was motivated by the problem in thermal combustion mentioned briefly in section 1. For slab geometry and infinite Biot number the steady state equation for the dimensionless temperature in an exothermic reaction can be reduced to,[1],[2],[7]

$$x'' + \lambda \exp[x/(1+\mu x)] = 0, \quad x(0) = x(1) = 0. \tag{5.2}$$

The solution (bifurcation) diagrams are shown in figure 1 and exhibit two simple quadratic turning points for $0<\mu<\mu^o$, and a simple cubic turning point when $\mu=\mu^o$. There are no singular (critical) points for $\mu>\mu^o$. The point (x^o,λ^o,μ^o) corresponds to a loss of criticality in the reaction and is the point which needs to be computed.

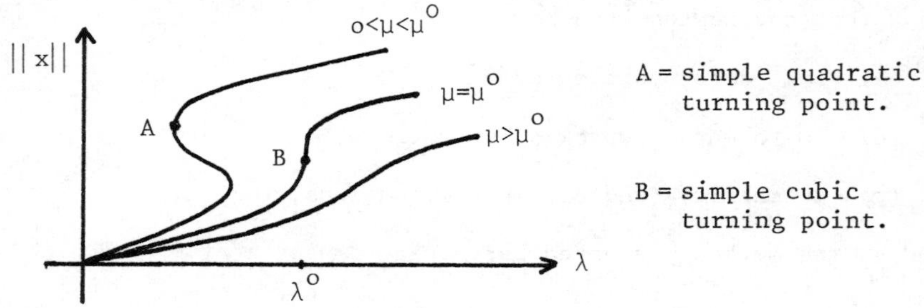

FIGURE 1. Bifurcation diagram for (5.2) as λ and μ vary.

A numerical method for the calculation of a simple cubic turning point (x^o,λ^o,μ^o) of (5.1) is presented in [20] (though the term *double point* was used instead of *simple cubic* turning point) and numerical results were given for equation (5.2). In this paper we present a different approach for such problems based on extending the ideas in sections 3 and 4. Again the fundamental idea is to derive an extended system which has (x^o,λ^o,μ^o) as an isolated solution.

In this section we use a notation similar to that in sections 2-4, e.g. $f_x^o\phi^o \equiv f_x(x^o,\lambda^o,\mu^o)\phi^o, \psi^o f_\lambda^o$ etc., and we assume that f satisfies (2.2),(2.7) and (2.12) at (x^o,λ^o,μ^o). To compute a cubic turning point of f we add the condition (2.12) to (3.1) to obtain

$$F(y) \equiv \begin{bmatrix} f(x,\lambda,\mu) \\ \psi f_x(x,\lambda,\mu)\phi \\ \psi f_{xx}(x,\lambda,\mu)\phi\phi \end{bmatrix} = 0, \qquad y = (x,\lambda,\mu)\in X\times\mathbb{R}^2 \tag{5.3}$$

where ψ and ϕ satisfy conditions similar to (3.2)-(3.4). We have the following theorem.

THEOREM 5.4 Assume that, at (x^o,λ^o,μ^o), f satisfies (2.2), (2.7),(2.12) and (2.15) (i.e. (x^o,λ^o,μ^o) is a simple cubic turning point of (5.1)), and that the generic condition (see remark 5.6 below)

$$\begin{vmatrix} \psi^o f_\lambda^o & \psi^o f_{\lambda x}^o\phi^o + \psi^o f_{xx}^o\phi^o v^1 \\ \psi^o f_\mu^o & \psi^o f_{\mu x}^o\phi^o + \psi^o f_{xx}^o\phi^o v^2 \end{vmatrix} \neq 0 \tag{5.5}$$

holds, where (recall (2.3))

$$f_x^o v^1 = -Qf_\lambda^o \qquad v^1\in V$$

$$f_x^o v^2 = -Qf_\mu^o \qquad v^2\in V$$

with Q the projection onto V defined by $Qy = y-(\psi_o y)\phi_o$. Then $y^o=(x^o,\lambda^o,\mu^o)$ is an isolated solution of $F(y)=0$.

PROOF The proof follows the lines of the proof of Theorem 3.5 though we omit the details. However we remark that the equation

$$g_x(x(s),\lambda(s))\phi(s) = \sigma(s)\phi(s)$$

may be used to prove that

$$\psi_x^o g_{xx}^o \phi^o \phi^o = \psi^o g_{xx}^o \phi^o \phi_x^o = \psi^o g_{xx}^o \phi^o v^o$$

where v^o is given by (2.14).

We note that the conditions (2.14) and (5.5) are precisely those required in the theory given in [20].

REMARK 5.6 The condition (5.5) is introduced in [4,§§10,11] and can be interpreted as the condition by which $f(x,\lambda,\mu)=0$ is a universal unfolding of $f(x,\lambda,\mu^o)=0$. This link with catastrophe theory and the connection between cubic turning points and the cusp catastrophe is explored in [2] and [20].

We end this section with some remarks about the implementation of Newton's method for $\underline{F}(\underline{y})=0$, the vector form of (5.3). Newton's method requires at each stage the solution of the linear system

$$\underline{F}_y(\underline{y}_k)\ \Delta\underline{y}_k = -\underline{F}(\underline{y}_k)$$

where, dropping the subscripts k for convenience,

$$F_y(y) \equiv \begin{bmatrix} \underline{f}_x & \underline{f}_\lambda & \underline{f}_\mu \\ \underline{\psi}_x\underline{f}_x\underline{\phi}+\underline{\psi}^T\underline{f}_x\underline{\phi}_x & \underline{\psi}_\lambda^T\underline{f}_x\underline{\phi}+\underline{\psi}^T\underline{f}_x\underline{\phi}_\lambda & \underline{\psi}_\mu^T\underline{f}_x\underline{\phi}+\underline{\psi}^T\underline{f}_x\underline{\phi}_\mu \\ +\underline{\psi}^T\underline{f}_{xx}\underline{\phi} & +\underline{\psi}^T\underline{f}_{x\lambda}\underline{\phi} & +\underline{\psi}^T\underline{f}_{x\mu}\underline{\phi} \\ E_1 & E_2 & E_3 \end{bmatrix} \qquad (5.6)$$

where

$$E_1 = \underline{\psi}_x\underline{f}_{xx}\underline{\phi}\underline{\phi} + 2\underline{\psi}^T\underline{f}_{xx}\underline{\phi}_x\underline{\phi} + \underline{\psi}^T\underline{f}_{xxx}\underline{\phi}\underline{\phi}$$

$$E_2 = \underline{\psi}_\lambda^T\underline{f}_{xx}\underline{\phi}\underline{\phi} + 2\underline{\psi}^T\underline{f}_{xx}\underline{\phi}_\lambda\underline{\phi} + \underline{\psi}^T\underline{f}_{xx\lambda}\underline{\phi}\underline{\phi}$$

with a similar expression for E_3. Now if one normalises $\underline{\psi}^T$ and $\underline{\phi}$ as in (2.2c) the first two terms in each of the components in the second row of $\underline{F}_y(\underline{y})$ vanish using (4.4). Also one can show that, in E_1,

$$\underline{\psi}_x \underline{f}_{xx} \underline{\phi}\underline{\phi} = \underline{\psi}^T \underline{f}_{xx} \underline{\phi}_x \underline{\phi}$$

with similar results for E_2 and E_3. It only remains to calculate the vectors $\underline{\phi}_x \underline{\phi}, \underline{\phi}_\lambda$ and $\underline{\phi}_\mu$. It can be shown that

$$(\underline{f}_x - \sigma \underline{I}) \underline{\phi}_x \underline{\phi} = -\underline{f}_{xx} \underline{\phi}\underline{\phi} + \left[\underline{\psi}^T \underline{f}_{xx} \underline{\phi}\underline{\phi} / \underline{\psi}^T \underline{\phi} \right] \underline{\phi} \tag{5.7}$$

which could be solved for $\underline{\phi}_x \underline{\phi}$. Now this would probably be regarded as too expensive, since a completely new LU decomposition of $(\underline{f}_x - \sigma \underline{I})$ would be needed. However an approximation to $\underline{\phi}_x \underline{\phi}$, $\underline{z}_1$ say, could be obtained more easily from

$$\underline{f}_x \underline{z}_1 = -\underline{f}_{xx} \underline{\phi}\underline{\phi}, \tag{5.8}$$

and similarly one could approximate $\underline{\phi}_\lambda$ and $\underline{\phi}_\mu$ by $\underline{z}_1, \underline{z}_2$ where

$$\underline{f}_x \underline{z}_2 = -\underline{f}_{x\lambda} \underline{\phi}$$

$$\underline{f}_x \underline{z}_3 = -\underline{f}_{x\mu} \underline{\phi}$$

respectively. Thus one could approximate $\underline{F}_y(\underline{y})$ by

$$B_k = \begin{bmatrix} \underline{f}_x & \underline{f}_\lambda & \underline{f}_\mu \\ \underline{\psi}^T \underline{f}_{xx} \underline{\phi} & \underline{\psi}^T \underline{f}_{x\lambda} \underline{\phi} & \underline{\psi}^T \underline{f}_{x\mu} \underline{\phi} \\ \underline{\psi}^T \underline{f}_{xxx} \underline{\phi}\underline{\phi} & \underline{\psi}^T \underline{f}_{xx\lambda} \underline{\phi}\underline{\phi} & \underline{\psi}^T \underline{f}_{xx\mu} \underline{\phi}\underline{\phi} \\ +3\underline{\psi}^T \underline{f}_{xx} \underline{z}_1 & +3\underline{\psi}^T \underline{f}_{xx} \underline{\phi} \underline{z}_2 & +3\underline{\psi}^T \underline{f}_{xx} \underline{\phi} \underline{z}_3 \end{bmatrix}$$

and obtain a quasi-Newton method

$$B_k \Delta \underline{y}_k = -\underline{F}(\underline{y}_k)$$

to compute $\underline{y}^o = (\underline{x}^o, \lambda^o, \mu^o)$.

ACKNOWLEDGEMENT

A. Jepson acknowledges support from the National Science Foundation, the Office of Naval Research, the Army Research, and the Air Force Office of Scientific Research.

REFERENCES

1. BAZLEY, N.W. and WAKE, G.C., The disappearance of criticality in the theory of thermal ignition. *ZAMP* 29, 971-976, (1978).

2. BODDINGTON, T., GRAY, P. and ROBINSON, C., Thermal explosions and the disappearance of critically at small activation energies: exact results for the slab. *Proc. Roy. Soc. Lond.* A. 368, 441-468, (1979).

3. BREZZI, F., RAPPAZ, J. and RAVIART, P.A., Finite dimensional approximations of nonlinear problems. Part I: Branches of Nonsingular solutions, *Numer Math* 36, 1-25, (1980); Part II: Limit points, *Numer Math* 37, 1-28, (1981); Part III: Simple Bifurcation points, *Numer Math* 38, 1-30, (1981).

4. CHOW, S., HALE, J. and MALLET-PARET, J., Applications of generic bifurcation I. *Arch. Rat. Mech. Anal.* 59, 159-188, (1975).

5. CRANDALL, M.G. and RABINOWITZ, P.H., Bifurcation, perturbation of simple eigenvalues and linearised stability. *Arch. Rat. Mech. Anal.* 52, 161-180, (1973).

6. DENNIS, J.E. and MORÉ, J.J., Quasi-Newton methods, motivation and theory. *SIAM Rev.* 19, 46-89, (1977).

7. FRADKIN, L.J. and WAKE, G.C., The critical explosion parameter of thermal ignition. *J. Inst. Math. Appl.* 20, 471-484, (1977).

8. HYERS, D.H., Some nonlinear integral equations of hydrodynamics, p.319-344, in P.M. Anselone (Ed.) *Nonlinear Integral Equations*.

9. JEPSON, A. and SPENCE, A., Folds in solutions of two parameter systems and their calculation: Part I. Stanford University Technical Report (to appear) (1981).

10. KEENER, J.P. and KELLER, H.B., Perturbed Bifurcation Theory. *Arch. Rat. Mech. Anal.* 50, 159-175, (1973).

11. KELLER, H.B., Approximation methods for nonlinear problems with applications to twopoint boundary value problems. *Math. of Comp.* 29, 464-474, (1975).

12. KELLER, H.B., Numerical solution of bifurcation and nonlinear eigenvalue problems, pp.359-384 of P.H. Rabinowitz (Ed.). *Applications of Bifurcation Theory*. Academic Press, New York (1977).

13. KELLER, H.B., Singular systems, inverse iteration and least squares (Private communication).

14. MOORE, G., The numerical treatment of nontrivial bifurcation points. *Numer. Funct. Anal. and Optimiz.* 2, 441-472 (1980).

15. MOORE, G. and SPENCE, A., The calculation of turning points of nonlinear equations. *SIAM J. Numer. Anal.* 17, 567-576 (1980).

16. MOORE, G. and SPENCE, A., The convergence of operator equations at turning points. *IMA J. of Numer. Anal.* 1, 23-38 (1981).

17. NOBLE, B., The numerical solution of nonlinear integral equations and related topics, pp.215-318 of P.M. Anselone (Ed.) *Nonlinear Integral Equations*. The University of Wisconsin Press, Madison (1964).

18. ORTEGA, J.M. and RHEINBOLDT, W.C., *Iterative solution of nonlinear equations in several variables*. Academic Press, New York, (1970).

19. SEYDEL, R., Numerical computation of branch points in ordinary differential equations. *Numer. Math.* 33, 339-352, (1979).

20. SPENCE, A. and WERNER, B., Nonsimple turning points and cusps. (Submitted to *IMA J. of Numer. Anal.*).

21. WARNOCK, R.L., A nonlinear integral equation from the Ball-Zachariasen Model of diffractive scattering: Numerical solution near a singularity of the Fréchet derivative. *J. Comp. Phys.* 33, 45-69, (1979).

22. WEISS, R., On the approximation of fixed points of nonlinear compact operators. *SIAM J. Numer. Anal.* 11, 550-553, (1974).

APPLICATIONS OF RESULTS OF VAINIKKO TO VOLTERRA INTEGRAL EQUATIONS

P. J. Taylor

University of Stirling

1. INTRODUCTION

Functional analysis of discretization methods has been applied to Fredholm integral equations (Anselone [1], Sloan [4], Vainikko [6][7])but little has been written regarding Volterra equations particularly of the first kind.

In analysing quadrature methods for linear Volterra equations, most authors use the invariance of stability under small perturbations to prove convergence. We can also prove convergence using a compact perturbation approach but we need some additional constraints on the quadrature formulae. For Volterra equations of the second kind, we find that a sufficient condition for compact convergence is that the asymptotic row repetition factor is one. Similar but stronger conditions are derived for Volterra equations of the first kind. A necessary condition for equicontinuity of approximations is also obtained. It would seem desirable to use compact convergence where possible in discretizations because of its smoothing effect. The results suggest that there may be a link between "numerical stability" and compact convergence. Wolkenfelt [8] has shown that for second kind Volterra equations, an asymptotic row repetition factor of one is a sufficient condition for numerical stability.

2. VAINIKKO'S ANALYSIS

Vainikko [6][7] extended the work of earlier authors including Anselone [1]. (A full list of references is given in [7].) A short description follows.

We have the well known diagram

$$\begin{array}{ccc} E & \xrightarrow{\quad A \quad} & F \\ p_n \Big\downarrow & & \Big\downarrow q_n \\ E_n & \xrightarrow{\quad A_n \quad} & F_n \end{array}$$

where $E, F, E_n, F_n, n \in \mathbb{N}$ are (real) Banach spaces. For brevity, we avoid full generality and assume that E_n and F_n are finite dimensional of equal dimensions and that A, A_n, p_n, q_n are bounded linear operators. We also require

$$\|p_n x\|_{E_n} \to \|x\|_E \quad \text{and} \quad \|q_n y\|_{F_n} \to \|y\|_F \quad \text{as} \quad n \to \infty$$

for all $x \in E$, $y \in F$.

Definition 1 A sequence $\{x_n \in E_n\}_{n=0}^{\infty}$ <u>p-converges</u> to $x \in E$ if

$$\|p_n x - x_n\|_{E_n} \to 0 \quad \text{as} \quad n \to \infty .$$

We write $x_n \xrightarrow{p} x$.

Definition 2 A sequence $\{A_n\}$ of operators <u>pq-converges</u> to A if

$$x_n \xrightarrow{p} x \quad \Rightarrow A_n x_n \xrightarrow{q} Ax$$

We write $A_n \xrightarrow{pq} A$.

Theorem 1 The following are equivalent

(i) $A_n \xrightarrow{pq} A$

(ii) $\|A_n\| \le$ constant $\quad \forall n \in \mathbb{N}$

and

$$\|A_n p_n x - q_n Ax\|_{F_n} \to 0 \qquad \forall x \in E'$$

where E' is a dense subspace of E.

Note that $A_n p_n x - q_n Ax$ is the consistency error.

Definition 3 We write

$$A_n \to A \ \underline{\text{stable}}$$

if $A_n \xrightarrow{pq} A$ and there exists $n_0 \in \mathbb{N}$ such that A_n^{-1} exists with $\|A_n^{-1}\|$ uniformly bounded for $n \ge n_0$.

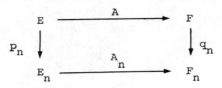

Theorem 2 "consistency + stability $\Longrightarrow$ convergence".
Given that range of A, $R(A) = F$

$$A_n \to A \text{ stable, and } y_n \xrightarrow{q} y$$

then

(i) $Ax = y$ has a unique solution x^*,

(ii) $\exists\, n_0 \in \mathbb{N}$ such that $A_n x_n = y_n$ has a unique solution x_n^* for each $n \geq n_0$,

(iii) $x_n^* \xrightarrow{p} x$.

Definition 4 A sequence $\{x_n \in E_n\}_{n\in\mathbb{N}}$ is said to be <u>p-compact</u> if each subsequence $\{x_n\}_{n\in\mathbb{N}'\subseteq\mathbb{N}}$ contains a p-convergent subsequence $\{x_n\}_{n\in\mathbb{N}''\subseteq\mathbb{N}'}$.

Definition 5 We write

$$A_n \to A \ \underline{\text{compact}}$$

if $A_n \xrightarrow{pq} A$ and

$\|x_n\|$ uniformly bounded $\Rightarrow$ $\{A_n x_n\}$ is q-compact.

Theorem 3 "Stability is invariant under compact perturbations".
Given that $S_n \to S$ stable, $R(S) = F$

$T_n \to T$ compact

$N(S + T) = \{0\}$

then

$R(S + T) = F$ and $(S_n + T_n) \to (S + T)$ stable.

We also use the following theorem which is in essence the result $\|A\| < 1 \Rightarrow \|(I + A)^{-1}\| \leq (1 - \|A\|)^{-1}$.

Theorem 4 "Stability is invariant under small perturbations".
Given that $S_n \to S$ stable, $T_n \xrightarrow{pq} T$ and there exist non-singular matrices $D_n : F_n \to E_n$ and $k \in [0, 1)$ such that

$$\|D_n T_n\| \leq k\, \|(D_n S_n)^{-1}\|^{-1} \quad \forall n \geq n_0 \in \mathbb{N}$$

then

$$(S_n + T_n) \to (S + T) \text{ stable.}$$

3. VOLTERRA EQUATIONS OF THE SECOND KIND

We consider

$$x(s) + \int_a^s K(s, t)x(t)dt = f(s) \quad s \in [0, a] \tag{3.1}$$

where $K \in C(D)$, $f \in C[0, a]$ and $D = \{(s,t) : 0 \le t \le s \le a\}$. We let

$$M = \max_D |K(s, t)| . \tag{3.2}$$

To consider small perturbations we take $E = F = C[0, a]$ with norms

$$\| x\|_E := \| x\|_F := \max_{0\le s\le a} |x(s)e^{-\alpha s}| \tag{3.3}$$

where $\alpha \ge 0$ will be chosen later. Write (3.1) in the form

$$(I + T)x = f$$

where I is the identity operator and T is defined by

$$(Tx)(s) = \int_0^s K(s, t)x(t)dt \quad . \tag{3.4}$$

We can easily show that if $\alpha > M$ then $\| T\| < 1$ and thus (3.1) has a unique solution (c.f. Bielecki's method for ordinary differential equations). We also note that T is a compact operator.

In the discretization, we introduce a mesh $s_i = ih$, $i = 0, \ldots, n$, $h = a/n$ for $n \in \mathbb{N}$ $(n > 0)$ and replace the integral in (3.1) by quadrature formulae

$$\int_0^{s_i} g(t)dt = h \sum_{j=0}^{i} w_{ij} g_j + e_{n,i}(g) \quad . \tag{3.5}$$

We will impose various conditions on the quadrature formulae.

Condition 1 There exists a dense subspace $X \subseteq C(D)$ such that

$$g \in X \qquad \max_{1\le i\le n} |e_{n,i}(g(s_i, \cdot))| \to 0 \quad \text{as} \quad n \to \infty \tag{3.6}$$

Condition 2

$$|w_{ij}| \le \text{constant} \quad j = 0, \ldots, i \quad i = 1, \ldots, n \tag{3.7}$$

where the constant is independent of n .

We take $E_n = F_n = \mathbb{R}^{n+1}$ with norms

$$\| \underline{x}\|_{E_n} = \| \underline{x}\|_{F_n} := \max_{0\le i\le n} |x_i e^{-\alpha s_i}| \tag{3.8}$$

where $\alpha \geq 0$ is to be chosen. We take $p_n = q_n$ with

$$p_n f := \{f(s_0), \ldots, f(s_n)\} \qquad f \in C[0, a] ,$$

$$S_n := I_n \quad (n + 1) \times (n + 1) \text{ unit matrix,}$$

$$T_n := h \begin{bmatrix} 0 & & \\ & & 0 \\ w_{10}K_{10} & w_{11}K_{11} & \\ \cdot & \ddots & \\ \cdot & & \ddots \\ \cdot & & \\ w_{n0}K_{n0} & \cdots & w_{nn}K_{nn} \end{bmatrix} \text{ lower triangular matrix,}$$

where $K_{ij} = K(s_i, s_j)$. Conditions 1 and 2 ensure that $T_n \xrightarrow{pq} T$. Also

$$\| T_n \| = \max_{\| \underline{x} \|_{E_n} = 1} \; \max_{1 \leq i \leq n} \left| h \left(\sum_{j=0}^{i} w_{ij} K_{ij} x_j \right) e^{-\alpha s_i} \right|$$

$$\leq \max_{1 \leq i \leq n} h \sum_{j=0}^{i} |w_{ij}| |K_{ij}| e^{-\alpha(s_i - s_j)} < \frac{MC}{\alpha} e^{\alpha h}$$

where C is the constant in (3.7). Thus $\| T_n \| < 1$ if α is sufficiently large and h sufficiently small. The conditions of Theorem 4 are satisfied with $D_n = I_n$.

To consider compact perturbations we take $\alpha = 0$ in (3.3) and (3.8) so that we have the usual Chebyshev norms. Suppose $\underline{x}_n \in E_n$, $\| \underline{x}_n \| \leq$ constant, $\underline{x}_n = \{x_{0n}, x_n, \ldots, x_{nn}\}$ and construct $z_n \in F$ as follows.

$$z_n(s_i) = h \sum_{j=0}^{i} w_{ij} K_{ij} x_{jn} \qquad i = 0, \ldots, n$$

$$z_n \text{ linear on } [s_i, s_{i+1}] \qquad i = 0, \ldots, n - 1 .$$

We have effectively introduced a prolongation operator from F_n to F . We show that the z_n are equicontinuous under certain conditions and from the Arzelà-Ascoli Theorem we deduce that $\{z_n : n \in \mathbb{N}\}$ is compact.

For $s_i > s_k$, consider

$$z_n(s_i) - z_n(s_k) = h \sum_{j=0}^{k} (w_{ij} K_{ij} - w_{kj} K_{kj}) x_{jn} + h \sum_{j=k+1}^{i} w_{ij} K_{ij} x_{jn}$$

$$= h \sum_{j=0}^{k} w_{ij} (K_{ij} - K_{kj}) x_{jn} + h \sum_{j=0}^{k} (w_{ij} - w_{kj}) K_{kj} x_{jn}$$

$$+ h \sum_{j=k+1}^{i} w_{ij} K_{ij} x_{jn} . \qquad (3.9)$$

To proceed we need a further condition on the quadrature formulae.

Condition 3a

$$\sum_{j=0}^{i} |w_{i+1,j} - w_{ij}| \leq \text{constant} \quad i = 1, \ldots, n \quad \text{and all} \quad n \in \mathbb{N} \,. \tag{3.10}$$

Under Conditions 1, 2, 3a, it follows from (3.9) that

$$|z_n(s_i) - z_n(s_k) \leq c_1 \, \omega(|s_i - s_k|) + c_2|s_i - s_k|$$

where ω is the modulus of continuity of $K(s, t)$ on D and c_1, c_2 are constants independent of i, k and n. Hence $\{z_n\}$ is compact and as $T_n\underline{x}_n = q_n z_n$ where the q_n are linear bounded operators, we deduce that $\{T_n\underline{x}_n\}$ is q-compact.

We make the following remarks regarding Condition 3a.

(a) It is a special case of the condition for collectively compact convergence derived by Sloan [4] for (Fredholm) equations of the second kind with weakly singular kernels.

(b) It is in essence equivalent to an asymptotic row repetition factor of one as defined by Wolkenfelt [8].

(c) We note that if

$$h \sum_{j=0}^{i} |w_{i+1,j} - w_{ij}| \geq \text{constant} > 0 \tag{3.11}$$

for all $i \geq n_0$, $n_0 \in \mathbb{N}$, then the functions z_n are not equicontinuous when $K(s,t) \equiv 1$. Only rather artificial quadrature rules can be found which satisfy neither (3.10) nor (3.11).

4. VOLTERRA EQUATIONS OF THE FIRST KIND - SMALL PERTURBATION ANALYSIS

We now consider

$$\int_0^s K(s, t)x(t)dt = f(s) \qquad s \in [0, a] \tag{4.1}$$

where

$$K, K_s \in C(D), \ D = \{(s,t) = 0 \leq t \leq s \leq a\}, \ K_s = \partial K(s,t)/\partial s, \tag{4.2}$$

$$K(s, s) = 1 \qquad \forall s \in [0, a] \,, \tag{4.3}$$

$$f \in C^1[0, a] \quad \text{with} \quad f(0) = 0 \,. \tag{4.4}$$

Under these conditions it is well known that (4.1) has a unique solution $x \in C[0, a]$ with $x(0) = f'(0)$. ((4.3) is equivalent to requiring $K(s, s) \neq 0$, $s \in [0, a]$). Let

$$K(s, t) = 1 + G(s, t)$$

where $G, G_s \in C(D)$ and $G(s, s) = 0 \quad \forall s \in [0, a]$. Take

$$E = C[0, a] \qquad F = \{y \in C^1[0, a] : y(0) = 0\}$$

and define $S:E \to F$ and $T:E \to F$ by

$$Sx := \int_0^s x(t)dt \qquad Tx := \int_0^s G(s, t)x(t)dt \,. \tag{4.5}$$

Define norms by

$$\|x\|_E := \max_{0\le s\le a} |x(s)e^{-\alpha s}| \tag{4.6}$$

$$\|y\|_F := \max\{\max_{0\le s\le a} |y(s)e^{-\alpha s}|,\ \max_{0\le s\le a} |y'(s)e^{-\alpha s}|\} \tag{4.7}$$

It is easily shown that if we choose $\alpha > \max\{M, M_s\}$ where

$$M = \max_D |G(s, t)| \quad \text{and} \quad M_s = \max_D |G_s(s, t)|$$

then $\|T\| < 1 \le \|S^{-1}\|^{-1}$ so that T is a "small" perturbation of S.

We now consider the discrete analogue. Take $E_n = F_n = \mathbb{R}^{n+1}$ and define $S_n, T_n : E_n \to F_n$, $p_n : E \to E_n$, $q_n : F \to F_n$ by

$$S_n\underline{x} := \begin{cases} x_i \\ h\sum_{j=0}^{i} w_{ij}x_j \end{cases} \qquad T_n\underline{x} := \begin{cases} 0 & i = 0, \ldots, m-1 \\ h\sum_{j=0}^{i} w_{ij}G_{ij}x_j & i = m, \ldots, n \end{cases}$$

$$p_n x := \{x(s_0), \ldots, x(s_n)\}$$

$$q_n f := \{L_0(f), \ldots, L_{m-1}(f), f(s_m), f(s_{m+1}), \ldots f(s_n)\}$$

where $s_j = jh$, $h = a/n$, $G_{ij} = G(s_i, s_j)$. Starting values are determined by $L_0, \ldots, L_{m-1}$, bounded linear functionals on $C^1[0, a]$ such that $L_i(f) \to f'(0)$ as $n \to \infty$, $i = 0, \ldots, m-1$, for all $f \in C^1[0, a]$. We choose norms $\|\underline{x}\|_{E_n}$ as in (3.8) and

$$\|\underline{y}\|_{F_n} = \max\{\max_{0\le i\le n} |y_i e^{-\alpha s_i}|,\ \max_{m\le i\le n} |\partial y_i e^{-\alpha s_i}|\} \tag{4.8}$$

where

$$\partial y_i = \begin{cases} \frac{1}{ih}\, y_i & i = m, \ldots, m+\rho-1 \\ \frac{1}{\rho h}\,(y_i - y_{i-\rho}) & i = m+\rho, \ldots, n \end{cases} \tag{4.9}$$

and $\rho \in \mathbb{N}$ is the asymptotic row repetition factor of the quadrature formulae.

Let D_ρ be an $(n+1) \times (n+1)$ "differentiation" matrix with elements

$$d_{ii} = \begin{cases} 1 & i = 0, \ldots, m-1 \\ \frac{1}{ih} & i = m, \ldots, m+\rho-1 \\ \frac{1}{\rho h} & i = m+\rho, \ldots, n \end{cases} \qquad d_{i,i-\rho} = -\frac{1}{\rho h} \quad i = m+\rho, \ldots, n$$

and all other $d_{ij} = 0$. As in Holyhead, McKee and Taylor [3], we find that for suitable quadrature formulae and ρ , $\|(D_\rho S_n)^{-1}\|_\infty$ are uniformly bounded as $n \to \infty$. However, we require a bound on $\|(D_\rho S_n)^{-1}\|_{(\alpha)}$, the operator norm corresponding to the norm (3.8) on E_n . If A_n is the $(n+1) \times (n+1)$ diagonal matrix with diagonal

$$a_{ii} = e^{-\alpha s_i} \qquad i = 0, \ldots, n$$

then

$$\|(D_\rho S_n)^{-1}\|_{(\alpha)} = \|A_n(D_\rho S_n)^{-1}A_n^{-1}\|_\infty \le \|(D_\rho S_n)^{-1}\|_\infty$$

as $D_\rho S_n$ is lower triangular.

We use two further conditions on the quadrature formulae (3.5).

Condition 3b Given $\eta \in (0, a)$, there exists $n_0 \in \mathbb{N}$ such that for $n \ge n_0$,

$$|w_{ij} - w_{i-\rho,j}| \le h \cdot \text{constant} \tag{4.10}$$

$$j = 0, \ldots, i - r_n , \quad i = \max\{r_n, m+\rho\}, \ldots, n$$

where r_n = integer part of η/n . (We ignore a strip of weights "near" the diagonal.)

Condition 4 There exists a dense subspace X of $Y = \{f = f(s, t) : f, f_s \in C(D)\}$ such that

$$g \in X \implies \max_{m \le i \le n} |\partial e_{n,i}(g(s_i, \cdot))| \to 0 \quad \text{as} \quad n \to \infty \tag{4.11}$$

where ∂ is as defined in (4.9).

Under Conditions 1, 2, 3b and 4, it is easily verified using Theorem 1 that $T_n \xrightarrow{pq} T$. After lengthy analysis it can also be shown that these conditions imply that given any $\varepsilon > 0$, there exist $\alpha \ge 0$ and $n_0 \in \mathbb{N}$ such that $\|D_\rho T_n\|_{(\alpha)} < \varepsilon$ for $n \ge n_0$. If $S_n \to S$ stable, with $\alpha = 0$, so that $\|(D_\rho S_n)^{-1}\|_\infty \le \gamma$ as $n \to \infty$ where γ is a constant, we deduce that for a suitable $\alpha \ge 0$

$$\|D_\rho T_n\|_{(\alpha)} < \frac{1}{\gamma} \le \|(D_\rho S_n)^{-1}\|_\infty^{-1} \le \|(D_\rho S_n)^{-1}\|_{(\alpha)}^{-1}$$

and Theorem 4 on small perturbations may be applied.

Of course the difficult task is to show that $S_n \to S$ stable, for a particular choice of quadrature formulae.

5. VOLTERRA EQUATIONS OF THE FIRST KIND - COMPACT PERTURBATION ANALYSIS

We take the same spaces E, F, E_n, F_n and operators S, T, S_n, T_n, p_n, q_n as in Section 4. We put $\alpha = 0$ so that we have the usual Chebyshev norms and take $\rho = 1$ in (4.9). Using $G(s, s) = 0$, it is easy to show that T is a compact operator and so we may expect to apply compact perturbation analysis. We impose a further condition on (3.5) but again ignore weights "near" the diagonal.

Condition 5 Given $\eta \in (0, a)$ and $\varepsilon > 0$, there exist $\delta > 0$ and $n_0 \in \mathbb{N}$ such that for $\delta + s_k > s_i > s_k \geq \eta$, $n \geq n_0$

$$\sum_{j=0}^{k-r_n} |\Delta w_{ij} - \Delta w_{kj}| < \varepsilon \tag{5.1}$$

where r_n = integer part of η/n and $\Delta w_{ij} = w_{ij} - w_{i-1,j}$.

Under Conditions 1, 2, 3a, 4 and 5 we can show that $T_n \to T$ compact. We need to show that $\partial T_n \underline{x}_n$ are "equicontinuous" on the grid points for bounded $\underline{x}_n$ and then apply the Arzelà-Ascoli Theorem. Write $\underline{x}_n = \{x_{0n}, \ldots, x_{nn}\}$ and $(\partial T_n \underline{x}_n)_i$ for the ith component of $\partial T_n \underline{x}_n$. For $i > k$,

$$(\partial T_n \underline{x}_n)_i - (\partial T_n \underline{x}_n)_k = \sum_{j=0}^{i} \Delta(w_{ij} G_{ij}) x_{jn} - \sum_{j=0}^{k} \Delta(w_{kj} G_{kj}) x_{jn}$$

$$= \sum_{j=0}^{k} w_{ij} (\Delta G_{ij} - \Delta G_{kj}) x_{jn} + \sum_{j=0}^{k} (w_{ij} - w_{kj})(\Delta G_{kj}) x_{jn}$$

$$+ \sum_{j=0}^{k} (\Delta w_{ij} - \Delta w_{kj}) G_{i-1,j} x_{jn} + \sum_{j=0}^{k} (\Delta w_{kj})(G_{i-1,j} - G_{k-1,j}) x_{jn}$$

$$= \sum_{j=k+1}^{i} w_{ij} (\Delta G_{ij}) x_{jn} + \sum_{j=k+1}^{i} (\Delta w_{ij}) G_{i-1,j} x_{jn}$$

Using the conditions on the quadrature formulae and the continuity of G and G_s (with $G(s, s) = 0$), it may be shown that each of the above summations has a bound which is independent of the choice of n and (bounded) $\underline{x}_n$, and which converges to zero as $s_k \to s_i$.

Note that Conditions 2, 3a and 5 will be satisfied by quadrature rules with an (exact) row repetition factor of one. Conditions 1 and 4 require the truncation error to be sufficiently smooth. All the conditions and $S_n \to S$ stable are satisfied by (ρ, σ)-reducible quadrature methods of Wolkenfelt [9] (when both ρ and σ are simple von Neumann polynomials). The methods include those based on "inverted differentiation" formulae (Taylor [5]) which therefore give compact convergence. The latter rules have been found numerically stable for a variety of Volterra equations of the first kind.

6. OTHER TYPES OF VOLTERRA EQUATIONS

This analysis can be extended to other types of equation, for example, Volterra integrodifferential equations, Abel equations and equations of the type considered by Cameron and McKee [2]. For Abel equations, the conditions on the approximation may involve both the weights and the kernel (c.f. Sloan [4]).

ACKNOWLEDGEMENTS

I wish to express my gratitude to Dr Wolf-Jürgen Beyn, Mathematisches Institut der Universität Konstanz for introducing me to Vainaikko's results. I am grateful to Dr Beyn for invaluable assistance and discussions. This work was started during sabbatical leave and I am grateful to the University of Calgary and the National Research Council of Canada for financial assistance.

REFERENCES

1. ANSELONE, P.M. *Collectively Compact Operator Approximation Theory*. Prentice-Hall, New Jersey (1971).

2. CAMERON, R.F. and McKEE, S., The Direct Numerical Solution of a Volterra Integral Equation Arising out of Visco-elastic Stress in Materials. *Comp. Meth. in Appl. Mech. and Eng.* 29, 219-232 (1981).

3. HOLYHEAD, P.A.W., McKEE, S. and TAYLOR, P.J., Multistep Methods for Solving Linear Volterra Integral Equations of the First Kind. *SIAM J. Numer. Anal.* 12, 698-711)1975).

4. SLOAN, I.H., Analysis of General Quadrature Methods for Integral Equations of the Second Kind. *Numer. Math.* 38, 263-278 (1981).

5. TAYLOR, P.J., The Solution of Volterra Integral Equations of the First Kind using Inverted Differentiation Formulae. *B.I.T.* 16, 416-425 (1976).

6. VAINIKKO, G. *Functionalanalysis der Diskretisierungsmethoden.* Teubner Verlagsgesellschaft, Leipzig (1976).

7. VAINIKKO, G., Approximate Methods for Nonlinear Equations (Two Approaches to the Convergence Problem). *Nonlinear Anal., Theory Methods, Applics.* 2, 647-687 (1978).

8. WOLKENFELT, P.H.M. *The Numerical Analysis of Reducible Quadrature Methoes for Volterra Integral and Integro-Differential Equations.* Doctoral Thesis. Mathematical Centre Amsterdam (1981).

9. WOLKENFELT, P.H.M., Reducible quadrature rules for Volterra integral equations of the first kind. *BIT* 21, 232-241 (1981).

SUPERCONVERGENCE AND THE GALERKIN METHOD FOR INTEGRAL EQUATIONS OF THE SECOND KIND

Ian H. Sloan

University of New South Wales

1. INTRODUCTION

This paper is concerned with the use of the Galerkin method and its iterative variants for the solution of integral equations of the second kind. In the case of differential equations it is well known that the Galerkin solution can exhibit 'superconvergence', in the sense of convergence that is faster than expected at first sight. For integral equations the study of superconvergence (see [11, 12, 13, 2, 3, 4, 5, 6, 7, 8, 15, 14]) is of more recent vintage, the first work on the subject apparently appearing only in 1976. Yet the superconvergence results for integral equations, especially for the iterative variants of the Galerkin method, are even more striking than those for differential equations.

I shall begin with a brief review of known results, and then outline some recent results (see [14]) obtained jointly with Vidar Thomée, of the Chalmers University of Technology in Gothenburg, Sweden.

Two aspects of superconvergence not covered in this paper are the eigenvalue problem (see [13, 5]) and the additional errors introduced into the Galerkin method and its iterative variants by the use of approximate quadratures (see [2, 4, 15]).

2. THE GALERKIN METHOD

We shall apply the Galerkin method and its variants to the equation

$$y(t) = f(t) + \int_{\Omega} k(t,s)y(s)d\sigma(s), \quad t \in \Omega, \tag{2.1}$$

where Ω is either a bounded open set in $\mathbb{R}^m$ or a suitable smooth m-dimensional surface in $\mathbb{R}^{m+1}$, and $d\sigma(s)$ is the element of surface area. Writing the equation as

$$y = f + Ky, \tag{2.2}$$

we shall assume that f belongs to a Banach space X, that the integral operator K is a compact operator on X, and that the equation $x = Kx$, $x \in X$, has only the trivial solution. Then the Fredholm theory applies, and the exact solution is $y = (I-K)^{-1}f$, with $(I-K)^{-1}$ a bounded operator on X and I the identity.

In the Galerkin method one defines a suitable linearly independent set of functions $u_1, \ldots, u_n$ belonging to X, and approximates y by

$$y_n = \sum_{i=1}^{n} a_i u_i.$$

The coefficients $a_1, \ldots, a_n$ are determined by

$$(y_n - Ky_n - f, u_j) = 0, \quad j = 1, \ldots, n,$$

where the inner product is

$$(v,u) = \int_\Omega v(s)\overline{u(s)}d\sigma(s). \tag{2.3}$$

Thus the coefficients satisfy a set of n equations in n unknowns, with matrix $((I-K)u_i, u_j)$.

How good is the Galerkin method? Usually it is very good indeed, in the sense that for sufficiently large n

$$\|y - y_n\|_X \le c \inf_{u \in U_n} \|y - u\|_X, \tag{2.4}$$

where $U_n = \text{span}(u_1, \ldots, u_n)$, and c is independent of n. (The symbol c denotes a generic constant, which may take different values at its different occurrences.)

For (2.4) to hold it is sufficient (see, for example, [9]) that P_n, defined to be the orthogonal projection operator onto U_n with respect to the inner product (2.3), is a bounded operator on X, which satisfies $\|f - P_n f\|_X \to 0$ and $\|K - P_n K\|_X \to 0$ as $n \to \infty$, and also $\|P_n\|_X \le M < \infty$, where M is independent of n. For the case $X = L_2$ the constant c in (2.4) can be replaced by $1 + \varepsilon$, where ε is any positive number.

3. AN EXAMPLE

As an introductory example let us consider

$$y(t) = t^2 + \int_0^1 \frac{1}{1+(t-s)^2}\, y(s)ds,$$

which is a very well behaved integral equation in one dimension, and take U_n to be the space of continuous piecewise-linear functions on $[0,1]$ with breakpoints at $0, h, 2h, \ldots, 1$. For this example we may take $X = L_\infty(0,1)$. Then (2.4) holds, and it follows from a standard result on approximation by piecewise-linear functions that the Galerkin error, in the uniform norm, is

$$\|y - y_n\|_{L_\infty} \le ch^2 \|D^2 y\|_{L_\infty}, \tag{3.1}$$

where D^2y denotes the second derivative of y.

Richter [10] has shown that one extra power of h may be gained by looking at $|y(t) - y_n(t)|$ at special points t; for the piecewise-linear case the special points are the zeros of the shifted second-degree Legendre polynomial on each sub-interval. I shall not pursue that subject here, because it turns out that convergence of even higher order may be obtained, and moreover at all points of the interval, by iterating the Galerkin solution to obtain a new approximate solution.

4. ITERATED GALERKIN SOLUTION

The first iterate of y_n is defined by

$$y_n^{(1)} = f + Ky_n = f + \sum_{i=1}^{n} a_i K u_i. \tag{4.1}$$

One attraction of $y_n^{(1)}$ is that it generally reflects much more faithfully the main analytic features of y than does y_n — for example, for the example of Section 3, $y_n^{(1)}$, like y, is infinitely differentiable, whereas y_n itself is merely piecewise-linear. A second attraction of $y_n^{(1)}$ is that it exhibits global superconvergence: under the conditions stated in Section 2 I showed in 1976 (see [11, 12]) that

$$\|y - y_n^{(1)}\|_{L_2} \le c\beta_n \|y - y_n\|_{L_2}, \tag{4.2}$$

where $\beta_n = \|K - KP_n\|_{L_2} \to 0$ as $n \to \infty$. The proof, indicated in Section 8, is quite easy. The corresponding result with L_2 replaced by the Banach space X holds if $\|K - KP_n\|_X \to 0$.

Quantitative superconvergence estimates for the case of one-dimensional integral equations and piecewise-polynomial spaces U_n have been given by Chandler [2, 3, 4] for the case of smooth kernels and Green's function kernels, and by Graham [7] for weakly-singular kernels, and have been extended to smooth kernels in higher dimensions by Hsiao and Wendland [8]. These results are very striking, especially for the case of smooth kernels. For example, for the piecewise-linear example of Section 3 it follows from Chandler's results that

$$\|y - y_n^{(1)}\|_{L_\infty} \le ch^4 \|D^2 y\|_{L_\infty} . \tag{4.3}$$

Similarly, for piecewise-polynomials of order r (i.e. of degree $\le r-1$) the Galerkin error is of order h^r, whereas the error in $y_n^{(1)}$ is of order h^{2r}.

Chandler's argument is based on the observation that

$$y(t) - y_n^{(1)}(t) = \int_\Omega k(t,s)[y(s)-y_n(s)]\,d\sigma(s) = (y-y_n,\bar{k}_t), \tag{4.4}$$

where $k_t(s) = k(t,s)$. But Chandler shows that every inner product of the form $(y - y_n, \phi)$ exhibits superconvergence if ϕ is sufficiently smooth, and in particular is of order h^4 for the piecewise-linear case. The proof is not repeated here, because a different approach, based on the smoothing properties of integral operators rather than the smoothness of the kernel, is used here and in [14]. That approach, which is in the spirit of (4.2), seems to extend more easily to complicated integral equations in higher dimensions. On the other hand, it does not yield Chandler's results [2, 3, 4] on superconvergence at special points for the case of Green's function kernels in one dimension.

The remainder of this paper gives a brief outline of the work of [14].

5. DERIVATIVES OF $y_n^{(1)}$

It turns out that the derivatives of $y_n^{(1)}$ have even more striking superconvergence properties. Let us begin with the example of Section 3. Because the Galerkin approximation y_n is piecewise-linear, its derivative Dy_n is piecewise-constant, and hence the best result we could hope to obtain for the error in Dy_n is

$$\|Dy - Dy_n\|_{L_\infty} \le ch\|D^2 y\|_{L_\infty} . \tag{5.1}$$

Yet for this example we can show that

$$\|Dy - Dy_n^{(1)}\|_{L_\infty} \le ch^4 \|D^2 y\|_{L_\infty}, \tag{5.2}$$

so that for the approximation $y_n^{(1)}$ no powers of h are lost in going to the first derivative. Similarly, for this example *all* derivatives of $y_n^{(1)}$ converge with the same h^4 order.

This result can be understood by the argument outlined in the previous section: from (4.4) we have

$$Dy(t) - Dy_n^{(1)}(t) = (y - y_n, \overline{\kappa_t}),$$

where

$$\kappa_t(s) = \kappa(t,s) = \frac{\partial}{\partial t} k(t,s) = \frac{\partial}{\partial t}\left[\frac{1}{1+(t-s)^2}\right],$$

and the result then follows from the observation that κ, like k, is a smooth kernel. In Section 8 the result will be proved by a different argument, which extends straightforwardly to more complicated integral equations in higher dimensions.

6. SOBOLEV SPACES

The arguments of [14] depend on the mapping properties of K and its adjoint from one Sobolev space to another. If Ω is a bounded open subset of $\mathbb{R}^m$, then the Sobolev space $W_p^k(\Omega) = W_p^k$, with k a non-negative integer and $1 \le p \le \infty$, is the space of functions $g \in L_p$ such that all the weak partial derivatives $D^\alpha g$ with $|\alpha| \le k$ belong to L_p. Here we use the multi-index notation: α stands for $(\alpha_1,\ldots,\alpha_m)$, with $\alpha_1, \ldots, \alpha_m$ non-negative integers, $|\alpha| = \alpha_1 + \cdots + \alpha_m$, and

$$D^\alpha g(t) = \frac{\partial^{\alpha_1+\cdots+\alpha_m}}{\partial t_1^{\alpha_1} \ldots \partial t_m^{\alpha_m}} g(t), \qquad t \in \mathbb{R}^m.$$

The space W_p^k is a Banach space under the norm

$$\|g\|_{W_p^k} = \sum_{|\alpha| \le k} \|D^\alpha g\|_{L_p}.$$

For example, in one dimension

$$\|g\|_{W_\infty^2} = \|g\|_{L_\infty} + \|Dg\|_{L_\infty} + \|D^2 g\|_{L_\infty}.$$

Thus the error bounds (3.1), (5.1) and (5.2) can be expressed in terms of the Sobolev norms as

$$\|y - y_n\|_{L_\infty} \le ch^2 \|y\|_{W_\infty^2}, \tag{6.1}$$

$$\|y - y_n\|_{W^1_\infty} \leq ch\|y\|_{W^2_\infty}, \tag{6.2}$$

and

$$\|y - y_n^{(1)}\|_{W^1_\infty} \leq ch^4\|y\|_{W^2_\infty}. \tag{6.3}$$

The Sobolev space definitions need modification if Ω is a smooth m-dimensional surface in $\mathbb{R}^{m+1}$; see [14] for details.

7. A SECOND EXAMPLE

To motivate the theory in the following section, consider the two-dimensional integral equation

$$y(t_1,t_2) = 1 + i\lambda \int_\Omega \log|t - s|\; y(s_1,s_2)ds_1 ds_2,$$

where

$$|t - s| = [(t_1 - s_1)^2 + (t_2 - s_2)^2]^{\frac{1}{2}},$$

with λ a real number, and Ω a bounded open region in the plane. This equation describes the distribution of alternating current in a conducting bar of cross section Ω. (For further details see, for example, [6].)

Suppose that Ω is divided in a quasi-uniform manner into triangles of maximum diameter h. (The quasi-uniformity is an annoying technical requirement which can probably be weakened.) Suppose also that U_n is the space of functions that are continuous on Ω and linear on each subregion. Then it can be shown, for p in $1 < p < \infty$, that

$$\|y - y_n\|_{L_p} \leq ch^2\|y\|_{W^2_p}, \tag{7.1}$$

$$\|y - y_n^{(1)}\|_{L_p} \leq cph^4\|y\|_{W^2_p}, \tag{7.2}$$

and

$$\|y - y_n^{(1)}\|_{W^1_p} \leq cph^3\|y\|_{W^2_p}. \tag{7.3}$$

The proofs are sketched in the following section. Note that one power of h is now lost in going from $y_n^{(1)}$ to its first derivative, but that the order of convergence is still very satisfactory.

The above estimates are unsatisfactory when p is large, because the right-hand sides diverge as $p \to \infty$. However, the following uniform error estimates can be squeezed from the theory:

$$\|y - y_n\|_{L_\infty} \le ch^2 \log(1/h)\|y\|_{L_\infty}, \tag{7.4}$$

$$\|y - y_n^{(1)}\|_{L_\infty} \le ch^4 \log^2(1/h)\|y\|_{L_\infty}, \tag{7.5}$$

$$\|y - y_n^{(1)}\|_{W_\infty^1} \le ch^3 \log^2(1/h)\|y\|_{L_\infty}. \tag{7.6}$$

The final orders of convergence for $y_n^{(1)}$ might be thought very encouraging.

8. OUTLINE OF THEORY

For simplicity, the theory is here specialized to the case of piecewise-linear approximating functions; the generalization to piecewise polynomials of higher degree is straightforward.

Corresponding to the piecewise-linear assumption, we assume that for every p in $1 \le p \le \infty$

$$\|g - P_n g\|_{L_p} \le ch^s\|g\|_{W_p^s}, \qquad s = 0,1,2. \tag{8.1}$$

In particular, this property holds for the piecewise-linear constructions in Sections 3 and 7. Moreover, we assume that for some fixed p we have $f \in W_p^2$ and K is a bounded operator from L_p to W_p^2. (For the example of Section 3 the latter holds for all p in $1 \le p \le \infty$, while for the example in Section 7 it holds for p in $1 < p < \infty$, by the Calderón-Zygmund theory [1].) Then $y \in W_p^2$, and it follows from (2.4) and (8.1) that

$$\|y - y_n\|_{L_p} \le c\|y - P_n y\|_{L_p} \le ch^2\|y\|_{W_p^2}. \tag{8.2}$$

The defining equation of the Galerkin method is $y_n = P_n f + P_n K y_n$, thus from (4.1) we have $y_n = P_n y_n^{(1)}$, and hence

$$y_n^{(1)} = f + KP_n y_n^{(1)}.$$

If $\|K - KP_n\|_{L_p} \to 0$ as $n \to \infty$, then for n sufficiently large the inverses $(I - KP_n)^{-1}$ are uniformly bounded operators on L_p, and

$$\begin{aligned} y - y_n^{(1)} &= (I - K)^{-1}f - (I - KP_n)^{-1}f \\ &= (I - KP_n)^{-1}(K - KP_n)y \\ &= (I - KP_n)^{-1}(K - KP_n)(y - y_n), \end{aligned}$$

where the last step follows because $P_n y_n = y_n$. Hence

$$\|y - y_n^{(1)}\|_{L_p} \le \|(I - KP_n)^{-1}\|_{L_p} \|K - KP_n\|_{L_p} \|y - y_n\|_{L_p},$$

or if $y_n \neq y$

$$\frac{\|y - y_n^{(1)}\|_{L_p}}{\|y - y_n\|_{L_p}} \le c\|K - KP_n\|_{L_p}. \tag{8.3}$$

This is essentially the superconvergence result of [11, 12].

It may also be shown that

$$\frac{\|y - y_n^{(1)}\|_{W_p^1}}{\|y - y_n\|_{L_p}} \le c\|K - KP_n\|_{L_p \to W_p^1} \equiv c \sup_{g \in L_p} \frac{\|(K-KP_n)g\|_{W_p^1}}{\|g\|_{L_p}}. \tag{8.4}$$

This follows from the identity

$$y - y_n^{(1)} = (I - K)^{-1}(K - KP_n)(y^{(1)} - y_n),$$

which leads to

$$\begin{aligned} \|y - y_n^{(1)}\|_{W_p^1} &= \|(I-K)^{-1}(K-KP_n)(y_n^{(1)} - y + y - y_n)\|_{W_p^1} \\ &\le \|(I-K)^{-1}\|_{W_p^1} \|K-KP_n\|_{L_p \to W_p^1} (\|y_n^{(1)} - y\|_{L_p} + \|y - y_n\|_{L_p}) \\ &\le c\|K - KP_n\|_{L_p \to W_p^1} \|y - y_n\|_{L_p}, \end{aligned} \tag{8.5}$$

where we have used (8.3) and the fact (see [14]) that $(I - K)^{-1}$ is a bounded operator on the space W_p^1. (It can be shown that the constants in (8.3) and (8.5) can be chosen independently of p.)

The task that remains is to estimate the superconvergence

factors on the right-hand sides of (8.3) and (8.4). For the first it follows by a simple duality argument (see [14]) that

$$\|K - KP_n\|_{L_p} = \|K^* - P_n K^*\|_{L_q} \equiv \sup_{v \in L_q} \frac{\|(I - P_n)K^*v\|_{L_q}}{\|v\|_{L_q}}, \tag{8.6}$$

where $q = p/(p-1)$ if $1 < p < \infty$ and $q = \infty$ or 1 if $p = 1$ or ∞ respectively, and K^* is the adjoint of K defined by

$$K^*v(t) = \int_\Omega \overline{k(s,t)}v(s)d\sigma(s).$$

Now suppose that K^* is a bounded operator from L_q to W_q^2. (For the example of Section 3 this is true for all q in $1 \le q \le \infty$, and hence for all p in $1 \le p \le \infty$; for the example of Section 7 it is true for all q in $1 < q < \infty$, and hence for all p in $1 < p < \infty$.) Then from (8.6) and (8.1) it follows that

$$\|K - KP_n\|_{L_p} \le ch^2 \sup_{v \in L_q} \frac{\|K^*v\|_{W_q^2}}{\|v\|_{L_q}} = ch^2 \|K^*\|_{L_q \to W_q^2}.$$

For the first example the desired result (4.3) then follows from (8.3) and (8.2) on setting $p = \infty$ and hence $q = 1$. For the example of Section 7 the Calderón-Zygmund theory of singular integral operators (see [1]) gives $\|K^*\|_{L_q \to W_q^2} \le cp$ for large p, and (7.2) then follows.

In a similar way it can be shown (see [14] for details) that

$$\|K - KP_n\|_{L_p \to W_p^1} = \sum_{|\alpha| \le 1} \sup_{v \in L_q} \frac{\|(I-P_n)(D^\alpha K)^*v\|_{L_q}}{\|v\|_{L_q}}, \tag{8.7}$$

and that for the example in Section 7 this yields

$$\|K - KP_n\|_{L_p \to W_p^1} \le ch\|K^*\|_{L_q \to W_q^2} \le cph,$$

from which (7.3) follows. On the other hand for the example in Section 3 it follows that $\|K-KP_n\|_{L_\infty \to W_\infty^1} \le ch^2$, because $(DK)^*$ is itself an integral operator with a smooth kernel, and so maps from L_1 to W_1^2. Hence (6.3) follows.

The extension of the error estimates (7.1) - (7.3) to the uniform error estimates (7.4) - (7.6) is a non-trivial matter, for which we must refer to [14].

9. CONCLUSION

It is clear that iteration of the Galerkin approximation often significantly improves the order of convergence, and that the improvement is particularly useful if good derivative information is desired.

The theory as presented here assumes that all integrals are evaluated exactly. In practice numerical integration will usually be necessary, and in that case it becomes important to choose the integration rule in such a way as to preserve the order of convergence. Because $y_n^{(1)}$ converges with a higher order than does y_n itself, this condition is more stringent for $y_n^{(1)}$ than for y. With that qualification, the calculation of $y_n^{(1)}$ requires in principle no extra work over that needed for y_n, because the necessary ingredients, namely the coefficients a_i and the integrals Ku_i (the latter perhaps only at discrete quadrature points) are already required for the Galerkin calculation itself.

The theory in [14] also considers higher iterates of the Galerkin approximation, and demonstrates that in some cases higher iteration is desirable, at least in principle. See [14] for details.

REFERENCES

1. CALDERÓN, A.P., and ZYGMUND, A., On the Existence of Certain Singular Integrals. *Acta Math.* 88, 85-139 (1952).
2. CHANDLER, G.A., Global Superconvergence of Iterated Galerkin Solutions for Second Kind Integral Equations. Technical Report, Australian National University, Canberra (1978).
3. CHANDLER, G.A., Superconvergence for Second Kind Integral Equations, pp.103-117 of R.S. Anderssen, F.R. de Hoog and M.A. Lukas (Eds.), *The Application and Numerical Solution of Integral Equations*. Sijthoff and Noordhoff, Alphen aan den Rijn (1980).
4. CHANDLER, G.A., Superconvergence of Numerical Solutions to Second Kind Integral Equations. Ph.D. thesis, Australian National University (1979).
5. CHATELIN, F., Sur les Bornes d'Erreur a posteriori pour les Éléments Propres d'Opérateurs Linéaires. *Numer. Math.* 32, 233-246 (1979).
6. GRAHAM, I.G., Some Application Areas for Fredholm Integral Equations of the Second Kind, pp.75-102 of R.S. Anderssen, F.R. de Hoog and M.A. Lukas (Eds.), *The Application and Numerical Solution of Integral Equations*. Sijthoff and Noordhoff, Alphen aan den Rijn (1980).
7. GRAHAM, I.G., Galerkin Methods for Second Kind Integral Equations with Singularities. *Math. Comp.*, to appear.
8. HSIAO, G.C., and WENDLAND, W.L., The Aubin-Nitsche Lemma for Integral Equations. *J. Integral Eqns.* 3, 299-315 (1981).

9. KRASNOSEL'SKII, M.A., VAINIKKO, G.M., ZABREIKO, P.P., RUTITSKII, Ya.B., and STETSENKO, V.Ya. *Approximate Solution of Operator Equations*. Wolters-Noordhoff, Groningen (1972).
10. RICHTER, G.R., Superconvergence of Piecewise Polynomial Galerkin Approximations for Fredholm Integral Equations of the Second Kind. *Numer. Math.* 31, 63-70 (1978).
11. SLOAN, I.H., Error Analysis for a Class of Degenerate-Kernel Methods. *Numer. Math.* 25, 231-238 (1976).
12. SLOAN, I.H., Improvement by Iteration for Compact Operator Equations. *Math. Comp.* 30, 758-764 (1976).
13. SLOAN, I.H., Iterated Galerkin Method for Eigenvalue Problems. *SIAM J. Numer. Anal.* 13, 753-760 (1976).
14. SLOAN, I.H., and THOMÉE, V., Superconvergence of the Galerkin Iterates for Integral Equations of the Second Kind, to be submitted.
15. SPENCE, A., and THOMAS, K.S., On Superconvergence Properties of Galerkin's Method for Compact Operator Equations. Technical Report, University of Southampton (1981).

THE ACCURATE COMPUTATION OF CERTAIN INTEGRALS ARISING IN INTEGRAL EQUATION METHODS FOR PROBLEMS IN POTENTIAL THEORY

D.M. Hough

Polytechnic of the South Bank

1. INTRODUCTION

Let Γ be an analytic arc in the complex z-plane with parametric equation

$$z = \zeta(s), \ a \leq s \leq b, \tag{1.1}$$

where s is any appropriate real parameter. Let an arbitrary mesh, whose nodal points are denoted by s_k, $k = 0(1)K$, with

$$a = s_0 < s_1 < \ldots < s_{K-1} < s_K = b,$$

be placed on Γ and define

$$L_k = \{s: s_{k-1} \leq s \leq s_k\},$$

$$\Gamma_k = \{z: z = \zeta(s), \ s \in L_k\}, \ k = 1(1)K. \tag{1.2}$$

This report is concerned with the accurate evaluation of the integral

$$A_k(z,w) = \int_{L_k} w(s)\log(z - \zeta(s))ds \tag{1.3}$$

where w is a density function defined over L_k, z denotes an arbitrary point in the plane,

$$\log(z - \zeta(s)) = \log|z - \zeta(s)| + i \arg(z - \zeta(s)) \tag{1.4}$$

and $\arg(z - \zeta(s))$ is a continuous argument as defined by Henrici [6, p230]; see also Jaswon and Symm [10, §11.7].

Integrals of the form (1.3) arise in the numerical solution of integral equations in potential theory and conformal mapping. The purpose of this paper is to consider the evaluation of the following two special cases which, in particular, occur in the numerical conformal mapping technique of Hough and Papamichael [9]:

$$\text{(a)}\quad A_k(z,(s-s_{k-1})^\beta) = \int_{L_k} (s-s_{k-1})^\beta \log(z-\zeta(s))ds, \qquad (1.5)$$

where $\beta>-1$ is a rational number;

$$\text{(b)}\quad A_k(z,M_{nj}(s)) = \int_{L_k} M_{nj}(s)\log(z-\zeta(s))ds,\ j=k(1)k-1+n, \qquad (1.6)$$

where $M_{nj}(s)$ is the B-spline of order n (degree n-1) based on the knots $s_{j-n}, s_{j-n+1},\ldots, s_j$ and where the knots outside the interval $[s_0,s_K]$ may be defined in any convenient manner.

The conformal mapping technique of [9] involves the numerical solution of an integral equation of the first kind with a logarithmic kernel, where the unknown source density ν is approximated by cubic splines and singular functions. The singular functions are used for overcoming the difficulties associated with corner singularities of ν. The integrals which arise in [9] correspond to (1.5) with $\beta>-\frac{1}{2}$ and (1.6) with n=4. However, the integration formulae derived in the present paper can be used to evaluate the integrals associated with other integral equation methods. For example, the methods of Symm [12] and Christiansen [1] require the evaluation of (1.5)-(1.6) with $\beta=0$ and n=1. Similarly, the method of Hayes, Kahaner and Kellner [5] essentially requires the evaluation of integrals of the form (1.5)-(1.6) with β an integer, $0\leq\beta\leq 2$, and $n\leq 3$. Finally the integrals in the method of Hough and Papamichael [8] correspond to the special case where Γ is a straight line, $\beta>-\frac{1}{2}$ and $n\leq 6$.

2. QUADRATURE SCHEMES

The technique used for evaluating any of the integrals $A_k(z,w)$, given by (1.5)-(1.6), depends on the position of z relative to the arc Γ_k. If the point z is "far" from Γ_k then the integral A_k is non-singular and may be evaluated accurately by Gaussian quadrature. However, as z approaches Γ_k all derivatives with respect to s of $\log(z-\zeta(s))$ acquire relatively large magnitudes, and in the extreme case, when $z\in\Gamma_k$, $\log(z-\zeta)$ is unbounded at $z=\zeta$. For these reasons, special quadrature techniques must be used in order to approximate A_k accurately when z lies in the vicinity of Γ_k. Here we adopt the following general strategy.

We assume that for each arc Γ_k, k=1(1)K, the z-plane can be divided into two mutually exclusive point sets F_k and N_k, containing respectively "far" and "near" points with respect to Γ_k. That is, for our quadrature purposes, we regard the kernel

$\log(z-\zeta(s))$ as being non-singular or singular according as to whether $z \varepsilon F_k$ or $z \varepsilon N_k$, and employ different quadrature schemes for each of the two cases. The criteria for defining the regions F_k and N_k are based on the quadrature schemes we use and also on the accuracy required. For these reasons, we first describe our two quadrature schemes and postpone the definition of the regions F_k and N_k until section 3.

2.1 The "Far" Quadrature Scheme

We consider separately the evaluation of each of the integrals (1.5)-(1.6).

2.1.1 The evaluation of $A_k(z,(s-s_{k-1})^\beta)$ *when* $z \varepsilon F_k$ With the change of variable

$$s = s_{k-1} + h_k x, \; h_k = s_k - s_{k-1} \qquad (2.1)$$

the integral (1.5) is transformed to

$$A_k(z,(s-s_{k-1})^\beta) = h_k^{\beta+1} \int_0^1 x^\beta \log(z-\zeta(s_{k-1}+h_k x))dx \qquad (2.2)$$

and is evaluated by a standard low order q-point Gauss-Jacobi quadrature formula. Full theoretical details and Fortran subroutines for determining the abscissae and weights for Gauss-Jacobi quadrature are given in Stroud and Secrest [11]. We note that the numerical results of [9] correspond to the choice q=4.

2.1.2 The evaluation of $A_k(z,M_{nj}(s))$ *when* $z \varepsilon F_k$ For any $s \varepsilon L_k$, the B-splines $M_{nj}(s)$, $j=k(1)k-1+n$, are polynomials of degree n-1 and, for this reason, a standard q*-point Gauss-Legendre quadrature formula is used for the evaluation of the integral $A_k(z,M_{nj}(s))$ given by (1.6). In order to minimize computational effort, the n integrals $A_k(z,M_{nj}(s))$, $j=k(1)k-1+n$, are evaluated concurrently. That is, the values of $\log(z-\zeta(s))$ at the Gaussian points on Γ_k are only computed once and are stored for use in the evaluation of each integral. Also, for a particular Gaussian point $s \varepsilon L_k$, the n B-splines $M_{nj}(s)$ are computed concurrently using standard B-spline techniques; see for example Cox [2] and De Boor [4].

Regarding the values q* and q corresponding to the two Gaussian rules used above, we observe the following. The q*-point Gauss-Legendre rule for the evaluation of

$$\int_{L_k} M_{nj}(s) f(s) ds$$

is exact if f is a polynomial of degree 2q*-n, whereas the q-point Gauss-Jacobi rule evaluates

$$\int_{L_k} (s-s_{k-1})^{\beta} f(s)ds$$

exactly when f is a polynomial of degree 2q-1. For this reason the values of q* and q are chosen so that

$$q + \tfrac{1}{2}(n-1) \leq q^* \leq q + \tfrac{1}{2}n. \tag{2.3}$$

For example, in [9], where n=4 and q=4, the value of q* used is q*=6.

2.2 The "Near" Quadrature Scheme

As before, we let

$$z = \zeta(s) \tag{2.4}$$

be the parametric equation of Γ. Since each arc Γ_k is analytic there exists a domain $D_k \supset L_k$ in the complex ξ=s+it-plane such that

$$z = \zeta(\xi) \tag{2.5}$$

is one-one analytic in D_k and defines a conformal map of D_k onto some domain $\zeta(D_k) \supset \Gamma_k$. It is convenient to assume that the "near" region N_k is such that $N_k \subset \zeta(D_k)$. Then, given any $z \in N_k$ there exists a unique point $\xi \in D_k$ which satisfies (2.5), and the logarithmic kernel in (1.3) may be written in the form

$$\log(z-\zeta(s)) = \log(\zeta(\xi)-\zeta(s))$$

$$= \log\zeta[\xi,s] + \log(\xi-s). \tag{2.6}$$

In (2.6),

$$\zeta[\xi,s] = \frac{\zeta(\xi) - \zeta(s)}{\xi - s}, \quad \xi \neq s, \tag{2.7}$$

and

$$\zeta[s,s] = \lim_{\xi \to s} \zeta[\xi,s]$$

$$= \zeta'(s) \neq 0; \tag{2.8}$$

i.e. $\zeta[\xi,s] \neq 0$ for all $s \in L_k$ and $z \in N_k$. (2.9)

Using (2.6) the integral $A_k(z,w)$ of (1.3) may be written in the form

$$A_k(z,w) = B_k(\xi,w) + C_k(\xi,w), \tag{2.10}$$

where

$$B_k(\xi,w) = \int_{L_k} w(s)\log\zeta[\xi,s]\,ds, \tag{2.11}$$

$$C_k(\xi,w) = \int_{L_k} w(s)\log(\xi-s)\,ds. \tag{2.12}$$

In view of (2.9), the logarithmic kernel of the integral B_k is non-singular for all $z\varepsilon N_k$. Hence, depending on the form of w, B_k is evaluated by using one of the Gaussian rules described in sections 2.1.1 or 2.1.2.

The integral $C_k(\xi,w)$ does not depend explicitly on the curve Γ and can be evaluated exactly for the two functions

$$w(s) = (s-s_{k-1})^{\beta} \text{ and } w(s) = M_{nj}(s)$$

of interest. The full details concerning the exact evaluation of $C_k(\xi,(s-s_{k-1})^{\beta})$ are given in Hough [7] and, for this reason, are not repeated here. Clearly, the integral $C_k(\xi,M_{nj}(s))$ is a linear combination of integrals of the type $C_k(\xi,(s-s_{k-1})^{\beta})$, where β is a positive integer. However, because of the properties of B-splines, simplifications occur when $w(s)=M_{nj}(s)$ and we prefer to consider the integrals $C_k(\xi,M_{nj}(s))$ separately, as follows.

Let the functions $u_r(\xi,s)$, $r\geq 0$, be defined recursively by

$$u_0(\xi,s) = \log(\xi-s),\ \xi\neq s, \tag{2.13}$$

$$u_r(\xi,s) = \{(s-\xi)u_{r-1}(\xi,s) - s^r/r!\}/r,\ r\geq 1, \tag{2.14}$$

so that

$$\frac{\partial u_r}{\partial s}(\xi,s) = u_{r-1},\ r\geq 1, \tag{2.15}$$

and, by continuity,

$$u_r(s,s) = \frac{-s^r}{r(r!)},\ r\geq 1. \tag{2.16}$$

Also, let

$$I_{rj}(\xi) = \int_{L_k} M_{rj}(s)u_{n-r}(\xi,s)\,ds,\ 1\leq r\leq n, \tag{2.17}$$

so that

$$I_{nj}(\xi) = C_k(\xi,M_{nj}(s)). \tag{2.18}$$

(Because of the compact support of the B-splines, $I_{rj}(\xi)\neq 0$ only if $k\leq j\leq k-1+r$). Then, by applying integration by parts to (2.17) and using (2.15) in conjunction with the differentiation formula

$$\frac{dM_{rj}}{ds} = -\frac{(r-1)}{s_j - s_{j-r}}\{M_{r-1,j}(s) - M_{r-1,j-1}(s)\},\ r \geq 2, \tag{2.19}$$

we obtain the recurrence relation

$$I_{1j}(\xi) = \begin{cases} \{u_n(\xi,s_k) - u_n(\xi,s_{k-1})\}/(s_k - s_{k-1}), & j=k, \\ 0 & , j \neq k, \end{cases} \tag{2.20}$$

$$I_{rj}(\xi) = M_{rj}(s_k)u_{n-r+1}(\xi,s_k) - M_{rj}(s_{k-1})u_{n-r+1}(\xi,s_{k-1})$$
$$+ \frac{(r-1)}{s_j - s_{j-r}}\{I_{r-1,j}(\xi) - I_{r-1,j-1}(\xi)\},\ 2 \leq r \leq n, \tag{2.21}$$

for the evaluation of $I_{nj}(\xi)$, $j=k(1)k-1+n$. It can be shown easily that (2.21) remains valid for the singular case where $\xi \varepsilon L_k$.

To summarise, the computational sequence for the evaluation of $C_k(\xi, M_{nj}(s))$ is as follows.

(i) By means of (2.13), (2.14) or (2.16) evaluate and store $u_r(\xi, s_{k-1})$ and $u_r(\xi, s_k)$, $r=1(1)n$.

(ii) Evaluate and store the B-spline values $M_{rj}(s_{k-1})$ and $M_{rj}(s_k)$, $r=2(1)n$, $j=k(1)k-1+r$. (These evaluations are performed by using the stable B-spline algorithm; see [2], [4].)

(iii) Evaluate $I_{nj}(\xi)$, $j=k(1)k-1+n$, by means of the recurrence formula (2.20)-(2.21).

3. DEFINITIONS OF THE "NEAR" AND "FAR" REGIONS

In order to define the "near" region N_k we assume that z lies close to the arc Γ_k, and consider the integral $A_k(z, (s-s_{k-1})^\beta)$ expressed in the form (2.10), i.e.

$$A_k(z, (s-s_{k-1})^\beta) = B_k(\xi, (s-s_{k-1})^\beta) + C_k(\xi, (s-s_{k-1})^\beta), \tag{3.1}$$

where B_k and C_k are as defined by (2.11) and (2.12). In particular we consider the integral C_k and, by using the change of variables

$$\sigma = (\xi - s_{k-1})/h_k,\ x = (s-s_{k-1})/h_k,\ h_k = s_k - s_{k-1}, \tag{3.2}$$

we express it as

$$C_k(\xi, (s-s_{k-1})^\beta) = h_k^{\beta+1}\{\int_0^1 x^\beta \log h_k dx + S(\sigma)\}, \tag{3.3}$$

where

$$S(\sigma) = \int_0^1 x^\beta \log(\sigma - x)\ dx. \tag{3.4}$$

We recall that in the "near" quadrature scheme B_k is always evaluated by using a q-point Gauss-Jacobi rule. Assume now that the same q-point rule is used for the evaluation of the integral C_k in (3.1) and let $\tilde{C}_k$ be the resulting approximation. Then it follows at once from (3.3) that

$$|C_k-\tilde{C}_k| = h_k^{\beta+1}|S-\tilde{S}|, \tag{3.5}$$

where $\tilde{S}$ is the q-point Gauss-Jacobi approximation to S. Furthermore, the error analysis of [11, §4.1] shows that

$$|S-\tilde{S}| \leq e_{2q}(\beta)B_{2q}, \tag{3.6}$$

where

$$B_{2q} = \max_{0\leq x\leq 1}\left|\frac{d^{2q}\log(\sigma-x)}{dx^{2q}}\right|, \quad \sigma\notin[0,1], \tag{3.7}$$

$$e_{2q}(\beta) = \frac{1}{(2q)!}\int_0^1 x^\beta\{P_q^{(0,\beta)}(x)\}^2dx, \tag{3.8}$$

and $P_q^{(0,\beta)}(x)$ is the Jacobi polynomial of degree q with leading coefficient unity based on the interval [0,1]. As is shown in Davis [3, §10.3], the integral (3.8) is given explicitly by

$$e_{2q}(\beta) = \frac{1}{(2q)!(2q+\beta+1)}\left[\frac{q!\Gamma(q+\beta+1)}{\Gamma(2q+\beta+1)}\right]^2$$

$$= \frac{1}{(2q)!(2q+\beta+1)}\left[\frac{q!}{(2q+\beta)(2q+\beta-1)..(q+\beta+1)}\right]^2. \tag{3.9}$$

Since

$$\frac{d^{2q}\log(\sigma-x)}{dx^{2q}} = -\frac{(2q-1)!}{(\sigma-x)^{2q}}, \quad q>0,$$

it follows that

$$B_{2q} = (2q-1)!\{h_k/d_k(\xi)\}^{2q}, \quad q>0, \tag{3.10}$$

where

$$d_k(\xi) = \min_{s\in L_k}|\xi-s| \tag{3.11}$$

and, as before,

$$L_k = \{s: s_{k-1}\leq s\leq s_k\}.$$

That is,

$$d_k(\xi) = \begin{cases} |\xi-s_k| \ , & Re\{\xi\}>s_k, \\ |\xi-s_{k-1}|, & Re\{\xi\}<s_{k-1}, \\ |Im\{\xi\}| \ , & s_{k-1}\leq Re\{\xi\}\leq s_k. \end{cases} \tag{3.12}$$

Hence, from (3.6), (3.9) and (3.10),

$$|S-\tilde{S}| \leq E_{2q}\{h_k/d_k(\xi)\}^{2q}, \ q>0, \tag{3.13}$$

where E_{2q} is a constant independent of β. For example, in the case where $\beta>-\frac{1}{2}$, which covers all the integrals arising in the numerical conformal mapping technique of [8,9], we may choose

$$E_{2q} = (2q-1)!e_{2q}(-\tfrac{1}{2}) = \frac{1}{2q(2q+\frac{1}{2})}\left[\frac{q!}{(2q-\frac{1}{2})\ldots(q+\frac{1}{2})}\right]^2, \ q>0. \tag{3.14}$$

Given any $\delta>0$ it follows from (3.13) that a sufficient condition for $|S-\tilde{S}|\leq\delta$ is

$$d_k(\xi) \geq \rho h_k, \tag{3.15}$$

where

$$\rho = (E_{2q}/\delta)^{1/2q}, \ q>0. \tag{3.16}$$

Values of ρ corresponding to $\beta>-\frac{1}{2}$ and to various choices of q and δ are given in Table 1.

TABLE 1

Values of ρ

δ	q = 2	3	4	5	6
10^{-3}	1.3	0.71	0.53	0.44	0.40
10^{-5}	4.1	1.5	0.94	0.70	0.58
10^{-7}	13	3.3	1.7	1.1	0.86
10^{-9}	41	7.1	3.0	1.8	1.3

Naturally, if $\tilde{C}_k$ is used to estimate C_k then the whole integral A_k in (3.1) is effectively computed by Gaussian quadrature, i.e. by the "far" quadrature scheme. This means that the "far" region F_k should be defined as the region where sufficiently accurate Gaussian approximations to C_k can be computed, i.e. the region where (3.15) is satisfied for a sufficiently small δ. This leads to the following definition of the "near" region N_k.

Definition 3.1 The "near" region N_k is the set of all points z for which

$$d_k(\xi) < \rho h_k, \tag{3.17}$$

where ξ is the solution, closest to the interval L_k, of the equation

$$\zeta(\xi) = z. \tag{3.18}$$

Clearly, an analysis similar to the one used above can be applied for the purpose of determining a separate definition of a "near" region associated with the integral $A_k(z,M_{nj}(s))$. Unfortunately however, such an analysis does not lead to a practically useful result. For this reason we adopt definition 3.1 as the general definition of N_k, irrespective of the form of w(s). This can be justified to some extent by observing that when q* satisfies (2.3) the q*-point Gauss-Legendre approximation to $C_k(\xi,M_{nj}(s))$ is of the same order of accuracy as the q-point Gauss-Jacobi approximation to $C_k(\xi,(s-s_{k-1})^{\beta})$.

It is of interest to observe that the equation

$$d_k(\xi) = \rho h_k \tag{3.19}$$

defines a closed curve in the ξ=s+it-plane which consists of the two straight line segments $t=\pm\rho$, $s_{k-1}\leq s\leq s_k$ and the two semi-circular arcs of radius ρ centred on s_{k-1} and s_k. The "near" region N_k is the image under the transformation $z=\zeta(\xi)$ of the interior of the curve described by (3.19).

In principle the process of testing whether a given point z lies in N_k may be carried out in either the ξ-plane or the z-plane. If the test is performed in the ξ-plane then the process requires the determination of a root of equation (3.18), for each value of z. Clearly, the computational effort involved in determining a solution of (3.18) depends strongly on the nature of the function ζ which describes Γ and, for arbitrary z, this may be a non-trivial problem. If Γ is a segment of any conic section, as is the case in all examples considered in [9], then (3.18) can always be solved analytically. For this reason, in all the applications considered in [9], the testing for $z\epsilon N_k$ is performed in the ξ-plane using the parameters q=4, $\rho=2.9, \delta\simeq 10^{-9}$; see Table 1. In other applications it may be necessary to carry out the testing in the z-plane. A strict interpretation of the definition 3.1 may then present practical difficulties in so far as it may be computationally expensive to test whether a given point z lies within the image contour of the curve described by (3.19). In this case, it is preferable to take as N_k a simple region which approximates the region of definition 3.1. One such possibility is to take as near region the disc

$$N_k = \{z: |z-\zeta(s_{k-\frac{1}{2}})|<(\rho+\tfrac{1}{2})|\zeta(s_k)-\zeta(s_{k-1})|\}, \tag{3.20}$$

where $s_{k-\frac{1}{2}} = s_k - \frac{1}{2}h_k$. If (3.20) is adopted then clearly it is a simple matter to test whether or not $z\varepsilon N_k$ and, if $z\varepsilon N_k$, to solve (3.18) numerically.

4. NUMERICAL EXAMPLE

As a simple example to illustrate the accuracy of the proposed method we consider the evaluation of

$$A(z,w) = \int_{s_0}^{s_0+h} w(s)\log(z-\zeta(s))ds$$

where Γ is the circular arc

$$\zeta(s) = e^{is}, \tag{4.1}$$

with

$$s_0 = -0.02,\ h = 0.1, \tag{4.2}$$

for the three densities

(i) $w(s) = (s-s_0)^{-\frac{1}{2}}$
(ii) $w(s) = M_{2j}(s);\ j=1,2$
(iii) $w(s) = M_{4j}(s);\ j=1(1)4.$

The knots of the B-splines in (ii) and (iii) are taken to be the equispaced points $s_j = s_0+jh$. For simplicity we compute $A(z,w)$ only at the real test points

$$z = 0(0.01)1\ . \tag{4.3}$$

The number of Gauss-Jacobi quadrature points and the accuracy parameter δ are taken to be $q=4$ and $\delta=10^{-7}$. Then Table 1 and equation (2.3) give respectively

$$\rho = 1.7 \text{ and } q^* = \begin{cases} 5 \text{ when } w(s) = M_{2j}(s), \\ 6 \text{ when } w(s) = M_{4j}(s). \end{cases}$$

Definition 3.1 for the "near" region requires the solution ξ of (3.18) which, for the parametric function (4.1), is

$$\xi = -i\log z\ . \tag{4.4}$$

It follows from (3.11), (3.17), (4.2) and (4.4) that z is "near" to the circular arc if $-\log z<\rho h$; i.e., z lies in the "near" region if

$$z > 0.84\ .$$

The numerical results are presented graphically in Fig. 1.

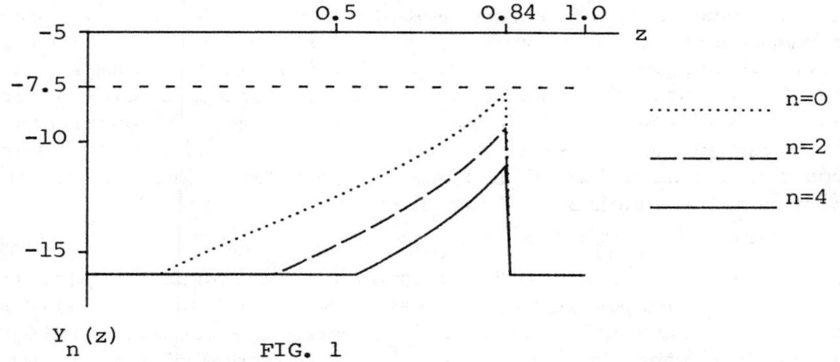

FIG. 1

The meaning of this graph is as follows. The absolute errors

$$E_0(z) = |\tilde{A}(z,(s-s_0)^{-\frac{1}{2}}) - A(z,(s-s_0)^{-\frac{1}{2}})|,$$

$$E_n(z) = \max_{j=1(1)n} |\tilde{A}(z,M_{nj}(s)) - A(z,M_{nj}(s))|;\ n=2,4,$$

where $\tilde{A}$ denotes the numerical approximation to the integral A, are computed at each test point (4.3). (In the above, if z is in the "near" region, i.e. $z>0.84$, then the "exact" value of $A(z,w)$ is computed by evaluating the non-singular component $B(\xi,w)$ of (2.11) by a sequence of appropriate Gaussian rules until successive estimates agree to within an absolute difference of 10^{-16}. A similar sequence of Gaussian rules is used to compute $A(z,w)$ itself when $z\leq 0.84$). The graph of Fig. 1 is then the plot of the values of

$$Y_n(z) = \max\{-16, \log_{10}E_n(z)\};\ n=0,2,4,$$

against z. The horizontal dashed line in Fig. 1 marks the reference error value $\log_{10}(h^{\frac{1}{2}}\delta) = -7.5$.

5. DISCUSSION

In the numerical example of section 4, the quadrature error does not exceed the value $h^{\frac{1}{2}}\delta$ which is the maximum error in $\tilde{C}(\xi,(s-s_0)^{-\frac{1}{2}})$ permitted by the method; see (3.5). This same error behaviour is observed in other numerical examples involving arcs of moderate curvature. However, if the curvature of the arc under consideration is large, then the error in $\tilde{B}(\xi,w)$ may contribute more significantly to the overall quadrature error. The numerical results of section 4 also show that the value of q^* is unnecessarily large. This suggests that it may be possible to replace (2.3) by a less conservative requirement which relates q^* directly to δ and ρ.

As was previously remarked, the integration methods described in this paper have been implemented successfully in [9], where it was essential to compute integrals of the form (1.3) accurately. Techniques similar to those descibed in sections 2 and 3 can also be used for dealing with density functions w(s) other than rational powers of $(s-s_{k-1})$ and B-splines. In particular, the treatment of corner singularities in integral equation methods for the solution of problems in potential theory might involve density functions of the form

$$w(s) = (s-s_{k-1})^{\beta}\log^{m}(s-s_{k-1}) \qquad (5.1)$$

where $\beta>-\frac{1}{2}$ is rational and $m>0$ is an integer. For example, the functions corresponding to the cases $\beta=1,3$ and $m=1$ of (5.1) occur in the numerical method of [9], where the resulting integrals are computed by techniques similar to those of the present paper; i.e. by using Gaussian formulae appropriate to the weight w(s) for the "far" quadrature scheme in conjunction with Gaussian and semi-analytic formulae for the "near" quadrature scheme.

The author wishes to thank Dr. N. Papamichael for his valuable advice in the preparation of this paper.

REFERENCES

1. CHRISTIANSEN, S., Numerical Solution of an Integral Equation with a Logarithmic Kernel. *BIT* 11, 276-287 (1971).
2. COX, M.G., Numerical Evaluation of B-Splines. *JIMA* 10, 134-149 (1972).
3. DAVIS, P.J., *Interpolation and Approximation*. Dover Publications, New York (1975).
4. DE BOOR, C., On Calculating with B-Splines. *J. Approx. Theory* 6, 50-62 (1972).
5. HAYES, J.K., KAHANER, D.K. & KELLNER, R.G., An Improved Method for Numerical Conformal Mapping. *Math. Comput.* 26, 327-334 (1972).
6. HENRICI, P., *Applied and Computational Complex Analysis, Volume 1*. Wiley, New York (1974).
7. HOUGH, D.M., Exact Formulae for Certain Integrals Arising in Potential Theory. *IMAJNA* 1, 223-228 (1981).
8. HOUGH, D.M. & PAPAMICHAEL, N., The Use of Splines and Singular Functions in an Integral Equation Method for Conformal Mapping. *Numer. Math.* 37, 133-147 (1981).
9. HOUGH, D.M. & PAPAMICHAEL, N., An Integral Equation Method for the Numerical Conformal Mapping of Interior, Exterior and Doubly-Connected Domains. *Numer. Math.* (to appear).
10. JASWON, M.A. & SYMM, G.T., *Integral Equation Methods in Potential Theory and Elastostatics*. Academic, London (1977).
11. STROUD, A.H. & SECREST, D., *Gaussian Quadrature Formulas*. Prentice-Hall, Englewood Cliffs N.J. (1966)
12. SYMM, G.T., An Integral Equation Method in Conformal Mapping. *Numer. Math.* 9, 250-258 (1966).

EXTENDED VOLTERRA-RUNGE-KUTTA METHODS

E. Hairer

University of Heidelberg

1. INTRODUCTION

Consider the (nonlinear) regular Volterra integral equation of the second kind

$$y(x) = f(x) + \int_a^x K(x,s,y(s))ds , \quad x\in I:=[a,b] \qquad (1.1)$$

and assume that a unique continuous solution exists on I. A direct discretization of (1.1) yields the *quadrature method*

$$y_n = f(x_n) + h \sum_{j=0}^{n} w_{nj} K(x_n,x_j,y_j) , \quad n\geq n_o . \qquad (1.2)$$

Here $x_n=a+nh$, y_n is a numerical approximation to $y(x_n)$ and w_{nj} are the weights of some quadrature formula.

We will turn our attention to another class of methods. Observe first that with the help of the function

$$F_n(x) = f(x) + \int_a^{x_n} K(x,s,y(s))ds , \quad x\in[x_n,b]. \qquad (1.3)$$

equation (1.1) can be rewritten as

$$y(x) = F_n(x) + \int_{x_n}^{x} K(x,s,y(s))ds, \quad x\in[x_n,b]. \qquad (1.4)$$

A *one step method* for (1.1) is then given by

$$y_{n+1} = \tilde{F}_n(x_{n+1}) + h\Phi_n(x_{n+1},\tilde{F}_n,h) , \quad n\geq 0 \qquad (1.5)$$

where $\tilde{F}_n(x)$ is an approximation to $F_n(x)$ and Φ_n is some increment function that approximates the integral in (1.4). Examples for Φ_n are the socalled Volterra-Runge-Kutta methods (VRK-methods) where

$$\Phi_n(x,\tilde{F}_n h) = \sum_{i=1}^{m} b_i K(x+(e_i-1)h,\ x_n+c_i h,\ Y_i^{(n)}(\tilde{F}_n))$$
$$Y_i^{(n)} = \tilde{F}_n(x_n+c_i h) + h \sum_{j=1}^{m} a_{ij} K(x_n+d_{ij}h,\ x_n+c_j h,\ Y_j^{(n)}) \tag{1.6}$$
$$i=1,\dots,m.$$

We always assume that

$$c_i = \sum_{j=1}^{m} a_{ij}\ , \quad i = 1,\dots,m. \tag{1.7}$$

There are two inportant special cases of (1.6) which are due to Pouzet [7] and Bel'tyukov [2], respectively. *Pouzet-type methods* (PRK-methods) are characterized by

$$d_{ij} = c_i,\ e_i = 1 \quad \text{for all } i,j \tag{1.8}$$

and are therefore completely determined by the coefficients a_{ij} and b_i of an ordinary Runge-Kutta method. *Bel'tyukov-type methods* (BRK-methods) satisfy the relations

$$d_{ij} = e_j \quad \text{for all } i,j. \tag{1.9}$$

Besides the increment function Φ_n also the tail approximation $\tilde{F}_n(x)$ has to be specified for a one-step method (1.5). There are several possibilities. For *mixed VRK-methods* the integral in (1.3) is approximated by a quadrature formula in the same way as for quadrature methods. For *extended VRK-methods* $\tilde{F}_n$ is defined recursively by

$$\tilde{F}_o(x) = f(x)$$
$$\tilde{F}_{n+1}(x) = \tilde{F}_n(x) + h\Phi_n(x,\tilde{F}_n,h) \tag{1.10}$$

or equivalently by

$$\tilde{F}_n(x) = f(x)+h \sum_{j=0}^{n-1} \sum_{i=1}^{m} b_i K(x+(e_i-1)h,x_j+c_i h,Y_i^{(j)}(\tilde{F}_j)) \tag{1.11}$$

For this approach, which seems to be the more appropriate one, no starting procedure is necessary.

The aim of this paper is to present the theory for extended VRK-methods in a compact form. First we state a convergence theorem (section 2) and show then in section 3 how VRK-methods can be constructed. After some remarks on the implementation (section 4) we finally investigate the stability of such methods in section 5.

2. CONVERGENCE OF EXTENDED VRK-METHODS

For the study of convergence of one-step methods one has first to look at the error caused by the increment function Φ_n only. Consider therefore the value

$$\bar{y}_{n+1} = F_n(x_{n+1}) + h\Phi_n(x_{n+1}, F_n, h)$$

which can be interpreted as the numerical result when one step of (1.5) is applied to (1.4). The one-step method (1.5) is said to be of *local order* p if for any integral equation the local error satisfies

$$y(x_{n+1}) - \bar{y}_{n+1} = O(h^{p+1}).$$

For VRK-methods the concept of local order has been extensively studied in [3]. There the local order is characterized by algebraic conditions for the coefficients of the method. E.g. for a 4-th order BRK-method the following 21 conditions have to be satisfied:

$$\sum_{i=1}^{m} b_i e_i^{\alpha} c_i^{\beta} = \frac{1}{(\beta+1)} \qquad \text{for } 0 \leqq \alpha+\beta \leqq 3 \qquad (2.1)$$

$$\sum_{i=1}^{m} \sum_{j=1}^{m} b_i e_i^{\alpha} c_i^{\beta} a_{ij} e_j^{\gamma} c_j^{\delta} = \frac{1}{(\beta+\gamma+\delta+2)(\delta+1)} \text{ for } 1 \leqq \gamma+\delta \leqq \alpha+\beta+\gamma+\delta \leqq 2 \qquad (2.2)$$

$$\sum_{i=1}^{m} \sum_{j=1}^{m} \sum_{k=1}^{m} b_i a_{ij} a_{jk} e_k^{\gamma} c_k^{\delta} = \frac{1}{(\gamma+\delta+3)(\gamma+\delta+2)(\delta+1)} \text{ for } \gamma+\delta=1 \qquad (2.3)$$

($\alpha, \beta, \gamma, \delta$ are nonnegative integers).

Since extended one-step methods are fully determined by the increment functions Φ_n the question arises whether local order p already implies global convergence of the same order. At a first look this seems unlikely since the $Y_i^{(j)}(\tilde{F}_j)$ which are involved in the computation of the integral in (1.3) (compare formula (1.11)) are in general only low order approximations to $y(x_j+c_ih)$. However we have

<u>Theorem 1</u> (Convergence of extended one-step methods)
Let the numerical values y_n be calculated by an extended one-step method. If the local order is p then we have convergence of order p, i.e.

$$|F_k(x) - \tilde{F}_k(x)| \leqq C\, h^p \ , \quad x \in [x_k, b]$$

and

$$|y(x_k) - y_k| \leqq C\, h^p.$$

A detailed proof of this theorem can be found in [5]. Observe that the second statement follows immediately from the first one since

$$y(x_k) = F_k(x_k) \quad \text{and} \quad y_k = \tilde{F}_k(x_k). \tag{2.4}$$

3. CONSTRUCTION OF VRK-METHODS

In this section we construct a 4-th order BRK-method and give the coefficients of a 5-th order PRK-method. We intend to give a feeling of the complexity of the order conditions and to show how they can be solved.

3.1 An explicit BRK-method of order 4

In order that $Y_i^{(n)}$ can be calculated explicitely from (1.6) we assume $a_{ij} = 0$ for $i \leqq j$. Since the parameters of a BRK-method satisfy in addition (1.7) and (1.9) there remain $m(m+3)/2$ parameters a_{ij}, b_i, e_i. For a 4-th order BRK-method these parameters must be a solution of the 21 equations (2.1) - (2.3). It is known (see [3]) that this system is contradictory for m=4. We therefore try to solve these order condition with m=5.

Since the right hand sides of the order conditions (2.1) and (2.2) do not depend on α, the addition of

$$b_i(1-e_i) = 0 \qquad i = 1,\ldots,m \tag{3.1}$$

causes that the equations with $\alpha > 0$ need no longer be considered. In fact one can prove that in our case condition (3.1) is also necessary.

Similar as for ordinary Runge-Kutta methods we next propose to consider the simplifying assumptions

$$\sum_{j=1}^{i-1} a_{ij} c_j = c_i^2/2 \tag{3.2a}$$

for i=3,4,5.

$$\sum_{j=1}^{i-1} a_{ij} e_j = c_i^2 \tag{3.2b}$$

Observe that (3.2a) cannot be satisfied for i=2, otherwise all c_i would be zero. Some trivial considerations on the system

(2.1) - (2.3) show that the addition of (3.2) implies

$$b_2 = 0 \, , \quad \sum_{i=1}^{m} b_i a_{i2} = 0 \, . \tag{3.3}$$

On the other hand (3.2) together with (3.3) effect that all order conditions with $\gamma+\delta = 1$ can be thrown away.

Besides the order conditions

$$\sum_{i=1}^{m} b_i c_i^{k-1} = 1/k \quad \text{for} \quad k = 1,2,3,4 \tag{3.4}$$

which indicate that the underlying quadrature formula is of order 4 only three equations remain, namely (2.2) with $\alpha=\beta = 0$, $\gamma+\delta = 2$. In order to manage these last conditions we need

Lemma 2. Under the assumption that (1.7) and (3.1) - (3.4) hold the three conditions (2.2) ($\alpha=\beta= 0$, $\gamma+\delta= 2$) are equivalent to

$$(e_4-e_3)c_3(1-2c_3)=(c_4-c_3)[c_3c_4(4-6e_1)+(e_3-e_1)(1-2c_3-2c_4)] \tag{3.5a}$$

$$(e_4-e_3)(2-4c_3)=(c_4-c_3)(3-4e_3) \tag{3.5b}$$

$$(e_4-e_3)[3-4e_1-4c_3(2-3e_1)]=(c_4-c_3)[6-8e_1-4e_3(2-3e_1)]. \tag{3.5c}$$

Proof. Observe first that the assumptions together with the three conditions among (2.2) imply

$$\begin{pmatrix} c_3 & c_4 \\ e_3 - e_1 & e_4 - e_1 \\ c_3^2 & c_4^2 \\ c_3 e_3 & c_4 e_4 \\ e_3(e_3-e_1) & e_4(e_4-e_1) \end{pmatrix} \begin{pmatrix} \sum b_i a_{i3} \\ \sum b_i a_{i4} \end{pmatrix} = \begin{pmatrix} 1/6 \\ 1/3 - e_1/2 \\ 1/12 \\ 1/8 \\ 1/4 - e_1/4 \end{pmatrix}$$

The necessity of (3.5) now follows from the solvability of this overdetermined system. To prove sufficiency is more tedious and therefore omitted.

Summarizing we can state

Proposition 3. Any solution of the reduced system, consisting of (1.7) and (3.1)-(3.5) is also a solution of the 21 order conditions (2.1)-(2.3).

Next the reduced system has to be solved. Let

$$b_1 = 0,\ b_2 = 0,\ b_3 = 1/2,\ b_4 = 0,\ b_5 = 1/2,$$

$$c_1 = 0,\ c_3 = (3-\sqrt{3})/6,\ c_5 = (3+\sqrt{3})/6$$

so that the BRK-method is based on the Gaussian quadrature. Because of (3.1) this implies

$$e_3 = 1,\quad e_5 = 1.$$

Condition (3.5) then constitutes of 3 nonlinear equations in the 3 unknowns e_1, c_4, e_4, the solution of which is

$$e_1 = (3-\sqrt{3})/4,\ c_4 = (9+2\sqrt{3})/23,\ e_4 = (57+5\sqrt{3})/92.$$

(1.7) and (3.2) for i=3 are three linear equations for a_{31} and a_{32}. This system can be solved if

$$e_2 = e_1 - c_2.$$

Now $c_2=c$ can be taken as free parameter and the a_{ij} can easily be computed from (1.7), (3.2) and (3.3) One gets

$$a_{21}=c,\ a_{32}=(2-\sqrt{3})/(12c),\ a_{31}=(3-\sqrt{3})/6-a_{32},$$

$$a_{42}=(2544-807\sqrt{3})/(13754c),\ a_{43}=(-90+1245\sqrt{3})/6877,$$

$$a_{41}=(2781-647\sqrt{3})/6877-a_{42},\ a_{52}=(-2+\sqrt{3})/(12c),$$

$$a_{51}=(-3+2\sqrt{3})/9-a_{52},\ a_{53}=1/5,\ a_{54}=(57-5\sqrt{3})/90.$$

In order that the derived BRK-method satisfies the *kernel condition*, i.e. all arguments of K lie in $\{(x,s);\ a\leq s\leq x\leq b\}\times\mathbb{R} = S\times\mathbb{R}$, the parameter c has to be restricted to the interval $0<c\leq(3-\sqrt{3})/8$.

3.2 *An explicit PRK-method of order 5*

We give here the coefficients of a 5-th order PRK-method (m=7). It is based on the 6-th order Gaussian quadrature formula , which is an advantage for the computation of (1.11) (see also section 4).

$$b_1=b_2=b_4=b_6= 0,\ b_3=b_7= 5/18,\ b_5= 4/9,$$

$$c_1= 0,\ c_2=c_3=(5-\sqrt{15})/10,\ c_4=c_5= 1/2,\ c_6=c_7= (5+\sqrt{15})/10,$$

$$a_{21}=c_2,\ a_{31}=c_2/2,\ a_{32}= c_2/2,\ a_{41}= -(1+\sqrt{15})/8$$

$$a_{42}=u,\ a_{43}=(5+\sqrt{15})/8-u,\ a_{51}=-1/8,$$

$a_{52}=-(5-\sqrt{15})/16$, $a_{53}=(5+\sqrt{15})/16$, $a_{54}=(5-\sqrt{15})/8$,

$a_{61}=(11+\sqrt{15})/20$, $a_{62}=c_2/2$, $a_{63}=c_2-1$,

$a_{64}=(1+\sqrt{15})/5-v$, $a_{65}=v$, $a_{71}=-(1-\sqrt{15})/20$,

$a_{72}=c_2/2$, $a_{73}=0$, $a_{74}=-2c_2$, $a_{75}=4/5$, $a_{76}=c_2$.

Here u and v are free parameters. A reasonable choice, implying small truncation errors, is

$$u=(145-11\sqrt{15})/904, \quad v=(84+20\sqrt{15})/75.$$

4. SOME IMPLEMENTATIONAL REMARKS

The VRK-methods presented in section 3 are both based on *Gaussian quadrature*, so that the number of nonzero b_i is as small as possible. This implies that the evaluation of $\tilde{F}_n(x)$ given by (1.11) is not too expensive.

In Fig. 1 we have drawn the points of S where the kernel has to be evaluated (crosses and circles). One observes a certain lack of symmetry and one would like to evaluate the kernel only at the points marked with a cross (x).

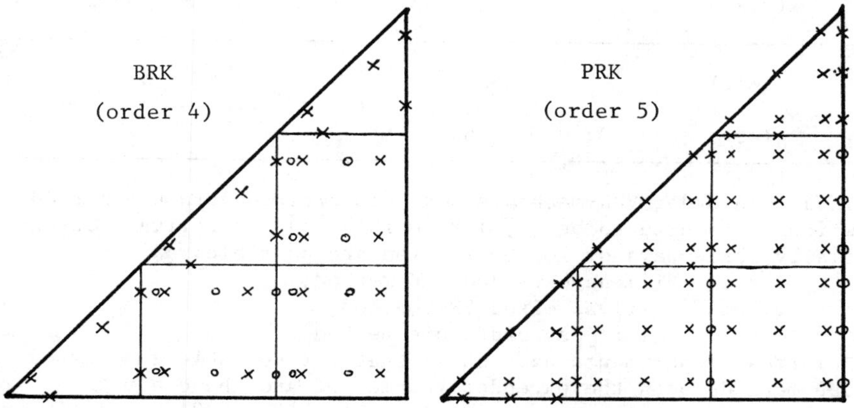

FIG. 1. Kernel evaluations

A way how this can be achieved is to replace $\tilde{F}_n(x)$ locally by the polynomial of degree p-1 that interpolates $\tilde{F}_n(x)$ at

$\xi_1^{(n)},\ldots,\xi_p^{(n)}\in M$. For the two methods above one can take

$$M_{BRK}=\{x_{n-1}+c_5h,\ x_n+c_3h,\ x_n+c_5h,\ x_{n+1}+c_3h\},$$

$$M_{PRK}=\{x_n+c_2h,\ x_n+c_4h,\ x_n+c_6h,\ x_{n+1}+c_2h,\ x_{n+1}+c_4h\}.$$

Using known techniques one can prove that this process does not affect the convergence results of section 2.

Finally we consider a class of integral equations which often appears in applications, the *Hammerstein integral equation of convolution type*

$$y(x) = f(x) + \int_a^x k(x-s)g(s,y(s))ds. \qquad (4.1)$$

Since the function k is constant on lines parallel to the diagonal x=s and g is constant on horizontal lines, the number of kernel evaluations may be reduced considerably. For the methods of section 3 this reduction can be seen in Table 1 (N denotes the number of steps).

TABLE 1

Amount of work for the methods of section 3

Evaluations of	BRK-order 4	PRK-order 5
$K(x,s,y)$	$2N^2 + 3N$	$(9N^2 + 21N)/2$
$k(t)$	$3N + 2$	$5N + 1$
$g(s,y)$	$5N$	$7N$

If a p-th order VRK-method and a p-th order quadrature method - which is assumed to be (ρ,σ)-reducible [9] - are given, essentially three ways of implementation are possible:

(i) as extended VRK-method;
(ii) as mixed VRK-method;
(iii) as quadrature method.

Numerical experiments have shown that for suitable choices of the methods both the extended VRK-method and the quadrature method are equally competitive (see also [9], p.91). The higher amount of work for extended methods is compensated by the larger stepsize that can be taken in order to achieve the same accuracy.

However, the mixed VRK-method is in general less competitive than the corresponding quadrature method. This can be seen as follows: Application to the simple integral equation

$$y(x) = y_o + \lambda \int_o^x y(s)ds \tag{4.2}$$

yields in both cases a linear recurrence relation. For the quadrature method its characteristic polynomial is

$$\rho(\zeta) - h\lambda\sigma(\zeta), \tag{4.3}$$

and for the mixed VRK-method it is [1]

$$\zeta\rho(\zeta) - h\lambda R(h\lambda)\sigma(\zeta). \tag{4.4}$$

Here $R(z)$ denotes the stability function of the underlying ordinary RK-method. A simple analysis shows that the essential roots of (4.3) and (4.4) satisfy

$$\zeta_1(h\lambda) = \exp(h\lambda) + C(h\lambda)^{p+1} + O((h\lambda)^{p+2})$$

with the *same* error constant in each case. For a fixed stepsize the mixed VRK-method yields thus only the same accuracy as the quadrature method but needs much more kernel evaluations.

5. STABILITY CONSIDERATIONS

Consider first the basic test equaiton

$$y(x) + y_o + \lambda \int_o^x y(s)ds \tag{5.1}$$

which is equivalent to the initial value problem $y'=\lambda y$, $y(0)=y_o$. Application of an extended VRK-method yields

$$y_{n+1} = R(h\lambda)y_n$$

where $R(z)$ is the stability function of the underlying ordinary RK-method (this is not the case for mixed VRK-methods). The stability properties of the ordinary RK-methods (e.g. A-stability, ... etc.) thus carry directly over to the test equation (5.1).

Already more intructive is the integral equation

$$y(x) = f(x) + \lambda \int_o^x e^{-\alpha(x-s)} y(s)ds, \quad \text{Re}\lambda<0, \alpha>0 \tag{5.2}$$

which satisfies for $x\to\infty$

$$y(x) \to 0 \qquad \text{if} \qquad f(x) \to 0$$

A nice property of A-stable extended PRK-methods is that the numerical solution also satisfies

$$y_n \to 0 \quad \text{if} \quad f(x) \to 0.$$

The reason is the following: With the notation $z(x)=e^{\alpha x}y(x)$, $g(x)=e^{\alpha x}f(x)$ the integral equation (5.2) is equivalent to

$$z(x) = g(x) + \lambda \int_0^x z(s)ds. \tag{5.3}$$

Extended PRK-methods now satisfy

$$z_n = e^{\alpha x_n} y_n$$

where y_n and z_n are the numerical solutions of (5.2) and (5.3) respectively. Thus the stability properties for the basic test equation carry over to the integral equation (5.2). For the γ-modified mixed PRK-methods (see [8]) this is not true if γ is chosen independently of (1.1).

At present the stability behaviour of extended PRK-methods for general convolution equations

$$y(x) = f(x) + \lambda \int_0^x k(x-s)\, y(s)ds$$

with Re $\lambda<0$ and k positive definite, is under investigation [4]. It is the aim to extend the results of [6], which are proved there for multistep methods, to one-step methods.

REFERENCES

1. BAKER, C.T.H. & WILKINSON, J.C., Stability Analysis of Runge-Kutta Methods Applied to a Basic Volterra Integral Equation. *J.Austral.Math.Soc.(Series B)* 22, 515-538 (1981).

2. BEL'TYUKOV, B.A., An Analogue of the Runge-Kutta Method for the Solution of Nonlinear Integral Equations of Volterra Type. *Differential Equations* 1, 417-433 (1965).

3. BRUNNER, H., HAIRER, E. & NØRSETT, S.P., Runge-Kutta Theory for Volterra Integral Equations of the Second Kind. *Math. of Comput.* (1982)

4. HAIRER,E. & LUBICH, Ch., On the Stability of Volterra-Runge-Kutta Methods. In preparation

5. HAIRER, E., LUBICH, Ch. & NØRSETT, S.P., Order of Convergence of One-Step Methods for Volterra Integral Equations of the Second Kind. *SIAM J. Num. Analysis* (1982)

6. LUBICH, Ch. On the stability of linear multistep methods for Volterra integral equations of the second kind. These proceedings.

7. POUZET, P., Etude en Vue de Leur Traitement Numérique des Équations Intégrales de Type Volterra. *Rev. Francais Traitement Information (Chiffres)* 6, 79-112 (1963).

8. VAN DER HOUWEN, P.J., WOLKENFELT, P.H.M. & BAKER, C.T.H., Convergence and Stability Analysis for Modified Runge-Kutta methods in the Numerical Treatment of Second-Kind Volterra Integral Equations. *IMA J. Num. Anal.* 1, 303-328 (1981)

9. WOLKENFELD, P.H.M., The Numerical Analysis of Reducible Quadrature Methods for Volterra Integral and Integro-Differential Equations. *Ph.D.Thesis*, Mathematisch Centrum, Amsterdam (1981)

ON THE STABILITY OF LINEAR MULTISTEP METHODS FOR VOLTERRA INTEGRAL EQUATIONS OF THE SECOND KIND

Ch. Lubich
Universität Heidelberg

1. INTRODUCTION

When solving a Volterra integral equation

$$y(x) = f(x) + \int_0^x K(x,s,y(s))ds \qquad (x \geq 0) \tag{1.1}$$

numerically the question arises as to whether the qualitative behaviour of the discretized equation reflects that of the original problem. In particular, will the numerical solution converge to zero or remain bounded if the exact solution does ? Naturally, a positive answer over the whole class of integral equations (1.1) cannot be expected. One therefore tries to investigate the indicated stability problem for special classes of test equations which should, of course, be as near to applications as possible. Much work in this field has been done for the "*basic test equation*"

$$y(x) = y_o + \lambda \int_0^x y(s)ds \qquad (\mathrm{Re}\ \lambda < 0), \tag{1.2}$$

see e.g. Baker and Keech [1]. Note that (1.2) reduces to the ordinary differential equation

$$y' = \lambda y \quad , \quad y(0) = y_o \ .$$

Other equations (see for example van der Houwen [9]) have also been proposed. So far test equations with non-separable kernels have been considered only in Nevanlinna [6],[7]. The author gratefully appreciates the influence of these two papers on the present work. Further references can be found in the survey article of Brunner [2].

In applications one often encounters convolution equations

$$y(x) = f(x) + \int_0^x a(x-s)g(s,y(s))ds \qquad (x \geq 0) \ . \tag{1.3}$$

In the subsequent stability analysis we shall consider the

linear case where

$$g(s,y) = y \quad .$$

It is well known (see Wolkenfelt [10], Matthys [5]) that a linear multistep method (ρ,σ) applied to (1.3) leads to a discrete convolution equation

$$y_n = f_n + h \sum_{j=0}^{n} \omega_{n-j} a_{n-j} g(x_j, y_j) \qquad (n \geq 0) \; . \tag{1.4}$$

Here $h > 0$ is a fixed step-size, $x_j = x_0 + jh$, y_n is an approximation to $y(x_n)$, $a_n^j = a(nh)$, and $f_n = f(x_n) + h \sum_{j=-k}^{-1} w_{nj} a_{n-j} g(x_j, y_j)$ where $y_{-k},\ldots,y_{-1}$ are given starting values.
The weights ω_{n-j} in (1.4) depend only on (ρ,σ).

In section 2 we discuss the asymptotic stability of discrete linear convolution equations. In Section 3 we investigate if and to what extent the classical stability concepts for linear multistep methods, which are related to the "basic test equation" (1.2) only, carry over to more general situations.

Proofs and extensions of the results are given in [4].

2. A THEOREM OF PALEY AND WIENER

In [8], p.59 Paley and Wiener have given the following result on the asymptotic stability of Volterra integral equations of convolution type:

Consider the linear Volterra equation

$$y(x) = f(x) + \int_0^x a(x-s) y(s) ds \qquad (x \geq 0)$$

where the kernel $a(x)$ belongs to $L^1(0,\infty)$. Then we have

$$y(x) \to 0 \quad \text{whenever} \quad f(x) \to 0 \quad (x \to \infty)$$

if and only if

$$\int_0^\infty e^{-wx} a(x) dx \neq 1 \qquad \text{for } \operatorname{Re} w \geq 0 \; .$$

The discrete version of this theorem, where we have put $\zeta = e^{-w}$, reads

Theorem 2.1. Consider the discrete linear Volterra equation

$$y_n = f_n + \sum_{j=0}^{n} a_{n-j} y_j \qquad (n \geq 0) \tag{2.1}$$

where the kernel $\{a_n\}_0^\infty$ belongs to ℓ^1. Then we have

$$y_n \to 0 \quad \text{whenever} \quad f_n \to 0 \quad (n \to \infty) \tag{2.2}$$

if and only if

$$\sum_{n=0}^{\infty} a_n \zeta^n \neq 1 \qquad \text{for} \quad |\zeta| \leq 1 . \tag{2.3}$$

The assumption on the summability of the kernel is weakened in

Corollary 2.2. Consider the discrete Volterra equation (2.1) where $\{a_n - a_\infty\}_0^\infty$ belongs to ℓ^1 for some $a_\infty \neq 0$. Then

$$y_n \to 0 \quad \text{whenever} \quad \Delta f_n \to 0 \quad (n \to \infty)$$

if and only if (2.3) is valid.

There is also a boundedness result.

Corollary 2.3. Consider the discrete Volterra equation (2.1) where $\{a_n - a_\infty\}_0^\infty$ belongs to ℓ^1 for some $a_\infty \in \mathbb{C}$. Then

y_n is bounded whenever f_n is bounded

if and only if (2.3) is valid.

For some kernels the Paley-Wiener condition (2.3) is obviously satisfied:

i) $\Sigma_0^\infty a_n \zeta^n$ lies in a half-plane that does not contain the positive real axis,

ii) $\Sigma_0^\infty a_n \zeta^n$ is contained in a small disk near the origin, and

iii) sums of both types *i)* and *ii)*.

This observation is the key for the results of the following section.

3. STABILITY RESULTS FOR LINEAR MULTISTEP METHODS

We investigate the asymptotic behaviour of the numerical solutions obtained by applying linear multistep methods with a fixed step-size h to the following class of Volterra integral eqations:

$$y(x) = f(x) + \lambda \int_0^x a(x-s)y(s)ds \qquad (x \geq 0) \tag{3.1}$$

where $\mathrm{Re}\,\lambda < 0$ and the continuous kernel $a(x) \in L^1(0,\infty)$ with $\{a(nh)\}_0^\infty \in \ell^1$ is

(PD) *positive definite* ,

that is (see, e.g., [3], p.137) ,

$$\sum_{j,k=1}^{n} a(\xi_j-\xi_k)z_j\bar{z}_k \geq 0 \quad \text{for any choice of } \xi_1,\ldots,\xi_n \text{ and}$$

complex numbers $z_1,\ldots,z_n$. Here $a(x)$ is extended to the negative axis by $a(-x) = \overline{a(x)}$.

This is equivalent to

$$\mathrm{Re} \int_0^\infty e^{-wx} a(x)dx \geq 0 \quad \text{for } \mathrm{Re}\, w \geq 0 .$$

It is easily seen that $a(0) > 0$ if $a(x)$ is not identically zero. Therefore we may assume $a(0) = 1$ without restricting generality.

For example, the following functions are positive definite: any convex, non-increasing, non-negative function, $\cos x$, $\exp(-x^2)$. Obviously linear combinations with positive coefficients preserve positive definiteness, and also the pointwise product of positive definite functions is again positive definite.

By the Paley-Wiener theorem, positive definite L^1-functions form the largest class of L^1-kernels for which the solution of (3.1) satisfies

$y(x) \to 0$ (is bounded) whenever $f(x) \to 0$ (is bounded,resp.) for any λ with $\mathrm{Re}\,\lambda < 0$.

Before we turn to see if this property can be obtained also for the discretized equation, recall the following:

The *stability region* D of a linear multistep method (ρ,σ) is the set of all $z = h\lambda$ for which the numerical solution y_n of the basic test equation (1.2) tends to zero as $n \to \infty$.

The method is called *A-stable* if the left half-plane $\mathrm{Re}\, z < 0$

is contained in D.

Theorem 3.1. Apply an A-stable linear multistep method to an integral equation (PD). Then the numerical solution satisfies

$y_n \to 0$ (is bounded) whenever $f(x_n) \to 0$ (is bounded, resp.).

A linear multistep method is called *strongly stable* if a disc in the left half-plane, touching the origin, is contained in D.

Theorem 3.2. Apply a strongly stable linear multistep method to an integral equation (PD). Assume $a(0) = 1$. Then the numerical solution satisfies

$y_n \to 0$ (is bounded) whenever $f(x_n) \to 0$ (is bounded, resp.)

and $h\lambda$ is contained in the stability disc of the method.

Note, however, that the class (PD) does not preserve the stability region D.

A linear multistep method is called *A(α)-stable* if D contains the sector $|\arg z - \pi| < \alpha$.

A(α)-stability is not preserved within the class (PD). However, in general A(α)-stability does carry over to the restricted class of integral equations (3.1) where the continuous L^1-kernel $a(x)$ is

(CM) *completely monotonic* ,

i.e., $(-1)^k . a^{(k)}(x) \geq 0$ for $k = 0,1,2,\ldots$ and $x \geq 0$.

Theorem 3.3. Apply an A(α)-stable method to an integral equation (CM). Assume that the method satisfies a certain non-pathology condition. Then the numerical solution satisfies

$y_n \to 0$ (is bounded) whenever $f(x_n) \to 0$ (is bounded, resp.)

and $|\arg \lambda - \pi| < \alpha$.

ACKNOWLEDGEMENT. The author wishes to thank E. Hairer for helpful discussions.

REFERENCES

1. BAKER, C.T.H. and KEECH, M.S., Stability regions in the numerical treatment of Volterra integral equations. SIAM J. Numer. Anal. 15, 394-417 (1978).
2. BRUNNER, H., A survey of recent advances in the numerical treatment of Volterra integral and integro-differential equations. To appear in J. Comput. Appl. Math. (1982).
3. KATZNELSON, Y., *An introduction to harmonic analysis.* Wiley Inc., New York (1968).

4. LUBICH, Ch., On the stability of linear multistep methods for Volterra equations. Doctoral thesis, Univ. Innsbruck (in preparation).
5. MATTHYS, J., A-stable linear multistep methods for Volterra integro-differential equations. Numer. Math. 27, 85-94 (1976).
6. NEVANLINNA, O., Positive quadratures for Volterra equations. Computing 16, 349-357 (1976).
7. NEVANLINNA, O., On the numerical solution of some Volterra equations on infinite intervals. Report Inst. Mittag-Leffler, Djursholm (1976).
8. PALEY, R.E.A.C. and WIENER, N., *Fourier Transforms in the Complex Domain*. Amer. Math. Soc., Providence, R.I. (1934).
9. VAN DER HOUWEN, P.J., On the numerical solution of Volterra integral equations of the second kind. Report NW 42/77, Mathematisch Centrum, Amsterdam (1977).
10. WOLKENFELT, P.H.M., Linear multistep methods and the construction of quadrature formulae for Volterra integral and integro-differential equations. Report NW 76/79, Mathematisch Centrum, Amsterdam (1979).

BOUNDARY INTEGRAL EQUATION METHODS FOR A VARIETY OF CURVED CRACK PROBLEMS

J.C. Mason and R.N.L. Smith

Royal Military College of Science, Shrivenham.

1. INTRODUCTION

Different mathematical formulations of crack problems in linear elasticity have led to a number of distinct types of numerical methods. These include boundary collocation methods based on a complex variable formulation, finite element methods based on a partial differential equation formulation, and boundary element methods based on an integral equation formulation (and sometimes called boundary integral equation or BIE methods).

The boundary collocation and BIE methods share the advantage of effectively reducing the dimension of the problem (e.g. solving a one-dimensional boundary problem in a two-dimensional region), but unlike the finite element method they lead to full systems of linear equations which can sometimes be ill-conditioned. However, this does not seem to present too much difficulty in practice. Of the two methods we believe the BIE method to be the more versatile and robust.

There are three classes of BIE methods: indirect, semi-direct, and direct. The indirect method solves an integral equation for a "density function" over the boundary and hence evaluates stresses and displacements (see Watson [11] for example). The semi-direct method solves integral equations for certain unknown functions related directly to stress functions (see Jaswon [6] for example). The direct method determines and solves an integral relationship between the known and unknown tractions and displacements on the boundary (see Cruse [2] and Lachat and Watson [7]. One disadvantage of the indirect and semi-direct methods is that they give poor resolutions of stresses near the surface (see Van Buren [10], which is the region of major importance in fracture mechanics. The direct method is therefore commonly used for crack problems, and we adopt it in the present work.

2. INTEGRAL EQUATION FORMULATION

The formulation of the direct method, which is summarised below, follows the discussion and notation of Lachat and Watson [7] and is restricted to homogeneous linearly elastic regions in the absence of body forces. As is customary, vectors such as the points x and y are not underlined, and tensor notation is adopted.

Suppose that $u_i(x)$ and $t_i(x)$ are displacement and traction fields at x in a region R with boundary S enclosing a volume V. Then

$$t_i(x) = \sigma_{ij}\,[u]\;.\;n_j(x)$$

where $\sigma_{ij}\,[u]$ is the stress field and $n_j(x)$ is the outward normal. To start with, elementary solutions $U_{ij}(x,y)$ and $T_{ij}(x,y)$ for displacements and tractions, respectively, are obtained from the governing equations for a point force in an infinite region. (See [1] for formulae.) Now, by the divergence theorem for continuously differentiable functions, it follows that

$$\int_V \left\{ b_j(y)\,\frac{\partial\sigma_{js}}{\partial y_s}\,[a] - a_j(y)\,\frac{\partial\sigma_{js}}{\partial y_s}\,[b] \right\} dV_y$$

$$= \int_S \left\{ b_j(y)\,\sigma_{js}\,[a] - a_j(y)\,\sigma_{js}\,[b] \right\} n_s(y)\,dS_y \qquad (2.1)$$

which is known as Betti's equation and is analogous for the biharmonic operator to the symmetric Green's identity for the Laplace operator.

Replacing $a_j(y)$ by $u_j(y)$ and $b_j(y)$ by $U_{ij}(x,y)$ (for x on S), noting that for a source free region $\frac{\partial\sigma_{ij}}{\partial x_j} = 0$, and taking for V the Region $R' = R - R_\varepsilon$, where R_ε isolates the point x, we obtain

$$\int_{S'} \left\{ U_{ij}(x,\,y)\,t_j(y) - T_{ij}(x,y)\,u_j(y) \right\} dS'_y = 0. \quad (2.2)$$

Here R_ε is conventionally taken to be the part of R contained within a sphere of radius ε, centre x (on S), and hence by letting $\varepsilon \to 0$ (see [2])

$$c_{ij}(x)\ u_j(x) + \int_S T_{ij}(x,y)\ u_j(y)\ dS_y$$

$$= \int_S U_{ij}(x,y)\ t_j(y)\ dS_y \tag{2.3}$$

where $c_{ij} = \lim_{\varepsilon\to 0} \int_{S_\varepsilon} T_{ij}(x,y)\ dS_y$. (2.4)

If the tangent plane is continuous at x then

$$c_{ij} = \tfrac{1}{2}\ \delta_{ij}$$

Thus (2.3) expresses $u_j(x)$ at one point x on S in terms of values of $u_j(y)$ and $t_j(y)$ at all points y of S.

The value of $u_i(x)$ at an interior point x may be obtained from Betti's theorem[1] by a similar limiting process in the form

$$u_i(x) = \int_S U_{ij}(x,y)\ t_j(y)\ dS_y - \int_S T_{ij}(x,y)\ u_j(y)\ dS_y. \tag{2.5}$$

A corresponding expression to (2.5) may also be obtained for the stress tensor at an interior point.

Equation (2.3) is similar in form to the Fredholm equations for boundary value problems of the first and second kinds. Although the left hand side includes a term in r^{-1} in 2 dimensions (or r^{-2} in 3 dimensions) the equation is regular (for admissible values of Poisson's ratio) and the index of the kernel is zero (see [8]). Hence the Fredholm alternatives apply, and all normal problems are solvable.

3. INCORPORATION OF CRACKS

There has been significant progress over recent years in extending the accuracy and applicability of the BIE method, especially for crack problems.

In early work it was noted that the standard direct method produced a singular algebraic system for an idealised crack (where both surfaces lie in the same plane), and so crack geometries had to be approximated by separating the surfaces by some small distance. This crude approach led to results of rather limited accuracy.

However, if the problem happens to be symmetric about the crack, then only half the region and hence only one side of the crack needs to be considered, and the resulting linear algebraic system may be solved successfully (Cruse [3]).

In the present paper we adopt the recent and more general technique of "stitching", which permits the two sides of the crack to be assigned to distinct subregions. This appears to overcome problems encountered by other methods and has been used with some success on various problems by Blandford et al [1], although these have all up to now involved straight cracks.

4. BOUNDARY ELEMENT IMPLEMENTATION

Applying the stitching method, the body is divided into two subregions (1) and (2), joined by a line I which includes all crack surfaces. Tractions and displacements must then satisfy the continuity conditions

$$u_I = u^{(1)} = u^{(2)}, \quad t_I = t^{(1)} = - t^{(2)}.$$

The boundary is represented by isoparametric quadratic elements, as also are tractions and displacements, but special crack-tip elements are also needed (see §5). At each point of the boundary, including crack surfaces, one of the variables t and u is specified while the other is unknown. By combining sub-regions, and taking u_I and t_I as unknowns, the integral equation (2.3) is hence modelled as a linear algebraic system for determining particular values of u and t. Four practical aspects of the solution are now discussed.

4.1 Cauchy Principal Values

The quantities c_{ij} defined as limits of integrals in (2.4) represent Cauchy principal values and are difficult to calculate for non-smooth boundaries. However, they do not, in fact, need to be calculated as we now show.

The discretisation and approximation of the boundary integral-equation (2.3) results in a linear system for $\underline{u}$ of the form

$$H\underline{u} = G\underline{t},$$

where the diagonal elements of H include the Cauchy principal value integral c_{ij}. If 1_k is a vector specifying a unit rigid body displacement with zero tractions on the boundary then

$H1_k = 0$ and hence $\sum_j h_{ij} = 0$.

The diagonal terms of H may thus be determined from the formulae

$$h_{ii} = -\sum_{j \neq i} h_{ij}$$

and do not need to be calculated by using (2.4).

4.2 *Weakly Singular Integrals*

The second possible complication which occurs is that some of the element integrals, based on (2.3), include a log r singularity. Denoting the element parameter by ξ $(0 \leqq \xi \leqq 1)$, the quadratic interpolation function by $\phi_i(\xi)$, and the boundary Jacobian by $G(\xi)$, such integrals take the form

$$I = \int_0^1 \log r(\xi) \,.\, \phi_i(\xi)\, G(\xi)\, d\xi \qquad (4.1)$$

There are 3 obvious ways of treating this integral. Firstly, the singularity is weak and might be ignored. However, only fair results are obtained by using Standard Gauss quadrature. Secondly, since r vanishes at $\xi = 0$, we might write

$$I = \int_0^1 \log \xi \; [(\log r/\log \xi)\; \phi_i(\xi)\; G(\xi)]\; d\xi$$

and use Gauss quadrature with weight $\log \xi$. However this method will only give accurate results if $(\log r/\log \xi)$ is non-singular. Finally, the singularity can be additively removed, since

$$r = \xi\sqrt{a\xi^2 + b\xi + c}.$$

Thus $I = I_1 + I_2$, where

$$I_1 = \int_0^1 \log \xi \,.\, \phi_i \,.\, G\, d\xi \text{ and}$$

$$I_2 = \tfrac{1}{2}\int_0^1 \log\,(a\xi^2 + b\xi + c) \,.\, \phi_i \,.\, G\, d\xi \qquad (4.2)$$

In our implementation (4.2) was used, and I_1 and I_2 were calculated by 6 point Gauss quadrature with weights log ξ and 1, respectively, and accurate results were obtained. Note that, on any straight line elements, r is $k\xi$ (k constant), G is constant, and so the integral is obtained exactly by Gauss quadrature with 2 or more points.

4.3 *Singular Algebraic Systems*

It is nontrivial to solve the linear algebraic system for displacements and tractions, when all boundary conditions are traction conditions. Indeed the system is singular, since it embodies an arbitrary rigid body motion. In the case of non-symmetric two-dimensional problems, this normally means that 3 equations are superfluous.

An obvious tactic is to fix 3 of the unknowns (e.g. 3 zero displacements), delete 3 apparently superfluous equations, and solve the remainder. This approach was found to be successful for simple and well-conditioned problems, namely for long straight cracks, but it is not to be recommended for most problems. Indeed it was found that for somewhat ill-conditioned problems, the solution varied considerably according to which constraint equations were chosen.

A better approach in practice is to add 3 extra equations to eliminate rigid body motions and solve the resulting over-determined system by least squares. This not only avoids exacerbating ill-conditioning, but also eliminates the need to decide which equations to delete. It is thus suitable as an automatic procedure.

4.4 *Error Estimation*

Rigorous error analysis is particularly difficult in the BIE method, and indeed there is very little established theory. (However, the reader is referred to the work of Wendland [12], who gives convergence results for BIE techniques based on Galerkin methods.) It does not even appear to be known in general how the error behaves as a function of h (some measure of element width), although the results in the present paper would appear to indicate that the overall error in the stress intensity factor in our method behaves like h^p where $p \cong 1$.

There are essentially three sources of truncation error to be taken into account: the error in evaluating integrals, the error in quadratic approximations to boundaries, stresses, and displacements, and the error in the crack-tip element approximation. We do not attempt any rigorous analysis here, but observe that we have found in practice that the choice of 6 point quadrature formulae, quadratic elements, and traction-singular quarter point elements provides a compatible balance of accuracies.

It is nevertheless important to estimate errors, and this is done here by comparisons of results over different element widths.

5. CRACK-TIP AND STRESS INTENSITY FACTOR MODELLING

The stresses around a straight crack tip are known to behave as $r^{-\frac{1}{2}}$ for r small, while displacements behave as $r^{\frac{1}{2}}$ (by integrating stress expansions). Thus a change of variable from r to $r^{\frac{1}{2}}$ is suggested, and Henshell and Shaw [5] have shown that this can effectively be achieved by shifting the mid point node of each 3-node element $0 \leqq \xi \leqq 1$ to the "quarter point" ($\xi = \frac{1}{2}$, $r = \frac{1}{4}$). (See Figure 1).

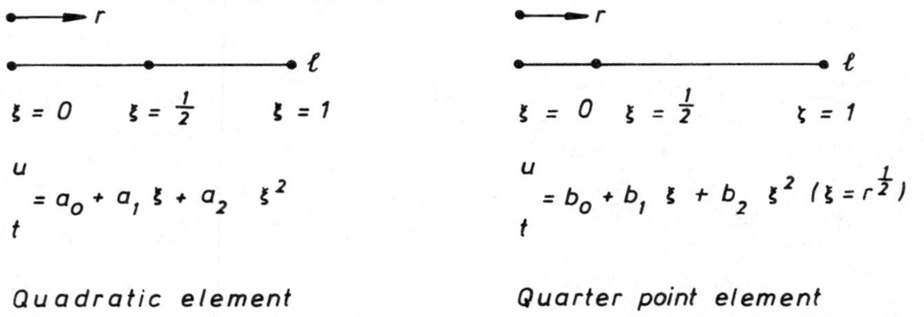

FIG. 1. Boundary Elements

The stress (traction) variation of $r^{-\frac{1}{2}}$ is achieved by simply multiplying the crack tip element coefficients by $(\ell/r)^{\frac{1}{2}}$, where ℓ is the element length (Cruse and Wilson [4]):

$$t = (b_0 + b_1 r^{\frac{1}{2}} + b_2 r)(\ell/r)^{\frac{1}{2}} = c_0 r^{-\frac{1}{2}} + c_1 + c_2 r^{\frac{1}{2}}.$$

For any curved crack, the crack tip still approximates a straight crack tip as $r \to 0$, and hence we conjecture that the leading term in the series expansions for displacements is always unchanged. This has been confirmed in the case of the circular arc crack by Panasyuk and Berezhnitskiy (see Savin [9]), and in all cases the following formulae are used for the x, y components of displacement

$$u_x = \frac{1}{8\mu}\left(\frac{2r}{\pi}\right)^{\frac{1}{2}}\Big\{K_I[(2\kappa-1)\cos\frac{\theta}{2} - \cos\frac{3\theta}{2}]$$

$$+ K_{II}[(2\kappa+3)\sin\frac{\theta}{2} + \sin\frac{3\theta}{2}]\Big\} + o(r^{\frac{1}{2}}) \qquad (5.1)$$

$$u_y = \frac{1}{8\mu}\left(\frac{2r}{\pi}\right)^{\frac{1}{2}}\Big\{K_I[(2\kappa+1)\sin\frac{\theta}{2} - \sin\frac{3\theta}{2}]$$

$$- K_{II}[(2\kappa-3)\cos\frac{\theta}{2} + \cos\frac{3\theta}{2}]\Big\} + o(r^{\frac{1}{2}}) \qquad (5.2)$$

where $\mu = E/[2(1+\nu)]$, E = Young's Modulus; ν = Poisson's ratio. Here K_I and K_{II} denote the mode I and II stress intensity factors, and κ is $3-4\nu$ for plane strain and $(3-\nu)/(1+\nu)$ for plane stress. The x, y coordinates are measured, respectively, from the crack tip and the tangent to it, so that the crack itself occurs at $\theta = \pi$. (See Figure 2.) By taking $\theta = \pi$ and r small in (5.1), (5.2), we deduce that on the crack surface adjacent to the tip,

$$u_x = \frac{\kappa+1}{4\mu} K_{II} \left(\frac{2r}{\pi}\right)^{\frac{1}{2}}, \quad u_y = \frac{-(\kappa+1)}{4\mu} K_I \left(\frac{2r}{\pi}\right)^{\frac{1}{2}} \qquad (5.3)$$

and hence that for r small

$$K_I \cong \frac{2\mu}{\kappa+1}\left(\frac{\pi}{2r}\right)^{\frac{1}{2}}(u_{yD} - u_{yB}), \quad K_{II} \cong \frac{2\mu}{\kappa+1}\left(\frac{\pi}{2r}\right)^{\frac{1}{2}}(u_{xD} - u_{xB}) \qquad (5.4)$$

where D, B are adjacent points on distinct crack surfaces.

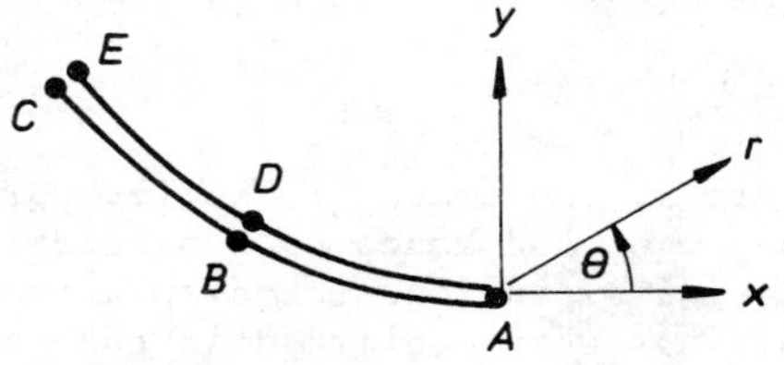

FIG. 2. Crack Tip

Although Blandford et al [1] determine K_I, K_{II} values from formulae based on two pairs of points (e.g. D, B; E, C in Figure 2) on the crack-tip, their approximations are not valid for curved cracks and indeed we have obtained worse results by using them. Moreover the formulae (5.3), (5.4) are certainly valid provided that large values of r are avoided.

6. PARTICULAR PROBLEMS AND NUMERICAL RESULTS

The general method described above was used to solve three particular curved crack problems for square plates under simple tension.

A circular arc crack is shown in Fig. 3 joining up two points A and B inside a square plate CDEF, with artificial boundaries PA and QB introduced to define an obvious subdivision

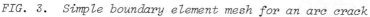

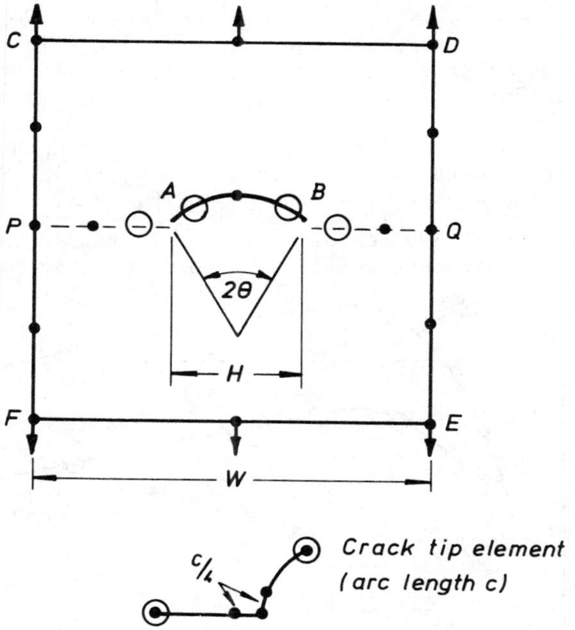

FIG. 3. Simple boundary element mesh for an arc crack

of the plate. A simple boundary element mesh is also shown in Fig. 3, and for each problem the number of crack elements (apart from the crack-tip element) and boundary elements could be increased proportionately to achieve a required accuarcy. Initially

TABLE 1

Computed K values for a Circular Arc Crack

θ (degrees)	K_I			K_{II}		
	Infinite plate	BEM	% error	Infinite plate	BEM	% error
5	5.57	5.49	- 1.5	0.49	0.48	- 0.8
15	5.28	5.18	- 1.8	1.43	1.43	0.4
30	4.36	4.28	- 1.9	2.62	2.65	1.2
45	3.05	2.99	- 2.0	3.41	3.47	1.8
60	1.58	1.54	- 2.5	3.71	3.81	2.6
75	0.17	0.15	- 14.1	3.54	3.66	3.4
90	- 0.99	- 1.01	- 1.6	2.97	3.09	3.8

the ratio H/W, which measures the relative size of the plate, was set equal to 0.1 to permit comparisons with known results for an infinite plate (see [9]).

For an arc of length 2a, an optimum length c of the crack-tip element was determined for the choice $\theta = 30^o$, by comparing errors in K_I and K_{II} obtained with a 24-point mesh with different choices of c/a. These errors, calculated from the infinite plate results, are shown in Figure 4, and it can

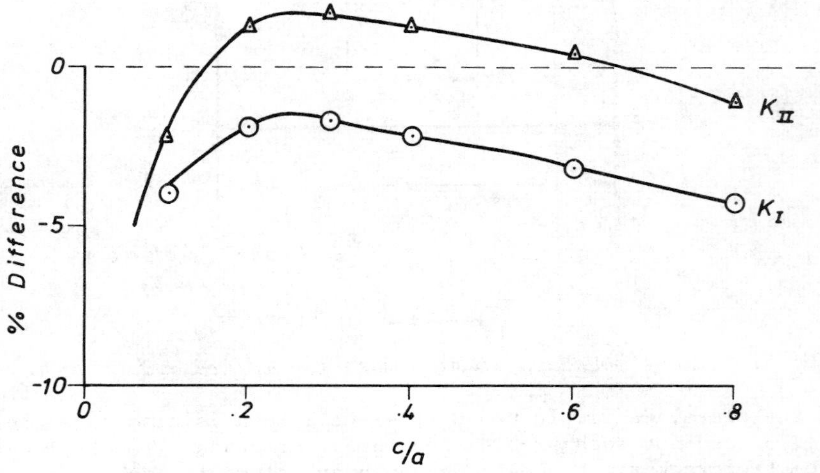

FIG. 4. Percentage difference in K_I and K_{II} for an arc crack ($\theta = 30^o$)

be seen that the average of the errors in K_I and K_{II} is about 3% for all c/a between 0.15 and 0.3. In particular it is clear that c/a = 0.2 is close to an optimal choice. (The program was also tested for a problem involving a double edge-cracked plate, and the choice c/a = 0.2 was again found to be near-optimal.) The value of c, which corresponded to c/a = 0.2, was therefore used as a fixed length for the crack tip element in all three curved crack problems. Numerical values of K_I and K_{II} (for c/a = 0.2) are given in Table 1 for θ up to 90°, together with errors calculated from the infinite plate solutions. It can be seen that no K value is in error by more than 4 per cent of its maximum value (over the range of θ values).

Values of K_I and K_{II} were also calculated for a variety of ratios H/W between 0.1 and 0.5, and in each case these values tended to the corresponding known values of K_I and K_{II} for a straight crack as θ approached zero.

The two other types of curved cracks considered were a parabolic arc (for which quadratic boundary elements are exact) and an S-shaped arc consisting of two circular arcs joined together antisymmetrically. In each case, a square plate of the same dimensions as for the circular arc was adopted, with the distance AB between crack tips fixed as before, and each crack tip making an angle of θ with AB. The values which were obtained for K_I for $\theta \cong 30^o$ for a variety of numbers N of crack elements (and with N elements for every boundary element shown in Figure 3) are shown in Figure 5 for the three types of curved crack. Taking into account the results of Table 1, it would appear that in all three cases the values of K_I are converging like h^p for $p \cong 1$, where h = 1/N is taken as a measure of element width. Indeed by plotting a graph of ln (error) against ln h, based on the "exact" values of K_I for an infinite plate, we estimated a value of 1.1 for p. However, we must point out that the length c of the crack tip element is <u>fixed</u> in Fig. 5, and so h = 1/N does not represent a true measure of element width. In the case of the S-shaped crack, the results for odd N do not quite keep up with the general trend of convergence, but this is to be expected since the quadratic element at the centre of the crack degenerates into a straight line. Similar results were also obtained for K_{II}. As far as the actual accuracies of the results are concerned, it is clear, by comparing the results in Figure 5, with those in Table 1, that the K_I and K_{II} values are unlikely to be in error by more than 2 per cent for the case $\theta = 30^o$ with N = 4.

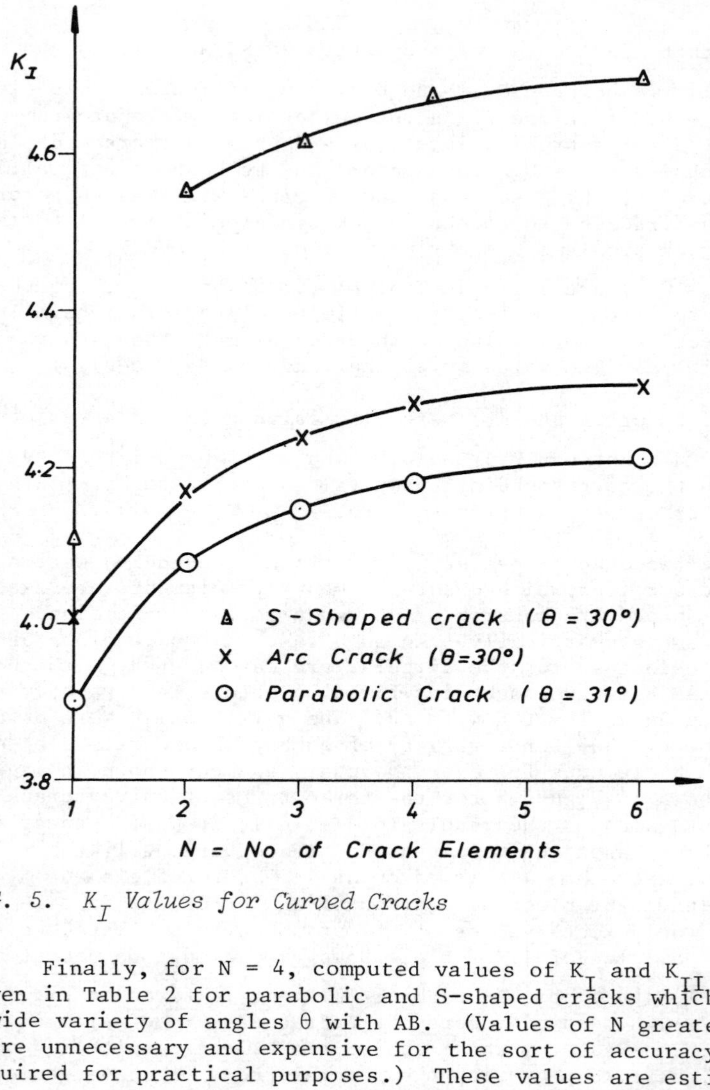

FIG. 5. K_I *Values for Curved Cracks*

Finally, for N = 4, computed values of K_I and K_{II} are given in Table 2 for parabolic and S-shaped cracks which make a wide variety of angles θ with AB. (Values of N greater than 4 are unnecessary and expensive for the sort of accuracy that is required for practical purposes.) These values are estimated to have errors comparable with the corresponding values in Table 1, as a consequence of the similar behaviours of K_I in Figure 5.

TABLE 2

Computed K Values for other Curved Cracks

θ (degrees)	K_I		K_{II}	
	Parabolic crack	S-shaped crack	Parabolic crack	S-shaped crack
5	5.49	5.49	0.48	0.38
15	5.19	5.28	1.43	1.13
30	4.24	4.67	2.66	2.11
45	2.96	3.78	3.50	2.85
60	1.48	2.74	3.76	3.25
75	0.15	1.73	3.26	3.26
90	-	0.88	-	2.91

7. CONCLUSIONS

Results, which are accurate for practical purposes, have been obtained for a variety of curved crack problems by combining and extending a number of modern BIE and finite element techniques and by using a mesh with remarkably few elements.

8. ACKNOWLEDGEMENTS

We should like to thank Mr D P Rooke of the Royal Aircraft Establishment, Farnborough (Materials Department) for his helpful advice.

9. REFERENCES

1. BLANDFORD, G.E., INGRAFFEN, A.R. and LIGGETT, J.A. Two-Dimensional Stress Intensity Factor Computations Using the Boundary Element Method. *Int. J. Num. Meth. Engg.* 17, 387-404 (1981).

2. CRUSE, T.A. Numerical Solutions in Three-Dimensional Elastostatics, *Int. J. Solids Structures* 5, 1259-1274 (1969).

3. CRUSE, T.A. Boundary Integral Equation Fracture Mechanics Analysis. *Proceedings ASME Conference*, Troy, New York (1975), pp 31-46.

4. CRUSE, T.A. and WILSON, R.B. Boundary Integral Equation Method for Elastic Fracture Mechanics. *AFSOR-TR-78-0355* (1977), pp 10-11.

5. HENSHELL, R.D., and SHAW, K.G. Crack Tip Finite Elements are Unnecessary. *Int. J. Num. Meth. Engg.* 9, 495-507 (1975).

6. JASWON, M.A., Integral Equation Methods in Potential Theory - I, *Proc. Roy. Soc.* A275, 23-32 (1963).

7. LACHAT, J.C., and WATSON, J.O. Effective Numerical Treatment of Boundary Integral Equations: A Formulation for Three-Dimensional Elastostatics. *Int. J. Num. Meth. Engg.* 10, 991-1005 (1976).

8. MIKHLIN, S.G. *Multidimensional Singular Integrals and Integral Equations*, Pergamon Press, Oxford (1965).

9. SAVIN, G.N. Stress Distribution Around Holes. *NASATT F-607* (1970) Ch. VIII, pp 638-642.

10. VAN BUREN, W. The Indirect Potential Method for Three-Dimensional Boundary Value Problems of Classical Elastic Equilibrium. *Res. Report 68-ID7-MEKMA-RZ*, Westinghouse Research Labs., Pittsburg (1968).

11. WATSON, J.A. *Analysis of Thick Shells With Holes by Using Integral Equation Methods*. Ph.D. Thesis, University of Southampton (1972).

12. WENDLAND, W. On Galerkin Collocation Methods for Integral Equations of Elliptic Boundary Value Problems. *Numerische Behandlung von Integralgleichungen ISNM 53*, Collatz. L. and Albrecht J. (Eds) Birkhauser Verlag, Basel (1980), pp S244-275.

COLLOCATION METHODS FOR A BOUNDARY INTEGRAL EQUATION ON A WEDGE

*Kendall E. Atkinson and †Frank de Hoog

**University of Iowa,* †*DMS, CSIRO, Canberra*

1. INTRODUCTION

Consider the Dirichlet problem

$$\begin{aligned} &\Delta u(A) = 0, \quad A \in D \\ &u(P) = f(P), \quad P \in S \equiv \partial D. \end{aligned} \tag{1.1}$$

D is a simply-connected region in the plane with a piecewise smooth boundary S. Representing u as a double layer potential,

$$u(A) = \int_S \varphi(Q)\frac{\partial}{\partial \nu_Q}[\ln|A-Q|]dS, \quad A \in D \tag{1.2}$$

the density function $\varphi(Q)$ is determined from

$$-\pi\varphi(P) + \int_S \varphi(Q)\frac{\partial}{\partial \nu_Q}[\ln|P-Q|]ds - [\pi-\Omega(P)]\varphi(P) = f(P), \quad P \in S. \tag{1.3}$$

$\Omega(P)$ denotes the interior angle of S at P, and ν_Q is the inner normal to S at Q. At all but a finite number of points on S, we assume $\Omega(P) = \pi$; elsewhere we assume

$$0 < \Omega(P) < 2\pi, \tag{1.4}$$

disallowing cusps on the boundary.

Let the equation (1.3) be written as

$$(-\pi + \mathcal{K})\varphi = f, \tag{1.5}$$

implicitly defining $\mathcal{K}$. This operator $\mathcal{K}$ is bounded from $C(S)$ to $C(S)$ and from $L^\infty(S)$ to $L^\infty(S)$. But if S has vertices at which $\Omega(P) \neq \pi$, then $\mathcal{K}$ is not compact; and thus the theory for the smooth boundary case, using the Fredholm alternative, does not apply to (1.5). A theory of the unique solvability of (1.5) is given in Radon [9], and a survey of the problem is given in [6]. A corresponding theory for three dimensional problems is given in [7] and [10].

The numerical solution of (1.3) has usually been based on collocation, using piecewise polynomial approximations to $\varphi(Q)$; for example, see [4], [5], [6], and [10]. Regarding collocation as a projection method, it can be written in the form

$$(-\pi + \mathcal{P}_m \mathcal{K})\varphi_m = \mathcal{P}_m f. \tag{1.6}$$

For a discussion of this notation and of projection methods, see [1, pp. 50-70]. The most general analysis of (1.6) given to date is that of Bruhn and Wendland [5]. It uses the general schema of [9], and this leads them to the assumption

$$(\overline{\mathrm{Sup}_m}\|\mathcal{P}_m\|) \operatorname*{Max}_{P \in S}|\pi - \Omega(P)| < \pi. \tag{1.7}$$

If this is satisfied, then the solutions of (1.6) converge to φ, with an error proportional to $\|\varphi - \mathcal{P}_m\varphi\|$, for all continuous data f.

The condition (1.7) is quite restrictive, and empirically it seems unnecessary. In this paper, we give a summary of our investigations of the operators $\mathcal{K}$ and $\mathcal{P}_m\mathcal{K}$. We seek to determine whether (1.7) is really necessary for the convergence of (1.6). This is done by considering the essential difficulty of the problem on a piecewise smooth boundary, namely the behavior of the integral operator $\mathcal{K}$ in the vicinity of corner points. We restrict our attention, initially, to the wedge boundary $S = W$ pictured in Figure 1. The arms have length 1, and the

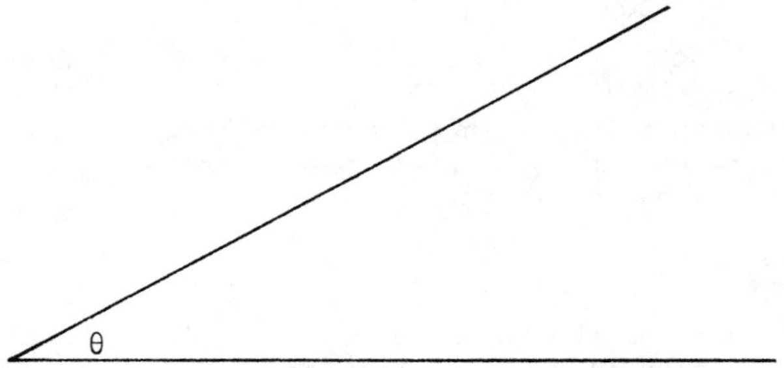

FIG. 1. The wedge boundary W.

angle θ satisfies $0 < \theta < \pi$. The integral equation (1.3) for W is discussed in section 2, and the collocation method for its solution is given in section 3. A short discussion of the relation of this material to the general case of (1.3) is given in section 4. This paper is a summary of results which will appear elsewhere in a more complete form [2].

2. THE INTEGRAL EQUATION ON A WEDGE

The integral equation (1.3) simplifies greatly when $S = W$. Let

$$\varphi_1(x) = \varphi(x,0), \quad \varphi_2(x) = \varphi(x\cos\theta, x\sin\theta), \quad 0 \le x \le 1$$

and define $f_1(x)$ and $f_2(x)$ analogously. Then (1.3) becomes

$$-\pi\varphi_1(x) - \int_0^1 \frac{x\sin\theta\, \varphi_2(y)dy}{x^2-2xy\cos(\theta)+y^2} = f_1(x), \quad 0 < x \le 1$$

$$-\pi\varphi_2(x) - \int_0^1 \frac{x\sin\theta\, \varphi_1(y)dy}{x^2-2xy\cos(\theta)+y^2} = f_2(x), \quad 0 < x \le 1 \qquad (2.1)$$

$$\varphi_1(0) = \varphi_2(0) = \varphi(0,0) = \frac{-1}{2\pi-\theta} f(0,0).$$

Let $\psi_1 = \varphi_1+\varphi_2$, $\psi_2 = \varphi_1-\varphi_2$, $F_1 = f_1+f_2$, and $F_2 = f_1-f_2$. Then adding and subtracting in (2.1) yields

$$-\pi\psi_j(x) \pm \int_0^1 \frac{x\sin\theta\, \psi_j(y)dy}{x^2-2xy\cos\theta+y^2} = F_j(x), \ 0<x\le 1, \ j=1,2. \qquad (2.2)$$

ψ_1 uses the negative sign, and ψ_2 the positive. Let $\mathcal{L}$ denote this integral operator, with

$$\mathcal{L}\psi(0) = (\pi-\theta)\psi(0), \quad \psi \in C[0,1].$$

Easily the spectrum of $\mathcal{K}$ satisfies

$$\sigma(\mathcal{K}) = \sigma(\mathcal{L}) \cup \sigma(-\mathcal{L}). \qquad (2.3)$$

Also, $\mathcal{L}$ maps $C[0,1]$ to $C[0,1]$, and

$$\|\mathcal{K}\| = \|\mathcal{L}\| = \pi-\theta \qquad (2.4)$$

relative to $C(W)$ and $C[0,1]$, respectively.

To analyze (2.2), we note it is equivalent to a Wiener-Hopf equation. Let $x = e^{-s}$, $y = e^{-t}$, $0 \le s,t < \infty$. Then (2.2) becomes

$$-\pi\hat{\psi}_j(s) \pm \int_0^1 \frac{\sin\theta\, \hat{\psi}_j(t)dt}{e^{s-t}-2\cos\theta+e^{t-s}} = \hat{F}_j(s), \quad 0 \le s < \infty \qquad (2.5)$$

with $\hat{\psi}(s) = \psi(e^{-s})$, etc. Let this new integral operator be denoted by $\hat{\mathcal{L}}$. Then from [8],

$$\sigma(\hat{\mathcal{L}}) = [0,\pi-\theta]. \tag{2.6}$$

Moreover, if $\mathcal{L}$ is regarded as an operator on $C[0,1]$, then none of the points λ in $(0,\pi-\theta)$ are eigenvalues of $\hat{\mathcal{L}}$. For $0 < \lambda < \pi-\theta$, the associated eigenfunctions of $\hat{\mathcal{L}}$ can be shown to have the form

$$\hat{\psi}(s) = A_\lambda \cos(C_\lambda s) + B_\lambda \sin(C_\lambda s) + o(1) \tag{2.7}$$

as $s \to \infty$. Letting $s = -\log(x)$ to transform back to $[0,1]$, it can be seen that the eigenfunction of $\mathcal{L}$ will not be continuous at $x = 0$ on $[0,1]$.

3. THE COLLOCATION METHOD

In the error analysis of Bruhn and Wendland [5], the restriction (1.7) arises from using the bound

$$\|\mathcal{P}_m \mathcal{K}_\epsilon\| \le \|\mathcal{P}_m\|\|\mathcal{K}_\epsilon\|, \tag{3.1}$$

where $\mathcal{K}_\epsilon$ is derived from $\mathcal{K}$ on S in such a way that

$$\|\mathcal{K}_\epsilon\| \doteq \operatorname*{Max}_{P\in S}|\pi - \Omega(P)|. \tag{3.2}$$

The authors needed $\|\mathcal{P}_m\mathcal{K}_\epsilon\| < \pi$, and thus the restriction (1.7) arises from (3.1) and (3.2).

We ask the more direct question: What is the size of $\|\mathcal{P}_m\mathcal{K}_\epsilon\|$? The reason for asking this is that (1.7) is quite restrictive on the allowable angles $\Omega(P)$ when $\|\mathcal{P}_m\|$ is larger than 1. For example, piecewise quadratic collocation functions lead to $\|\mathcal{P}_m\| = 5/4$, thus yielding the restriction

$$\frac{\pi}{5} < \Omega(P) < \frac{9\pi}{5}.$$

Empirically, this appears unnecessary.

To examine the size of $\|\mathcal{P}_m\mathcal{K}\|$, we consider both piecewise linear and piecewise quadratic interpolation in defining the collocation method. Let

$$h = \frac{1}{m}, \quad t_j = jh, \quad j = 0,1,\cdots,m.$$

Let $0 < a \le \frac{1}{2}$ be constant, and define the interpolation nodes by

$$x_{2j-1} = (j - \tfrac{1}{2} - a)h, \quad x_{2j} = (j - \tfrac{1}{2} + a)h, \quad 1 \le j \le m.$$

For arbitrary $\varphi \in C[0,1]$, let $\mathcal{P}_m\varphi(x)$ be the piecewise linear interpolating function

$$\mathcal{P}_m\varphi(x) = \frac{1}{2ah}\left[(x_{2j}-x)\varphi(x_{2j-1})+(x-x_{2j-1})\varphi(x_{2j})\right],$$

$$t_{j-1} < x < t_j. \tag{3.3}$$

For $a = \frac{1}{2}$, this leads to the usual continuous piecewise linear interpolating function. Its analysis is covered by [5], and thus we consider only $0 < a < \frac{1}{2}$. Because (3.3) is only piecewise continuous, we shift our analysis to $L^\infty(0,1)$. To extend $\mathcal{P}_m$ to $L^\infty(0,1)$ with the preservation of the norm and of the standard properties, see [3]. Then

$$\|\mathcal{P}_m\| = \frac{1}{2a}, \tag{3.4}$$

which becomes as large as desired by taking a close to zero.

If (3.3) is applied to (2.1), it is easily reduced to the application of $\mathcal{P}_m$ to the equivalent formulation (2.2). Thus we consider only the size of $\|\mathcal{P}_m\mathcal{L}\|$. Let $\ell_1(x),\cdots,\ell_{2m}(x)$ be the usual cardinal basis functions for the interpolation in (3.3), so that

$$\mathcal{P}_m\varphi(x) = \sum_1^{2m}\varphi(x_j)\ell_j(x).$$

Then the solution of

$$(-\pi \pm \mathcal{P}_m\mathcal{L})\psi_m = \mathcal{P}_m F, \quad \psi_m \in L^\infty(0,1) \tag{3.5}$$

is equivalent to solving the linear system

$$-\pi\psi_m(x_i) \pm \sum_{j=1}^{2m}\int_{t_{j-1}}^{t_j}\frac{x_i\sin(\theta)\ell_j(y)dy}{x_i^2-2x_iy\cos(\theta)+y^2} = F(x_i), \quad 1 \le i \le 2m. \tag{3.6}$$

We denote this system by

$$(-\pi \pm A_m)\underline{\psi}_m = \underline{F}. \tag{3.7}$$

The operator $\mathcal{P}_m\mathcal{L}$ is a degenerate kernel integral operator, and this can be used to prove the following.

<u>Theorem 1</u>. Let $\theta = \frac{\pi}{2}$, $0 < a < \frac{1}{2}$. Then the kernel of $\mathcal{P}_m\mathcal{L}$ is nonnegative for all $m \ge 1$, and

$$\|\mathcal{P}_m\mathcal{L}\| = \frac{\pi}{2} - \eta(m), \tag{3.8}$$

where $\eta(m)$ is positive and tends to zero as $m \to \infty$.

This result leads directly to the existence and uniform boundedness of $(-\pi \pm \mathcal{P}_m\mathcal{L})^{-1}$, for all $m \geq 1$ and $0 < a < \frac{1}{2}$. For comparison, the restriction (1.7) in this case would require

$$\frac{\pi}{4a} < \pi,$$

or $\frac{1}{4} < a < \frac{1}{2}$. The result (3.8) also generalizes to $\frac{\pi}{2} \leq \theta < \pi$.

The proof of (3.8) breaks down when $0 < \theta < \frac{\pi}{2}$. Instead we look at the matrix A_m in (3.7) and examine its norm.

Theorem 2. There is a value $\hat{a}$ for which A_m has only nonnegative elements for $\hat{a} \leq a < .5$, $0 < \theta < \frac{\pi}{2}$, and $m \geq 1$. It then follows that

$$\|A_m\| < \pi - \theta \tag{3.9}$$

using the matrix row norm. The critical value $\hat{a}$ satisfies $\hat{a} \leq a^*$, with $a^* \doteq .24500$.

The condition (3.9) implies the existence and uniform boundedness of $(-\pi \pm A_m)^{-1}$ for all $0 < \theta < \frac{\pi}{2}$, $m \geq 1$, and $\hat{a} \leq a < \frac{1}{2}$. This can be used to give a complete error analysis for the collocation method (3.5). The results (3.8) and (3.9) also bound the spectrum of $\mathcal{P}_m\mathcal{L}$, showing its spectral radius is less than $\pi-\theta$, independent of θ, m, and a, subject to the limitations in the two theorems.

The results for quadratic interpolation follow a similar pattern. Define $\mathcal{P}_m\varphi(x)$ to be the usual continuous piecewise quadratic function interpolating to $\varphi(x)$ at $2m+1$ evenly spaced points on $[0,1]$. For $\theta = \frac{\pi}{2}$, we are able to prove a result corresponding to (3.9), again showing that the bound (3.1) overestimates the true size of the spectrum of $\mathcal{P}_m\mathcal{L}$.

We also investigated the behavior of A_m numerically, looking at both the sign of the elements and the eigenvalues of the matrix. Theorem 2 was found to be true, with $\hat{a} = a^*$ as the critical value of a at which some matrix entries become negative. More importantly, it was found that

$$\sigma(\mathcal{P}_m\mathcal{L}) = \sigma(A_m) \subset \sigma(\mathcal{L}) = [0, \pi-\theta] \tag{3.10}$$

for $0 < \theta < \pi$, $0 < a \leq \frac{1}{2}$, and $m \geq 1$. The same was also true with piecewise quadratic collocation, with all θ and m. These results are quite surprising, as they would seem to imply that some kind of maximum principle is valid for $\mathcal{P}_m\mathcal{L}$. But the operators are not in the usual context or of the usual form for such a principle to be valid.

4. THE GENERAL EQUATION

The material in sections 2 and 3 can be used to study the equation (1.3) on a general piecewise smooth curve S. Following ideas in [5], we are able to split $\mathcal{K} = \mathcal{W} + \mathcal{V}$. The operator $\mathcal{V}$ is compact, and $\mathcal{W}$ is essentially the sum of wedge operators as in section 2, one for each corner point of S.

Applying the collocation methods of section 3 to equation (1.3), the results of Theorems 1 and 2 can be combined with the methods of [5] to give a complete error analysis. This extends the results for the wedge case, and it shows that the earlier restriction (1.7) is unnecessary, at least for these particular numerical methods.

To our knowledge, no general analysis of numerical methods for (1.3) is known. But it appears from the present work that any such analysis will be closely related to numerical methods for Wiener-Hopf equations.

ACKNOWLEDGEMENTS

The work of the first author was supported in part by NSF grant MCS-8002422 and by a visiting appointment to the Division of Mathematics and Statistics, CSIRO, Canberra.

REFERENCES

1. ATKINSON, K., *A Survey of Numerical Methods for the Solution of Fredholm Integral Equations of the Second Kind*, SIAM, Philadelphia (1976).

2. ATKINSON, K. and F. deHOOG, The numerical solution of Laplace's equation on a wedge, submitted for publication.

3. ATKINSON, K., I. GRAHAM, and I. SLOAN, Piecewise continuous collocation for integral equations, *SIAM J. Numer. Anal.*, to appear.

4. BENVENISTE, J., Projective solutions of the Dirichlet problem for boundaries with angular points, *SIAM J. Appl. Math.* 15, 558-568 (1967).

5. BRUHN, G. and WENDLAND, W., Uber die naherungsweise Losung von linearen Funktionalgleichungen, pp. 136-164 of L. Collatz, G. Meinardus, and H. Unger (Eds.), *Functionalanalysis, Approximationstheorie, Numerische Mathematik*, Birkhauser, Basel (1967).

6. CRYER, C., The solution of the Dirichlet problem for Laplace's equation when the boundary data is discontinuous and the domain has a boundary which is of bounded rotation by means of the Lebesgue-Stieljes integral equation for the double layer potential, *Computer Sciences Tech. Rep.* #99, University of Wisconsin, Madison (1970).

7. KRAL, J., *Integral Operators in Potential Theory*, Lecture Notes in Mathematics #893, Springer-Verlag, Berlin (1980).

8. KREIN, M., Integral equations on a half-line with kernel depending on the difference of the arguments, *AMS Transl.*, Series 2, Vol. 22, 163-288 (1963).

9. RADON, J., Uber die Randwertaufgaben beim logarithmischen Potential, *Stizungsberichte der akademie der Wissenschaften Wien* 128 Abt. IIa, 1123-1167 (1919).

10. WENDLAND, W., Die Behandlung von Randwertaufgaben im R_3 mit Hilfe von Einfach- und Doppelschichtpotentialen, *Numer. Math.* 11, 380-404 (1968).

PROJECTION METHODS FOR CAUCHY SINGULAR INTEGRAL EQUATIONS WITH CONSTANT COEFFICIENTS ON [-1,1]

Michael A. Golberg

University of Nevada, Las Vegas

1. INTRODUCTION

In the past 10 years there has been an increasing interest in the numerical solution of Cauchy singular integral equations (CSIE). Although work in the Soviet Union has been going on since the 1950's, current Western interest seems to have been stimulated by the publication of a series of papers by Erdogan, Gupta, Cook and Krenk between 1969 and 1975 [7,8,15].

Since numerical work on CSIE has been done by groups of people in different disciplines, there has been little contact between them, and the literature is widely scattered in engineering, numerical analysis, mathematics and Soviet journals.

In this paper, we will try to bring together some of this widely dispersed material (over 200 papers since 1969) which we hope will be of use to both experts and novices in the field.

Since equations on intervals of the line are currently of most interest we will confine our attention to these, and in particular to projection methods for linear equations with constant coefficients. For these equations the basic numerical analysis is reasonably complete and there is ample numerical evidence that the existing algorithms work.

In Section 2 we will present a brief outline of some of the theory of CSIE and an indication of the special problems that are expected to occur in their numerical solution. Sections 3 and 4 are devoted to a discussion of Galerkin and collocation methods, particularly those using orthogonal polynomial bases. In Section 5 we will discuss the applicability of these procedures for the solution of equations of the first kind with a discontinuous right hand side, a practical problem which we have found particularly difficult to solve numerically.

2. L_2 THEORY FOR CSIE

We consider the CSIE

$$av(x)+\frac{b}{\pi}⨍_{-1}^{1}\frac{v(t)}{t-x}dt+\int_{-1}^{1}K(x,t)v(t)dt=f(t), \tag{2.1}$$

where for convenience we use the normalization $a^2+b^2=1$ and all functions are real. When $a=0$ we get an equation of the first kind and for $a\neq 0, b\neq 0$ an equation of the second kind. When $K(x,t)=0$ (2.1) is called the *dominant equation* and for $K(x,t) \neq 0$ the *complete equation*.

Because of its widespread occurrence in aerodynamics, we refer to the case $a=0$ as the generalized airfoil equation. In the Soviet literature this special case is often discussed in the trigonometric form that results by setting $t=\cos\tau$, $x=\cos\xi$. This leads naturally to trigonometric approximations analogous to the polynomial ones we discuss.

In general, the solution to (2.1) will be of the form $v(x)=\rho(x)u(x)$ where $\rho(x)=(1-x)^{\alpha}(1+x)^{\beta}$ and $(\alpha,\beta)>-1$ are given by

$$\alpha=\{\log[(a-ib)/(a+ib)]/2\pi i\}+N,$$
$$\beta=-\{\log[(a-ib)/(a+ib)]/2\pi i\}+M, \tag{2.2}$$

where N and M are integers chosen so that the index $\nu=-(\alpha+\beta)=-(N+M)$ is restricted to $\nu=\pm 1,0$. For $\nu=0$ solutions are generally unique, for $\nu=1$ there will be a one dimensional kernel, while for $\nu=-1$ solutions will exist only if f is orthogonal to the solution of the homogeneous adjoint equation [14].

Using $v(x)=\rho(x)u(x)$ we will rewrite (2.1) considering $u(x)$ the new unknown. This opens the way for a functional analytic approach to (2.1). Let $L_\rho, L_{1/\rho}$ be the Hilbert spaces of functions square integrable with respect to ρ and $1/\rho$ respectively, with the norms and inner products defined in the usual way. (For convenience, we will denote the norms and inner products in both spaces by $\|\ \|$ and $<\ ,\ >$.) Assume that

$$\int_{-1}^{1}\int_{-1}^{1}[K^2(x,t)\rho(t)/\rho(x)]dxdt<\infty \tag{2.3}$$

and that $f\in L_{1/\rho}$. It follows from the results in [6,9] that the dominant part of (2.1) can be extended to a bounded linear operator from $L_\rho \to L_{1/\rho}$ and by virtue of (2.3) that

$$Ku=\int_{-1}^{1}\rho(t)K(x,t)u(t)dt \tag{2.4}$$

defines a compact operator. Thus we may write (2.1) in the form Hu+Ku=f where

$$Hu=a\rho u+\frac{b}{\pi}⨍_{-1}^{1}\frac{\rho u(t)}{t-x}dt \tag{2.5}$$

and solutions are to be found in $L\rho$.

Since H is bounded its adjoint H* is defined. H* has the property that for $\nu=0$ $H^*Hu=u$, $HH^*u=u$; for $\nu=1$ $H^*Hu=u+c\Phi_0$ $(H\Phi_0=0)$, $HH^*u=u$; and for $\nu=-1$ $H^*Hu=u$. Thus $H^*=H^{-1}$ for $\nu=0$, H* is a right inverse for $\nu=1$ and a left inverse for $\nu=-1$. If we multiply (2.1) on the left by H* we find that u satisfies the Fredholm equation $u+H^*Ku=H^*f+b_\nu$, where $b_\nu=0$, $\nu=(-1,0)$.

By the Fredholm alternative, this equation has a unique solution iff $N(I+H^*K)=0$ which we assume holds from now on. When $\nu=1$ the arbitrary constant needs to be determined and in practice this is done by requiring u to satisfy the additional condition

$$\langle l,u\rangle=c, \tag{2.6}$$

$l\varepsilon L_\rho$. Our assumptions guarantee that (2.1), (2.6) have a unique solution.

An important role in the numerical solution of (2.1) is played by the polynomials $\{\Phi_n\}$,$\{\psi_n\}$, n=0,1,..., which are orthonormal with respect to ρ and $1/\rho$ respectively. These polynomials are suitably normalized Jacobi polynomials with the property that $H\Phi_n=\psi_{n-\nu}$, $(\psi_{-1}=0)$.

The two features that must be accounted for in solving (2.1) numerically are that fact that v(x) is generally singular near x= 1 and the fact that ν may not be 0. The first problem is usually accounted for by looking for solutions of the form

$v_N = \rho(x)u_N$, $u_N \varepsilon L_\rho$ while for the second we must be careful how (2.1) is approximated. For $\nu=0$, we should have a system with the same number of equations as unknowns, for $\nu=1$ there will be one less equation and for $\nu=-1$ one more equation than unknowns. This will become apparent with the specific algorithms discussed in Sections 3 and 4.

3. GALERKIN'S METHOD

From our discussion in Section 2 it seems appropriate to seek approximations u_N to u in L_ρ. For Galerkin's method with u_N a polynomial this approach appears to have been first discussed by Erdogan in [7] and later analyzed for equations of the first kind with index 1 by Linz [17]. Further work appears in [4,6,8,9,11,12,13,18]. Spline approximations for equations of index 0 are examined by Thomas in [19].

To set up Galerkin's method we assume that $\{\phi_n\}$ is a basis for L_ρ, $\{\chi_n\}$ is a basis for $L_{1/\rho}$ and look for approximations to u of the form

$$u_N = \sum_{n=0}^{N} c_n \phi_n . \tag{3.1}$$

The expansion coefficients $\{c_n\}$ are obtained by forming the residual

$$r_N = (H+K)u_N - f \tag{3.2}$$

and setting

$$\langle r_N, \chi_k \rangle = 0, \quad k=0,1,2,\ldots,N-\nu. \tag{3.3}$$

Writing (3.3) out in full gives the $N+1-\nu$ linear equations

$$\sum_{n=0}^{N} c_n \langle H\phi_n, \chi_k \rangle + \sum_{n=0}^{N} c_n \langle K\phi_n, \chi_k \rangle = \langle f, \chi_k \rangle,$$

$$k=0,1,2,\ldots,N-\nu. \tag{3.4}$$

If $\nu = -1$ then (3.5) is an overdetermined system and one equation may be deleted and the resulting square system solved in the usual way. When $\nu = 0$ we get N+1 equations in N+1 unknowns and (3.4) may be solved as they stand. For $\nu = 1$ the system is underdetermined. In physical problems u is usually required to satisfy the additional condition (2.6) and this condition may be approximated by

$$<l, u_N> = \sum_{n=0}^{N} c_n <l, \phi_n> = c, \tag{3.5}$$

which when added to (3.3) will generally give a uniquely solveable system. This approach, rather than trying to find the general solution to (3.3), seems to be preferred in applications [8].

As for Fredholm equations the bulk of the computation time involves the calculation of the double integrals $\{<H\phi_n, \chi_k>\}$, $\{<K\phi_n, \chi_k>\}$. If one uses the orthogonal polynomials $\{\Phi_n\}$ for ρ as a basis then $H\Phi_n = \psi_{n-\nu}$, so that choosing $\chi_k = \psi_k$ gives

$$<H\Phi_n, \psi_k> = <\psi_{n-\nu}, \psi_k> = \delta_{k,n-\nu}. \tag{3.6}$$

In this case only $\{<K\Phi_n, \psi_k>\}$ need to be evaluated.

Because of (3.6) polynomial expansions seem to be preferred and most of our further discussion will be restricted to these.

If the kernel K(x,t) and f are continuous then the integrals $<K\Phi_n, \psi_k>$ are probably best evaluated by Gaussian quadrature as has been done in [12]. If K(x,t) is discontinuous, as is typical of problems in aerodynamics, then the integration may have to be done in a different fashion if adequate rates of convergence are to be achieved [18]. In this regard, it is important to point out that certain strategies for numerical integration will yield an algorithm which is not Galerkin's method at all.

To see this, suppose that the inner products in (3.4) are evaluated by

$$<(H+K)\phi_n, \chi_k> \simeq \sum_{j=0}^{N-\nu} w_j (H+K)\phi_n(x_j)\chi_k(x_j), \tag{3.7}$$

and

$$\langle f,\chi_k\rangle \simeq \sum_{j=0}^{N-\nu} w_j f(x_j)\chi_k(x_j), \tag{3.8}$$

and these approximations substituted into (3.4). Then the system to be solved numerically is $\chi WG\underline{c}=\chi W\underline{f}$ where $\chi=[\chi_k(x_j)]$, $W=\text{diag}(w_j)$, $G=[(H+K)(\phi_n(x_j)]$ and $\underline{f}=[f(x_j)]$. If the functions $\{\chi_k\}$, $k=0,1,\ldots,n-\nu$ form a Haar system (they will in the polynomial case) then det $(\chi W)\neq 0$ and the system becomes $G\underline{c}=\underline{f}$ which is the linear system to be solved for the collocation method discussed in Section 4.

In the case of orthogonal polynomial expansions, this system also arises even if one uses the exact values $\delta_{k,n-\nu}$ for $\langle H\phi_n,\chi_k\rangle$ and $N+1-\nu$ point Gaussian quadrature is used to evaluate the inner products. This follows because $H\Phi_n\psi_k=\psi_{n-\nu}\psi_k$ is a polynomial of degree at most $2N-2\nu$ and Gaussian quadrature is exact for polynomials of degree $2N-2\nu+1$.

3.1 Convergence

The convergence of Galerkin's method has been the subject of a number of papers [4,6,9,11,17,19] and a typical method of proof is to regard it as a projection method and then to proceed in a fashion analogous to that for Fredholm equations [9,19]. We will illustrate this approach with an analysis of the orthogonal polynomial expansion described by equations (3.4) and (3.5) for $\nu=1$.

To simplify matters a little, we will assume that $l=\Phi_0$ and $c=0$ so that $c_0=0$.

Let $P_{N-1}:L_{1/\rho}\to L_{1/\rho}$ be the operator of orthogonal projection onto $X_{N-1}=\text{span}\ \{\psi_0,\psi_1,\ldots,\psi_{N-1}\}$ then it follows from (3.3) that u_N satisfies

$$P_{N-1}Hu_N+P_{N-1}Ku_N=P_{N-1}f. \tag{3.9}$$

Now $Hu_N=\Sigma_{n=0}^{N}c_nH\Phi_n=\Sigma_{n=0}^{N}c_n\psi_{n-1}=\Sigma_{n=0}^{N-1}c_{n+1}\psi_n\varepsilon X_{N-1}$ so that $P_{N-1}Hu_N=Hu_N$ and (3.9) becomes

$$Hu_N+P_{N-1}Ku_N=P_{N-1}f. \tag{3.10}$$

In this case H has a bounded right inverse $H^*:L_{1/\rho}\to L_\rho$ so that solving (3.10) along with $<\Phi_0,u_N>=0$ is equivalent to solving the equation of the second kind

$$u_N+H^*P_{N-1}Ku_N=H^*P_{N-1}f. \tag{3.11}$$

An easy calculation shows that $H^*P_{N-1}=Q_NH^*$ where Q_N is the operator of orthogonal projection onto span$\{\Phi_1,\Phi_2,\dots,\Phi_N\}$. Using this in (3.11) it becomes

$$u_N+Q_NH^*Ku_N=Q_NH^*f, \tag{3.12}$$

which may be viewed as the "classical" Galerkin method applied to the regularized equation $u+H^*Ku=H^*f$. The convergence of u_N to u in L_ρ now follows from standard theorems [2]. Moreover, the error estimate

$$\|u-u_N\| \le (1+\gamma_N)\ \|u-Q_Nu\|,\quad \gamma_N\to 0, \tag{3.13}$$

holds [2]. If $u\varepsilon C^r[-1,1]$ then (3.13) gives $\|u-u_N\|=0(1/N^r)$[2].

In some applications one may wish to approximate various inner products of the solution and for these we have a "super-convergence" result of the type familiar in finite element methods. A generalization of the result obtained in [12] is given in Theorem 3.1.

Theorem 3.1 Assume that u_N is an approximation to the solution of (2.1) obtained via (3.3). Let $P_{N-\nu}$ be the operator of orthogonal projection onto $X_{N-\nu}=$span$\{\chi_0,\chi_1,\dots,\chi_{N-\nu}\}$. If $g\varepsilon L_\rho$ let the approximation to $<g,u>$ be given by $<g,u_N>$. Assume that u* is a solution to the adjoint equation $(H^*+K^*)u^*=g$ then

$$|<g,u>-<g,u_N>|\le c\ \|H^*+K^*\|\ \|u^*-P_{N-\nu}u^*\|\ \|u-u_N\|. \tag{3.14}$$

Proof If we let A=H+K then $<g,u>-<g,u_N>=<g,u-u_N>=$ $<A^*u^*,u-u_N>=<u^*,A(u-u_N)>=<u^*,f-Au_N>=<u^*,r_N>=<(I-P_{N-\nu})u^*$ $+P_{N-\nu}u^*,r_N>=<(I-P_{N-\nu})u^*,r_N>$, since by definition of Galerkin's method $<P_{N-\nu}u^*,r_N>=0$. Thus $<g,u-u_N>=<A^*(I-P_{N-\nu})u^*,u-u_N>$ and

(3.14) follows by applying the Cauchy-Schwarz inequality.

Further application of this result will be given in Section 5.

4. COLLOCATION

As for Galerkin's method, we begin by looking for approximations of the form in (3.1) where again we choose $\{\phi_n\}$ to be complete in L_ρ. To obtain $\{c_n\}$ we form the residual r_N and set

$$r_N(x_k)=0, \quad k=0,1,\ldots,N-\nu\,, \tag{4.1}$$

where $\{x_k\}$ are $N+1-\nu$ distinct collocation points in $[-1,1]$. If the collocation points are chosen as the nodes of the integration rule in (3.7) then it is easily seen that (4.1) coincides with $G\underline{c}=\underline{f}$ a fact alluded to in Section 3. This relation indicates that sets of integration nodes which achieve high accuracy should be prime candidates for collocation points-the most important set being the nodes for Gaussian quadrature. As is well known, these are the zeros of the polynomial $\psi_{N-\nu+1}$.

When the basis elements are chosen to be $\{\Phi_n\}$ we arrive at a method which generalizes the well known technique for solving equations of the first kind with $\nu=0$ [3,5,9,10].

The principle advantage of a collocation method over a Galerkin method is that only single integrals $H\phi_n$ and $K\phi_n$ need be obtained numerically. Moreover, if one uses $\{\Phi_n\}$ as a basis then $H\Phi_n=\psi_{n-\nu}$ and then only $K\phi_n$ needs to be evaluated. We have had considerable success in applying this technique to equations of the first kind with kernels of the form $K(x,t)=a(x-t)\log|x-t|+b(x-t)$, where $a(x)$ and $b(x)$ are analytic [9].

4.1 Convergence

With the exception of the paper by Lifanov and Polonskii [16] the only collocation method whose convergence has been proved is the one with $\{\Phi_n\}$ as a basis and Gaussian nodes as collocation points [3,5,10]. (The trigonometric analogue has also been studied in [3].) As for Galerkin's method, we take a projection approach. The mathematics here is slightly more complicated since the necessary projections are unbounded in L_ρ. This difficulty may be overcome by a technique introduced by Vainikko to study the convergence of polynomial collocation methods for Fredholm equations.

Again we will consider the case $\nu=1$ since that for $\nu=0$ has been considered in detail previously [3,10]. The case $\nu=-1$ may be treated in a similar fashion [4].

To simplify matters, we will assume that $K(x,t)$ and $f(x)$ are continuous. If $w\varepsilon C^\circ$ let $P_{N-1}w$ be the polynomial that interpolates to w at the collocation points $\{x_k\}$. This defines a bounded projection $P_{N-1}:C^\circ\to L_{1/\rho}$ by virtue of properties of Gaussian integration [10]. Using a classical theorem of Erdos and Turan on the mean square convergence of interpolation polynomials we have $\lim N\to\infty \|w-P_{N-1}w\|=0$ [10] and by the principle of uniform boundedness (or by direct calculation) $\|P_{N-1}\|\le M, N\ge 1$.

Using the definition of P_{N-1} it follows that (4.1) and (3.5) can be expressed as

$$P_{N-1}Hu_N+P_{N-1}Ku_N=P_{N-1}f, \langle l,u_N\rangle=c. \tag{4.2}$$

Since $P_{N-1}Hu_N=Hu_N$ we find that u_N satisfies

$$Hu_N+P_{N-1}Ku_N=P_{N-1}f, \langle l,u_N\rangle=c. \tag{4.3}$$

For simplicity we assume that $l=\Phi_0$ and $c=0$ and then (4.3) is equivalent to

$$Hu_N+P_{N-1}Ku_N=P_{N-1}f. \tag{4.4}$$

Since $K:L_\rho\to C^\circ$ is compact it follows from the properties of $\{P_{N-1}\}$ that $\lim N\to\infty\|K-P_{N-1}K\|=0$ [10]. Again a standard convergence theorem shows that [2] $\lim N\to\infty u_N=u$ and for an error estimate we have

$$\|u-u_N\|\le c\|u-Q_Nu\| \tag{4.5}$$

where $Q_N=H^*P_{N-1}H$ is a projection onto span $\{\Phi_1,\ldots,\Phi_N\}$.

4.2 *Collocation Points*

For equations of the first kind the collocation points for the polynomial method may be found as the zeros of Chebyshev polynomials, and so are known explicitly. For equations of the second kind these will generally have to be determined numerically for each equation. Consequently, it is of some interest to see to what extent a fixed set of nodes may be used. A crude analysis for $\nu=1$ follows.

If one examines the convergence proof it essentially depends on showing that $P_{N-1}w$ converges in $L_{1/\rho}$ for every $w\varepsilon C^{\circ}$ and this certainly is true if we choose the Gaussian nodes for $1/\rho$. Now let $\{x_k'\}$ be the nodes of a Gaussian rule for the weight $1/\rho'$ where ρ'/ρ is bounded and let $P_{N-1}'w$ be the interpolation polynomial of w on these nodes. Then $\|w-P_{N-1}'w\|^2=\int_{-1}^{1}[(\rho'/\rho)(w-P_{N-1}'w)^2/\rho')]dx\leq\max(\rho'/\rho)x$ $\int_{-1}^{1}[(w-P_{N-1}'w)^2/\rho']dx=c\|w-P_{N-1}'w\|^2$. By the Erdos-Turan theorem $\lim N\to\infty\|w-P_{N-1}'w\|^2=0$ so that using $\{x_k'\}$ gives a convergent algorithm as well.

For $\nu=1$ $\rho=(1-x)^{\sigma}(1+x)^{-(1+\sigma)}$, $-1<\sigma<0$ so that $1/\rho$ is already bounded and we may choose $\rho'=1$. This gives $\{x_k'\}$ as the nodes for classical Gaussian integration. Since there is no explicit formula for these, perhaps a better choice is $\rho'=(1-x^2)^{\frac{1}{2}}$ so that $\{x_k'\}$ may be chosen as the zeros of $T_N(x)$ the Nth Chebyshev polynomial.

As a last comment on polynomial collocation with Gaussian nodes we note that it reduces to the currently popular quadrature method if the integrals $K\Phi_n(x_k)$ are evaluated by N+1 point Gaussian quadrature.

5. THE FLAP PROBLEM

Our motivation for studying a variety of methods for solving CSIE has been in the hope of finding an efficient algorithm for solving the generalized airfoil equation where $f(x)$ is a step function. In aerodynamics this problem arises in the study of flows over thin airfoils and in mechanics it occurs in crack problems with discontinuous loadings. What makes this problem difficult to solve is the presence of a logarithmic singularity at a point of discontinuity of f. Consequently the methods discussed in Sections 3-4 can be expected to converge slowly, if they converge at all. For Galerkin's method it can be shown that $u-u_N=0(1/\sqrt{N})$ and nothing better can be expected from collocation or quadrature. Numerical computations show this rate with oscillatory approximations. Various convergence accelerating devices have been examined but do not seem computationally efficient.

In aerodynamic applications one may often be satisfied with knowing various moments of the solution which may be expressed as inner products. In this case Theorem 3.1 shows that these may be obtained efficiently even if u cannot. Typical inner products are proportional to the Fourier coefficients $\{c_n\}$ and numerical results obtained in [12] show that one may obtain 6 figures for c_0 using fewer than 10 basis elements in the approximation. At present, Galerkin's method appears to be the most reliable method we have found for calculating these quantities and further work is continuing in order to improve the efficiency of the algorithm.

ACKNOWLEDGEMENT

I would like to acknowledge the help of Gloria Golberg in the preparation of this paper.

REFERENCES

1. ABD-ELAL,L.F., A Galerkin Algorithm for Solving Cauchy-Type Singular Integral Equations. *Indian J. Pure and Appl. Math.* 11, 699-709 (1980).
2. BAKER,C.T.H. *The Numerical Treatment of Integral Equations*. Oxford University Press, Oxford (1977).
3. DUSKOV,P.N,B.G. GABDULHAEV, Direct Methods of Solution of Singular Integral Equations of the First Kind. *Izv. Vyss. Ucebn.Zaved.Mat.* 7, 112-124 (1973).
4. DZHISHKARIANI, A.V., The Solution of Singular Integral Equations by Approximate Projection Methods, *USSR Comp. Math. and Math.Phys.* 19, 61-74, (1979).
5. DZHISHKARIANI,A.V., On the Solution of Singular Integral Equations by Collocation Methods, *Zh. Vychisl. Mat.i Mat. Fiz.* 21,355-362 (1981).
6. ELLIOTT, D., Orthogonal Polynomials Associated With Singular Integral Equations Having a Cauchy Kernel, *SIAM J. of Math. Anal.*(to appear).
7. ERDOGAN, F., Approximate Solutions of Systems of Singular Integral Equations, *SIAM J. on Appl. Math.* 17, 1041-1059 (1969).

8. ERDOGAN, F., G.D. GUPTA and T.S. COOK. Numerical Solution of Singular Integral Equations. pp. 368-425 of G.C. Sih (Ed.), *Methods of Analysis and Solutions of Crack Problems, Vol 1.* Noordhoff Int. Pub. Co, Leyden (1973).
9. FROMME, J.A., M.A. GOLBERG, *Unsteady Two Dimensional Airloads Acting on Oscillating Thin Airfoils in Subsonic Ventilated Wind Tunnels,* NASA CR 2967, Washington, D.C. (1978).
10. FROMME, J.A., M.A. GOLBERG, Convergence and Stability of a Collocation Method for the Generalized Airfoil Equation, *Appl. Math. and Comp.* 8, 281-292, (1981).
11. GABADEVA, T.V., Approximate Solution of Systems of Singular Integral Equations, *SSSR. Mech. Akad. Math. Inst. Srom.* 52, 39-48 (1976).
12. GOLBERG, M.A., M. LEA, G. MIEL, A Superconvergence Result for a Class of Singular Integral Equations, *J. Int. Equations*,(to appear).
13. IOAKIMIDIS, N.I., On the Weighted Galerkin Method of Numerical Solution of Cauchy Type Singular Integral Equations, *SIAM J. On Num. Anal.* 18, 1120-1127, (1981).
14. IVANOV, V.V. *The Theory of Approximate Methods and Their Application to the Numerical Solution of Singular Integral Equations.* Noordhoff Int. Publ. Co., Leyden (1976).
15. KRENK, S., On Quadrature Formulas For Singular Integral Equations of the First and Second Kind, *Quart. Appl. Math.* 33, 225-282 (1975).
16. LIFANOV, I.K., IA. E. POLONSKII, Proof of the Method of Discrete Vortices for Solving Singular Integral Equations, *J. Appl. Math. and Mech. (PMM)* 39, 713-718 (1974).
17. LINZ, P., An Analysis of a Method for Solving Cauchy Singular Integral Equations, *BIT* 17, 329-337 (1977).
18. MOSS, W.F., The Two Dimensional Oscillating Airfoil: A New Implementation of Galerkin's Method, *SIAM J. Num. Anal.*,(to appear).
19. THOMAS, K.S., Galerkin Methods for Singular Integral Equations, *Math. of Comp.* 36, 193-207 (1981).

SOME RESULTS FOR ABEL-VOLTERRA INTEGRAL EQUATIONS OF THE SECOND KIND

D. Kershaw

University of Lancaster

1. INTRODUCTION

In this contribution we shall prove the convergence of a numerical procedure for the approximate solution of the scalar Abel-Volterra integral equation

$$f(x) = g(x) + \frac{1}{\Gamma(\alpha)} \int_0^x K[x,t,f(t)] \frac{dt}{(x-t)^{1-\alpha}} \tag{1.1}$$

where $0 < \alpha < 1$, $0 \leqslant x \leqslant X$. The function $g \in C[0,X]$ is given and K will be assumed to satisfy certain conditions which will be imposed in section 3.

The paper falls into three main parts. In the first we present a test equation which has for its solution the Mittag-Leffler function. This arises naturally in a study of (1.1) as demonstrated in the next section which is devoted to a proof of the existence of a unique solution of (1.1) by a method which is based on Bielecki's existence theorem for ordinary differential equations.

The numerical method is derived next and in the final section is shown to be convergent.

2. TEST EQUATION

We suggest that the appropriate equation for testing and comparing numerical methods for the solution of (2.1) is

$$f_0(x) = 1 + \frac{\lambda}{\Gamma(\alpha)} \int_0^x \frac{f_0(t)}{(x-t)^{1-\alpha}} dt, \quad 0 < \alpha < 1. \tag{2.1}$$

In terms of the Abel-Liouville fractional integral [5], this can be written

$$f_0 = 1 + \lambda I_\alpha f_0, \tag{2.2}$$

and since $I_\alpha . I_\beta = I_{\alpha+\beta}$ we see that the Neumann series solution of (2.1) is given by

$$f_0(x) = 1 + \sum_{n=1}^{\infty} \frac{\lambda^n}{\Gamma(n\alpha)} \int_0^x (x-t)^{n\alpha-1}\, dt. \tag{2.3}$$

This series can be summed to give the solution of (2.1) as

$$f_0(x) = E_\alpha(\lambda x^\alpha) \tag{2.4}$$

where E_α is the Mittag-Leffler function:

$$E_\alpha(z) = \sum_{n=0}^{\infty} \frac{z^n}{\Gamma(n\alpha+1)} . \tag{2.5}$$

E_α is clearly a generalization of the exponential function to which it reduces for $\alpha = 1$. A convenient reference for this function is [4] chapter 18.

E_α is an entire function for $\alpha > 0$ and the following asymptotic result holds:

$$E_\alpha(z) = -\sum_{n=1}^{N-1} \frac{z^{-n}}{\Gamma(1-\alpha n)} + O(|z|^{-N}) \text{ as } z \to \infty \tag{2.6}$$

for $|\arg(-z)| < (1 - \tfrac{1}{2}\alpha)\pi$.

Finally we note that if unity in (2.1) is replaced by the function g then the solution of this more general equation is

$$\frac{d}{dx} \int_0^x g(t)\, E_\alpha[\lambda(x-t)^\alpha]dt. \tag{2.7}$$

3. EXISTENCE THEOREM

The existence of a unique solution of (1.1) can be proved by setting up the Picard iteration given by

$$f_{n+1}(x) = T\, f_n(x) \tag{3.1}$$

where $f_0(x) = g(x)$ and the operator T is defined by

$$T\,F(x) = g(x) + \frac{1}{\Gamma(\alpha)} \int_0^x K[x,t,F(t)] \frac{dt}{(x-t)^{1-\alpha}}, \quad x>0 . \tag{3.2}$$

The usual argument (see for example [8] chapter 2) which uses a fixed point theorem leads to the existence of a solution only over a short enough range of the independent variable for T to be a contraction. The existence for a larger range has to be established by a continuation argument. In [1] Bielecki showed how this could be avoided when proving the existence of a system of ordinary differential equations. (A more convenient reference is probably [2] p.154). We shall adapt his method to deal with (1.1).

Theorem 3.1

Let $K[x,t,F]$ be continuous for $0 \leqslant t \leqslant x \leqslant X$, $-\infty < F < \infty$, and let there exist a positive scalar L such that

$$|K[x,t,F_1] - K[x,t,F_2]| \leqslant L|F_1 - F_2|$$

for $0 \leqslant t \leqslant x \leqslant X$.

Then there exists a unique solution F of (1.1) which is continuous for $0 \leqslant x \leqslant X$.

Proof

Define the norm $||\cdot||$ on $C[0,X]$ by

$$||F|| = \sup_{0 \leqslant x \leqslant X} \frac{1}{\phi(x)} |F(x)| \tag{3.3}$$

where

$$\phi(x) = E_\alpha[2\,L\,x^\alpha].$$

Now we have seen that ϕ satisfies

$$\phi(x) = 1 + \frac{2\,L}{\Gamma(\alpha)} \int_0^x \frac{\phi(t)}{(x-t)^{1-\alpha}}\,dt, \quad x > 0,$$

consequently, since $\phi(x) > 0$,

$$\frac{1}{\phi(x)} \frac{L}{\Gamma(\alpha)} \int_0^x \frac{\phi(t)}{(x-t)^{1-\alpha}} = \frac{1}{2} - \frac{1}{2\phi(x)} < \frac{1}{2}, \quad 0 \leqslant x \leqslant X. \tag{3.4}$$

It is not difficult to show that T defined by (3.2) is a continuous mapping of $C[0,X]$ into itself. Moreover, with the aid of the Lipschitz condition,

$$|TF_1(x)-TF_2(x)| \leqslant \frac{1}{\Gamma(\alpha)} \int_0^x \Big| K[x,t,F_1(t)]-K[x,t,F_2(t)] \Big| \frac{dt}{(x-t)^{1-\alpha}}$$

$$\leqslant \frac{L}{\Gamma(\alpha)} \int_0^x |F_1(t)-F_2(t)| \frac{dt}{(x-t)^{1-\alpha}}\,. \tag{3.5}$$

It follows that

$$\frac{1}{\phi(x)} |TF_1(x)-TF_2(x)| \leqslant \frac{1}{\phi(x)} \frac{L}{\Gamma(\alpha)} \int_0^x \phi(t)\, ||F_1-F_2|| \frac{dt}{(x-t)^{1-\alpha}}$$

which, with the aid of (3.4), can be replaced by

$$\frac{1}{\phi(x)} |TF_1(x) - TF_2(x)| < \frac{1}{2} ||F_1-F_2||. \tag{3.6}$$

It remains to take the supremum of the left hand side to give the result that

$$||TF_1 - TF_2|| < \frac{1}{2}\,||F_1 - F_2|| , \tag{3.7}$$

in other words the mapping T is a contraction and the result follows from Banach's fixed point theorem.

Note that although this theorem is sufficient for our purpose the proof can be adapted to cover the situation when (1.1) is replaced by a system of equations.

4. NUMERICAL APPROXIMATION

We shall describe in this section a simple numerical method for the approximate solution of (1.1). This is based on Lagrange's two point interpolation formula. It is clear that more elaborate and more accurate schemes can be devised with the aid of higher order interpolation formulae. However these will require special methods for the initial steps and so in order to make this paper self contained as far as possible we restrict ourselves to the following scheme.

Take a fixed step length $h > 0$, and write

$$x_n = nh, \quad n = 0,1,\ldots . \tag{4.1}$$

Then (1.1) can be written

$$f(x_n) = g(x_n)+\frac{1}{\Gamma(\alpha)}\sum_{r=0}^{n-1}\int_{x_r}^{x_{r+1}} K[x_n,t,f(t)]\,\frac{dt}{(x_n-t)^{1-\alpha}}, \quad n=1,2,\ldots \tag{4.2}$$

In each interval $[x_r,x_{r+1}]$, $r = 0,1,\ldots,n-1$ we use Lagrange's linear interpolation formula with remainder, namely

$$K[x_n,t,f(t)] = \frac{(x_{r+1}-t)}{h}\,K[x_n,x_r,f(x_r)]+\frac{(t-x_r)}{h}K[x_n,x_{r+1},f(x_{r+1})]+ \\ + \; e_r(x_n,t), \quad x_r \leqslant t \leqslant x_{r+1}. \tag{4.3}$$

In (x_r,x_{r+1}) replace t by x_r+ph to give the result that (4.2) can be rewritten as

$$f(x_n)=g(x_n)+\frac{h^\alpha}{\Gamma(\alpha)}\sum_{r=0}^{n-1}\left\{K[x_n,x_r,f(x_r)]\int_0^1\frac{(1-p)}{(n-r-p)^{1-\alpha}}\,dp + \right. \\ \left. K[x_n,x_{r+1},f(x_{r+1})].\int_0^1\frac{p}{(n-r-p)^{1-\alpha}}\,dp\right\} + \\ \frac{1}{\Gamma(\alpha)}\sum_{r=0}^{n-1}\int_{x_r}^{x_{r+1}} e_r(x_n,t)\,\frac{dt}{(x_n-t)^{1-\alpha}} . \tag{4.4}$$

To obtain an approximate solution set $e_r \equiv 0$ and replace $f(x_r)$ by f_r, $r = 1,2,\ldots,$ then

$$f_n = g(x_n) + \frac{h^\alpha}{\Gamma(\alpha)} \sum_{r=0}^{n-1} \left\{ K[x_n,x_r,f_r] \int_0^1 \frac{(1-p)}{(n-r-p)^{1-\alpha}}\, dp + K[x_n,x_{r+1},f_{r+1}] \cdot \int_0^1 \frac{p}{(n-r-p)^{1-\alpha}}\, dp \right\}, \quad n = 1,2,\ldots . \tag{4.5}$$

Replace f_0 by g(0) and rearrange this to give

$$f_n = g(x_n) + \frac{h^\alpha}{\Gamma(\alpha)} K[x_n,0,g(0)] \int_0^1 \frac{(1-p)}{(n-p)^{1-\alpha}}\, dp + \frac{h^\alpha}{\Gamma(\alpha)} \sum_{r=1}^{n-1} K[x_n,x_r,f_r] \left\{ \int_0^1 \frac{(1-p)}{(n-r-p)^{1-\alpha}}\, dp + \int_0^1 \frac{p}{(n-r+1-p)^{1-\alpha}}\, dp \right\} + \frac{h^\alpha}{\Gamma(\alpha)} K[x_n,x_n,f_n] \int_0^1 \frac{p}{(1-p)^{1-\alpha}}\, dp, \quad n = 1,2,\ldots .$$

The integrals can be evaluated to produce the scheme

$$f_n = g(x_n) + \frac{h^\alpha}{\Gamma(\alpha+2)} \left\{ (n-1)^{\alpha+1} - n^{\alpha+1} + (\alpha+1)n^\alpha \right\} K[x_n,0,g(0)] + \frac{h^\alpha}{\Gamma(\alpha+2)} \sum_{r=1}^{n-1} K[x_n,x_r,f_r][(n-r+1)^{\alpha+1} - 2(n-r)^{\alpha+1} + (n-r-1)^{\alpha+1}] + \frac{h^\alpha}{\Gamma(\alpha+2)} K[x_n,x_n,f_n], \quad n = 1,2,\ldots . \tag{4.6}$$

At each stage (4.6) has to be solved for f_n, but in view of the Lipschitz condition this can be achieved by a bootstrap iteration for small enough h.

5. CONVERGENCE

It will have been noticed that the remainder in Lagrange's interpolation formula in $[x_r,x_{r+1}]$ was not written out in its usual form in terms of the second derivative of K and, as a consequence, of f. This is because in general f will not be differentiable at the origin; for further details see [3] or [7]. Away from the origin f will usually have a continuous second order derivative and then the usual form for the remainder can be used. It is possible to avoid the problem at x = 0 by adopting an alternative procedure for the first interval, but for our purposes it will be sufficient to show that uniformly

$$|e_r(x_n,t)| \leqslant \omega(K[x_n,\cdot,f],h) \tag{5.1}$$

where ω is the modulus of continuity.

This follows from the fact that we can write, for $x_r \leqslant t \leqslant x_{r+1}$,

$$e_r(x_n,t) = \frac{(t-x_r)}{h}\,(K[x_n,t,f(t)] - K[x_n,x_{r+1},f(x_{r+1})]) +$$

$$+\frac{(x_{r+1}-t)}{h}\,(K[x_n,t,f(t)] - K[x_n,x_r,f(x_r)])\,, \qquad (5.2)$$

and so

$$\begin{aligned}|e_r(x_n,t)| &\leqslant \max\{|K[x_n,t,f(t)]-K[x_n,x_r,f(x_r)]|,\\ &\qquad |K[x_n,t,f(t)]-K[x_n,x_{r+1},f(x_{r+1})]|\}\\ &\leqslant \omega(K[x_n,\cdot,f],h).\end{aligned}$$

Since $K[x_n,\cdot,f]$ is continuous we see that with x_n fixed

$$\lim_{h\to 0} \omega(K[x_n,\cdot,f],h) = 0. \qquad (5.3)$$

We now prove the following theorem for the convergence of the numerical scheme.

Theorem 5.1

Let $\varepsilon(x_n,h) = f(x_n)-f_n$,

then

$$|\varepsilon(x_n,h)| \leqslant \omega(K[x_n,\cdot,f],h)\,\frac{x_n^{\alpha}}{\Gamma(\alpha+1)}\cdot\frac{1}{(1-\mu h^{\alpha})}\,E_{\alpha}\left(\frac{\lambda x_n^{\alpha}}{1-\mu h^{\alpha}}\right)$$

where $\lambda = L(\alpha+3)/(2\alpha+2)$, $\mu = L/\Gamma(\alpha+2)$, and

$$\mu h^{\alpha} < 1.$$

In view of (5.3) this has the immediate consequence:

Corollary

If $x = nh$ then

$$|\varepsilon(x,h)| \to 0 \text{ as } h \to 0.$$

Proof

Subtract (4.5) from (4.4), take the modulus of the left hand side, use the triangle inequality on the right hand side followed by an imposition of the Lipschitz condition in each subinterval to give the inequalities

$$|\varepsilon_n| \leqslant \frac{h^\alpha . L}{\Gamma(\alpha)} \sum_{r=0}^{n-1} |\varepsilon_r| \cdot \int_0^1 \frac{(1-p)}{(n-r-p)^{1-\alpha}} dp + \frac{h^\alpha . L}{\Gamma(\alpha)} \sum_{r=0}^{n-1} |\varepsilon_{r+1}| \cdot \int_0^1 \frac{p}{(n-r-p)^{1-\alpha}} dp +$$

$$+ \frac{1}{\Gamma(\alpha)} \sum_{r=0}^{n-1} \int_{x_r}^{x_{r+1}} |e_r(x_n,t)| \frac{dt}{(x_n-t)^{1-\alpha}}, \quad n = 1,2,\ldots \ . \tag{5.4}$$

(For brevity we have written ε_n for $\varepsilon(x_n,h)$.)

Since $\varepsilon_0 = f(0)-g(0) = 0$ this can be rearranged to give

$$|\varepsilon_n| \leqslant \frac{h^\alpha . L}{\Gamma(\alpha)} \sum_{r=1}^{n-1} |\varepsilon_r| . a_{n-r} + \frac{h^\alpha . L}{\Gamma(\alpha+2)} |\varepsilon_n| + E(x_n) \tag{5.5}$$

where

$$a_r = \int_0^1 \left[\frac{1-p}{(r-p)^{1-\alpha}} + \frac{p}{(r+1-p)^{1-\alpha}} \right] dp, \quad r = 1,2,\ldots \tag{5.6}$$

and

$$E(x_n) = \frac{1}{\Gamma(\alpha)} \sum_{r=0}^{n-1} \int_{x_r}^{x_{r+1}} |e_r(x_n,t)| \frac{dt}{(x_n-t)^{1-\alpha}}$$

$$\leqslant \frac{1}{\Gamma(\alpha)} \omega(K[x_n,\cdot,f],h) \int_0^{x_n} \frac{dt}{(x_n-t)^{1-\alpha}}$$

$$= \frac{1}{\Gamma(\alpha+1)} \omega(K[x_n,\cdot,f],h) x_n^\alpha . \tag{5.7}$$

Rewrite (5.6) as

$$a_r = r^{\alpha-1} \left[\int_0^1 \frac{(1-p)}{(1-p/r)^{1-\alpha}} dp + \int_0^1 \frac{p}{(1+(1-p)/r)^{1-\alpha}} dp \right].$$

Since $r \geqslant 1$ the first of these integrals is bounded above by

$$\int_0^1 \frac{1-p}{(1-p)^{1-\alpha}} dp = \frac{1}{\alpha+1},$$

and the second is bounded above by

$$\int_0^1 p \, dp = \frac{1}{2} .$$

Consequently

$$a_r \leqslant \left(\frac{\alpha+3}{2\alpha+2} \right) r^{\alpha-1} . \tag{5.8}$$

It was shown by W. Gautschi in [6] that if $r \geq 1$ and $0 \leq \alpha \leq 1$ then

$$\frac{\Gamma(r+\alpha)}{\Gamma(r+1)} \leq r^{\alpha-1} \leq \frac{\Gamma(r+\alpha-1)}{\Gamma(r)}, \tag{5.9}$$

consequently

$$a_r \leq \frac{(\alpha+3)}{(2\alpha+2)} \frac{\Gamma(r+\alpha-1)}{\Gamma(r)}$$

which when used in (5.5) gives, for $n = 1,2,\ldots,$

$$|\varepsilon_n| \leq h^\alpha L \cdot \frac{(\alpha+3)}{(2\alpha+2)} \cdot \sum_{r=1}^{n-1} |\varepsilon_r| \cdot \frac{\Gamma(n-r+\alpha-1)}{\Gamma(\alpha)\Gamma(n-r)} + \frac{h^\alpha \cdot L}{\Gamma(\alpha+2)} \cdot |\varepsilon_n| + E(x_n). \tag{5.10}$$

We shall now show that if $\frac{h^\alpha . L}{\Gamma(\alpha+2)} < 1$ then

$$|\varepsilon_n| \leq \max_{r \leq n} E(x_r) . \sum_{r=0}^{n} \frac{\lambda^r h^{\alpha r}}{(1-\mu h^\alpha)^{r+1}} \frac{\Gamma(n+r\alpha-r+1)}{(n-r)!\Gamma(r\alpha+1)}, \tag{5.11}$$

where $\lambda = L(\alpha+3)/(2\alpha+2)$ and $\mu = L/\Gamma(\alpha+2)$.

To do this consider the equations

$$\gamma_n = \lambda h^\alpha \sum_{r=1}^{n-1} \gamma_r \cdot \frac{\Gamma(n-r+\alpha-1)}{\Gamma(\alpha)\Gamma(n-r)} + \mu h^\alpha \gamma_n + \delta_n, \qquad n = 1,2,\ldots . \tag{5.12}$$

Define the functions γ and δ formally by

$$\gamma(x) = \sum_{n=1}^{\infty} \gamma_n x^n \quad \text{and} \quad \delta(x) = \sum_{n=1}^{\infty} \delta_n x^n, \tag{5.13}$$

then, with the aid of (5.12), it is not difficult to show that

$$\gamma(x) = \lambda h^\alpha x(1-x)^{-\alpha}\gamma(x) + \mu h^\alpha \gamma(x) + \delta(x). \tag{5.14}$$

It follows that γ_n is the coefficient of x^n in the expansion in ascending powers of x of $b(x)\,\delta(x)$ where

$$b(x) = [1-\mu h^\alpha - \lambda h^\alpha x(1-x)^{-\alpha}]^{-1}. \tag{5.15}$$

Write $b(x) = \sum_{r=0}^{\infty} b_r x^r$. It is easy to see that if $\lambda > 0$ and $\mu h^\alpha < 1$ then $b_r > 0$, $r = 0,1,\ldots$. Consequently since $\gamma_n = \sum_{r=1}^{n} b_{n-r} . \delta_r$, $n = 1,2,\ldots$ we can assert that

$$\gamma_n \leqslant \sum_{r=1}^{n} b_{n-r} \cdot \max_{r \leqslant n} \delta_r, \tag{5.16}$$

which, for simplicity later on will be replaced by the slightly coarser inequality

$$\gamma_n \leqslant \left(\sum_{r=0}^{n} b_r \right) \cdot \max_{r \leqslant n} \delta_r. \tag{5.17}$$

Thus we require $\sum_{r=0}^{n} b_r$ which is the coefficient of x^n in the expansion of

$$(1-x)^{-1}[1-\mu h^{\alpha}-\lambda h^{\alpha}x(1-x)^{-\alpha}]^{-1}. \tag{5.18}$$

It is not difficult to show that this is

$$\sum_{r=0}^{n} \frac{\lambda^r h^{r\alpha}}{(1-\mu h^{\alpha})^{r+1}} \frac{\Gamma(n+r\alpha-r+1)}{(n-r)!\,\Gamma(r\alpha+1)}. \tag{5.19}$$

In the case under consideration we have $\delta_r \leqslant E(x_r)$, $r = 1,2,\ldots$ which gives (5.11)

The next step is to show that if $x > 0$ then

$$\sum_{r=0}^{n} \left(\frac{x}{n^{\alpha}}\right)^r \frac{\Gamma(n-r+r\alpha+1)}{(n-r)!\,\Gamma(r\alpha+1)} < \sum_{r=0}^{n} \frac{x^r}{\Gamma(r\alpha+1)}. \tag{5.20}$$

To accomplish this we prove that if

$$a_r = \frac{\Gamma(n-r+r\alpha+1)}{n^{\alpha r}(n-r)!} \tag{5.21}$$

then

$$a_r \leqslant 1, \quad 0 \leqslant r \leqslant n. \tag{5.22}$$

First we note that $a_0 = 1$ and consider

$$\frac{a_{r+1}}{a_r} = \frac{(n-r)}{n^{\alpha}} \cdot \frac{\Gamma(n-r+r\alpha+\alpha)}{\Gamma(n-r+r\alpha+1)}. \tag{5.23}$$

Now Gautschi's inequalities (5.9) are valid for non integral values and we have for $z \geqslant 1$, $0 \leqslant \alpha \leqslant 1$

$$\frac{\Gamma(z+\alpha)}{\Gamma(z+1)} < \frac{1}{z^{1-\alpha}}, \tag{5.24}$$

then with $z = n-r+r\alpha$ we can replace (5.22) by the inequality

$$\frac{a_{r+1}}{a_r} < \frac{(n-r)}{n^\alpha}\frac{1}{(n-r+r\alpha)^{1-\alpha}}$$

$$= \left(\frac{n-r}{n-r+r\alpha}\right).\left(\frac{n-r+r\alpha}{n}\right)^\alpha < 1.$$

Hence $\{a_r\}$ is a decreasing sequence which proves (5.22), and, as a consequence, (5.20). Replace h by x_n/n in (5.11) and use (5.20) to give

$$|\varepsilon_n| \leqslant \max_{r\leqslant n} E(x_r).\frac{1}{(1-\mu h^\alpha)}\sum_{r=0}^{n}\frac{1}{\Gamma(r\alpha+1)}\left(\frac{\lambda x_n^\alpha}{1-\mu h^\alpha}\right)^r.$$

$$< \frac{1}{\Gamma(\alpha+1)}\,\omega(K[x_n,\cdot,f],h)x_n^\alpha.\frac{1}{(1-\mu h^\alpha)}\;E_\alpha\left(\frac{\lambda x_n^\alpha}{1-\mu h^\alpha}\right),$$

which is the inequality stated in the theorem.

ACKNOWLEDGEMENTS

The material of section 5 is based on some results presented in seminars at University College, London and the universities of Oxford and Manchester.

The author has great pleasure in acknowledging his indebtedness to Ben Noble for his stimulating interest.

REFERENCES

1. BIELECKI,A., *Bull. Acad. Polon. sci.IV*,pp.261-268 (1956).

2. EDWARDS,R. *Functional Analysis.* Holt, Reinhart and Wilson (1965).

3. de HOOG,F., and WEISS,R., *SIAM J. of Math. Anal.* 4, pp.561-573 (1973).

4. ERDÈLYI,A., et al., *Higher Transcendental Functions vol.III.*

5. ERDÈLYI,A., et al., *Integral Transforms vol.II.*

6. GAUTSCHI,W., *J. of Maths and Physics*.39, pp.77-81 (1959).

7. LUBICH,Ch., Runge-Kutta theory for Volterra and Abel integral equations of the second kind. *Stochastische Mathematische Modelle,* Preprint 154, Universitat Heidelberg. (1982).

8. MILLER,R.K., *Nonlinear Volterra Integral Equations.* Benjamin. (1971).

COMPUTATIONAL METHODS FOR GENERALIZED CROSS-VALIDATION WITH LARGE DATA SETS

Douglas M. Bates and Grace Wahba

University of Wisconsin-Madison
Department of Statistics

Abstract

The use of Generalized Cross Validation (GCV) has proven to be an effective method for choosing the smoothing or regularization parameters in data smoothing problems and in the approximate solution of mildly and moderately ill-posed problems. The main drawback of the method, in problems for which it is suitable, is its relatively high computational cost when very large data sets are involved. In meteorological applications in particular, it is desirable to be able to handle large data sets. Several computational tricks and shortcuts are presented to reduce the cost. In particular, a method for cheaply and effectively truncating the calculation of the singular value decomposition just as it reaches negligible singular values is presented.

1. INTRODUCTION

Generalized Cross Validation (GCV) has proven to be an excellent method for smoothing noisy data with spline functions (Craven and Wahba, [4]) choosing ridge parameters in ridge regression (Golub, Heath, and Wahba, [8]) and choosing the smoothing parameter in regularized solutions of Fredholm integral equations of the first kind (Wahba, [16, 18, 20]). In the case of smoothing data values, the method provides a means of choosing a smoothing parameter λ to balance the requirements of fidelity of the fitted function to the data, and smoothness.

Meteorological applications include the analysis (smoothing and interpolation) of height and wind fields. See Wahba and Wendelberger, [22] and Wahba [21].

In the the context of Fredholm integral equations of the first kind, the method provides an estimate of a regularization parameter with certain optimality properties which has also been shown to be effective in various applications. In this context, the setup is as follows: Data $\underline{y} = (y_1, \ldots y_n)'$ is assumed to be of the form

$$y_i = \int_0^1 K(t_i,s)\, f(s)ds + \varepsilon_i, \quad i = 1,2,\ldots n,$$

where K is known and the ε_i are independent zero mean errors with common (but unknown) variance. A regularized estimate of f is obtained by finding f in the Sobolev Space W_2^m to

minimize

$$\frac{1}{n} \sum_{i=1}^{n} \left((Kf)(t_i)-y_i\right)^2 + \lambda \int_0^1 \left(f^{(m)}(s)\right)^2 ds \,. \tag{1.1}$$

Under suitable conditions, a unique minimizer f_λ will exist and is linear in the data vector $\underline{y}$. Let $A(\lambda)$ be the n x n influence matrix satisfying

$$\begin{bmatrix} (Kf_\lambda)(t_1) \\ \cdot \\ \cdot \\ \cdot \\ (Kf_\lambda)(t_n) \end{bmatrix} = A(\lambda)\underline{y}. \tag{1.2}$$

The GCV estimate $\hat{\lambda}$ for λ is the minimizer of $V(\lambda)$ given by

$$V(\lambda) = \frac{\frac{1}{n} \| (I-A(\lambda))\, \underline{y} \|^2}{\left(\frac{1}{n} \text{Trace } (I-A(\lambda)\right)^2} \tag{1.3}$$

where $\|\cdot\|$ is the Euclidean norm. For the reasoning behind and some of the properties of this method, see Wahba, [16], Craven and Wahba, [4], Golub, Heath and Wahba, [8], Utreras, [15], Speckman, [14], Lukas, [10]. Some examples have been tried by Merz, [11], Lukas, [10], Crump and Seinfeld, [2], and others.

Although the exact minimizer of (1.1) can be written explicitly (possibly modolo some quadrature), manipulation of full matrices of the order of n x n are required to compute the solution in the general cases that we have in mind. (In special cases, such as convolution equations with equally-spaced data, the special structure of the problem can be exploited in other ways. See Davies, this volume and Wahba, [20].) If n is much larger than, say 200, one is likely (for any fixed λ) to get a good approximation to f_λ by finding $f_{\lambda,p}$, the minimizer of (1.1) in the span of p suitably defined basis functions, for example, B-splines of degree 2m-1 with p<n. Letting $B_1, \ldots B_p$ be the basis functions, then

$$f_{\lambda,p} = \sum_{j=1}^{p} c_j B_j$$

where $\underline{c} = (c_1, \ldots c_p)'$ is the minimizer of

$$\frac{1}{n} \| \underline{y} - X\underline{c} \|_n^2 + \lambda \underline{c}' \Sigma \underline{c} \quad ,$$

where X is the n x p matrix with ij th entry $\int_0^1 K(t_i,s)B_j(s)ds$ and Σ is the p x p matrix with jk th entry $\int_0^1 B_j^{(m)}(s)B_k^{(m)}(s)\, ds$. We have $\underline{c} = (X'X+n\lambda\Sigma)^{-1}X'\underline{y}$

and $A(\lambda) = A(\lambda,p) = X(X'X+n\lambda\Sigma)^{-1}X'$. If p is small, p will also act as a smoothing parameter. The value of p can also be chosen by GCV; however, the point of view here is to choose p large enough so that $f_{\lambda,p}$ is a good approximation to f_λ, and let the choice of λ do the smoothing. This point of view is buttressed by the character of f_λ as a Bayesian estimate, see Wahba, [17]. (One could, of course, go to the extreme of setting $\lambda = 0$ thus doing regression on $B_1, \ldots B_p$. It is believed that this procedure is suboptimal except for extremely large data sets. Otherwise, detectable fine structure in the solution may be obscured if p is too small. If λ is fixed at 0, the optimal p is of the order $n^{1/5}$ under certain assumptions, see Agarwal and Studden, [1].) Having set p moderately large, the numerical problem is then to minimize $V(\lambda)$ of (1.2), where $A(\lambda)$ is given by (1.3), and to compute $\underline{c}$ efficiently. Although we have described this problem for $s \varepsilon [0,1]$, the same approach can be used for s in a bounded subset of the plane or higher dimensions (see Wahba, [19]); however, as the dimension goes up, n and p will generally be larger and it will be even more important to have efficient computational techniques.

Recently, we have developed an approach to minimizing (1.1) subject to $f \varepsilon C$, where C is a closed convex set, for example, $C = \{f: \; f(s) \geq 0, \; 0 \leq s \leq 1,\}$ and an extension of the method of GCV for choosing λ in this case (see Wahba, [18], [20], also Rutman, [13].) However, we only consider the unconstrained case here.

The calculations required for GCV, say as given in Wahba, [18], are often expressed in terms of iterative matrix decompositions such as the eigenvalue-eigenvector decomposition or the singular value decomposition (SVD). These decompositions can be very computationally intensive so we present a computational method to replace all but one of the iterative decompositions with direct decompositions such as the Cholesky decomposition and QR decomposition. This makes GCV feasible for very large data sets.

Evaluating the GCV function, even with these modifications, still requires the evaluation of the singular values and left singular vectors of a matrix X which is of size n by p where n is the number of data points and p is the number of basis functions (or parameters) being used. When working with large data sets, n could quite conceivably be in the thousands. Also, the matrix X tends to be very ill-conditioned. In such cases, even one evaluation of the singular value decomposition (SVD) can be very time consuming. It would not be unusual for this single step in the overall computation to consume 90% or more of the total c.p.u. time for the GCV problem.

Even worse than to have the SVD consume large amounts of computer time is to have the SVD fail to converge altogether. Since the matrix X is large and ill-conditioned, the SVD algorithm will converge slowly. As an example, while the usual settings for the SVD subroutine in the Linpack package (Dongarra et al., [7]) allow a maximum of 30 iterations of the singular value algorithm for each singular value, we have seen cases of matrices used in GCV where the SVD algorithm will not converge even after allowing 90 iterations for each singular value.

For n large and data of "engineering accuracy," the manner in which the singular values and singular vectors are used in GCV, described in Section 3, frequently shows that only the largest 30% to 50% of the singular values and the corresponding singular vectors are needed. Thus we introduce a truncated singular value decomposition (TSVD) which provides a very close approximation to the larger singular values and their singular vectors while using smaller, better conditioned matrices that are derived from X.

In Section 2 we present the background of the matrix decompositions and in Section 3 present the calculation method for GCV. The TSVD is introduced in Section 4 and the results are discussed in Section 5. Other short cut results related to the SVD and regularization may be found in Golub, Luk and Overton, [9], O'Leary and Simmons, [12], Cuppen, [5].

2. MATRIX DECOMPOSITIONS

Matrix decompositions can be classified as iterative decompositions, such as the eigenvalue-eigenvector decomposition and the singular value decomposition, or direct decompositions, such as the QR decomposition and the Cholesky decomposition. Since the iterative decompositions can be very computationally intensive and can even fail to converge on some poorly conditioned matrices, it is an advantage to use direct decompositions whenever possible. In this section we introduce and compare four different matrix decompositions, two of which are iterative and two of which are direct.

A singular value decomposition is defined for any matrix, either rectangular or square. We will be primarily concerned with the singular value decomposition for rectangular matrices where the number of rows is at least as great as the number of columns. The SVD of such an n by p matrix X can be written as

$$X = UDV' \tag{2.1}$$

where U is n by p with orthonormal columns, D is p by p diagonal, and V is p by p orthogonal. In the terminology of Dongarra et al., [6], this is the "singular value factorization" as opposed to a similar decomposition where U is n by n. We will continue to refer to this form as the singular value decomposition, however.

An alternative to the SVD is the QR decomposition which is a direct calculation and represents the matrix X as

$$X = QR = [Q_1 Q_2] \begin{bmatrix} R_1 \\ 0 \end{bmatrix} = Q_1 R_1 \tag{2.2}$$

where Q is n by n orthogonal and R is n by p with zeroes below the main diagonal. Q_1 consists of the first p columns of Q. Q_2 is the next n -p columns, and R_1 is the first p rows of R. Because the matrix Q is orthogonal, the columns of Q_1 form an orthonormal basis for the column span of X and the columns of Q_2 form an orthonormal basis for the orthogonal subspace to this column span. Similar results hold for the matrix U in the SVD but there is a space advantage for the QR decomposition. From the way that the orthogonal matrix Q is formed using Householder transformations, it is possible to store the information necessary to multiply a vector or matrix by Q or Q', along with the matrix R_1, in an array of size n by p + 1. This can represent a considerable storage space saving when n is very large since it is not necessary to store any matrices of size n by n.

In the process of forming the QR decomposition using Householder transformations, a pivoting scheme can be introduced which is equivalent to producing a p by p permutation matrix E such that

$$XE = QR = Q_1R_1 \tag{2.3}$$

and R has the property that

$$r_{k,k}^2 \geq \sum_{i=k}^{j} r_{i,j}^2 \quad (j = k,k+1,\ldots,p) \tag{2.4}$$

where $r_{i,j}$ is the i,jth entry of R_1. This means that not only are the elements of the diagonal of R non-increasing in absolute value, but also each diagonal element dominates the length of every sub-column to the right and below it in R. In ill-conditioned matrices the pivoting scheme results in entries in the lower parts of R_1 being quite small in comparison to entries in the upper part.

With square matrices, an eigenvalue-eigenvector decomposition can be used. In particular, a p by p symmetric matrix B can be written

$$B = VDV' \tag{2.5}$$

where V is p by p orthogonal and D is p by p diagonal. An alternative for a positive definite, symmetric matrix B is the Cholesky decomposition

$$B = R'R \tag{2.6}$$

where R is p by p and upper triangular. The eigenvalue-eigenvector decomposition is iterative while the Cholesky decomposition is direct.

The Cholesky decomposition is closely related to the QR decomposition since R_1 from the QR decomposition is identical to the Cholesky decomposition of the matrix X'X. There is also a pivoted version of the Cholesky decomposition where a p by p permutation matrix E and an upper triangular matrix R are determined so that

$$E'BE = R'R \tag{2.7}$$

3. GENERALIZED CROSS-VALIDATION

We will be considering the use of GCV in the choice of a smoothing parameter or ridge parameter λ when we are fitting a parameter vector $\underline{c}$ for a given set of basis functions to an observed data vector $\underline{y}$. In general, for each value of λ, we will choose $\underline{c}$ to minimize

$$Q_\lambda(\underline{c}) = \frac{1}{n}\|\underline{y}-X\underline{c}\|^2 + \lambda\underline{c}'\Sigma\underline{c} \tag{3.1}$$

where $\underline{y}$ is the observed n dimensional response vector, X is the n by p matrix whose t'th row is the value of the basis functions at the t'th experiment (in regression terms, it is the design matrix), $\underline{c}$ is the p dimensional coefficient vector, Σ is the p by p positive semi-definite symmetric matrix that defines the smoothing penalty, and λ is the smoothing parameter.

We wish to determine from the data a value of λ which will provide a solution with good fidelity to the data, as measured by $\|\underline{y} - X\underline{c}\|^2$, and good smoothness, as measured by $\underline{c}'\Sigma\underline{c}$.

For the special case where Σ is the identity matrix, as considered in Golub, Heath and Wahba, [8], we can proceed in the following way. First form the singular value decomposition of X

$$X = UDV' \tag{3.2}$$

and

$$\underline{z} = U'\underline{y} \;.$$

We can then write the predicted responses for any given λ as

$$\hat{\underline{y}}_\lambda = A(\lambda)\underline{y} = \sum_{j=1}^{p} \underline{u}_j \frac{d_j^2 z_j}{(d_j^2 + n\lambda)} \tag{3.3}$$

where u_j is the j'th column of U. The matrix $A(\lambda)$ corresponding to (1.2) is

$$A(\lambda) = UD^2(D^2+n\lambda I)^{-1}U' , \tag{3.4}$$

yielding, after some simplification,

$$V(\lambda) = \frac{n\|(I-A(\lambda))\underline{y}\|^2}{(tr(I-A(\lambda)))^2}$$

$$= \frac{n\left\{\|\underline{y}\|^2-\|\underline{z}\|^2+\sum_{j=1}^{p}\left[\frac{n\lambda}{d_j^2+n\lambda}\right]^2 z_j^2\right\}}{\left(n-p+\sum_{j=1}^{p}\frac{n\lambda}{d_j^2+n\lambda}\right)^2} \tag{3.5}$$

and the value of λ which minimizes $V(\lambda)$ is used to provide the coefficient estimates.

The calculations outlined above are very specific to the case where Σ is an identity matrix but we want to consider more general Σ matrices. Particularly in smoothing applications, we want to allow Σ to be any positive semi-definite p by p matrix with rank p-m and still produce a generalized cross-validation function of the form of eqn. (3.5).

We begin by dividing the parameter vector into a part that lies in the null space of Σ and a part orthogonal to this null space. First a pivoted Cholesky decomposition of Σ is formed as

$$E'\Sigma E = S'S \tag{3.6}$$

where S is p-m by p with zeroes below the main diagonal. A QR decomposition of S' gives

$$S' = QR \tag{3.7}$$

with the columns of Q_2 spanning the null space of Σ and the columns of Q_1 spanning the orthogonal to this null space.

We can define transformed parameters $\underline{\gamma}$ and $\underline{\delta}$ where $\underline{\delta}$ is in the null space of Σ by

$$\begin{bmatrix} \underline{\gamma} \\ \hline \underline{\delta} \end{bmatrix} = \begin{vmatrix} R_1' & 0 \\ 0 & I \end{vmatrix} Q'E'\underline{c} \tag{3.8}$$

and the corresponding transformation of the X matrix as

$$Y = [Y_1 Y_2] = XEQ \begin{vmatrix} (R_1')^{-1} & 0 \\ 0 & I \end{vmatrix} \tag{3.9}$$

The function of $\underline{\gamma}$ and $\underline{\delta}$ that corresponds to $nQ_\lambda(c)$ is

$$g_\lambda(\underline{\gamma}, \underline{\delta}) = \left\| \underline{y} - Y \begin{vmatrix} \underline{\gamma} \\ \hline \underline{\delta} \end{vmatrix} \right\|^2 + n\lambda \underline{\gamma}'\underline{\gamma} \tag{3.10}$$

so the penalty part of the objective function is in the desired form. However, the "deviation from the data" part involves $\underline{\delta}$ as well as $\underline{\gamma}$. We can isolate the dependence of the deviation term upon $\underline{\delta}$ by taking a QR decomposition of Y_2 as

$$Y_2 = FG = [F_1 \; F_2] \begin{bmatrix} G_1 \\ 0 \end{bmatrix} \tag{3.11}$$

and pre-multiplying everything in the deviation term by F'. This does not affect the value of the deviation term since F is orthogonal.

This produces

$$g_\lambda(\underline{\gamma}, \underline{\delta}) = \|\underline{w}_1 - T_1\underline{\gamma} - G_1\underline{\delta}\|^2 + \|\underline{w}_2 - T_2\underline{\gamma}\|^2 + n\lambda\underline{\gamma}'\underline{\gamma} \tag{3.12}$$

where

$$\underline{w} = \begin{bmatrix} \underline{w}_1 \\ \underline{w}_2 \end{bmatrix} = F'\underline{y} \tag{3.13}$$

and

$$T = \begin{bmatrix} T_1 \\ T_2 \end{bmatrix} = F'Y_1 \tag{3.14}$$

The leading term in the right hand side of eqn. (3.12) can be made zero for any choice of $\underline{\gamma}$ by setting

$$\underline{\delta} = G_1^{-1}(\underline{w}_1 - T_1\underline{\gamma}) \tag{3.15}$$

so we can concentrate on the second and third terms and apply the method of Golub, Heath and Wahba, [8] to them. That is, we form the SVD

$$T_2 = UDV' \tag{3.16}$$

and calculate

$$\underline{z} = U'\underline{w}_2 \tag{3.17}$$

to get

$$V(\lambda) = \frac{n\left\{\|\underline{w}_2\|^2 + \|\underline{z}\|^2 + \sum_{j=1}^{p-m}\left[\frac{n\lambda}{d_j^2+n\lambda}\right]^2 z_j^2\right\}}{\left\{n - (p-m) + \sum_{j=1}^{p-m}\left[\frac{n\lambda}{d_j^2+n\lambda}\right]\right\}^2} \tag{3.18}$$

Once the optimal λ is chosen, $\underline{\gamma}$ can be calculated as

$$\underline{\gamma} = V(D^2 + n\lambda I)^{-1}D\underline{z} \tag{3.19}$$

and $\underline{\delta}$ from eqn. (3.15) and thence the coefficients $\underline{c}$.

4. THE TRUNCATED SVD

The purpose of the SVD is to factor the matrix X into a product of orthogonal and diagonal matrices but, as mentioned in section 1, this can be very time consuming with large, ill-conditioned matrices. In many cases, adequate information about X can be furnished by providing a QR decomposition instead of the SVD.

If we take a pivoted QR decomposition

$$XE = QR \tag{4.1}$$

of X and form the SVD of R_1 as

$$R_1 = KDL' \tag{4.2}$$

we can produce the SVD of X as

$$X = Q_1 KDL'E' = UDV' \tag{4.3}$$

where

$$U = Q_1 K$$

and

$$V = EL .$$

This is related to the method of Chan [3] for producing the SVD but uses the pivoted QR decomposition instead of the unpivoted version. This does not, however, provide better conditioning for the SVD step since the condition number of R_1 is the same as the condition number of X.

To provide better conditioning, we truncate the matrix R_1 after the k'th row and take the SVD of the resulting k by p matrix R_k ($k \leq p$) as

$$R_k = K_k D_k L_k \tag{4.4}$$

where K_k is k by k and L_k is k by p. The diagonal elements of D_k are no longer the singular values of X but now represent the singular values of a matrix

$$X_k = Q_1 \begin{bmatrix} R_k \\ 0 \end{bmatrix} E' \tag{4.5}$$

which is different from X.

It is possible, however, to measure the difference between X and X_k in the sense of the Frobenius norm (the square root of the sum of the squares of all the elements of a matrix) of $X - X_k$ without forming X_k because

$$\|X-X_k\|_F = \Big(\sum_{i=k+1}^{p} \sum_{j=i}^{p} r_{i,j}^{\,2} \Big)^{1/2} \tag{4.6}$$

so we can choose k to be as small as possible subject to the constraint that

$$\frac{\|X-X_k\|_F}{\|X\|_F} \leq \varepsilon \tag{4.7}$$

where ε is a very small number, say 10^{-8}. The double sum on the right hand side of (4.6) can be easily evaluated a row at a time starting with the p'th row until the constraint (4.7) is violated and the smallest k is determined.

Numerical experiments with the truncated singular value decomposition have shown that the larger singular values of X and the corresponding singular vectors are well determined while the singular values close to the k'th singular value (singular values are usually written in decreasing size) are poorly determined. From equation (3.18) it can be seen that singular values where squares are small compared to $n\hat{\lambda}$ have neglible effect on the value of $n\hat{\lambda}$ and $\hat{\underline{y}}_{\hat{\lambda}}$. If it is discovered after using a TSVD that the square of the estimate of the k'th singular value is not small compared to $n\hat{\lambda}$, the TSVD should be re-run with a smaller value of ε.

5. DISCUSSION

We have presented methods of streamlining the calculations involved in using GCV with large data sets. These techniques are based on replacing expensive iterative decompositions with less expensive direct decompositions and providing an approximation to the singular value decomposition which is easier to compute than the full decomposition but provides the information needed for GCV.

ACKNOWLEDGMENTS

This research was supported by the Office of Naval Research under Contract N00014-77-G-0675, by the National Aeronautics and Space Administration under Contract NAG5-128, and by the National Science Foundation under research Grant MCS-8102732.

REFERENCES

1. AGARWAL, G., and STUDDEN, W. J., Asymptotic Integrated Mean Square Error Using Least Squares and Bias Minimizing Splines. *Ann. Statist.* 8,6, 1307-1325 (1980).
2. CRUMP, J. G., and SEINFELD, J. H., A New Algorithm for Inversion of Aerosol Size Distribution Data. *Aerosol Science and Technology* 1, 15-34 (-982).
3. CHAN, T. F., An Improved Algorithm for Computing the Singular Value Decomposition. *ACM Transactions on Mathematical Software* 8, 72-83 (1982).
4. CRAVEN, P., and WAHBA, G., Smoothing Noisy Data with Spline Functions. *Numerische Mathematik* 31, 377-408 (1979).

5. CUPPEN, J. J. M., The Singular value decomposition in Product Form. *University of Amsterdam, Dept. of Mathematics Report 81-06*, (1981).
6. DAVIES, A. R., this volume (1982).
7. DONGARRA, J. J., BUNCH, J. R., MOLER, C. B., and STEWART, G. W., *Linpack User's Guide*, S.I.A.M., Philadelphia, (1979).
8. GOLUB, G., HEATH, M., and WAHBA, G., Generalized Cross Validation as a Method for Choosing a Good Ridge Parameter. *Technometrics* 31, 315-224 (1979).
9. GOLUB, G. H., LUK, F. T., and OVERTON, M. L., A Block Lanczos Method to Compute the Singular Values and Corresponding Singular Vectors of a Matrix. *ACM Trans. Math. Software* 7,2, 149-169 (1981).
10. LUKAS, M. A., Regularization of Linear Operator Equations, *Ph.D. Thesis, Australian National University*. (1981).
11. MERZ, P., Determination of Adsorption Energy Distribution by Regularization and a Characterization of Certain Adsorption Isotherms. *J. Comp. Physics* 38, 64-85 (1980).
12. O'LEARY, D. P., and SIMMONS, J. A., A Bidiagonalization-Regularization Procedure for Large Scale Discretization of Ill Posed Problems. *S.I.A.M. J. COMPUT.* 2,4, 474-487 (1981).
13. RUTMAN, R., this volume (1982).
14. SPECKMAN, P., Spline Smoothing and Optimal Rates of Convergence in Nonparametric Regression Models. *Preprint* (1981).
15. UTRERAS, F., Optimal Smoothing of Noisy Data Using Spline Functions. *S.I.A.M. J. Sci. Stat. Comput.* 2,3 (1981).
16. WAHBA, G., Practical Approximate Solutions to Linear Operator Equations when the Data are Noisy. *S.I.A.M. J. Numer. Anal.* 14,4 (1977).
17. WAHBA, G., Improper Priors, Spline Smoothing, and the Problem of Guarding Against Model Errors in Regression. *J. Roy. Stat. Soc. B* 49,3, 364-372 (1978).
18. WAHBA, G., Ill-Posed Problems: Numerical and Statistical Methods for Mildly, Moderately and Severely Ill-Posed Problems with Noisy Data. *Statistics Dept. UW-Madison, TR 595*. Prepared for the Proceedings of the International Symposium on Ill-Posed Problems held at Newark, Delaware Oct 2-6, 1979, (1980).
19. WAHBA, G., Spline Bases, Regularization, and Generalized Cross Validation for Solving Approximation Problems with Large Quantities of Noisy Data. In *Approximation Theory III*, E. W. Cheney, ed Academic Press. (1980).

20. WAHBA, G., Constrained Regularization for Ill-Posed Linear Operator Equations, with Applications in Meteorology and Medicine. In *Statistical Decision Theory and Related Topics, III*, Vol. 2, S. S. Gupta and J. O. Bergen, eds., Academic Press (1982).

21. WAHBA, G., Vector Splines on the Sphere, with Application to the Estimation of Vorticity and Divergence from Discrete, Noisy Data. *Statistics Dept. UW-Madison, TR 674* May 1982. To appear, *Multivariate Approximation Theory,* Vol. 2, W. Schempp and K. Zeller, eds. Birkhauser Verlag. (1982).

22. WAHBA, G., and J. WENDELBERGER, Some New Mathematical Methods for Variational Objective Analysis Using Splines and Cross-Validation. *Mon. Wea. Rev.* 108, 36-57 (1980).

THE NUMERICAL TREATMENT OF SINGULAR INTEGRAL EQUATIONS - A REVIEW

David Elliott

University of Tasmania

1. INTRODUCTION

The sort of singular integral equations I wish to discuss are those involving a Cauchy principal value integral taken over a finite interval. Such an integral will be denoted and defined by

$$\fint_{-1}^{1} \frac{\phi(\tau)d\tau}{\tau-t} = \lim_{\varepsilon\to 0+} \left(\int_{-1}^{t-\varepsilon} + \int_{t+\varepsilon}^{1} \right) \frac{\phi(\tau)d\tau}{\tau-t} , \qquad (1.1)$$

for $t \in (-1,1)$, provided that the limit exists. In his book [1] published in 1977, the convener of this symposium devoted just 8 pages, out of a total of 984, to such equations. This did not so much reflect the importance of such equations, as the state of the art of their approximate solution, circa 1975. The last few years, however, have seen an increasing number of papers devoted to the approximate solution of a variety of such singular integral equations, and in this lecture I should like to say something about the present state of the art and where progress may be expected in the future.

We shall consider, in some detail, the equation

$$a(t)\phi(t) + \fint_{-1}^{1} \frac{K(t,\tau)\phi(\tau)d\tau}{\tau-t} = f(t) , \qquad (1.2)$$

for $t \in (-1,1)$, which can be rewritten as

$$a(t)\phi(t) + \frac{b(t)}{\pi} \fint_{-1}^{1} \frac{\phi(\tau)d\tau}{\tau-t} + \int_{-1}^{1} k(t,\tau)\phi(\tau)d\tau = f(t) , \qquad (1.3)$$

where $b(t) = \pi K(t,t)$ and $k(t,\tau) = (K(t,\tau) - K(t,t))/(\tau - t)$. The real-valued functions a, f and K or a, b, f, and k are assumed to be given and it is required to determine the real function ϕ. (I shall refer to this equation as CSIE). I have chosen the interval of integration to be (-1,1) but the theory of CSIE as developed in Muskhelishvili's book [10] allows the integrals to be taken over unions of arcs or contours in the complex plane. But, for this lecture, the arc (-1,1) is of sufficient interest for our study.

Before considering approximate methods for the solution of (1.3) we must take a look at the theory of such equations. Although this theory has long been established it does not appear to be widely known and there are aspects of it which are quite different from the theory of Fredholm integral equations. We shall consider this theory in the next section.

2. SOME THEORY OF CSIE

We shall, following Muskhelishvili [10], assume that the functions a, b, f, and k are Hölder continuous in each independent variable on [-1,1]. That is, a function g is *Hölder continuous* on [-1,1] if there exist constants $C_1 \geq 0$ and α, $0 < \alpha \leq 1$, such that

$$|g(t_1) - g(t_2)| \leq C_1 \; |t_1 - t_2|^{\alpha} \tag{2.1}$$

for all t_1, $t_2 \in [-1,1]$.

For any given CSIE we must specify the class of functions in which its solution ϕ is to be found. To this end we define a class of functions, which we shall denote by $L\{p_1,p_2\}$, where $p_1,p_2 \geq 1$, as follows. We have $g \in L\{p_1,p_2\}$ if

(i) g is Hölder continuous on every closed interval $[c,d]$ of (-1,1);

(ii) $\displaystyle\int_{-1}^{-1+\delta_1} |g(\tau)|^{p_1} d\tau$ exists for some $\delta_1 > 0$;

(iii) $\displaystyle\int_{1-\delta_2}^{1} |g(\tau)|^{p_2} d\tau$ exists for some $\delta_2 > 0$.

First, let us consider the *dominant equation*, which is (1.3) with $k(t,\tau) \equiv 0$, and write it as $M\phi = f$ where

$$M\phi(t) = a(t)\phi(t) + \frac{b(t)}{\pi} \int_{-1}^{1} \frac{\phi(\tau)d\tau}{\tau - t} . \tag{2.2}$$

We shall look for solutions of $M\phi = f$ with $\phi \in L\{p_1,p_2\}$, for some $p_1,p_2 \geq 1$. (Throughout his book [10], Muskhelishvili looks for solutions in the class $L\{1,1\}$ which he denotes by H^*). It can be shown that if $\phi \in L\{p_1,p_2\}$ then $M\phi \in L\{p_1,p_2\}$ also. An operator closely related to M is its adjoint M^* defined by

$$M^*\phi(t) = a(t)\phi(t) - \frac{1}{\pi} \int_{-1}^{1} \frac{b(\tau)\phi(\tau)}{\tau - t} d\tau \ . \tag{2.3}$$

This is such that for every $\phi \in L\{p_1,p_2\}$ and $\psi \in L\{q_1,q_2\}$ where $1/p_1 + 1/q_1 = 1$, $1/p_2 + 1/q_2 = 1$, we have

$$\int_{-1}^{1} \psi(\tau)M\phi(\tau)d\tau = \int_{-1}^{1} \phi(\tau)M^*\psi(\tau)d\tau \ . \tag{2.4}$$

Let us commence our analysis by looking for solutions $\phi \in L\{p_1,p_2\}$ of the homogeneous dominant equation $M\phi = 0$. To solve it we must move off the interval $[-1,1]$ into the complex plane and define, for all $z \in C\backslash[-1,1]$ (i.e. the complex plane with the closed interval $[-1,1]$ deleted), the function

$$\Phi(z) = \frac{1}{2\pi i} \int_{-1}^{1} \frac{\phi(\tau)d\tau}{\tau - z} \ . \tag{2.5}$$

This function is analytic in the deleted plane and tends to zero as $|z| \to \infty$. If we define, for $t \in (-1,1)$,

$$\Phi^{\pm}(t) = \lim_{\varepsilon \to 0+} \Phi(t \pm i\varepsilon)$$

then the Sokhotski-Plemelj equations (see, for example, [10]) give

$$\left.\begin{aligned} \Phi^+(t) - \Phi^-(t) &= \phi(t) \ , \\ \Phi^+(t) + \Phi^-(t) &= \frac{1}{\pi i} \int_{-1}^{1} \frac{\phi(\tau)d\tau}{\tau - t} \ , \end{aligned}\right\} \tag{2.6}$$

for $t \in (-1,1)$. Substituting from (2.6) into $M\phi = 0$ we find that our problem can be reformulated as the Riemann boundary value problem of finding Φ such that

$$\Phi^+(t) = \frac{a(t)-ib(t)}{a(t)+ib(t)} \Phi^-(t) \text{ for } -1 < t < 1 \ . \tag{2.7}$$

To solve this problem we introduce, on $[-1,1]$, a continuous function θ which is such that

$$\theta(t) = -\frac{1}{2\pi i} \operatorname{Ln} \left(\frac{a(t)-ib(t)}{a(t)+ib(t)}\right) = \frac{1}{\pi} \arctan \frac{b(t)}{a(t)} + N(t) \ , \tag{2.8}$$

where N takes only integer values and may have discontinuities at the zeros of a/b, and $-\pi/2 < \arctan x \leq \pi/2$. In terms of θ, a solution of (2.7) is given by

$$\Phi(z) = \exp\left(-\int_{-1}^{1} \frac{\theta(\tau)d\tau}{\tau - z}\right), \quad z \notin [-1,1] . \tag{2.9}$$

However, any function $p\Phi$ where p is analytic on (-1,1) will also satisfy (2.7). Applying the Sokhotski-Plemelj formulae again to $p\Phi$ we find that a solution of $M\phi = 0$ is given by

$$\phi(t) = -2i\sin(\pi\theta(t))p(t) \exp\left(-\fint_{-1}^{1} \frac{\theta(\tau)d\tau}{\tau - t}\right) . \tag{2.10}$$

Now, from (2.8), sin $(\pi\theta(t)) = b(t)/r(t)$ where $r = (a^2 + b^2)^{\frac{1}{2}}$ and we assume $r(t) > 0$ for all $t \in [-1,1]$, so that if we choose $p(t) = (iC_2/2)(1+t)^{n_1}(1-t)^{n_2}$ where C_2 is an arbitrary constant and n_1, n_2 are integers, then we find that a solution of the equation $M\phi = 0$ can be written as

$$\phi(t) = C_2\, b(t)Z(t)/r(t) , \tag{2.11}$$

where the so called *fundamental function* Z is defined by

$$Z(t) = (1+t)^{n_1}(1-t)^{n_2} \exp\left(-\fint_{-1}^{1} \frac{\theta(\tau)d\tau}{\tau - t}\right) , \tag{2.12}$$

for $t \in (-1,1)$. If we had proceeded in a similar manner with the homogeneous adjoint equation $M^*\psi = 0$ we would have found that a solution is given by

$$\psi(t) = C_3/Z(t)r(t) , \tag{2.13}$$

where C_3 is an arbitrary constant. These two solutions point up the importance of the fundamental function Z which is defined by (2.12) and is determined once we have fixed the integers n_1 and n_2. To do this let us recall that we shall require $Z \in L\{p_1,p_2\}$. From (2.12) we can rewrite Z as

$$Z(t) = (1-t)^{n_2-\theta(1)}\,(1+t)^{n_1+\theta(-1)}\,\Omega(t) \tag{2.14}$$

say, where

$$\Omega(t) = \exp\{(\theta(1) - \theta(t))\log(1-t) + (\theta(t) - \theta(-1))\log(1+t) - \int_{-1}^{1}\left(\frac{\theta(\tau)-\theta(t)}{\tau - t}\right)d\tau\} . \tag{2.15}$$

Consider first the behaviour of Z near the end point -1. If $\theta(-1)$ is an integer we have what Muskhelishvili calls a *special end* and we shall choose $n_1 = -\theta(-1)$. In this case, both Z and $1/Z$ are bounded at the point -1. Suppose, on the other hand, that $\theta(-1)$ is not an integer. Muskhelishvili calls such an end *non-special* and there is the possibility of a choice. If we choose n_1 so that $n_1 + \theta(-1) > 0$ then Z will be bounded (in fact zero) at -1 and it is certainly p_1-integrable at that point for all values of p_1. However, $1/Z$ will then be unbounded at that point and if we require it to be integrable then we must have $-(n_1 + \theta(-1)) > -1$. Thus if Z is to be bounded at the end point -1 then we try to choose the integer n_1 such that

$$0 < n_1 + \theta(-1) < 1 .$$

On the other hand, if $n_1 + \theta(-1) < 0$ then for Z to be p_1-integrable at -1 we must have $p_1(n_1 + \theta(-1)) > -1$ and in this case $1/Z$ will be zero at -1. So for Z unbounded we choose n_1 such that

$$-1/p_1 < n_1 + \theta(-1) < 0 .$$

A similar discussion can be given at the other end point. If $\theta(1)$ is an integer then it is a special end and $n_2 = \theta(1)$. If it is non-special then we try to choose n_2 such that either

$$0 < n_2 - \theta(1) < 1, \quad \text{if } Z \text{ is to be bounded, or}$$

$$-1/p_2 < n_2 - \theta(1) < 0, \quad \text{if } Z \text{ is to be unbounded.}$$

In all cases we define an *index* κ of M by

$$\kappa = -(n_1 + n_2) . \tag{2.16}$$

It turns out that M may have up to three values of κ depending upon whether Z is chosen to be bounded or unbounded at non-special ends. The largest of these values of κ we shall refer to as *the* index of M and it is such that, where possible, we choose Z to be unbounded at each non-special end. Returning to (2.14), if we define

$$\alpha = n_2 - \theta(1), \quad \beta = n_1 + \theta(-1) \tag{2.17}$$

then we shall always have

$$-1/p_2 < \alpha < 1, \quad -1/p_1 < \beta < 1 , \tag{2.18}$$

so that both Z and $1/Z$ are integrable on $(-1,1)$, a fact we shall exploit later.

Having defined the index it now remains for us to consider its significance and for this we return again to the homogeneous equations $M\phi = 0$ and $M^*\psi = 0$ where we look for solutions ϕ in $L\{p_1,p_2\}$. For $\kappa > 0$ it can be shown that $M\phi = 0$ has κ linearly independent solutions given by $bZ/r, btZ/r,\ldots,bt^{\kappa-1}Z/r$ whereas the adjoint equation $M^*\psi = 0$ has only the trivial solution $\psi = 0$. On the other hand, when $\kappa < 0$ the equation $M\phi = 0$ has only the trivial solution $\phi = 0$ while $M^*\psi = 0$ has $-\kappa$ linearly independent solutions given by $1/rZ,\ t/rZ,\ \ldots,\ t^{-\kappa-1}/rZ$. When $\kappa = 0$ both equations have only the trivial solution. We can combine these statements to give

$$\kappa(M) = \dim\ker(M) - \dim\ker(M^*)\ . \tag{2.19}$$

It can further be shown that if we add to M some compact operator K say, then

$$\kappa(M+K) = \kappa(M)\ . \tag{2.20}$$

After these comments on the solution of the homogeneous equation we find (see, for example [10]) that the solution of the inhomogeneous dominant equation $M\phi = f$ is given by

$$\phi = \hat{M}^I f + \frac{bZ}{r} P_{\kappa-1}\ , \tag{2.21}$$

where

$$\hat{M}^I f = \frac{af}{r^2} - \frac{bZ}{r\pi} \int_{-1}^{1} \frac{f(\tau)d\tau}{r(\tau)Z(\tau)(\tau-t)}\ , \tag{2.22}$$

and $P_{\kappa-1}$ denotes an arbitrary polynomial of degree $\leq(\kappa - 1)$ which is identically zero if $\kappa \leq 0$, provided that when $\kappa < 0$ the *consistency* conditions

$$\int_{-1}^{1} \frac{\tau^{k-1} f(\tau)d\tau}{r(\tau)Z(\tau)} = 0\ ,\ k = 1(1)(-\kappa) \tag{2.23}$$

are satisfied by f. Thus, when $\kappa > 0$, $M\phi = f$ has a solution for all f, the solution not being unique. When $\kappa = 0$ the equation has a unique solution for all f, and when $\kappa < 0$ again we have a unique solution but only for those functions f satisfying (2.23). The operator $\hat{M}^I$ is a one-sided inverse of M; to be more specific we have

$$\left.\begin{aligned} M\hat{M}^I\phi &= \phi,\quad \kappa \geq 0\ ,\\ \hat{M}^I M\phi &= \phi,\quad \kappa \leq 0\ ,\end{aligned}\right\} \tag{2.24}$$

and only when $\kappa = 0$ is $\hat{M}^I$ the inverse of M. Although we shall require some more theory later, let us take a look at the consequences of these results for finding approximate solutions of (1.3).

3. APPROXIMATE SOLUTION OF CSIE: DISCRETE SYSTEM OF EQUATIONS

Let us consider first the dominant equation $M\phi = f$. From (2.21) and (2.22) we have

$$\phi(t) = \frac{Z(t)}{r(t)}\left\{\frac{a(t)f(t)}{r(t)Z(t)} - \frac{b(t)}{\pi}\int_{-1}^{1}\frac{f(\tau)d\tau}{r(\tau)Z(\tau)(\tau-t)} + \right.$$
$$\left. + b(t)P_{\kappa-1}(t)\right\}, \tag{3.1}$$

and from (2.14), (2.15) we see that Z may have a complicated algebraic/logarithmic singularity at either or both of the end points and will, in general, be unbounded at both ± 1. This suggests that we define a new dependent variable ψ by

$$\phi = \frac{Z}{r}\psi \text{ or } \psi = \frac{r\phi}{Z}, \tag{3.2}$$

and try approximating to ψ which, so it turns out, will be Hölder continuous on $[-1,1]$. With this change of dependent variable we shall now rewrite our CSIE as $(A + K)\psi = f$ where

$$\left.\begin{aligned} A\psi(t) &= \frac{a(t)Z(t)\psi(t)}{r(t)} + \frac{b(t)}{\pi}\int_{-1}^{1}\frac{Z(\tau)\psi(\tau)d\tau}{r(\tau)(\tau-t)}, \\ K\psi(t) &= \int_{-1}^{1}\frac{Z(\tau)}{r(\tau)}k(t,\tau)\psi(\tau)d\tau . \end{aligned}\right\} \tag{3.3}$$

The adjoint operator A^* is chosen so that for all Hölder continuous functions ψ_1, ψ_2 we have

$$\int_{-1}^{1}\frac{\psi_1(\tau)A\psi_2(\tau)}{r(\tau)Z(\tau)}d\tau = \int_{-1}^{1}\frac{Z(\tau)}{r(\tau)}\psi_2(\tau)A^*\psi_1(\tau)d\tau, \tag{3.4}$$

and consequently

$$A^*\psi(t) = \frac{a(t)\psi(t)}{r(t)Z(t)} - \frac{1}{\pi}\int_{-1}^{1}\frac{b(\tau)\psi(\tau)d\tau}{r(\tau)Z(\tau)(\tau-t)}. \tag{3.5}$$

Finally we introduce the operator $\hat{A}^I$ which is defined by

$$\hat{A}^I\psi(t) = \frac{a(t)\psi(t)}{r(t)Z(t)} - \frac{b(t)}{\pi}\int_{-1}^{1}\frac{\psi(\tau)d\tau}{r(\tau)Z(\tau)(\tau-t)} \tag{3.6}$$

which (cf.(2.24)) is a right inverse of A when $\kappa \geq 0$ and a left inverse of A when $\kappa \leq 0$. The solution of the dominant equation $A\psi = f$ is given by

$$\psi = \hat{A}^I f + bP_{\kappa-1} \tag{3.7}$$

provided, as before, that f satisfies (2.23) when $\kappa < 0$. From (3.7) we see that one way to obtain approximate solutions of the dominant equation is by applying quadrature rules to evaluate $\hat{A}^I f$. This will require the approximate evaluation of Cauchy principal value integrals and an excellent review of quadrature rules for such integrals has been given by Rabinowitz [13]. We shall explore this problem no further here.

Let us consider now the complete equation $(A + K)\psi = f$. From the theory of CSIE it follows that this equation can be *regularized* so that it is equivalent to the Fredholm equation

$$\psi + \hat{A}^I K\psi = \hat{A}^I f + bP_{\kappa-1} , \tag{3.8}$$

provided that when $\kappa < 0$ we have

$$\int_{-1}^{1} \frac{(f(\tau)-K\psi(\tau))\tau^{k-1}}{r(\tau)Z(\tau)} d\tau = 0 , \quad k = 1(1)(-\kappa) , \tag{3.9}$$

a condition which can be verified once we have found ψ. This suggests that we use methods for Fredholm equations in order to solve (3.8). However, this is not quite as simple as it might seem. If we look at the term $\hat{A}^I K\psi$, it is given by

$$\hat{A}^I K\psi = \int_{-1}^{1} \frac{Z(\xi)}{r(\xi)} \left\{ \frac{a(t)k(t,\xi)}{r(t)Z(t)} - \frac{b(t)}{\pi} \int_{-1}^{1} \frac{k(\tau,\xi)d\tau}{r(\tau)Z(\tau)(\tau-t)} \right\} \psi(\xi)d\xi , \tag{3.10}$$

and is of a rather complicated nature. Some algorithms have been developed using this approach; see Dow and Elliott [2] for an attempt along these lines although it does not appear to have found much favour. Such methods are known as *indirect* in contrast to the so-called *direct* methods which attempt a discretization of the original equation $(A + K)\psi = f$, without performing the process of regularization. One can ask what sort of system of linear algebraic equations should one be looking for given that the equation $(A + K)\psi = f$ is of index κ?

Let A be any real $m \times n$ matrix and let $\rho \leq \min(m,n)$ be the rank of A_n. Now dim ker $(A_n) = n - \rho$ and since the adjoint of

A_n is A_n^T we have dim ker $(A_n^T) = m - \rho$. Recalling (2.19) for the index of M (or A) we can define the index of the matrix A_n by

$$\kappa(A_n) = \dim\ker(A_n) - \dim\ker(A_n^T) = n - m\ . \tag{3.11}$$

Again recalling that $\hat{A}^I$ is a right inverse of A when $\kappa \geq 0$ and is a left inverse when $\kappa \leq 0$, then we would like A_n to possess a right inverse when $\kappa \geq 0$ and a left inverse when $\kappa \leq 0$. From matrix theory we know that such inverses only exist provided that rank $(A_n) = \min(m,n)$. Finally, we might recall that if the system of m equations in n unknowns given by $A_n \underline{x}_n = \underline{y}_m$ is to possess a solution then $\underline{y}_m$ must be orthogonal to the null space of A_n^T.

To sum up; when we discretize an operator A of index κ, we should replace it by an $m \times n$ matrix A_n of rank $= \min(m,n)$ where $m = n - \kappa$. Further properties will of course be needed so that A_n is in some sense a good approximation to A. When $\kappa > 0$ the null spaces of A and A_n also need to correspond in some sense; and when $\kappa < 0$ a similar remark is relevant for the null spaces of A^* and A_n^T. If we look upon a Fredholm equation as a limit of n linear algebraic equations in n unknowns, then we might like to look upon a CSIE of index κ as a limit of $n - \kappa$ equations in n unknowns, as $n \to \infty$.

4. GLOBAL APPROXIMATION METHODS

In these methods we approximate to ψ on $[-1,1]$ by some suitably chosen polynomial. In this section I can outline only the method of *classical collocation*, see [4]. (The reader is referred to [6] for a description of a Galerkin method based on a similar global approximation). Many papers have now been published on these methods as applied to particular cases of (1.3) and we refer the reader to the references given in [4] and [6]. However, in this paper, we are attempting to be as general as possible.

The motivation for these methods must surely have come from a study of the *airfoil* equation which is a special case of (1.3) with $a = 0$, $b = 1$, $k = 0$. From the analysis of §2; if we look for a solution $\phi \in L\{1,1\}$ then we find that $Z(t) = (1 - t^2)^{-\frac{1}{2}}$ and $\kappa = 1$. In terms of ψ the airfoil equation is

$$\frac{1}{\pi}\int_{-1}^{1}\frac{\psi(\tau)d\tau}{(1-\tau^2)^{\frac{1}{2}}(\tau-t)} = f(t), \quad -1 < t < 1 . \tag{4.1}$$

A remarkable pair of equations, which will bring a gleam to the eye of any numerical analyst, is given by

$$\left.\begin{aligned} &\frac{1}{\pi}\int_{-1}^{1}\frac{T_n(\tau)d\tau}{(1-\tau^2)^{\frac{1}{2}}(\tau-t)} = U_{n-1}(t) , \quad n = 1,2,3,\ldots ; \\ &\frac{1}{\pi}\int_{-1}^{1}\frac{(1-\tau^2)^{\frac{1}{2}}U_n(\tau)}{\tau-t}\,d\tau = -T_{n+1}(t) , \quad n = 0,1,2,\ldots ; \end{aligned}\right\} \tag{4.2}$$

where T_n, U_n are Chebyshev polynomials of degree n of the first and second kind respectively. Thus in this special case we see that the particular operators A, $\hat{A}^I$ and A^* map polynomials into polynomials. Should we use Chebyshev polynomials in the more general case? The answer is in the negative since we see from equations (2.14), (2.17) and (2.18) that Z/r behaves like $(1 - t)^{\alpha}(1 + t)^{\beta}\Omega_1(t)$, where $\Omega_1(t)$ is strictly positive on $[-1,1]$. This weight function generates on $[-1,1]$ what Nevai [11] calls the *generalized Jacobi polynomials*, and we must now look for equations corresponding to (4.2) in this more general case. We can define two non-negative integrable weight functions on $(-1,1)$ by

$$w_1 = Z/r , \quad w_2 = 1/Zr . \tag{4.3}$$

Let $\{t_n\}$, $\{u_n\}$ be sequences of orthonormal polynomials generated on $(-1,1)$ by the weight functions w_1 and w_2 respectively. If we assume that A is of index κ and that b is a polynomial of degree μ, then we may establish the following relationships between $\{t_n\}$ and $\{u_n\}$ (see Elliott [5])

$$\left.\begin{aligned} &\text{(i)} \quad \text{for } n \geq \max(\mu,\kappa) , \quad At_n = (-1)^{\kappa}u_m , \\ &\text{(ii)} \quad \text{for } n \geq \max(\mu,\kappa) , \quad \hat{A}^I u_m = (-1)^{\kappa}t_n , \\ &\text{(iii)} \quad \text{for } n \geq \max(0,\kappa) , \quad A^*u_m = (-1)^{\kappa}t_n , \end{aligned}\right\} \tag{4.4}$$

where, as before, $m = n - \kappa$. Equations (4.4) are generalizations of (4.2). In addition, we can show that for $n \geq \max(1 + \kappa, 1 + \mu)$ and all z, both t_n and $u_{n-\kappa}$ satisfy a three term recurrence relation of the form

$$s_{n+1}(z) - (A_n z + B_n)s_n(z) + C_n s_{n-1}(z) = 0 \; . \tag{4.5}$$

Finally, for $n \geq \max(\kappa,\mu)$ and all z,

$$t_{n+1}(z)u_{n-\kappa}(z) - t_n(z)u_{n-\kappa+1}(z) = D_n b(z) \; , \tag{4.6}$$

where D_n depends only on n.

With these preliminaries established we can now consider the method of classical collocation as a direct method for the approximate solution of the complete equation $(A + K)\psi = f$. Let $\tau_{j,n}$, $j = 1(1)n$, denote the zeros of t_n. From the theory of orthogonal polynomials we know that these zeros are all simple and lie in $(-1,1)$. Let us approximate to ψ by ψ_n where

$$\psi_n(t) = \sum_{j=1}^{n} \frac{t_n(t)}{t'_n(\tau_{j,n})(t-\tau_{j,n})}\,\psi(\tau_{j,n}) \; , \tag{4.7}$$

this being the n point Lagrange interpolation polynomial approximation to ψ based on the zeros of t_n. Now for $n \geq \max(\mu,\kappa)$, and using (4.4) we find

$$A\psi_n(t) = (-1)^{\kappa}\,\frac{\psi_n(t)u_m(t)}{t_n(t)} + \frac{b(t)}{\pi}\sum_{j=1}^{n} \mu_{j,n}\,\frac{\psi(\tau_{j,n})}{\tau_{j,n}-t} \; , \tag{4.8}$$

provided that t is not a zero of t_n. Furthermore, we can write

$$K\psi_n(t) = \sum_{j=1}^{n} \mu_{j,n}\, k(t,\tau_{j,n})\,\psi(\tau_{j,n}) + \text{remainder} \tag{4.9}$$

where $\mu_{j,n}$, $j = 1(1)n$, are the Christoffel numbers for the polynomial t_n. The remainder term in (4.9) will be zero for all $t \in [-1,1]$ only if $k(t,\tau)$, considered as a function of τ, is a polynomial of degree $\leq n$. If we define a residual R_n by

$$R_n = (A + K)\psi_n - f \tag{4.10}$$

then in order to find an appropriate system of algebraic equations we shall both neglect the remainder in (4.9) and require that

$$R_n(t_{i,m}) = 0 \; , \quad i = 1(1)m \; , \tag{4.11}$$

where $u_m(t_{i,m}) = 0$ for $i = 1(1)m$. Again, the zeros $t_{i,m}$ are real, simple and lie in $(-1,1)$. From (4.6) we see that only at a zero of b can zeros of t_n and u_m coincide, but we shall ignore this case below although it is readily accounted for (see [4]). In this way we can construct a system of m linear algebraic equations in n unknowns given by

$$(A_n + K_n)\underline{\tilde{\psi}}_n = \underline{f}_m \ , \tag{4.12}$$

where the $m \times n$ matrices A_n and K_n are defined by

$$A_n = \left(\frac{b(t_{i,m})\mu_{j,n}}{\pi(\tau_{j,n}-t_{i,m})}\right) \ , \tag{4.13}$$

$$K_n = (\mu_{j,n}\ k(t_{i,m},\tau_{j,n})) \ , \tag{4.14}$$

for $i = 1(1)m$, $j = 1(1)n$. The $n \times 1$ vector $\underline{\tilde{\psi}}_n$ is defined by

$$\underline{\tilde{\psi}}_n^T = (\tilde{\psi}_n(\tau_{1,n}) \ ,\ldots,\ \tilde{\psi}_n(\tau_{n,n})) \tag{4.15}$$

and we observe that since we neglect the remainder in (4.9) we must replace $\psi(\tau_{j,n})$ by $\tilde{\psi}_n(\tau_{j,n})$, say. Finally the $m \times 1$ vector $\underline{f}_m$ is defined by

$$\underline{f}_m^T = (f(t_{1,m}) \ ,\ldots,\ f(t_{m,m})) \ . \tag{4.16}$$

The matrix A_n is such that we can find *explicitly* a left or right inverse according as $\kappa \le 0$ or $\kappa \ge 0$. If we denote this inverse by $\hat{A}_m^I$, this being an $n \times m$ matrix, then we find

$$\hat{A}_m^I = \left(\frac{b(\tau_{i,n})\nu_{j,m}}{\pi(\tau_{i,n}-t_{j,m})}\right) \ , \tag{4.17}$$

$i = 1(1)n$, $j = 1(1)m$, where $\nu_{j,m}$ are the Christoffel numbers of the polynomial u_m.

In order to carry out a convergence analysis of this particular method, we can apply a general convergence theorem for CSIE which has been developed by the author in [3]. This analysis makes use of the fact that the CSIE, $(A + K)\psi = f$ is equivalent to the Fredholm equation (3.8), and convergence theorems for such equations are readily available. Returning to this particular method it can be shown, making use of Nevai's results [11] for generalized Jacobi polynomials, that

$$||\hat{A}_m^I||_\infty \le C_4 + C_5 \log m \ , \tag{4.18}$$

for large enough m. If we further assume that ψ' is Hölder continuous on $[-1,1]$ then, given any solution ψ of $(A + K)\psi = f$, we can find a sequence of solutions $\{\tilde{\psi}_n\}$ of (4.12) such that

$$||\psi - \tilde{\psi}_n||_\infty \to 0 \text{ as } n \to \infty , \tag{4.19}$$

where $\tilde{\psi}_n$ is given by (4.7) with $\tilde{\psi}_n(\tau_{j,n})$ replacing $\psi(\tau_{j,n})$. Upper bounds can also be given for the rate of convergence, see [4]. It is conjectured that (4.19) is also true if ψ is Hölder continuous with exponent $> \max(-\alpha,-\beta)$. Various extensions of these results are possible if we replace the Gauss quadrature by Gauss-Radau or Gauss-Lobatto quadratures. Some discussion on the application of these quadrature rules has been given by Junghanns and Silbermann [8]. It should be noted that this recently published paper provides an excellent and comprehensive survey of approximate methods for the solution of the same equation that we are considering here.

5. LOCAL APPROXIMATION METHODS

By local approximation methods, I mean those in which we approximate to ϕ by means of piecewise continuous polynomials defined on a suitably chosen mesh. For the CSIE over the interval $(-1,1)$, these methods have not so far been considered as extensively as those of the previous section. The most comprehensive analysis that has appeared to date is that of a Galerkin method described by Thomas [14]. He considers (1.2) with $a \equiv 1$ and K smooth in each variable. The equation can be rewritten as

$$\phi + K_1\phi + K_2\phi = f , \tag{5.1}$$

where

$$\left.\begin{aligned} K_1\phi &= \frac{1}{2} \int_{-1}^{1} \frac{K(t,\tau)+K(\tau,t)}{\tau-t}\,\phi(\tau)d\tau \quad , \\ K_2\phi &= \frac{1}{2} \int_{-1}^{1} \frac{K(t,\tau)-K(\tau,t)}{\tau-t}\,\phi(\tau)d\tau \quad . \end{aligned}\right\} \tag{5.2}$$

We look for solutions ϕ of (5.1) in $L\{2,2\}$ and within this class of functions it turns out that the index of (5.1) is zero. If we look upon K_1 and K_2 as operators from $L_2(-1,1)$ into $L_2(-1,1)$ then it is readily shown that K_1 is skew-symmetric and

K_2 is symmetric and compact. Thomas considers first the Galerkin analysis for the equation

$$B\phi \equiv \phi + K_1\phi = f \, , \tag{5.3}$$

since it readily follows that B is positive definite. The analysis is then extended to the complete equation (5.1). Suppose we look for an approximate solution ϕ_n to ϕ out of a finite-dimensional subspace S_n, say, of $L_2(-1,1)$. If P_n denotes the projection operator from $L_2(-1,1)$ onto S_n then it can be shown that

$$||\phi - \phi_n||_{L_2} \le \{1 + C(n)\}||B|| \; ||(I - P_n)\phi||_{L_2} \tag{5.4}$$

where, if $P_n g \to g \; \forall \; g \in L_2(-1,1)$ then $C(n) \to 0$ as $n \to \infty$. Choosing as a finite dimensional subspace a set of piecewise continuous polynomials of degree k on an $(N + 1)$ point non-uniform mesh on $[-1,1]$, one attempts to choose the mesh so that $||\phi - \phi_n||_{L_2} = O(N^{-k})$. However, there are problems because of the behaviour of ϕ near the ends points $\pm$ 1. This can be overcome by introducing into the basis functions a pair of suitably chosen singular functions near the end points $\pm$ 1 which account for the known behaviour of ϕ near these points. Again one chooses the mesh points so that the error is $O(N^{-k})$ and Thomas provides criteria for this choice.

6. CONCLUSION

In this paper I have attempted to point out some of the problems associated with CSIE, and ways in which attempts have been made to overcome them. The ultimate aim is, I suppose, to design a procedure so that once the user has chosen the functions a, b, k and f and has prescribed the error that is acceptable in the computed solution, such a solution is then produced. We are still a long way from this goal. For example, with the method of classical collocation there are computational difficulties associated with finding the zeros and Christoffel numbers of the generalized Jacobi polynomials. In general one has neither analytic expressions for the moments of the weight functions w_1 and w_2, nor explicit expressions for the coefficients A_n, B_n and C_n in the three term recurrence relation (4.5). I have attempted to overcome this problem by calculating certain modified moments using appropriate Gauss-Jacobi quadrature rules and then using Wheeler's algorithm [16] for the numerical evaluation of the coefficients in (4.5). From the matrix of the recurrence relation we can readily compute the zeros and Christoffel numbers of t_n and u_m. This entire procedure has been described by Gautschi [7]. When $\kappa > 0$ we have, in general, a further κ conditions to incorporate in order to obtain a system of n equations in n unknowns which can then be solved by a standard procedure. When $\kappa < 0$, so that we can have more equations than unknowns, I have been experimenting with the singular value decomposition of the matrices. It is hoped to present the results of these numerical experiments in a future paper.

Having outlined some of the computational complexities of one of the global methods, one is tempted to consider methods based on piecewise polynomial approximation. Some of the elegant mathematics of the classical methods now appears to be lost, and there are other problems. We have mentioned in §5 the introduction of suitable singular functions into the polynomial basis and this raises the question as to whether one should solve directly for ψ rather than ϕ. However, there are now problems involved with the evaluation of the integrals which arise; these again cannot, in general, be given analytically. Thomas' analysis also required the solution to be in $L\{2,2\}$; how can it be generalized to larger classes of functions and to the case when $\kappa \neq 0$?

So there is still work to be done on the approximate solution of CSIE over (-1,1). Once we have solved these problems satisfactorily, we can consider CSIE over unions of arcs and contours. Incidentally I might mention here that much work has been done on CSIE taken over the unit circle on which a, b, k and f are Hölder continuous. The development of the approximate solution for these equations appears to have been more satisfactorily developed than the case we have been discussing here (see, for example, Prössdorf [12]). We should also consider systems of CSIE taken over various arcs and contours, the theoretical basis having been given by Vekua [15]. Finally there are multi-dimensional SIE for which the theory is to be found in the book by Mikhlin [9]. Very little work appears to have been done to date on the approximate solution of such equations.

REFERENCES

1. BAKER, C.T.H. *The Numerical Treatment of Integral Equations*. Oxford University Press, Oxford (1977).
2. DOW, M.L. and ELLIOTT, D. The Numerical Solution of Singular Integral Equations over [-1,1]. *SIAM Journal on Numerical Analysis* 16, 115-134 (1979).

3. ELLIOTT, D. A Convergence Theorem for Singular Integral Equations. *J. Austral. Math. Soc.* (*Series B*) 22, 539-552 (1981).
4. ELLIOTT, D. The Classical Collocation method for Singular Integral Equations. *SIAM Journal on Numerical Analysis* 19, 816-832 (1982).
5. ELLIOTT, D. Orthogonal Polynomials associated with Singular Integral Equations having a Cauchy kernel. *SIAM Journal on Mathematical Analysis* (to be published).
6. ELLIOTT, D. The Classical Galerkin Method for Singular Integral Equations. *University of Tasmania, Mathematics Department, Tech. Rpt. No.*173 (1982).
7. GAUTSCHI, W. On generating Gaussian Quadrature Rules. pp. 147-154 of G. Hämmerlin (Ed.) *Numerische Integration.* ISNM 45, Birkhauser Verlag (1979).
8. JUNGHANNS, P. and SILBERMANN, B. Zur Theorie der Näherungsverfahren für Singuläre Integralgleichungen auf Intervallen. *Math. Nachr.* 103, 199-244 (1981).
9. MIKHLIN, S.G. *Multidimensional Singular Integrals and Integral Equations.* Pergamon Press, Oxford (1965).
10. MUSKHELISHVILI, N.I. *Singular Integral Equations.* Nordhoff, Groningen (1953).
11. NEVAI, P.G. *Orthogonal Polynomials.* Mem. of American Math. Soc. 18 No.213, Providence (1979).
12. PRÖSSDORF, S. *Some Classes of Singular Equations.* North
13. RABINOWITZ, P. The Numerical Evaluation of Cauchy Principal Value Integrals. *Proceedings of the Fourth Symposium on Numerical Mathematics*, Durban, South Africa, 54-82 (1978).
14. THOMAS, K.S. Galerkin methods for Singular Integral Equations. *Math. of Computation* 36, 193-205, (1981).
15. VEKUA, N.P. *Systems of Singular Integral Equations.* Noordhoff, Groningen (1967).
16. WHEELER, J.C. Modified Moments and Gaussian Quadrature. *Rocky Mountain Journ. of Math.* 4, 287-296 (1974).

DESCRIPTIVE REGULARIZATION OF THE FREDHOLM INTEGRAL EQUATIONS OF THE FIRST KIND

*Roman S. Rutman and †Luiz M. Cabral

**Southeastern Massachusetts Univ.* *†Naval Underwater Systems Ctr.*

1. INTRODUCTION

The term "descriptive regularization" has been introduced recently [9] to designate various methods for solving the inverse problem

$$K\, x(s) = y(t), \qquad (1.1)$$

$$K: \mathscr{X} \to \mathscr{Y} \qquad a \le s \le b, \quad c \le t \le d,$$

in the Hilbert spaces $\mathscr{X}$ and $\mathscr{Y}$ of real-valued functions, with shape conditions imposed on the solution x(s): nonnegativity, monotonicity, convexity, the positions of the extrema, or some similar description by which information exogenous to (1.1) is provided. This term is retained here in order to describe a technique where, in addition to shape constraints, information pertaining to the smoothness of the solution may be available.

Introduction of the conditions of smoothness is the traditional way to regularize the ill-posed problem (1.1), i.e. to convert it into a related well-posed problem. The conditions of smoothness may be deterministic [8] or stochastic [10,12]. Both cases result in the equation

$$K^*K\, x^{\alpha}(s) + \alpha \boldsymbol{\Omega} x^{\alpha}(s) = K^*y(t), \qquad (1.2)$$

where the asterisk denotes the adjoint, the operator $\boldsymbol{\Omega}$ is positive definite and the constant $\alpha > 0$. The family of solutions $x^{\alpha}(s)$ defined by (1.2) is called the regularized solution. The selection of specific α and $\boldsymbol{\Omega}$ represents a problem much addressed though not yet satisfactorily solved.

Alternatively, shape constraints have been utilized to achieve a regularizing effect. The shape constraints may be specified by assigning the solution to one of the subsets of the parent Hilbert space $\mathscr{X}$ defined as the following E-sets:

1. set E^+ of nonnegative functions
   E^+: $\{x(s): x(s) \geq 0, a \leq s \leq b\}$;

2. set $E\downarrow$ of monotone (nonincreasing) functions
   $E\downarrow$: $\{x(s): x(s_1) \geq x(s_2), a \leq s_1 < s_2 \leq b\}$;

3. set $E^\cup$ of convex (downwards) functions
   $E^\cup$: $\{x(s): x(s_3) \leq \beta x(s_1) + (1-\beta) x(s_2)$,
   $s_3 = \beta s_1 + (1-\beta) s_2$, $0 \leq \beta \leq 1$,
   $a \leq s_1, s_2 \leq b$, $s_1 \neq s_2\}$.

That the set $E\downarrow$ (for bounded functions) defines a compact in the metric of the space $L_p(a,b)$ is an almost trivial fact [3]. Its mapping by the bounded operator of integration, i.e. the subset of bounded functions in $E^\cup$, is also a compact set. If the E-set is compact, the inverse problem becomes well-posed in the sense of Tikhonov [9]. Some methods for obtaining solutions in the sets $E\downarrow$ and $E^\cup$ were established in [2,6]; see also [4,5].

2. SHAPE CONSTRAINTS

Along with the function $x(s)$, its finite-dimensional approximation $\underline{x} \in \mathbb{R}^n$ is considered. The E-sets as above will be replaced by the corresponding F-sets:

$$F^+: \{\underline{x}: x_i \geq 0, i=1,2,\ldots,n\}, \quad (2.1)$$

$$F\downarrow: \{\underline{x}: x_{i+1}-x_i \leq 0, i=1,2,\ldots,n-1\}, \quad (2.2)$$

$$F^\cup: \{\underline{x}: x_{i-1}-2x_i+x_{i+1} \geq 0, i=2,3,\ldots,n-1\}. \quad (2.3)$$

These sets are specifications of the general type of constraints on $\underline{x}$:

$$(\underline{g}^i,\underline{x}) \geq 0, \quad (2.4)$$

where $\underline{g}^i \in \mathbb{R}^n$, $i \in \pi \subset \{1,2,\ldots,n\}$; $\underline{g}^i$ are linearly independent. Thus, for F^+,

$$\underline{g}^i = \underline{h}^i, \qquad \pi = \{1,2,\ldots,n\}; \quad (2.5)$$

for $F\downarrow$,

$$\underline{g}^i = \underline{h}^i - \underline{h}^{i+1}, \qquad \pi = \{1,2,\ldots,n-1\}; \quad (2.6)$$

for $F^\cup$,

$$\underline{g}^i = \underline{h}^i - 2\underline{h}^{i+1} + \underline{h}^{i+2}, \qquad \pi = \{1,2,\ldots,n-2\}, \tag{2.7}$$

where

$$\underline{h}^i := (0,\ldots,0,\overset{(i)}{1},0,\ldots,0)^T, \qquad \underline{h} \in \mathbb{R}^n. \tag{2.8}$$

We shall introduce here the constraints of selective nonnegativity

$$\underline{g}^i = \underline{h}^i, \quad i \in \pi \subset \{1,2,\ldots,n\}, \tag{2.9}$$

and show that the constraints (2.6) - (2.7) can be reduced to the basic case (2.9) by the transformation

$$\underline{x} = R\underline{z}. \tag{2.10}$$

Towards this end, we will compose the matrix B of the rows $(\underline{g}^i)^T$, $i \in \pi$, and assign arbitrary rows for $i \notin \pi$ in such a way that B is nonsingular. For example, for $F\downarrow$, $\underline{g}^n = \underline{h}^n$, and for $F^{\cup}$, $\underline{g}^{n-1} = \underline{h}^{n-1} - 2\underline{h}^n$, $\underline{g}^n = \underline{h}^n$. Then we define

$$R := B^{-1}. \tag{2.11}$$

Thus, for $F\downarrow$,

$$B = I - E, \tag{2.12}$$

$$R = \sum_{i=1}^{n} E^{i-1}, \tag{2.13}$$

and for $F^{\cup}$,

$$B = I - 2E + E^2, \tag{2.14}$$

$$R = \sum_{i=1}^{n} i\, E^{i-1}, \tag{2.15}$$

where

$$E = [\,\underline{h}^2\ \underline{h}^3 \ldots \underline{h}^n, \underline{0}\,]^T, \tag{2.16}$$

$\underline{0}$ is the zero in $\mathbb{R}^n$ and E^o is assumed equal to I_n, the identity in $\mathbb{R}^n$.

Constraints of the type (2.1) - (2.3) are active only on some set $\pi \subset \{1,2,\ldots,n\}$. Furthermore, different types of constraints can be applied on different sets π_i. Then each of the constraints is acted upon by its own transformation R_i, $i=1,2,\ldots,k$,

$$R = \text{diag}\ \{I_0, R_1, I_1, R_2, \ldots, R_k, I_k\}, \tag{2.17}$$

where the identity matrices I_{n_j}, $j=0,1,\ldots,k$, of the corresponding dimensions n_j are introduced for the intervals on which the constraints are not active. It is understood that $n_j=0$ in the case of adjacent intervals of active constraints.

As an example, the case of selective nonnegative and nonpositive constraints ($x_i \geq 0$, $i \in \pi_+$ and $x_i < 0$, $i \in \pi_-$) yields:

$$\underline{g}^i = \underline{h}^i,\ i \in \pi_+ \text{ or } \underline{g}^i = -\underline{h}^i,\ i \in \pi_-; \tag{2.18}$$

$$\pi = \pi_+ \cup \pi_-; \tag{2.19}$$

$$R = B = \text{diag}\ \{b_i\}; \tag{2.20}$$

$$b_i = -1 \text{ for } i \in \pi_- \text{ and } b_i = 1 \text{ for } i \notin \pi_-. \tag{2.21}$$

Another useful case is the constraints of unimodality. For a maximum at $j=q$, a combination of monotonicity constraints results in

$$B = \text{diag}\ \{B\uparrow_{(qXq)}, B\downarrow_{(n-q)X(n-q)}\} \tag{2.22}$$

$$(B\downarrow = I - E, \quad B\uparrow = E - I),$$

and

$$\pi = \{1,2,\ldots,q-1, q+1,\ldots,n\}. \tag{2.23}$$

For the cases considered above, the transformation $B=R^{-1}$ maps any of the corresponding F-sets into the basic set defined by the selective nonnegativity constraints

$$\underline{z}_i \geq 0,\ i \in \pi \subset \{1,2,\ldots,n\}. \tag{2.24}$$

3. COMBINED CONSTRAINTS ON SHAPE AND SMOOTHNESS

Consider now the finite-dimensional approximation to (1.1) where the values of the observed "data" vector $\underline{y}$ are corrupted by the additive "noise" vector $\underline{w}$:

$$K\underline{x} = \underline{y} + \underline{w}, \tag{3.1}$$

$$K:\ R_n \rightarrow R_m,\ n \geq m,\ \underline{x} \in \mathbb{R}^n,\ \underline{y},\underline{w} \in \mathbb{R}^m.$$

It is assumed that $\underline{w} \in N(0,\sigma)$, with $\underline{y}$ and $\underline{w}$ uncorrelated. The covariance matrix C of the solution vector $\underline{x}$ and the noise variance σ^2 are assumed given.

The problem of finding a consistent and asymptotically

unbiased estimate for the solution vector $\underline{x}$ under constraints (2.20) is now addressed. Following [11], we use here a standard Bayesian approach where the posteriori probability density $P(\underline{x}/\underline{y})$ of the solution $\underline{x}$ is sought via the conditional probability density $P(\underline{y}/\underline{x})=P_n(\underline{y}-K\underline{x})$ and the a priori probability density $P_a(\underline{x})$:

$$P(\underline{x}/\underline{y}) = \text{const } P_a(\underline{x})\ P(\underline{y}/\underline{x}). \tag{3.2}$$

The latter is "recovered" from the covariance matrix of the solution vector $C=[\int x_i x_j P_a(x)dx]$, $i,j=1,2,\ldots,n$, by setting the variational problem of minimization of the information integral $\int \ln P_a(\underline{x})\ P_a(\underline{x})d\underline{x}$. The solution to this problem is known to be

$$P_a(\underline{x}) = \text{const } \exp[-(\underline{x},C^{-1}\underline{x})]. \tag{3.3}$$

The estimate $\hat{\underline{x}}$ for the solution is found from the condition

$$P(\hat{\underline{x}}/\underline{y}) = \sup_{\underline{x}\in\mathbb{R}^n} P(\underline{x}/y). \tag{3.4}$$

If

$$P_n(K\underline{x}-\underline{y}) = \text{const } \exp\ [-\|\underline{y}-K\underline{x}\|^2], \tag{3.5}$$

then

$$P(\underline{x}/\underline{y}) = \text{const } \exp\ \{-[(\underline{x},(K^TK + \sigma^2C^{-1})\underline{x}) -2(K^T\underline{y},\underline{x})]\} \tag{3.6}$$

and

$$\hat{\underline{x}}:\ M(\hat{\underline{x}}) = \inf_{\underline{x}\in\mathbb{R}^n} M(\underline{x}), \tag{3.7}$$

where

$$M(\underline{x}) = (\underline{x},D\underline{x}) - 2(K^T\underline{y},\underline{x}) \tag{3.8}$$

$$D = K^TK + \sigma^2C^{-1} > 0. \tag{3.9}$$

The minimization problem (3.7) is to be solved under the constraints (2.4).

The transformation (2.10) reduces the minimization problem to the following one:

$$\text{find } \hat{\underline{z}}:\ M_R(\hat{\underline{z}}) = \inf_{z\in Q} M\ (\underline{z}), \tag{3.10}$$

$$Q: \{\underline{z} = (z_1,\ldots,z_n)^T;\ z_i \geq 0 \text{ for } i \in \pi \subset \{1,2,\ldots,n\}\}, \tag{3.11}$$

where

$$M_R(\underline{z}) := M(R\underline{x}) \tag{3.12}$$

$$= (\underline{z}, D_R\underline{z}) - 2(\underline{y}_R, \underline{z})$$

$$D_R = R^T(K^TK + \sigma^2C^{-1})R > 0, \tag{3.13}$$

$$\underline{y}_R = R^TK^T\underline{y}.$$

Minimization of $M_R(\underline{z})$ under constraints (3.10) is a well-posed problem of quadratic optimization. To solve this problem, a finite iteration procedure has been developed [7] which consists of the selection of an appropriate subspace based on the Kuhn-Tucker optimality check [1,11]. Since the objective function (3.12) is convex and the minimization is performed on a convex set, the Kuhn-Tucker optimality conditions are both necessary and sufficient.

4. NUMERICAL EXAMPLES

The first two experiments were set with the equation

$$\int_a^b (t-s)[(t-s)^2 + f(t)]^{-1/2} x(s)\,ds = y(t), \tag{4.1}$$

$$a \leq t,s \leq b,\ f(t) > 0.$$

For the numerical procedure, the vectors $\underline{x},\underline{y} \in \mathbb{R}^n$ were utilized instead of $x(s)$ and $y(t)$; Simpson's rule was used for the quadrature. The "data vector" $\underline{y}$ was corrupted by the noise which was assumed gaussian, with the covariance matrix $C=[c_{ij}]$ of the solution $\underline{x}$ assumed equal to

$$[c_{ij}] = [a_{ij}a_{ji}],\quad 1 \leq i,j \leq n, \tag{4.2}$$

where

$$a_{ij} = \begin{cases} \sqrt{5}(j-i+1), & j \geq i, \\ 0, & j<i, \quad i,j = 1,2,\ldots,61. \end{cases} \tag{4.3}$$

In the first case,

$$f(t) = 1.315[1 - (3t/5)^2], \quad b = -a = 5/3, \qquad (4.4)$$

$$n = 61, \; \sigma = 10^{-3}.$$

The exact solution $\bar{x}(s)$ to the problem was known to be represented by two impulses, positive and negative. It is shown in Fig. 1 along with the right-hand part $y(t)$ generated by it (no noise), the solution $x^{\alpha}(s)$ obtained with no shape constraints, and the solution $\hat{x}(s)$, which was obtained by the procedure of descriptive regularization under the assumption that $x(s) \geq 0$, $s \leq 0$, and $x(s) \leq 0$, $s > 0$.

In the second experiment,

$$f(t) = 5/3 - |t|, \quad b = -a = 5/3, \qquad (4.5)$$

$n=21$, and $\sigma=2.10^{-2}$. Fig. 2 illustrates the application of the constraints of selective nonnegativity. The assumption of the solution being nonnegative for $-13/15 \leq s \leq -2/3$ and $2/3 \leq s \leq 13/15$ (crosses) results in a considerable improvement over the case of no shape constraints (dotted line). The circles correspond to the nonnegativity constraints over the whole interval $[a,b]$.

The third experiment (Fig. 3) was set with the kernel

$$K(t,s) = \begin{cases} 1 + \cos(1.0472(t-s)), & |t-s| \leq 6, \\ 0 & , \text{ otherwise}, \end{cases} \qquad (4.6)$$

and $a=0$, $b=20$. The covariance matrix of the solution was chosen as in (4.2) - (4.3). The noise was gaussian with $\sigma=10^{-2}$ and $n=21$. Although the application of nonnegativity constraints improves the regularized solution, it still differs considerably from the exact solution $\bar{x}(s)$. If, however, the nonnegativity constraints are replaced by unimodality, the regularized solution becomes undistinguishable from the exact solution.

5. DISCUSSION

When a combination of specifications is utilized as is the case in the combined shape and smoothness regularizing technique, a question arises naturally: what are the relative contributions of those factors in the solution of the problem? In our case, this question is specifically as follows: which constraints--shape or smoothness--produce a stronger regularizing effect?

Having this question in mind, consider the first example

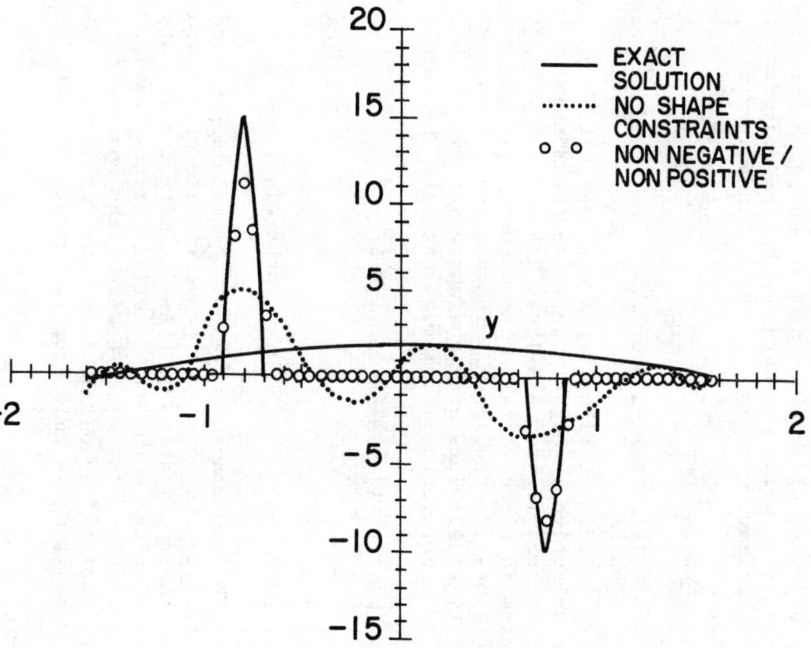

Fig. 1. Constraints of nonnegativity/nonpositivity

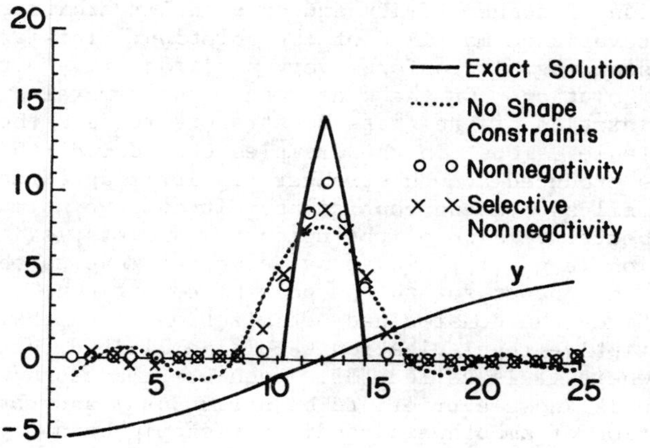

Fig. 2. Constraints of selective nonnegativity

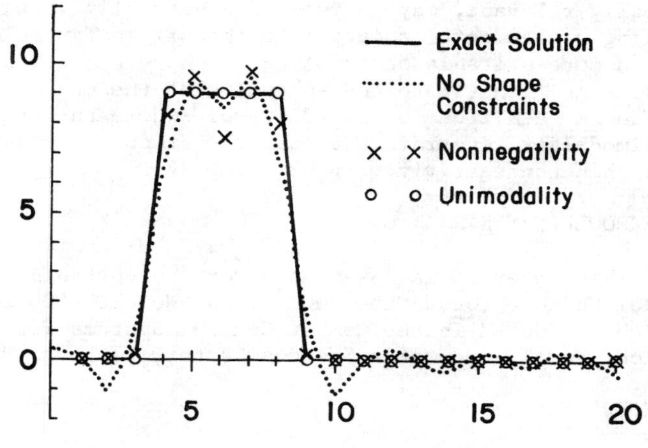

Fig. 3. Constraints of unimodality

of section 4. In that case the solution was obtained with the assumption of nonnegativity and on a rather arbitrary assumption on the covariance matrix C of the solution. This arbitrariness notwithstanding, it conforms very satisfactorily with the "exact" solution. On the other hand, the removal or a significant relaxation of the shape constraints renders the solution unacceptable. Thus, in the examples considered, the shape constraints produced a much stronger regularizing effect than the traditional smoothness conditions. It must be noted, however, that a better version of the unconstrained statistical regularization (e.g. [12]) may provide better results than those used in this paper for the purpose of comparison.

Since the constrained optimization is expensive, the use of descriptive regularization should be limited to the situations where it is called for. Such are the cases where the solution is known a priori to be not smooth, and therefore the application of smoothness conditions should be avoided or reduced to the level where it causes no harm. For other problems, the notion of the intrinsic rank of an ill-posed problem [13] may be helpful in detecting severely ill-posed problems which cannot be treated by the standard regularization alone and require the use of descriptive information whenever it is available. These considerations are all the more useful if one bears in mind that the exogenous information, rather than being always physically relevant, may be imposed arbitrarily in the hope of obtaining a meaningful solution to the original problem.

A more suitable description of shape conditions can contribute significantly to the enhancing of the quality of the solution as seen from the third example where the utilization of the unimodality constraints apparently produces a much stronger effect than nonnegativity.

6. ACKNOWLEDGEMENTS

This material is based upon work supported by the National Science Foundation under Grant No. ECS-8007108 and research conducted in the Combat Control Systems Technology Division, Naval Underwater Systems Center, Newport, RI.

REFERENCES

1. BEAL, E.M.L., Selecting an Optimum Subset. pp. 451-462 of J. Abadie (Ed.), *Integer and Non-Linear Programming*. North Holland Publ., Amsterdam (1970).

2. GONČARSKII, A.V. and V.V. STEPANOV, Algorithms for the Approximate Solution of Incorrectly Posed Problems on Certain Compact Sets. *Soviet Mathematics: Doklady* 20, 414-416 (1979).

3. GONČARSKII, A.V. and A.G. YAGODA, Uniform Approximation of Monotonic Solutions of Incorrect Problems. *Soviet Mathematics: Doklady* 10, 155-157 (1969).

4. GORENFLO, R. and M. HILPERT, On the Continuity of Convexly Constrained Interpolation. pp. 449-454 of E.W. Cheney (Ed.), *Approximation Theory III.* Academic Press (1980).

5. LARKIN, F.M. Estimation of Non-Negative Functions. *BIT* 9, 30-52 (1969).

6. MOROZOV, V.A., N.L. GOLDMAN and M.K. SAMARIN, Methods of Descriptive Regularization and Quality of Approximate Solutions. *Inzhenerno-Fizicheskii Zhurnal* 33, 1117-1124 (1977).

7. RUTMAN, R.S. and L.M. CABRAL, A Non-Linear Regularization Technique for Some Ill-Posed Problems of Identification and Estimation. *Proc. 19-th Conference on Decision and Control.* Albuquerque (1980).

8. TIKHONOV, A.N., Regularization of Incorrectly Posed Problems. *Soviet Mathematics: Doklady* 4, 1624-1627 (1963).

9. TIKHONOV, A.N. and V.A. MOROZOV, Methods of Regularization of Ill-Posed Problems. to appear in *Proc. of the International Symposium on Ill-Posed Problems.* University of Delaware, Newark, Delaware (1979).

10. TURCHIN, V.F., V.P. KOZLOV and M.S. MALKEVICH, The Use of Mathematical-Statistics Methods in the Solution of Incorrectly Posed Problems. *Soviet Physics: Uspekhi* 13, 681-840 (1971).

11. TURCHIN, V.F. and L.S. TUROVTSEVA, Restoration of Optical Spectra and Other Non-Negative Functions by the Statistical Regularization Method. *Optics and Spectroscopy* 36, 162-165 (1974).

12. WAHBA, G., Practical Approximate Solutions to Linear Operator When the Data Are Noisy. *SIAM J. Numer. Analysis.* 14, 651-667 (1977).

13. WAHBA, G., Ill Posed Problems: Numerical and Statistical Methods for Mildly, Moderately and Severely Ill Posed Problems with Noisy Data. to appear in *Proc. of the International Symposium on Ill Posed Problems.* University of Delaware, Newark, Delaware (1979).

ASYMPTOTIC PROPERTIES OF MATRICES ASSOCIATED WITH THE QUADRATURE METHOD FOR INTEGRAL EQUATIONS

K. Wright

University of Newcastle upon Tyne, Computing Laboratory

1. INTRODUCTION

This paper is concerned with matrices used in the quadrature method for the numerical solution of Fredholm Integral equations of the second kind. It is shown that the norm of the inverse of the matrix based on an n point formula tends to the norm of the operator associated with the given equation as $n \to \infty$ provided certain conditions are satisfied. The result is illustrated with a variety of equations and quadrature formulae.

Suppose that the given integral equation is

$$x(t) - \int_a^b k(t,s)x(s)ds = y(t), \tag{1}$$

where a and b are assumed finite. The kernel $k(t,s)$ is assumed to be continuous. This equation may be written in operator form as

$$(I-K)\ x = y, \tag{2}$$

with x and y elements of some Banach space such as $C[a,b]$ or $R[a,b]$ the space of Riemann integrable functions on $[a,b]$ with the infinity norm. It is assumed the $I-K$ is invertible. Since K is assumed continuous K will be a compact operator.

Approximate solutions to (1) are assumed to be obtained using n point quadrature formulae

$$\int_a^b z(t)dt = \sum_{i=1}^{n} w_i\ z(\xi_i) \tag{3}$$

where ξ_i, $i=1,\ldots n$ are some points in $[a,b]$ ordered so that $\xi_1 < \xi_2 < \cdots < \xi_n$, and w_i are corresponding weights assumed positive and satisfying $\sum_{i=1}^{n} w_i = b - a$.

It is also assumed that

$$\Delta_n = \max_{i=2,n} \{(\xi_i - \xi_{i-1}), \xi_1 - a, b - \xi_n\} \to 0 \tag{4}$$

as $n \to \infty$. Other assumptions will be introduced when required to emphasise why they are needed.

Applying the formula in the usual way gives a set of algebraic problems of the form

$$A \underline{x}_n = \underline{y} \tag{5}$$

where $\underline{x}_n^T = \{x_n(\xi_1), \ldots x_n(\xi_n)\}$, the values of the approximate solution x_n at the given points. Similar notation is used for $\underline{y}$ and y.

Here the Matrix A has elements

$$A_{ij} = \delta_{ij} - w_j\, K(\xi_i, \xi_j). \tag{6}$$

Having solved the algebraic equations the function $x_n(t)$ may be defined in a variety of ways using some form of extension or prolongation operator Ψ. In the derivation Ψ is chosen so that $\Psi \underline{x}$ is a piecewise constant function rather than a smoother extension which might seem more natural.

If the inverse of A is denoted by W_n, that is

$$W_n = A^{-1} \tag{7}$$

then the result to be derived is that

$$\|W_n\| \to \|(I-K)^{-1}\| \quad \text{as } n \to \infty \tag{8}$$

with appropriate infinity norms used, if either the quadrature rules are all Riemann sums for sufficiently large n, or the quadrature is an m times repeated q point rule with positive weights (so that $n = mxq$).

In view of convergence results for the solution of integral equations using quadrature methods, the result is not particularly surprising. However since the convergence is not uniform for all right hand sides with norm less than one, it is not trivial.

The result clearly has relevance to error estimation, since if we are given some approximate solution $x^*(t)$ and we can calculate the residual

$$R(x^*,t) = y(t) - x^*(t) + \int_a^b k(t,s)x^*(s)ds,$$

or

$$R = y - (I-K)x^*, \tag{9}$$

then using (2) the error is

$$e = x^* - x = -(I-K)^{-1}R, \tag{10}$$

which may be bounded using

$$\|e\| \leq \|(I-K)^{-1}\| \cdot \|R\|.$$

The result also indicates that if $(I-K)$ is a well conditioned operator then for n sufficiently large the algebraic problem will be well conditioned too.

Previous work in this area such as Anselone [1], Phillips [5] and Noble [4] have given bounds for $\|W\|$ in terms of $\|(I-K)^{-1}\|$ and vice versa. The result here is also related to previous work on collocation methods for ordinary differential boundary value problems, where similar results hold; details are given in Wright [6,7], Gerrard [2] and Gerrard and Wright [3].

2. OPERATOR EXPRESSION FOR W_n

To proceed further it is necessary to introduce some further notation, so that the approximate equation may be written in operator form.

Since the extension operator $\Psi: R^n \to Y$ will be chosen to give a piecewise constant function, Y is taken as $R[a,b]$ to allow for discontinuities in the approximate solution. The extension is also chosen so that

$$\|\Psi\| = 1.$$

A further evaluation (or restriction) operator is also needed; this will be denoted by $\Phi: Y \to R^n$, such that Φy is a vector whose components are the values of y at points ξ_i, $i=1,\dots n$. Clearly $\|\Phi\| = 1$ if the infinity norm is used in both spaces. The operator

$$P_n = \Psi\,\Phi \tag{11}$$

will be a projection operator onto the space $Y^n = \Psi R^n$.

Suppose the operator $K_n : Y \to Y$ is defined by

$$K_n z = \sum_{i=1}^{n} w_i k(s,\xi_i) z(\xi_i). \tag{12}$$

Then the equation (5) and (6) are equivalent to

$$\Phi(I-K_n)\ \Psi \underline{x}_n = \Phi y. \tag{13}$$

Defining $x_n = \Psi \underline{x}_n$ and applying the operator Ψ on the left gives

$$\Psi\Phi(I-K_n)x_n = \Psi\Phi y.$$

Since $x_n \in Y^n$, it follows that $\Psi\Phi x_n = P_n x_n = x_n$,

$$(I-P_n K_n)x_n = P_n y \tag{14}$$

and

$$x_n = (I-P_n K_n)^{-1} P_n y. \tag{15}$$

Applying Φ on the left gives

$$\underline{x}_n = \Phi x_n = \Phi(I-P_n K_n)^{-1} P_n y$$

$$= \Phi(I-P_n K_n)^{-1} \Psi \underline{y} \tag{16}$$

This in turn implies that the matrix W_n defined by (7) may be written as

$$W_n = \Phi(I-P_n K_n)^{-1}\Psi. \tag{17}$$

This expression is the starting point for the analysis of the behaviour of W_n. In the next section a comparison matrix Z_n is introduced, and conditions obtained for $\|W_n - Z_n\|$ to tend to zero. In section 4 $\|Z_n\|$ is shown to tend to $\|(I-K)^{-1}\|$ and in section 5 the remaining conditions are reduced to ones on the quadrature formula. Finally some illustrative examples and further comments on the observed values of $\|W_n\|$ are given.

3. THE COMPARISON MATRIX Z_n

In this section W_n is compared to the matrix

$$Z_n = \Phi\,(I-K)^{-1}\Psi. \tag{18}$$

Unlike W_n, this matrix will depend on the precise definition of Ψ.

From (17) and (18) it follows that

$$W_n - Z_n = \Phi(I-P_n K_n)^{-1}\{I-K-(I-P_n K_n)\}\,(I-K)^{-1}\Psi$$

$$= \Phi(I-P_n K_n)^{-1}(P_n K_n - K)(I-K)^{-1}\Psi$$

Expanding $(I-P_n K_n)^{-1}$ this becomes

$$\Phi\{I + (I-P_n K_n)^{-1} P_n K_n\}\,(P_n K_n - K)(I-K)^{-1}\Psi$$

$$= \Phi(P_n K_n - K)(I-K)^{-1}\Psi + W_n\,\Phi K_n (P_n K_n - K)(I-K)^{-1}\Psi,$$

using the expression (17) for W_n and the definition (11) $P_n = \Psi\Phi$.

Treating $(I-K)^{-1}$ in a similar way and collecting terms gives

$$W_n\{I-\Phi K_n(P_nK_n-K)\Psi - \Phi K_n(P_nK_n-K)K(I-K)^{-1}\Psi\} = Z_n + \Phi(P_nK_n-K)\Psi + \Phi(P_nK_n-K)K(I-K)^{-1}\Psi. \tag{19}$$

So if

$$\|Z_n\| \to \|(I-K)^{-1}\| \tag{20}$$

$$\|\Phi K_n(P_nK_n-K)\Psi\| \to 0$$

$$\|\phi K_n(P_nK_n-K)K(I-K)^{-1}\Psi\| \to 0$$

$$\|\Phi(P_nK_n-K)\Psi\| \to 0$$

and

$$\|\phi(P_nK_n-K)K(I-K)^{-1}\Psi\| \to 0 \quad \text{as} \quad n \to \infty$$

then $\|W_n\| \to \|(I-K)^{-1}\|$ also, and the slightly stronger result $\|W_n-Z_n\| \to 0$ will also hold. Now $\|\Psi\| = \|\Phi\| = 1$, both $\|K\|$ and $\|(I-K)^{-1}\|$ are fixed, and $\|K_n\|$ can easily be seen to be bounded using the definition (12), the assumption that $\sum_{i=1}^{n} w_i = b-a$ and the boundedness of $k(s,t)$. Noting also that $K_nP_n = K_n$ and $\Phi P_n = \Phi$, these conditions will hold if

$$\|\Phi(K_n-K)\Psi\| \to 0 \tag{21}$$

and

$$\|(K_n-K)K\| \to 0 \tag{22}$$

as $n \to \infty$ along with condition (20).

4. TREATMENT OF $\|Z_n\|$.

Here further conditions are found which ensure that (20) will hold. Since $\|\Phi\| = \|\Psi\| = 1$ it follows immediately that

$$\|Z_n\| \leq \|(I-K)^{-1}\| \tag{23}$$

By considering the resolvent kernel corresponding to $(I-K)^{-1}$, and the form of the function y which maximizes $\|(I-K)^{-1}y\|$, it follows that $\|(I-K)^{-1}\|$ is the same in both $C[a,b]$ and $R[a,b]$. This implies that given any $\varepsilon > 0$ there is a continuous function y_0 with $y_0 = 1$ such that

$$u_0 = (I-K)^{-1}y_0$$

satisfies

$$\|u_0\| \geq \|(I-K)^{-1}\| - \varepsilon. \tag{24}$$

Let

$$\underline{v} = Z_n \Phi y_0 = \Phi(I-K)^{-1} P_n y_0$$

$$= \Phi\{I + (I-K)^{-1}K\} P_n y_0$$

$$= \Phi y_0 + \Phi(I-K)^{-1} KP_n y_0.$$

Now if

$$\|K(P_n - I)z\| \to 0 \quad \text{as} \quad n \to \infty \tag{25}$$

for <u>any</u> fixed continuous z then

$$\|\underline{v} - \Phi y - \Phi(I-K)^{-1} Ky_0\| \to 0$$

so that

$$\|\underline{v} - \Phi(I-K)^{-1} y_0\| = \|\underline{v} - \Phi u_0\| \to 0,$$

and for n sufficiently large

$$\|\underline{v} - \Phi u_0\| < \varepsilon \tag{26}$$

Now since Δ_n defined by (4) tends to zero as $n \to \infty$ then

$$\|\Phi u\| \to \|u\|$$

for any fixed continuous u, and hence

$$|\ \|\Phi u_0\| - \|u_0\|\ | < \varepsilon$$

for n sufficiently large.

Together (24), (26) and (27) imply that

$$\|\underline{v}\| \geq \|(I-K)^{-1}\| - 3\varepsilon$$

for n sufficiently large.

But

$$\|\underline{v}\| = \|Z_n y,\| \leq \|Z_n\| \leq \|(I-K)^{-1}\|$$

using (23), and so the result follows as long as condition (25) is satisfied. This will be treated in the next section along with the other conditions

5. FURTHER REDUCTION OF THE CONDITIONS

In this section the remaining conditions (21), (22) and (25) are shown to hold if one further condition on the quadrature formula (3) is satisfied. In doing this the extension function will be given a precise definition.

First consider condition (21), this needs to be treated with care as $\|K_n - K\|$ does not tend to zero. As $\Phi(K_n - K)\Psi$ is a matrix we have the explicit expression

$$\|\Phi(K_n-K)\Psi\| = \max_j \sum_{i=1}^{n} |\{(K_n-K)\Psi\underline{e}_i\}(\xi_j)| \tag{28}$$

where $\underline{e}_i$ is the ith unit vector. Now

$$\{(K_n-K)\Psi\underline{e}_i\}(\xi_j) = \sum_{m=1}^{n} w_m k(\xi_j,\xi_m)(\Psi\underline{e}_i)(\xi_m)$$

$$- \int_a^b k(\xi_j,s)\ (\Psi\underline{e}_i)(s)ds$$

$$= w_i k(\xi_j,\xi_i) - \int_a^b k(\xi_j,s)(\Psi\underline{e}_i)(s)ds$$

since by definition $\Psi\underline{e}_i(\xi_m) = \delta_{im}$. To proceed further the definition of ψ needs to be completed, that is values at points other than the ξ_j, j=1,...n must be specified. First define the points η_i, $j = 0,\ldots n$ by

$$\eta_0 = a$$

$$\eta_i = \eta_{i-1} + w_i,\ i = 1,\ldots n,$$

so that $w_n = b$. Now define

$$(\Psi\underline{e}_j)(t) = \begin{cases} \delta_{jm}, & \text{if } t = \xi_m,\ m=1,\ldots n, \\ 1, & \text{otherwise and } \eta_{j-1} \leqslant t < \eta_j \\ 0, & \text{otherwise.} \end{cases} \tag{29}$$

with the choice of Ψ

$$\{(K_n-K)\Psi\underline{e}_i\}(\xi_j) = \int_{\eta_{i-1}}^{\eta_i} \{ k(\xi_j,\xi_i)-k(\xi_j,s)\}ds.$$

Now let

$$\Delta_n^* = \max_i |\xi_i - \eta_i|, \tag{30}$$

and suppose that the quadrature formula is such that

$$\Delta_n^* = \to 0 \quad \text{as} \quad n \to \infty, \tag{31}$$

as well as $\Delta_n \to 0$ which has already been assumed. Then since $k(t,s)$ is assumed continuous

$$|k(\xi_j,\xi_i) - k(\xi_j,s)| \to 0$$

for any s in $[\eta_{i-1},\eta_i]$. Since continuity implies uniform continuity we have for any given $\varepsilon > 0$ a δ such that if $\Delta_n < \delta$

and $\Delta_n < \delta$

$$|k(\xi_j,\xi_i) - k(\xi_j,s)| < \varepsilon$$

for any s in $[\eta_{i-1},\eta_i]$, independently of j. Hence

$$\sum_{i=1}^{n} |\{(K_n-K)\Psi \underline{e}_i\}(\xi_j)| < \varepsilon \sum_{i=1}^{n} (\eta_i-\eta_{i-1}) = \varepsilon(b-a).$$

As this is independent of j substitution in (28) gives the required result that

$$\|\Phi(K_n-K)\Psi\| \to 0 \quad \text{as} \quad n \to \infty.$$

With the assumed conditions on the quadrature formula propositions 2.1 and 1.7 of Anselone [1] imply that

$$\|(K_n-K)K\| \to 0 \quad \text{as} \quad n \to \infty. \tag{32}$$

The remaining condition is (25), that

$$\|K(P_n-K)z\| \to 0$$

for any continuous z. Here continuity of z implies that

$$\|(P_n-I)z\| \to 0$$

since

$$\{(P_n-I)z\}(s) = z(\xi_i) - z(s)$$

where i is chosen so that $\eta_{i-1} \leqslant s < \eta_i$.

6. ILLUSTRATIONS AND DISCUSSION

First note that condition (3) will be satisfied if the quadrature formula forms a Riemann sum, or is a repeated q point rule over m equal intervals so that $n = mq$. To illustrate the behaviour of $\|W_n\|$, a selection of values for a variety of problems and quadrature formulae are tabulated. Three kernels are considered

1. $\exp(st)$
2. $|s-t|$
3. $\begin{cases} \frac{1}{2}(t-1), & s \leqslant t \\ \frac{1}{2}(t+1), & s > t. \end{cases}$

In all cases $a = -1$ and $b = +1$.
The first kernel is smooth, the second has a discontinuous derivative on $t = s$, and the third has a discontinuity on $t = s$ and so does not satisfy the initial assumption.

TABLE 1 : KERNEL 1 – REPEATED FORMULAE

M	A	B	C	D	E
5	4.722 4.151	5.075 4.447	6.397 5.576	4.289 4.130	27.317 17.894
10	4.907 4.825	5.029 4.925	5.370 5.211	4.771 4.723	33.117 9.433
15	4.963 4.940	5.026 5.004	5.204 5.172	4.901 4.880	35.044 6.992
20	4.995 4.970	5.037 5.009	5.148 5.110	4.954 4.947	35.873 7.993

TABLE 2 : KERNEL2 – REPEATED FORULAE

M	A	B	C	D	E
5	4.047 3.077	3.961 2.993	3.760 2.814	5.512 4.855	149.908 27.011
10	4.887 4.058	4.884 4.008	4.867 3.891	5.535 4.972	64.549 9.252
15	5.204 4.184	5.215 4.156	5.233 4.088	5.618 5.084	51.497 6.910
20	5.370 4.253	5.383 4.233	5.410 4.187	5.674 5.154	48.223 7.984

TABLE 1* :
KERNEL 1 – HIGH ORDER FORMULAE

n	GAUSS	CHEBYSHEV
5	4.759 *	4.982 *
10	4.997 *	5.062 *
15	5.047 *	5.077 *
20	5.065 *	5.082 *
25	5.073 *	5.085 *
30	5.078 5.047	5.086 5.050

TABLE 2* :
KERNEL 2 – HIGH ORDER FORMULAE

n	GAUSS	CHEBYSHEV
5	3.906 3.787	4.602 4.110
10	5.287 4.340	5.562 4.426
15	5.612 4.320	5.749 4.401
20	5.734 3.664	5.815 3.715
25	5.793 3.793	5.846 3.841
30	5.825 3.882	5.863 3.928

Two types of quadrature formulae are used. The first consists of repeated two point interpolating quadrature formulae based on the following selections of points (normalized to –1,1) :

A Gauss points : ±.57735
B Chebyshev Zeros : ±.70711
C Trapezoidal rule : ± 1
D - .5, .1
E - .5, - .4

Choice D has positive weights and so satisfies the conditions but the quadrature does not form a Riemann sum. Choice E has weights -8 and 10 and so does not satisfy the conditions. The second type of formulae are interpolating quadratures based on n Gauss points and zeros of Chebyshev polynomials.

In addition to the calculated values of $\|W_n\|$ the tables include estimates obtained by a single forward and back substitution using the LU decomposition of A with the right hand side having elements ±1 with sign chosen to maximise the value calculated at each stage. This of course involves much less work than the full calculation of $\|W_n\|$.

Tables 1 and 1*, 2 and 2* give the values for kernel 1 and kernel 2. The lower values are the estimates of $\|W_n\|$ and an asterisk is used when the estimate is equal to the correct value. m indicates the number of subintervals used and n the number of points in the higher order formulae.

The results all clearly illustrate the predicted convergence. Results E are clearly not tending to the same limit, indicating that positive weights are a necessary condition. Note that the rate of convergence for most of the piecewise results is $O(1/m)$ while that for the global formulae is $O(1/n^2)$. These results can be explained by considering the function u* which maximzes $\|(I-K)^{-1} u\|$ subject to $\|u^*\| = 1$ and the corresponding function $v^* = (I-K)^{-1}u^*$.

The corresponding vector components will be approximations to $v^*(\xi_i)$ and so one component of the error will arise from the distance between the point t* where $v^*(t)$ attains its maximum and the nearest value of ξ_i.

Tables 3 and 3* illustrate the corresponding values for kernel 3.

For this problem the exact value is 2.313 to three decimal places. It is clear that convergence is taking place in all cases except E indicating that $k(s,t)$ does not need to be continuous. In fact it would not be too difficult to modify the derivation to allow for some discontinuities.

The estimates of the matrix norms show a number of values in precise agreement with the calculated values and others very close, but the results are somewhat erratic. However the values do usually at least give a rough order of magnitude and so might still be useful in error estimation. The larger differences occur with the choice E, two point repeated formula, which is unlikely to be used in practice. Note also that the conditions for the main result here are more stringent than would be needed for the convergence of the approximate solution, which would occur for example using points E above.

TABLE 3 : KERNEL 3 - REPEATED FORMULAE

M	A	B	C	D	E
5	1.809	1.813	1.790	1.982	55.401
	.987	.987	1.070	1.108	*
10	2.035	2.036	2.000	2.135	21.602
	.995	.995	1.052	1.060	13.335
15	2.121	2.121	2.091	2.192	17.248
	.998	.998	1.040	1.042	5.816
20	2.166	2.167	2.141	2.221	14.669
	.999	.999	1.038	1.032	3.415

TABLE 3* : KERNEL 3 - HIGH ORDER FORMULAE

n	GAUSS	CHEBYSHEV
5	1.654	1.735
	*	*
10	2.023	2.063
	*	*
15	2.138	2.159
	*	*
20	2.190	2.203
	*	*
25	2.219	2.228
	*	*
30	2.237	2.243
	*	*

7. CONCLUSIONS

The result derived in this paper helps to justify the use of $\|W_n\|$ as an estimate for $\|(I-K)^{-1}\|$ for use in the error analysis of quadrature methods for integral equations. The empirical results suggest that estimates of $\|W_n\|$ involving much less work than the full calculation may provide a useful though rougher estimate of $\|(I-K)^{-1}\|$.

8. REFERENCES

1. ANSELONE, P.M. *Collectively Compact Operator Approximation Theory*. Prentice Hall, Englewood Cliffs N.J. (1971).

2. GERRARD, C. *Computable Error Bounds for Approximate Solutions of Ordinary Differential Equations*. Ph.D. Thesis University of Newcastle upon Tyne (1979)

3. GERRARD, C. and WRIGHT, K. Asymptotic Properties of Collocation Matrix Norms 2: Piecewise Polynomial Approximation. Submitted for publication (1982).

4. NOBLE, B. Error Analysis of Collocation Methods for Solving Fredholm Integral Equations. pp. 211-232 of J.J.H. Miller (Ed.) *Topics in Numerical Analysis*. Academic Press, London and New York (1973).

5. PHILLIPS, J.L. Collocation as a projection method for solving integral and other Operator Equations. *S.I.A.M.J. Num. Anal. 9*, 14-28 (1972).

6. WRIGHT, K. *Asymptotic Properties of the Norms of Certain Collocation Matrices*. Technical Report. University of Newcastle upon Tyne, Computing Laboratory (1979).

7. WRIGHT, K. Asymptotic Properties of Collocation Matrix Norms 1: Global Polynomial Approximation. Submitted for publication (1982).

BOUNDARY INTEGRAL EQUATIONS AND MODIFIED GREEN'S FUNCTIONS

*R. E. Kleinman and †G. F. Roach

* *University of Delaware and* † *University of Strathclyde*

1. INTRODUCTION

It is well known that the Dirichlet and the Neumann problem can each be reduced to a boundary integral equation by using a fundamental solution of the equation under consideration in conjunction with either a Layer Theoretic Method or a Green's Theorem Method.

The boundary integral equations arising when dealing with the Helmholtz equation present a number of intriguing problems, mainly concerned with questions of uniqueness of solution and difficulties near eigenvalues, which have been the source of interest for a number of years. Of course, these problems can be removed entirely if the exact Green's function for the problem is known. Unfortunately the exact Green's function is only known for a few simple surfaces. Despite this however an attempt has been made [7] [9] to use, instead of the fundamental solution, a Green's function which is known exactly for some neighbouring region. This approximate Green's function technique leads to similar boundary integral equations to those obtained when the fundamental solution was used, which is not surprising since the approximate Green's function is itself a fundamental solution. Whilst the boundary integral equations obtained by using an approximate Green's function offer good prospects for numerical analysis, nevertheless a number of problems remain.

Recently Jones [7] introduced a theory of modified Green's functions in order to overcome the uniqueness problem arising in the boundary integral formulation of the exterior Dirichlet and Neumann problems for the Helmholtz equation. In this theory the fundamental solution, or free space Green's function, for the Helmholtz equation was modified by adding radiating spherical wave functions, that is outgoing solutions of the Helmholtz equation, and the coefficients of these added terms were chosen so as to ensure that the boundary integral formulation of the problem was uniquely solvable for all real values of the wave number. Ursell [9] simplified the proof of a key theorem in [7] but confined his remarks to the exterior Neumann problem in two dimensions.

In [4] the authors gave a systematic account of the boundary integral formulations of both the Dirichlet and the Neumann problems and presented a number of properties of the boundary integral operators arising in both the Layer Theoretic Method and the Green's Theorem Method. In particular it was shown that uniqueness of the boundary integral equation formulation of exterior problems could be retained even at eigenvalues of the corresponding adjoint interior problems by treating a pair of coupled equations. Similar results for the Robin problem are given in [2] where it is shown how classical problems for smooth boundaries may include boundary values in L_2.

In a recent paper [5] the authors have shown how Jones' modification can be incorporated into the boundary integral formulation presented in [4]. Ursell's simplification has been adapted to three dimensions and explicit results have been obtained for both the Dirichlet and Neumann problems. In particular it is shown that a single boundary integral equation is uniquely solvable in each case even at interior eigenvalues of the adjoint problems by suitably modifying the Green's function in the way suggested by Jones [7]. Furthermore the authors have shown that by abandoning the restriction to real coefficients in the modification which Jones and Ursell found sufficient to eliminate non-uniqueness of interior eigenvalues, the coeffcients may be chosen to be optimal with respect to certain specific criteria. This was motivated by a desire not only to ensure unique solvability but also to provide a constructive method of solving the boundary integral equations. In particular in [5] results are presented which show how to choose the coefficients so as to minimize the difference between modified and exact Green's functions for the Dirichlet and Neumann problems. In [6] it is shown that the coefficients can be chosen to minimize the norm of the modified boundary integral operator. This provides a bound on the spectral radius of the modified boundary integral operator and hence an indication when the associated boundary integral equation is solvable by iteration. In [6] an explicit definition of these "optimal" coefficients is given together with an alternative definition which, although not optimal, nevertheless simplifies the computation of the coefficients considerably whilst at the same time ensuring unique solvability. These results for the Dirichlet and Neumann problem have been extended to the Robin problem in [1] [2].

In this note we give a brief description of the modified Green's function technique together with a summary of results obtained so far.

The authors are very grateful to Dr. A. Kirsch of the University of Göttingen for carrying out the numerical work leading to Table 3. A more extensive numerical treatment of these problems will appear elsewhere [3].

2. NOTATION AND DEFINITIONS

Let B_- denote a bounded domain in $\mathbb{R}^3$ with closed boundary ∂B and simply connected exterior B_+. The boundary ∂B will be assumed to be a Lyapunov surface.

Let $P=(r_p,\theta_p,\phi_p)\in\mathbb{R}^3$ be a typical point in $\mathbb{R}^3$ where the coordinates are spherical polar coordinates relative to a Cartesian frame with origin in B_-.

We denote by $\hat{n}_p$ the unit outward normal to ∂B at a point $p\in\partial B$ ($\hat{n}_p$ points into B_+). By $\frac{\partial}{\partial n_p}$ we mean the derivative in the direction of the normal $\hat{n}_p$. Further we shall write $\frac{\partial}{\partial n_p^+}$ and $\frac{\partial}{\partial n_p^-}$ to denote normal derivatives as $P\to p\in\partial B$ from B_+ and B_- respectively.

Basic problem:

Determine $u(P)$ satisfying

$$(\Delta+k^2)u(P) = 0 \quad P\in B_+ \tag{2.1}$$

$$\lim_{r_p\to\infty}\{r_p[\frac{\partial u}{\partial r_p}(p) - iku(p)]\} = 0 \tag{2.2}$$

and either

$$u(p)=f(p)\ ,\quad p\in\partial B \qquad \text{Dirichlet Problem} \tag{2.3}$$

or

$$\frac{\partial u}{\partial n_p}(p)=f(p)\ ,\quad p\in\partial B \qquad \text{Neumann Problem} \tag{2.4}$$

or

$$\frac{\partial u}{\partial n_p}(p)+\sigma u(p) = f(p)\ ,\quad p\in\partial B \qquad \text{Robin Problem}. \tag{2.5}$$

Any solution of (2.1) and (2.2) is a radiating solution of the Helmholtz equation.

A fundamental solution of the Helmholtz equation is given by

$$\gamma_0(P,Q;k)\equiv\gamma_0(P,Q) := -\frac{e^{ikR}}{2\pi R},\quad R=|P-Q| \tag{2.6}$$

and if $g(P,Q)$ is a radiating solution of the Helmholtz equation in both P and Q then

$$\gamma_1(P,Q:k)\equiv\gamma_1(P,Q) := \gamma_0(P,Q) + g(P,Q) \tag{2.7}$$

is also a fundamental solution. We refer to γ_1 as the modified (free space) Green's function.

If $w \in L_2(\partial B)$ then we may obtain standard or modified forms of the single and double layer distributions of density w according to whether γ_o or γ_1 is used as fundamental solution. Specifically we have for $j = 0,1$

$$(S_j w)(P) := \int_{\partial B} \gamma_j(P,q)w(q)\,dS_q \qquad P \in \mathbb{R}^3 \tag{2.8}$$

$$(D_j w)(P) := \int_{\partial B} \frac{\partial \gamma_j}{\partial n_q}(P,q)w(q)\,dS_q \qquad P \in \mathbb{R}^3 . \tag{2.9}$$

Denote by K_j, j 0,1 the boundary integral operator

$$(K_j w)(p) := \int_{\partial B} \frac{\partial \gamma_j}{\partial n_p}(p,q)w(q)\,dS_q \tag{2.10}$$

and its $L_2(\partial B)$ adjoint

$$(K_j^* w)(p) := \int_{\partial B} \frac{\partial \overline{\gamma}_j}{\partial n_q}(p,q)w(q)\,dS_q . \tag{2.11}$$

In terms of K_j the jump conditions for the single and double layer distributions can be written

$$\frac{\partial}{\partial n_p}\left\{S_j w\right\}_{\pm}(p) = (\pm I + K_j)w(p) \qquad p \in \partial B \tag{2.12}$$

$$\lim_{p \to p^{\pm}} \left\{D_j w\right\}(p) = (\mp I + \overline{K}_j^*)w(p) \qquad p \in \partial B. \tag{2.13}$$

With this notation an appropriate form of Green's theorem for external problems can be written

$$(S_j u_n)(p) - (D_j u)(p) = 2u(P) \qquad P \in B_+ \qquad u_n = \frac{\partial u}{\partial n} \tag{2.14}$$

$$(S_j u_n)(p) - (\overline{K}^*_j u)(p) = u(p) \qquad p \in \partial B . \tag{2.16}$$

If now either a Layer Theoretic Method or a Green's Theorem Method is used to solve one or other of the problems (2.1) to (2.5) then we obtain the following representation of solutions and associated boundary integral equations.

Boundary Conditions	Representation in B_+	Boundary Integral Equations
Dirichlet	$u=\frac{1}{2}(S_j w-D_j f)$	$(I-K_j)w=-\frac{\partial}{\partial n}(D_j f)$
$u=f$ on ∂B	$u=-D_j w$	$(I-\bar{K}_j^*)w=f$
Neumann	$u=\frac{1}{2}(S_j f-D_j w)$	$(I+\bar{K}_j^*)w=S_j f$
$\frac{\partial u}{\partial n}=f$ on ∂B	$u=S_j w$	$(I+K_j)w=f$
Robin	$u=\frac{1}{2}(S_j f-\sigma S_j w-D_j w)$	$(I+\bar{K}_j^*+\sigma S_j)w=S_j f$
$\frac{\partial u}{\partial n}+\sigma u=f$ on ∂B	$u=S_j w$	$(I+K_j+\sigma S_j)w=f$

Table 1: Boundary Integral Equations

3. CONCERNING UNIQUENESS

Three boundary integral equations in Table 1 have the typical form

$$(I+A_j)w = g$$

where A_j denotes K_j or $(K_j+\sigma S_j)$ and g the known terms. Strictly $A_j=A_j(k)$ is an operator valued function of the frequency parameter k in the Helmholtz equation. Consequently consideration must be given to the influence of the so called characteristic or irregular values of $A_j(k)$ which are defined to be those real values of k for which $(I+A_j(k))^{-1}$ does not exist. It is these values of k which give rise to problems of non-uniqueness and difficulties near eigenvalues. Consequently we enquire into the possibility of modifying $\gamma_o(P,Q;k)$ in a systematic way so that the resulting modified fundamental solution $\gamma_1(P,Q:k)$ generates boundary integral equations for which there is no problem of non-uniqueness. In the modification (2.7) the choice of the function g is still at our disposal. We shall define

$$g(P,Q) = \sum_{\ell=o}^{\infty} \alpha_\ell V_\ell(P)V_\ell(Q) \tag{3.1}$$

where

$$V_\ell^{e,i} := \Lambda_{nm}\, z_n^{e,i}(kr) P_n^m(\cos\theta) \begin{cases} \cos m\phi & , \ \ell \text{ even.} \\ \sin m\phi & , \ \ell \text{ odd.} \end{cases} \tag{3.2}$$

with

$$z_n^e(kr) = h_n^{(1)}(kr), \qquad z_n^i(kr) = j_n(kr)$$

$$\Lambda_{nm} = \left\{ \frac{-ik}{2\pi} \epsilon_m (2n+1) \cdot \frac{(n-m)!}{(n+m)!} \right\}^{\frac{1}{2}}, \ \epsilon_0 = 1, \ \epsilon_m = 2, \ m > 0$$

$$\ell = \tfrac{1}{2} n(n+1) + m \quad , \ 0 \leq m \leq n .$$

The object now is to choose the coefficients α_ℓ in the definition of g so as to ensure that the boundary integral equations associated with the modified fundamental solution (2.7) are uniquely solvable.

We notice first the following result [4]:

Theorem 3.1

k is a characteristic value of K_o $(-K_o - \sigma S_o)$ if and only if k is an eigenvalue of the interior Neumann problem (interior Dirichlet problem).

This Theorem holds equally well when σ=0.

This result does not generalise to the operator K_1 because g is not defined throughout B_-. Nevertheless it is possible to establish the following [5]:

Theorem (3.2)

If g (P,Q) is defined by (3.1) so that

$|2\alpha_\ell + 1| < 1$ for all ℓ,

then K_1 and $-K_1 - S_1$ have no characteristic values.

Thus the modification of the fundamental solution as indicated by (3.1) removes the characteristic values of the associated boundary integral operators.

4. OPTIMAL CHOICE OF COEFFICIENTS.

In this section we present results based on different criteria for choosing the coefficients in the modification (3.1). These are obtained by a desire not only to ensure unique solvability but also to provide constructive methods of solving the integral equation. For instance we might wish to choose the coefficients so as to minimize the difference between the modified and exact Green's functions for the Dirichlet and Neumann problems. Alternatively we might take as a desideratum the minimization of the norm of the modified integral operator as this will provide a bound on the spectral radius of the operator and so give an indication of when the modified boundary integral equation can be solved by iteration. The derivation of the results which are simply presented in Table 2 can be found in [1], [5], [6].

Problem	Quantity to be minimized	Optimal Coefficients: General	Optimal Coefficients: Sphere
Dirichlet $\gamma_1^D=\gamma_o+\sum_\ell \alpha_\ell^D V_\ell(P)V_\ell(Q)$	$\int_{S_A}\int_{\partial B} \lvert\gamma_1^D(P,q)\rvert^2 dS_q dS_p$	$\alpha_\ell^D = \dfrac{-(V_\ell^i, V_\ell^e)}{\lVert V_\ell^e\rVert^2}$	$\alpha_{2\ell}^D = \alpha_{2\ell+1}^D = \dfrac{-j_n(ka)}{h_n^{(1)}(ka)}$
Neumann γ_1^N, α_ℓ^N	$\int_{S_A}\int_{\partial B} \left\lvert\dfrac{\partial\gamma_1^N}{\partial n}\right\rvert^2 dS_q\, dS_p$	$\alpha_\ell^N = -\dfrac{\left(\dfrac{\partial V_\ell^i}{\partial n}, \dfrac{\partial V_\ell^e}{\partial n}\right)}{\left\lVert\dfrac{\partial V_\ell^e}{\partial n}\right\rVert^2}$	$\alpha_{2\ell}^N = \alpha_{2\ell+1}^N = \dfrac{-j_n'(ka)}{h_n^{(1)'}(ka)}$
Robin γ_1^R, α_ℓ^R	$\int_{S_A}\int_{\partial B} \left\lvert\dfrac{\partial\gamma_1^R}{\partial n}+\sigma(q)\gamma_1^R\right\rvert^2 \cdot dS_q\, dS_p$	$\alpha_\ell^R = -\dfrac{\left(\dfrac{\partial V_\ell^i}{\partial n}+\sigma V_\ell^i, \dfrac{\partial V_\ell^e}{\partial n}+\sigma V_\ell^e\right)}{\left\lVert\dfrac{\partial V_\ell^e}{\partial n}+\sigma V_\ell^e\right\rVert^2}$	
Minimum Norm γ_1^D, α_ℓ^o	$\lVert K_1\rVert$	$\alpha_\ell^o = -\dfrac{\left(K_o\bar{V}_\ell^{\perp}, \dfrac{\partial V_\ell^e}{\partial n}\right)}{\left\lVert\dfrac{\partial V_\ell^e}{\partial n}\right\rVert^2}$	$\alpha_\ell^o = \frac{1}{2}(\alpha_\ell^D+\alpha_\ell^N)$

Table 2. Coefficient choices for optimal modifications.

In Table 2 the symbol S_A denotes an auxiliary sphere completely containing the given surface ∂B; furthermore $\{V_\ell^\perp\}$ denotes a dual basis to $\{V_m\}$ with the property that $(V_\ell^\perp V_m)=\delta_{\ell m}$.

Here $(\cdot,\cdot)$ and $||\cdot||$ denote the usual $L_2(\partial B)$ inner product and norm.

5. NUMERICAL RESULTS

If we restrict our attention to two dimensions the various quantities introduced in the previous sections assume the following particular form:

$$V_{2\ell}^{e}(p) := \sqrt{\frac{-i\epsilon_\ell}{2}}\, H_\ell^{(1)}(kr)\cos\ell\theta$$

$$V_{2\ell+1}^{e}(P) := \sqrt{\frac{-i\epsilon_\ell}{2}}\, H_\ell^{(1)}(kr)\sin\ell\theta$$

$$V_{2\ell}^{i}(P) := \sqrt{\frac{-i\epsilon_\ell}{2}}\, J_\ell(kr)\cos\ell\theta$$

$$V_{2\ell+1}^{i}(P) := \sqrt{\frac{-i\epsilon_\ell}{2}}\, J_\ell(kr)\sin\ell\theta$$

$$\gamma_o(P,q) = -\frac{i}{2} H_o^{(1)}(kR(P,q)) = \sum_{\ell=o}^{\infty} V_\ell^{i}(P_<)V_\ell^{e}(P_>)$$

where

$$P_< := \begin{cases} P, & r_p<r_q \\ q, & r_q<r_p \end{cases} \qquad P_> := \begin{cases} P, & r_p>r_q \\ q, & r_q>r_p \end{cases}$$

$$\gamma_1(P,q) = \gamma_o(P,q) + \sum_{\ell=o}^{\infty} \alpha_\ell V_\ell^{e}(P)V_\ell^{e}(q)$$

$$= \sum_{\ell=o}^{\infty} V_\ell^{e}(P_>)\,\{V_\ell^{i}(P_<) + \alpha_\ell V_\ell^{e}(P_<)\}.$$

We now consider the question of how to define the coefficients α_ℓ in the modifications of the Green's function so as to minimize $||K_1||$, because if we can ensure that $||K_1||<|$ then the associated boundary integral equations will be solvable by iteration. The main result in this direction is provided by the following [6]

Theorem 5.1

If

$$\partial B := \{P \in \mathbb{R}^2 : r_p = a + \epsilon f(\theta_p)\}$$

$$\alpha_\ell := (-\tfrac{1}{2}) \left\{ \frac{(v_\ell{}^i, v_\ell{}^e)}{||v_\ell{}^e||^2} + \frac{\left(\frac{\partial v_\ell{}^i}{\partial n}, \frac{\partial v_\ell{}^e}{\partial n}\right)}{\left|\left|\frac{\partial v_\ell{}^e}{\partial n}\right|\right|^2} \right\}$$

then $||K_1|| = O(\epsilon)$.

It is now natural to enquire if $||K_1||$ will be reduced if the modification contains only a finite number of non-zero coefficients. To demonstrate that this is indeed the case we shall assume that the coefficients are chosen as in Theorem 5.1 for $\ell \leq 2N+1$ and zero otherwise, and present numerical results corresponding to the case when ∂B is a sphere of radius a.

The spectral radius of the operator K_1 is denoted by

$$\sigma_{K_1}(k) := \sup_n \{\lambda_n(k) : (\lambda_n(k)I - K_1)w = o\} \leq ||K_1||.$$

In the particular case of a sphere of radius a it is known that

$$\lambda_n(k) = 1 - ika\pi J_n{}'(ka)\, H_n^{(1)}(ka) = -1 - ika\pi\, J_n(ka)\, H_n^{(1)'}(ka).$$

Furthermore, if

$$J_n{}'(ka) = 0 \quad \text{then } \lambda_n = 1$$

and k is an eigenvalue of the interior Neumann problem, and if

$$J_n(ka) = 0 \quad \text{then } \lambda_n = -1$$

and k is an eigenvalue of the interior Dirichlet problem.

If now we choose α_ℓ as in Theorem 5.1 for $\ell \leq 2N+1$ and zero otherwise then

$$\sigma K_1(k) = \sup_{n>N} \{|\lambda_n| : (\lambda_n I - K_1)w = 0\}.$$

The manner in which $\sigma_{K_1}(k)$ varies with N and k is illustrated in Table 3.

Number of terms	Spectral Radius $\sigma_{K_1}(k)$										
	k=0.1	0.5	1.0	1.5	2.0	2.5	3.0	3.5	4.0	4.5	5.0
0	1.01	1.20	1.35	1.31	1.14	1.20	1.21	1.13	1.18	1.14	1.16
1	0.11	0.18	0.49	0.82	1.07	1.20	1.21	1.13	1.18	1.14	1.16
3	0.11	0.15	0.23	0.33	0.51	0.76	0.98	1.13	1.18	1.14	1.16
5	0.11	0.15	0.23	0.33	0.42	0.50	0.54	0.74	0.93	1.10	1.16
7	0.11	0.15	0.23	0.32	0.42	0.50	0.54	0.55	0.53	0.72	0.90
9	0.11	0.15	0.23	0.32	0.42	0.50	0.53	0.55	0.53	0.48	0.54
11	0.11	0.15	0.23	0.32	0.42	0.50	0.53	0.55	0.53	0.47	0.41

Table 3.

These results clearly indicated that the number of terms in the modification required to reduce significantly the spectral radius increases with kq. Nevertheless, we note that even for ka=5 (well into the so-called resonance region) only 11 terms are required in the modification to reduce the spectral radius below 0.5. Results for non-circular boundaries will be given in [3].

ACKNOWLEDGEMENT

The first author (REK) was supported under NSF Grant MCS-820 2033.

6. REFERENCES

1. ANGELL, T.S. and KLEINMAN, R.E. Modified Green's Functions and the Exterior Robin Ploblem for the Helmholtz equation. *J. Math. Anal and Applics.* To appear.

2. ANGELL, T.S. and KLEINMAN, R.E. Boundary Integral Equations for the Helmholtz Equation: the Third Boundary Value Problem. *Math. Meth. in the App. Sci.* 4 (2) 164-193 (1982)

3. KIRSH, A. and KLEINMAN, R.E. On the Spectral Radius of Boundary Integral Operators with Modified Green's Function Kernels. To Appear.

4. KLEINMAN, R.E. and ROACH, G.F. Boundary Integral Equations for the Three Dimensional Helmholtz Equation, *SIAM Review*, 16 (2), 214-236, (1974).

5. KLEINMAN, R.E. and ROACH, G.F. On Modified Green's Functions in Exterior Problems. *Proc. Roy. Soc. To appear* (1982)

6. KLEINMAN, R.E. and ROACH, G.F. Operators of Minimal Norm via Modified Green's Functions. *Proc. Roy. Soc. Edin. To appear* (1982).

7. JONES, D.S., Integral Equations for the Exterior Acoustic Problem, *Q. Jl. Mech. App. Math. Vol.XXVII, 1,* (1974), 129-142.

8. ROACH, G.F., On the Approximate Solution of Elliptic Self-adjoint Boundary Value Problems. *Arch. Rational Mech. Anal.* 27, 3, 243-254, (1967).

9. URSELL, F. On the Exterior Problems of Acoustics: II *Math. Proc. Camb. Phil. Soc. Vol 84,* (1978) 545-548.

THE METHOD OF FUNDAMENTAL SOLUTIONS FOR PROBLEMS IN POTENTIAL THEORY

*Graeme Fairweather and †R.L. Johnston

**University of Kentucky*, †*University of Toronto*

1. INTRODUCTION

A useful technique for the numerical solution of certain elliptic boundary-value problems is the boundary integral equation method (BIEM), in any of its various forms. This method is applicable when a fundamental solution of the differential equation in question is known and is used to reformulate the boundary-value problem as an integral equation (or a coupled system of integral equations) on the boundary of the domain of the problem. A boundary method of more recent vintage is the method of fundamental solutions [14], which appears to have some of the characteristics of a BIEM, and possesses several advantages over such a method. In this paper, we briefly describe the MFS for the solution of interior boundary-value problems for Laplace's equation, and show how it can be formulated as a BIEM. Some numerical results, which demonstrate the adaptivity of a specific implementation of the MFS, are presented.

One of the drawbacks of the MFS has been its lack of generality due, mainly, to its initial formulation. Its use has been restricted to solving boundary-value problems for Laplace's equation in two and three dimensions [10,14] and for the Helmholtz equation [13]. By considering the MFS as a "discrete" simple-layer method, we indicate how the method may be extended to the biharmonic equation.

Before we describe the various attributes of the MFS, it is useful to review some of the basic boundary integral methods for Laplace's equation in the plane. This will provide the appropriate setting in which to give a new interpretation of the MFS. For simplicity, we restrict our attention to the Dirichlet problem in a bounded domain in the plane with a smooth boundary. The manner in which the MFS handles more general boundary conditions is indicated.

2. BOUNDARY METHODS FOR THE DIRICHLET PROBLEM

Consider the boundary-value problem

$$\Delta u(P) = 0 , \quad P \in \Omega , \tag{2.1}$$

$$u(P) = f(P) , \quad P \in \partial\Omega , \tag{2.2}$$

where Ω is a bounded domain in the plane with smooth boundary $\partial\Omega$, Δ denotes the Laplacian and f is a prescribed function.

2.1 Indirect BIEMs

It is well known that a solution to (2.1) can be represented in the form of a simple-layer potential, viz.,

$$u(P) = \int_{\partial\Omega} \sigma(Q) \log r(P,Q) dS_Q , \quad P \in \overline{\Omega} , \tag{2.3}$$

where $r(P,Q)$ denotes the distance between the points P and Q, and σ is the unknown source density function, which is determined so that (2.2) is satisfied. Thus

$$\int_{\partial\Omega} \sigma(Q) \log r(P,Q) dS_Q = f(P) , \quad P \in \partial\Omega , \tag{2.4}$$

which is a Fredholm integral equation of the first kind. An alternative representation of u is given by the double-layer potential

$$u(P) = \int_{\partial\Omega} \mu(Q) \frac{\partial}{\partial n_Q} \log r(P,Q) dS_Q , \quad P \in \overline{\Omega} , \tag{2.5}$$

where $\partial/\partial n_Q$ denotes the outward normal derivative at $Q \in \partial\Omega$. In this case, μ is the unknown density function defined by the second-kind integral equation

$$-\pi\mu(P) + \int_{\partial\Omega} \mu(Q) \frac{\partial}{\partial n_Q} \log r(P,Q) dS_Q = f(P) , \quad P \in \partial\Omega . \tag{2.6}$$

In the current literature, these methods are usually described as *indirect* BIEMs because, after the density functions (or approximations to them) are determined, a quadrature is required to calculate u and any of its derivatives at a point in $\overline{\Omega}$. Numerical methods for the solution of (2.4) and (2.6) are discussed in [12], for example.

2.2 Direct BIEMs

In most current applications, direct BIEMs are used. The key ingredients of such a method are a reciprocal theorem and a fundamental solution - Green's Second Identity and $G(P,Q) = -\log r(P,Q)$, respectively, in the case of (2.1) - the combination of which enables one to reformulate the boundary-value problem as an integral equation (or system of integral equations) on the boundary of the domain. In the case of (2.1), this combination yields Green's Third Identity, which gives

$$c(P)u(P) = \int_{\partial\Omega} \left[u(Q) \frac{\partial}{\partial n_Q} \log r(P,Q) - \frac{\partial u}{\partial n_Q} \log r(P,Q) \right] dS_Q \, , \qquad (2.7)$$

where

$$c(P) = \begin{cases} 2\pi \, , & P \in \Omega \\ \pi \, , & P \in \partial\Omega \\ 0 \, , & \text{otherwise} \, . \end{cases}$$

Equation (2.7) can be viewed as the combination of a simple-layer of density $\partial u/\partial n$, and a double-layer of density u. Of course, in a properly-posed boundary value problem, either u(P) or $(\partial u/\partial n)(P)$, $P \in \partial\Omega$, are prescribed and the problem is to determine, from (2.7) with $P \in \partial\Omega$, the remaining boundary data. For the Dirichlet problem we see that on choosing $P \in \partial\Omega$ in (2.7) and setting $u(P)=f(P)$, $P \in \partial\Omega$, the unknown data, $(\partial u/\partial n_P)(P)$, $P \in \partial\Omega$, satisfies the Fredholm integral equation of the first kind

$$\int_{\partial\Omega} \frac{\partial u}{\partial n_Q} \log r(P,Q) dS_Q =$$

$$\int_{\partial\Omega} f(Q) \frac{\partial}{\partial n_Q} \log r(P,Q) dS_Q - c(P)f(P) \, , \qquad P \in \partial\Omega \, , \qquad (2.8)$$

where $c(P) = \pi$.

This formulation is the basis of the *direct* BIEM so-called because, in contrast to the formulations described in Section 2.1, the solution of (2.8) is a meaningful physical quantity. Of course, once all the boundary data is determined, quadratures are required to compute u and its derivatives from (2.7) at points in Ω, should these quantities be desired.

In standard methods for the numerical solution of (2.8), the boundary $\partial\Omega$ is subdivided by a set of points Q_j, $j=1,\ldots,N$, with $\partial\Omega_j = \overline{Q_{j-1}Q_j}$, and a set of *observation points* P_i, $i=1,\ldots,N$, with $P_i \in \partial\Omega_i$, is chosen. (See Fig. 1.) If, for example, we approximate u and $\partial u/\partial n$ by piecewise constant functions, then successive evaluations of (2.8) at $P_i = \hat{Q}_i$, the mid-point of $\partial\Omega_i$, produces the system of equations

$$\sum_{j=1}^{N} u_{n_j} \int_{\partial\Omega_j} \log r(P_i,Q)\,dS_Q$$

$$= \sum_{j=1}^{N} u_j \int_{\partial\Omega_j} \frac{\partial}{\partial n_Q} \log r(P_i,Q)\,dS_Q - c(P_i)u_i \ , \ i=1,\ldots,N \ , \qquad (2.9)$$

where $u_i = f(\hat{Q}_i)$, $u_{n_i} \approx (\partial u/\partial n)(\hat{Q}_i)$, and $c(\hat{Q}_i) = \pi$, $i=1,\ldots,N$. On approximating the integrals by appropriate quadrature formulas, or by evaluating them analytically, we obtain an N×N linear system for the unknowns u_{n_i}, $i=1,\ldots,N$, (cf. [2,6]).

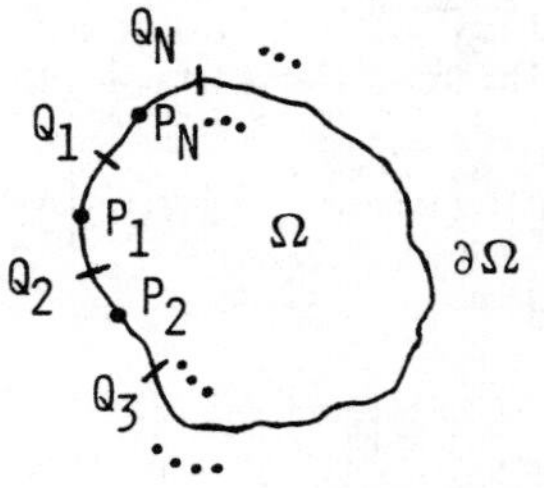

FIG. 1.

It should be mentioned that there are cases in which the Fredholm integral equations of the first kind (2.6) and (2.8) do not have unique solutions. This problem is addressed in [3] and [11], and remedies are proposed.

2.3 Methods Using An Auxiliary Boundary

2.3.1 Kupradze's Method

Since the integrands on the left-hand side of (2.9) are singular when j=i, special care must be taken when applying a quadrature rule to approximate the corresponding integrals. In Kupradze's method (see [4] and references therein), this difficulty is avoided by choosing the observation points P_i to lie on an *auxiliary boundary* $\partial\Omega_a$ enclosing the original region Ω. See Figure 2. Since each P_i lies outside $\bar{\Omega}$, $c(P)=0$ in (2.8). This approach appears attractive but it is not without difficulties of its own. Christiansen [4] has shown that the solution of (2.8), with $P \in \partial\Omega_a$, may not be unique, and even when this problem is eliminated, the location of $\partial\Omega_a$ must be chosen carefully to obtain accurate results. (See also [5].)

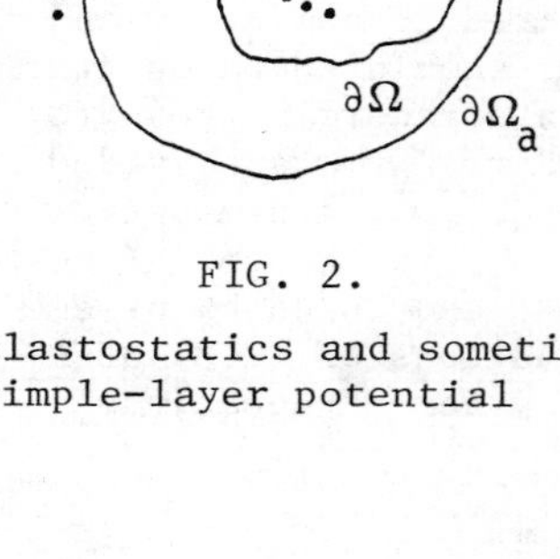

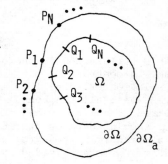

FIG. 2.

2.3.2 The Method of Oliveira

In this approach, proposed by Oliveira [15] for problems in plane elastostatics and sometimes called "point matching," we use the simple-layer potential representation with respect to an auxiliary boundary $\partial\Omega_a$

$$u(P) = \int_{\partial\Omega_a} \sigma_a(Q) \log r(P,Q) dS_Q , \quad P \in \bar{\Omega} . \tag{2.10}$$

Here, the unknown density σ_a is defined on $\partial\Omega_a$, but it is determined by the boundary condition (2.2) specified on $\partial\Omega$. A numerical method for solving (2.10) consists in choosing the observation points $P_i \in \partial\Omega$, $i=1,\ldots,N$, and approximating the integral over $\partial\Omega_a$ appropriately. For example, on subdividing $\partial\Omega_a$ by the points Q_j, $j=1,\ldots,N$, with $\partial\Omega_{a,j} = \overline{Q_{j-1},Q_j}$ (see Fig. 3), we obtain the N×N system of linear equations

$$\sum_{j=1}^{N} \sigma_{a,j} \int_{\partial\Omega_{a,j}} \log r(P_i,Q) dS_Q = f(P_i) , \quad i=1,\ldots,N , \tag{2.11}$$

where $\sigma_{a,j}$ is a (constant) approximation to $\sigma_a(Q)$, $Q \in \partial\Omega_{a,j}$. This method is, in a sense, a converse of Kupradze's method. In the latter, the observation points P_i are placed on $\partial\Omega_a$ and the quadrature is done on $\partial\Omega$, whereas the opposite obtains in Oliveira's method. One obvious advantage of this approach is that one can choose $\partial\Omega_a$ to be polygonal, thus simplifying the integrations required.

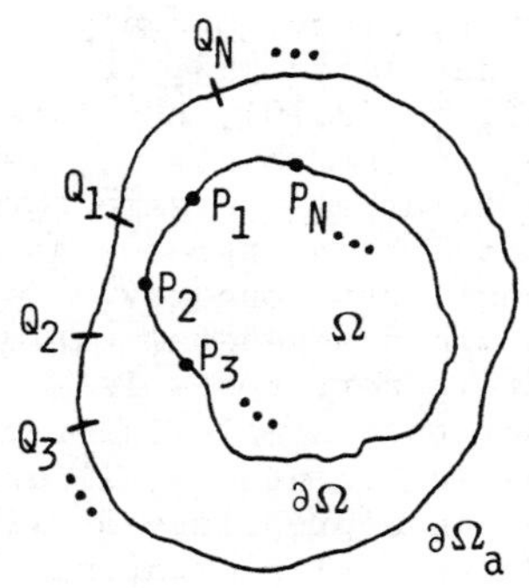

FIG. 3.

The numerical properties of both this method and Kupradze's method were investigated by Heise [8] for the solution of problems in plane elastostatics. He concluded that such methods can be used successfully for solving engineering problems. However, he emphasized that care must be exercised in the choice of $\partial\Omega_a$, as has Christiansen [4,5] with respect to Kupradze's method.

2.3.3 The Method of Fundamental Solutions

The method of fundamental solutions (MFS) is, in certain respects, a generalization of Oliveira's approach. In the MFS as defined in [13], the desired solution u is approximated by a function u_N of the form

$$u_N(\underline{c},\underline{Q};P) = \sum_{j=1}^{N} c_j K(P,Q_j) , \quad P \in \bar{\Omega} , \tag{2.12}$$

where $K(P,Q_j) = -\log r(P,Q_j)$, $Q_j \notin \bar{\Omega}$, is the fundamental solution of (2.1), $\underline{c} = (c_1,\ldots,c_N)^T$ and $\underline{Q}$ is a 2N-vector giving the coordinates of the singularities Q_j. Since u_N satisfies the differential equation (2.1), the unknowns $\underline{c}$ and $\underline{Q}$ are chosen so that u_N approximates the boundary data as well as possible. This is done by least squares. We select a set of points $P_i \in \partial\Omega$, i=1,...,M, and define $u_N^* \equiv u_N(\underline{c}^*,\underline{Q}^*;\cdot)$ by

$$\|Bu_N(\underline{c}^*,\underline{Q}^*;\cdot)-f\|_2^2 \equiv \min_{\underline{c},\underline{Q}} \|Bu_N(\underline{c},\underline{Q};\cdot)-f\|_2^2$$
$$\equiv \min_{\underline{c},\underline{Q}} \sum_{i=1}^{M} |Bu_N(\underline{c},\underline{Q};P_i)-f(P_i)|^2 , \quad (2.13)$$

where, in the case of (2.2), B is the identity operator. The main advantage of the MFS is its adaptivity. By permitting the singularities to move, the method is able to adapt the approximation u_N automatically to reflect any bad behaviour in u. The price one pays for this feature is that the least squares problem (2.13) becomes nonlinear in the coordinates of Q_j, j=1,...N. However, since good algorithms for solving this problem have been developed [9], it is not a serious drawback.

If, in (2.10), the integral is approximated using the values of σ_a at the points Q_j, j=1,...,N, then the resulting approximation to u has the same form as (2.12). Hence we see that the MFS can be interpreted as a discrete simple-layer potential using an auxiliary boundary. In this regard, it is clear that the MFS represents a generalization of Oliveira's method in two aspects. First it performs a least squares fit of the boundary data rather than simple point-matching, i.e., the number, M, of observation points P_i, is greater than N. Second, the MFS determines an appropriate auxiliary boundary $\partial\Omega_a$ automatically.

Software implementing the MFS has been developed for the solution of Laplace's equation in both two and three dimensions [10]. (In the latter case, $K(P,Q) = 1/r(P,Q)$ in (2.12), where $\underline{Q}$ is now a 3N-vector.) In the MFS algorithm implemented in [10], normalized fundamental solutions are used. Specifically, K(P,Q) is defined by

$$K(P,Q) = -\log \{r(P,Q) / (\rho+r(Q)\} ,$$

where ρ is the radius of the circumscribing circle of the region Ω and r(Q) is the distance of Q from the origin, which must lie inside Ω. This normalization improves the rate of convergence of the nonlinear least squares algorithm. To solve the nonlinear least squares problem, a slightly modified version of the subroutine VAØ7AD from the Harwell subroutine library [9] is used. This subroutine implements a Gauss-Newton-Marquardt process and the modification permits an optional line search.

The main advantage of an MFS algorithm is its ease of use. The user need only supply (a) the locations of the initial singularities; (b) the observation points on the boundary $\partial\Omega$; (c) the boundary data; (d) an error tolerance for the Gauss-Newton-Marquardt process. When specifying the number and locations of the singularities $\{Q_j\}_{j=1}^N$, the user should, of course, use all available information on the problem at hand. For example, one ought to increase the density of the singularities in the neighborhood of a reentrant corner or near a point where the form of the boundary conditions changes. As to the number of observation or least squares points, a rule of thumb appears to be that M should be chosen to be approximately three times the number of parameters to be determined.

It should be noted that the software described in [10] can handle not only the Dirichlet problem but also the Neumann problem, the Robin problem and problems with mixed boundary conditions. If the boundary conditions are written in the form $Bu(P)=f(P)$, $P \in \partial\Omega$, the corresponding MFS is defined by (2.13).

3. NUMERICAL EXPERIMENTS

The efficacy of the MFS is demonstrated by the numerical results presented in [13] and [14]. In the first paper, a particularly difficult problem in the computation of dipole fields was solved. This problem involved a coupled system of Helmholtz equations.

The two examples given here are intended to illustrate the adaptivity of the MFS in choosing appropriate locations for the singularities, or, equivalently, an appropriate auxiliary curve $\partial\Omega_a$. In each case, Laplace's equation is solved, subject to the boundary conditions specified in the following. In both problems, the initial singularities were placed at a uniform distance 0.1 from the boundary and spaced approximately evenly. The final locations of the singularities are denoted by × in Figs. 4 and 5.

In the first example, the polar equation of the outer boundary is

$$r = 0.7 + 0.3 \cdot \cos 3\theta ,$$

and on this boundary u=0 while on the inner boundary u=1. In the computations N=14 and M=60. Two of the singularities were placed in the hole, and, of the boundary points, 25 were on the inner boundary. Since there was a tendency for the singularities in the hole to move into Ω, it was necessary to have a sufficiently dense set of boundary points on the

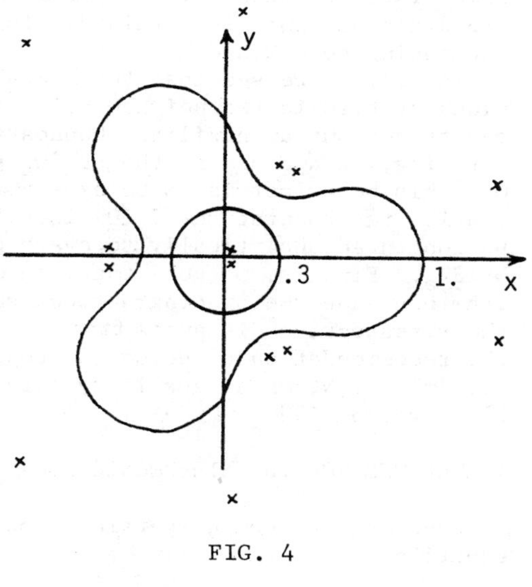

FIG. 4

circular boundary in order to serve as a barrier to such undesirable movements of the singularities. At points where the inner and outer boundaries are near each other, u changes quite rapidly. The MFS algorithm has placed singularities close to $\partial\Omega$ in these regions, see Fig. 4.

The second example, which arises in the analysis of certain crack problems, is a special case of one discussed in [1]. Fig. 5 shows one half of a cross-section of a rectangular bar with the segment AO representing the crack, the point O being the crack tip. In this problem the boundary conditions are $\partial u/\partial x = 0$ on AO, u=0 on OB, $\partial u/\partial y = 1/3\,u$ on BC, $u = y+2$ on CD, $\partial u/\partial y = u$ on AD. In this example, N=15 and the final locations of the singularities shown in Fig. 5 are those determined with 161 boundary points (M=161). Approximately half of these points were on AB to prevent the movement of the singularities into Ω. With this choice of M and N, we estimate that 2-3 digit accuracy is obtained. The accuracy can be improved by increasing both M and N.

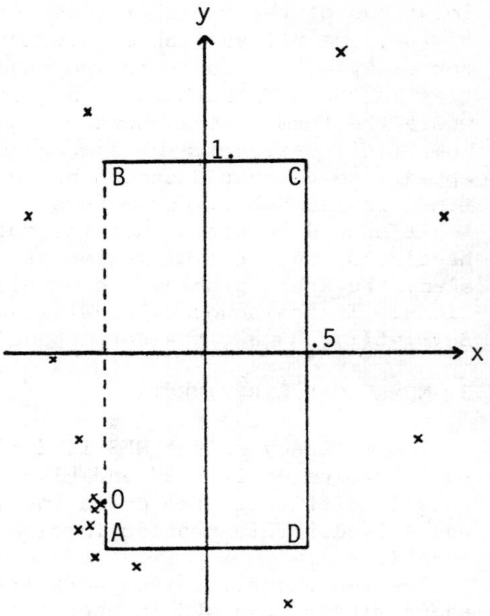

FIG. 5

In Fig. 5, we see that the MFS algorithm has placed several singularities in the neighborhood of the crack tip. In fact, if one traces out an auxiliary boundary $\partial\Omega_a$ defined by the singularities, it appears as though $\partial\Omega_a$ should touch $\partial\Omega$ at the point O. This behaviour is to be expected since the simple-layer potential representation (2.10) is valid if and only if u(P) can be continued analytically to every point of the region bounded by $\partial\Omega_a$. Finally, we note that, in each example, the MFS algorithm has made the spacing between successive Q_j greater than their respective distances from $\partial\Omega$. This is in agreement with the recommendation of Heise [8] concerning the location of the Q_j, j=1,...,N, on an auxiliary curve $\partial\Omega_a$, and with the results of Oliveira [15].

4. THE MFS FOR THE BIHARMONIC EQUATION

Consider the boundary-value problem for the biharmonic equation

$$\Delta^2 u(P) = 0 , \qquad P \in \Omega \subset \mathbb{R}^2, \tag{4.1}$$

$$B_1 u(P) = f_1(P) , \qquad B_2 u(P) = f_2(P) , \qquad P \in \partial\Omega . \tag{4.2}$$

Several boundary integral equation methods have been proposed for the solution of (4.1)-(4.2), see, for example, [12]. In this section we describe three formulations of the MFS which are currently under investigation. Two of these are based on the connection with layer-type integral equation methods.

The first, and rather obvious, formulation is to define u_N by (2.12), where $K(P,Q) = [r(P,Q)]^2 \log r(P,Q)$, a fundamental solution of (4.1). As in the case of Laplace's equation, we determine the solution u_N^* by matching the boundary conditions in the least squares sense, that is, we minimize

$$\| B_1 u_N(\underline{c},\underline{Q};\cdot) - f_1 \|_2^2 + \| B_2 u_N(\underline{c},\underline{Q};\cdot) - f_2 \|_2^2 . \tag{4.3}$$

Preliminary experimentation with this form of u_N has yielded somewhat disappointing results. An alternative formulation is based on the Almansi representation of a biharmonic function, [12]. Here the solution u of (4.1) can be expressed in the form

$$u(P) = r^2(P)\phi(P) + \psi(P) , \qquad P \in \overline{\Omega} , \tag{4.4}$$

where ϕ and ψ are harmonic in $\overline{\Omega}$ and $r(P)$ denotes the distance of P from the origin, which lies inside Ω. Using simple-layer potential representations of ϕ and ψ on an auxiliary boundary, we have

$$u(P) = r^2(P) \int_{\partial\Omega_a} \sigma_a(Q) \log r(P,Q) ds_Q + \int_{\partial\Omega_a} \mu_a(Q) \log r(P,Q) ds_Q , \qquad P \in \overline{\Omega} . \tag{4.5}$$

We now set

$$u_N(P) = r^2(P) \sum_{j=1}^{N} c_{\sigma,j} \log r(P,Q_j) + \sum_{j=1}^{N} c_{\mu,j} \log r(P,Q_j) , \qquad P \in \overline{\Omega} , \tag{4.6}$$

and determine u_N^* as before.

The third approach is to use the simple-layer potential representation proposed by Fichera [7], (see also [11] and [16])

$$u(P) = \int_{\partial\Omega_a} \left\{ \sigma(Q) \frac{\partial}{\partial x_Q} G(P,Q) + \mu(Q) \frac{\partial}{\partial y_Q} G(P,Q) \right\} ds_Q \ , \quad P \in \overline{\Omega} \ , \qquad (4.7)$$

where $G(P,Q) = -1/4\ r^2(P,Q) \log r^2(P,Q)$. Then we set

$$u_N(P) = \sum_{j=1}^{N} \left\{ c_{\sigma,j}(x_P - x_{Q_j})[\log r^2(P,Q_j)+1] \right.$$

$$\left. + c_{\mu,j}(y_P - y_{Q_j})[\log r^2(P,Q_j)+1] \right\} , \quad P \in \overline{\Omega} \ . \qquad (4.8)$$

MFS algorithms based on (4.6) and (4.8) are currently being investigated by the authors.

ACKNOWLEDGEMENTS

The authors wish to thank Professor F.J. Rizzo of the University of Kentucky for helpful discussions.

The work of the first author was supported in part by the National Science Foundation and that of the second by the Natural Sciences and Engineering Research Council of Canada.

REFERENCES

1. ATKINSON, C., L.S. XANTHIS and M.J.M. BERNAL. Boundary integral equation crack-tip analysis and applications to elastic media with spatially varying elastic properties. *Comput. Methods Appl. Mech. Engrg.* 29, 35-49 (1981).
2. BREBBIA, C.A. *The Boundary Element Method for Engineers.* Halsted Press, New York (1978).
3. CHRISTIANSEN, S. Integral equations without a unique solution can be made useful for some plane harmonic problems. *J. Inst. Maths Applics.* 16, 143-159 (1975).
4. CHRISTIANSEN, S. On Kupradze's functional equations for plane harmonic problems, pp. 205-243 of R.P. Gilbert and R.J. Weinacht (Eds), *Function Theoretic Methods in Differential Equations.* Pitman Publishing, London (1976).
5. CHRISTIANSEN, S. A comparison of various integral equations for treating the Dirichlet problem, in these proceedings.
6. FAIRWEATHER, G., F.J. RIZZO, D.J. SHIPPY and Y.S. WU. On the numerical solution of two-dimensional potential problems by an improved boundary integral equation method. *J. Comput. Phys.* 31, 96-112 (1979).
7. FICHERA, G. Linear elliptic equations of higher order in two independent variables and singular integral equations, with applications to anisotropic inhomogeneous elasticity, pp. 55-80

7. (contd) of R.E. Langer (Ed), *Partial Differential Equations and Continuum Mechanics*. Univ. of Wisconsin Press, Madison (1961).
8. HEISE, U. Numerical properties of integral equations in which the given boundary values and the sought solutions are defined on different curves. *Computers and Structures* 8, 199-205 (1978).
9. HOPPER, M.J. (Ed). Harwell Subroutine Library Catalogue, Theoretical Physics Division, AERE, Harwell (1973).
10. HO-TAI, S., R.L. JOHNSTON and R. MATHON. Software for solving boundary-value problems for Laplace's equation using fundamental solutions. Technical Report 136/79, Department of Computer Science, University of Toronto (1979).
11. HSIAO, G. and R.C. MACCAMY. Solution of boundary value problems by integral equations of the first kind. *SIAM Rev.* 15, 687-705 (1973).
12. JASWON, M.A. and G.T. SYMM. *Integral Equation Methods in Potential Theory and Elastostatics*. Academic Press, London (1977).
13. JOHNSTON, R.L. and R. MATHON. The computation of electric dipole fields in conducting media. *Internat. J. Numer. Methods. Engrg.* 14, 1739-1760 (1979).
14. MATHON, R. and R.L. JOHNSTON. The approximate solution of elliptic boundary-value problems by fundamental solutions. *SIAM J. Numer. Anal.* 14, 638-650 (1977).
15. OLIVEIRA, E.R. Plane stress analysis by a general integral method. *J. Engrg. Mech. Div. ASCE* 94, 79-101 (1968).
16. RICHTER, G.R. An integral equation method for the biharmonic equation, pp. 41-45 in R. Vichnevetsky (Ed), *Advances in Computer Methods for Partial Differential Equations-II*. IMACS (AICA) (1977).

NUMERICAL TREATMENT OF CERTAIN SINGULAR INTEGRAL EQUATIONS ARISING IN PROBLEMS OF ACOUSTIC RADIATION AND DIFFRACTION

Geoffrey F. Miller

The National Physical Laboratory
Teddington, Middlesex, UK

1. INTRODUCTION AND STATEMENT OF THE PROBLEM

The purpose of the present paper is twofold. First we propose a method for treating a class of mixed boundary value problems such as arise, for example, in diffraction by (or radiation from) a strip. This will involve the derivation of dual integral equations, their conversion to an integro-differential equation and subsequent reduction to a 'standard' form suitable for numerical treatment. Secondly we describe a very general method for the numerical solution of the reduced equation which may involve singular kernels of a variety of types. The latter method is applicable also to problems in potential theory, etc, in which derivatives of logarithmic kernels are present.

It is convenient (and, we hope, instructive) to describe the methods in the context of a particular problem concerning underwater acoustics which has been the subject of recent work at NPL. This problem is somewhat special and its main rôle here is to illustrate the use of the analytical and computational methods. We consider the problem in which an infinite flexible steel plate situated in the plane $z = 0$ is adjacent to a compressible fluid occupying the region $z > 0$. The upper side of the plate is assumed to be lined with pressure-release material everywhere *except* on the strip defined by $-L < x < L$, $-\infty < y < \infty$; this has the effect of insulating the plate from the fluid - so that the surface pressure is zero in the affected region. An oscillating force with magnitude $f(x)$ (i.e. independent of the y co-ordinate) and time factor $e^{-i\omega t}$ is applied to the lower surface of the plate. It is required to determine the resulting displacement of the plate and the pressure at infinity in the fluid. A main aim is the determination of the transform, $\bar{p}(s)$, of the surface fluid pressure, in terms of which the pressure at infinity may be readily evaluated. Without loss of generality we may suppose that $L = 1$, i.e. that all lengths are expressed in terms of the half-width of the strip.

The differential equations to be satisfied by the fluid

pressure, $p(x,z)$, and vertical displacement of the plate, $W(x)$, are given by

$$\nabla^2 p + k^2 p = 0, \quad z > 0, \tag{1.1}$$

$$(1-i\eta)\, D \frac{\partial^4 W}{\partial x^4} - \omega^2\, MW = -p(x,0) + f(x), \tag{1.2}$$

where

$k = \omega/c$ is the wavenumber for the fluid,
c = velocity of sound in the fluid,
M = mass per unit area of the plate,
$(1 - i\eta)D$ = (complex) flexural rigidity of plate.

The (positive) parameter η is included in order to permit a small amount of damping of the wave motion in the plate.

The boundary conditions representing continuity of the displacement between plate and fluid in the region of the strip, and the effect of the insulating coating for $|x| \geq 1$ take the form

$$(dp/dz)_{z=o} = \rho\omega^2 W, \quad |x| < 1, \tag{1.3}$$

$$p(x) = 0, \quad |x| \geq 1, \tag{1.4}$$

where ρ is the density of the fluid and $p(x)$ denotes the surface pressure $p(x,0)$.

In addition, to ensure uniqueness $p(x,z)$ must satisfy a radiation condition at infinity, namely

$$\lim_{r\to\infty} [r^{\frac{1}{2}} \frac{\partial}{\partial r} \{e^{-ikr} p(x,z)\}] = 0, \tag{1.5}$$

where $r = (x^2 + z^2)^{\frac{1}{2}}$, and $W(x)$ must vanish as $x \to \pm\infty$. The condition (1.5) signifies that the pressure disturbance in the fluid has the character of an outgoing wave.

2. DERIVATION OF THE DUAL INTEGRAL EQUATIONS

Let the Fourier transform of an arbitrary function $\phi(x)$ be defined by

$$\bar{\phi}(s) = \int_{-\infty}^{\infty} e^{-isx}\, \phi(x)dx. \tag{2.1}$$

Taking the transform of equations (1.1) and (1.2) we obtain

$$(d^2/dz^2)\, \bar{p}(s,z) + (k^2-s^2)\bar{p}(s,z) = 0, \quad z > 0, \tag{2.2}$$

$$D\, b(s)\, \bar{W}(s) = -\bar{p}(s) + \bar{f}(s), \tag{2.3}$$

where

$$b(s) = (1 - i\eta)s^4 - Y^4, \quad Y^4 = \omega^2 M/D.$$

Solution of (2.2) then yields

$$\left.\begin{aligned} \bar{p}(s,z) &= \bar{p}(s)\, e^{-\gamma(s)z}, \\ (d\bar{p}/dz)_{z=o} &= -\gamma(s)\bar{p}(s), \end{aligned}\right\} \tag{2.4}$$

where

$$\gamma(s) = \begin{cases} (s^2-k^2)^{\frac{1}{2}}, & |s| \geq k, \\ -i(k^2-s^2)^{\frac{1}{2}}, & |s| < k. \end{cases}$$

(This choice of the branch of the radical may be shown to be in accord with the radiation condition (1.5).) Also, from (2.3) we obtain

$$\bar{W}(s) = \{\bar{f}(s) - \bar{p}(s)\}/Db(s). \tag{2.5}$$

Taking inverse transforms and applying the boundary conditions (1.3) and (1.4) we derive the following dual pair of integral equations for the function $\bar{p}(s)$:

$$\frac{1}{2\pi}\int_{-\infty}^{\infty} \left\{-\gamma(s) + \frac{C}{b(s)}\right\} \bar{p}(s) e^{isx}\, ds = G(x), \quad |x| < 1, \tag{2.6}$$

$$\frac{1}{2\pi}\int_{-\infty}^{\infty} \bar{p}(s) e^{isx}\, ds = 0, \quad |x| \geq 1, \tag{2.7}$$

where $C = \rho\omega^2/D$ and

$$G(x) = \frac{C}{2\pi}\int_{-\infty}^{\infty} \frac{\bar{f}(s)}{b(s)}\, e^{isx}\, ds. \tag{2.8}$$

Particular interest attaches to the case when f(x) is a unit force concentrated on the line $x = \xi$, so that

$$f(x) = \delta(x-\xi), \quad \bar{f}(s) = e^{-is\xi}. \tag{2.9}$$

Then the function G(x) can be evaluated explicitly in the form

$$G(x) = C\beta(x-\xi), \tag{2.10}$$

where

$$\beta(x) = (iY'/4Y^4)\{e^{iY'|x|} + ie^{-Y'|x|}\},$$

$$Y' = Y(1-i\eta)^{-\frac{1}{4}}.$$

We note that if $\bar{p}(s)$ can be determined it is a simple matter to compute the pressure at infinity from the asymptotic formula

$$p(x,z) \sim \bar{p}(k \cos \theta) \left(\frac{k}{2\pi r}\right)^{\frac{1}{2}} \sin \theta \exp(ikr - \frac{1}{4} i\pi) \tag{2.11}$$

as $r \to \infty$, where $\theta = \cos^{-1}(x/r)$; see, for example, [3], Chapter 5.

3. DERIVATION OF INTEGRO-DIFFERENTIAL EQUATION FOR p(x)

Our next step will be to convert the dual pair (2.6) and (2.7) to an integro-differential equation for the surface pressure $p(x)$. If, following Noble [6], we employ the identity

$$\gamma(s) = (s^2-k^2)/\gamma(s)$$

and make use of the expression for the derivative of a Fourier integral, we obtain the following expression for the first term in (2.6):

$$-\frac{1}{2\pi}\int_{-\infty}^{\infty} \gamma(s)\, \bar{p}(s)\, e^{isx}\, ds = \left(\frac{d^2}{dx^2} + k^2\right) I_1(x;\ p), \tag{3.1}$$

where

$$I_1(x;\ p) = \frac{1}{2\pi}\int_{-\infty}^{\infty} \frac{1}{\gamma(s)}\, \bar{p}(s) e^{isx}\, ds. \tag{3.2}$$

Now the function $1/\gamma(s)$ has the inverse transform (see [6], p.67)

$$\frac{1}{2\pi}\int_{-\infty}^{\infty} \frac{1}{\gamma(s)}\, e^{isx}\, ds = \frac{1}{2}\, iH_0^{(1)}(k|x|), \tag{3.3}$$

where $H_0^{(1)}(r)$ is the Hankel function of the first kind of order zero, defined (in terms of the Bessel functions J_0 and Y_0) by

$$H_0^{(1)}(r) = J_0(r) + iY_0(r).$$

Therefore on applying the convolution theorem for Fourier transforms and making use of the condition (1.4), we obtain

$$I_1(x;\ p) = \int_{-1}^{1} \frac{1}{2}\, iH_o^{(1)}(k|x-y|)p(y)dy. \tag{3.4}$$

Similarly the second term on the left of equation (2.6) can be transformed to

$$I_2(x;\ p) = C\int_{-1}^{1} \beta(x-y)p(y)dy. \tag{3.5}$$

Equation (2.6) thus reduces to

$$(d^2/dx^2 + k^2)I_1(x;\ p) + I_2(x;\ p) = G(x), \quad -1 \le x \le 1, \tag{3.6}$$

a linear integro-differential equation for p(x), which is to be solved subject to the boundary conditions

$$p(x) = 0 \quad \text{at} \quad x = \pm 1. \tag{3.7}$$

4. REDUCTION OF THE INTEGRO-DIFFERENTIAL EQUATION

In order to solve the integro-differential equation numerically we need to transform the left-hand member of equation (3.6). Our aim will be to decompose it into the sum of certain singular terms of standard type plus a remainder consisting of the integral of a smooth function, which may be treated by routine methods.

As a first step we will remove the dominant part of the singularity in the kernel $H_o^{(1)}(k|x|)$. We have

$$\frac{1}{2} iH_o^{(1)}(k|x|) = -\frac{1}{\pi}\ln|x| + V(x), \tag{4.1}$$

where V(x) is a function whose second derivative is integrable across the origin. Performing the double differentiation we obtain

$$V''(x) = -\frac{1}{2} ik^2 H_o^{(1)}(k|x|) + \frac{ik}{2|x|} H_1^{(1)}(k|x|) - \frac{1}{\pi x^2}, \tag{4.2}$$

and hence, writing

$$L_x[p(y)] = \frac{1}{\pi}\int_{-1}^{1} \ln|x-y|\ p(y)dy, \tag{4.3}$$

we obtain after some algebra

$$\left\{\frac{d^2}{dx^2}+k^2\right\} I_1(x;p) = -\frac{d^2}{dx^2} L_x[p(y)] + \int_{-1}^{1} K_1(x-y)p(y)dy, \qquad (4.4)$$

where

$$K_1(x) = V''(x) + \frac{1}{2} ik^2 H_o^{(1)}(k|x|)$$

$$= \frac{ik}{2|x|} H_1^{(1)}(k|x|) - \frac{1}{\pi x^2} . \qquad (4.5)$$

As we presently show, the first term in the right-hand member of (4.4) can be treated analytically provided that we assume a suitable series representation for the unknown function p(y). The integrand of the second term still contains a logarithmic singularity at y = x, which it is desirable to remove before resorting to numerical integration. For small x we have

$$K_1(x) = \frac{k^2}{4}\left[i - \frac{2}{\pi}(\gamma_1 - \frac{1}{2} + \ln\frac{1}{2}k|x|) + O(k^2x^2 \ln k|x|)\right],$$

where γ_1 is the Euler-Mascheroni constant. Consequently we are led to define a new auxiliary kernel

$$K_2(x) = K_1(x) + (1/2\pi)k^2 \ln|x|$$

$$= k^2\left\{\frac{i}{2k|x|} H_1^{(1)}(k|x|) - \frac{1}{\pi(kx)^2} + \frac{1}{2\pi}\ln|x|\right\}, \qquad (4.6)$$

having the behaviour

$$K_2(x) = \text{const.} + O(k^2x^2 \ln k|x|) \quad \text{as} \quad x \to 0$$

(where the constant term depends on k). Rewriting (4.4) in terms of the kernel $K_2(x-y)$ we obtain the form

$$\left(\frac{d^2}{dx^2}+k^2\right) I_1(x;p) = -\frac{d^2}{dx^2} L_x[p] - \frac{k^2}{2} L_x[p] + \int_{-1}^{1} K_2(x-y)p(y)dy. \qquad (4.7)$$

On substituting this expression into (3.6) it is seen that the integro-differential equation has been reduced to the form

$$\frac{d^2}{dx^2} L_x[p] + \frac{1}{2}k^2 L_x[p] + \int_{-1}^{1} K(x-y)p(y)dy = - G(x), \qquad (4.8)$$

where the twice-differentiable function K(x) is defined by

$$K(x) = - K_2(x) - C\beta(x).$$

5. NUMERICAL SOLUTION OF THE INTEGRO-DIFFERENTIAL EQUATION

Here we consider two closely related methods for the numerical solution of equation (4.8). Both employ collocation and involve polynomial approximation of a suitable auxiliary function, and they are essentially equivalent. The choice between them depends on whether we prefer to evaluate a set of series coefficients or a set of function values at specified nodal points. For reasons of space we provide only a brief outline, omitting error analysis and details of the derivation of individual formulae - which will be reported elsewhere. We shall require the Chebyshev polynomials of the first and second kinds defined by

$$T_n(\cos \phi) = \cos n\phi, \quad U_n(\cos \phi) = \sin(n+1)\phi/\sin \phi.$$

Method A - Solution in series form

It can be deduced from the form of equation (4.8) (and also from physical considerations) that the pressure p(y) exhibits square-root singularities at $y = \pm 1$. Accordingly we set

$$p(y) = (1-y^2)^{\frac{1}{2}} \psi(y). \qquad (5.1)$$

Then the conditions $p(\pm 1) = 0$ are automatically satisfied provided that $\psi(y)$ is bounded. We will seek an approximation for $\psi(y)$ in the form of a finite sum:

$$\psi(y) = \sum_{r=1}^{N-1} a_r^{(N)} U_{r-1}(y). \qquad (5.2)$$

Substituting this expression into equation (4.8) we obtain

$$\sum_{r=1}^{N-1} a_r^{(N)} \{\alpha_r''(x) + \frac{1}{2} k^2 \alpha_r(x) + C_r(x)\} = - G(x), \qquad (5.3)$$

where

$$\alpha_r(x) = \frac{1}{\pi}\int_0^{\pi} \ln|x - \cos\phi| \sin\phi \sin r\phi \, d\phi, \tag{5.4}$$

$$C_r(x) = \int_0^{\pi} K(x - \cos\phi) \sin\phi \sin r\phi \, d\phi.$$

It can be shown with the aid of results in [1], p.785, that

$$\alpha_r(x) = \begin{cases} -\frac{1}{2}\ln 2 + \frac{1}{4}T_2(x), & r = 1, \\ \frac{1}{2}\left[\frac{T_{r+1}(x)}{r+1} - \frac{T_{r-1}(x)}{r-1}\right], & r \geq 2; \end{cases} \tag{5.5}$$

$$\alpha_r'(x) = T_r(x), \quad \alpha_r''(x) = rU_{r-1}(x). \tag{5.6}$$

It is convenient to choose as collocation points the set of N-1 zeros of the polynomial $U_{N-1}(x)$, i.e.

$$x_j = \cos j\pi/N, \quad j = 1,2, \ldots , N-1.$$

On setting $x = x_j$ in (5.3) we obtain a system of N-1 equations for the N-1 quantities $a_r^{(N)}$, which is readily solved. A good estimate of the accuracy achieved, and also of the convergence as N increases, may usually be obtained by observing the behaviour of the $a_r^{(N)}$.

The need for explicit evaluation of the oscillatory integrals $C_r(x_j)$ can, however, be avoided by using the following alternative approach.

Method B - Solution at nodal points

This method consists in applying formulae of Lagrangian type to represent the functionals occurring in equation (4.8). For completeness we will include also a further case - that of a Cauchy kernel coupled with the factor $(1-y^2)^{\frac{1}{2}}$ - which, though absent from the present problem, is of frequent occurrence in similar contexts. Specifically we construct formulae of the form

$$I(x) = \int_{-1}^{1} w(x,y)\psi(y)dy \approx \sum_{j=1}^{N-1} A_j^{(N)}(x)\psi(x_j), \tag{5.7}$$

(where the abscissae x_j coincide with the collocation points defined above) in the three cases

$$\text{(i)} \quad w(x,y) = \frac{1}{\pi}(1-y^2)^{\frac{1}{2}} \ln|x-y|,$$

$$\text{(ii)} \quad w(x,y) = \frac{1}{\pi}\frac{(1-y^2)^{\frac{1}{2}}}{x-y} = \frac{1}{\pi}\frac{d}{dx}\{(1-y^2)^{\frac{1}{2}}\ln|x-y|\},$$

$$\text{(iii)} \quad w(x,y) = \frac{1}{\pi}\frac{d^2}{dx^2}\{(1-y^2)^{\frac{1}{2}}\ln|x-y|\}.$$

In particular we are interested in the formulae which result from setting $x = x_j$.

We will seek formulae of type (5.7) which are exact when $\psi(y)$ is a polynomial of degree up to N-2. Such a polynomial may be expressed in the form (5.2). This is equivalent to

$$\psi(\cos\phi)\sin\phi = \sum_{r=1}^{N-1} a_r^{(N)} \sin r\phi, \tag{5.8}$$

and hence using orthogonality properties of the sine function we obtain

$$a_r^{(N)} = \frac{2}{N}\sum_{j=1}^{N-1} \psi(x_j)\sin\phi_j \sin r\phi_j, \tag{5.9}$$

where $\phi_j = \pi j/N$, $x = \cos\phi_j$. (Clearly the representation (5.8) will in general be only approximate and the coefficients will depend on N.) Substituting from (5.8) and (5.9) into (5.7) we obtain

$$I(x) = \sum_{r=1}^{N-1} a_r^{(N)} \int_0^{\pi} w(x,\cos\phi)\sin r\phi \, d\phi$$

$$= \sum_{j=1}^{N-1} A_j^{(N)}(x)\psi(x_j), \tag{5.10}$$

where

$$A_j^{(N)}(x) = \frac{2}{N}\sin\frac{j\pi}{N}\sum_{r=1}^{N-1}\sin\frac{rj\pi}{N} m_r(x), \tag{5.11}$$

$$m_r(x) = \int_0^\pi w(x, \cos\phi) \sin r\phi \, d\phi. \tag{5.12}$$

The formulae (5.11) and (5.12) provide a prescription for calculating the quadrature weights corresponding to an arbitrary weighting function $w(x,y)$. Here it may be remarked that the idea of constructing quadrature formulae of this type is not altogether new. A similar method was applied by Erdogan and Gupta [2] to the evaluation of Cauchy singular integrals. In [5] I indicated the extension to general kernels but did not give particular formulae. Formulae for logarithmic (and other) kernels have recently been given by Krenk [4]. The present treatment differs slightly from that of [5] in that the quadrature points are here taken to be the zeros of $U_{N-1}(y)$ rather than the zeros or turning values of $T_N(y)$; and the application to the improper integral of case (iii) perhaps represents a novel element.

In the cases of present interest considerable simplification results when we substitute the known expressions for $m_r(x)$ (cf. equations (5.5) and (5.6)) and set $x = x_i$. Here we will state only the *results* for the cases considered, omitting the derivation.

Case (i)

$$A_j^{(N)}(x_i) = -(1-x_j^2)(\beta_{i+j} + \beta_{i-j}), \tag{5.13}$$

where

$$\beta_m = \frac{1}{N}\left\{\frac{1}{2}\ln 2 + \sum_{r=1}^{N}{}^{*} \frac{1}{r}\cos\frac{mr\pi}{N}\right\},$$

in which the asterisk on the summation indicates that the last term is to be halved. Since

$$\beta_m = \beta_{-m} = \beta_{2N-m},$$

we need evaluate β_m only for $m = 0, 1, \ldots, N$. Thus the double array of coefficients can be computed in terms of a single array of length $N+1$.

Case (ii)

$$A_j^{(N)}(x_i) = \begin{cases} 0, & j-i \text{ even}, \\ \dfrac{2}{N}\dfrac{1-x_j^2}{x_i-x_j}, & j-i \text{ odd}. \end{cases} \tag{5.14}$$

Case (iii)

Here we obtain (after substantial algebra)

$$A_j^{(N)}(x_i) = \begin{cases} -\dfrac{2}{N}\dfrac{1-x_j^2}{(1-x_{j+i})(1-x_{j-i})}, & j-i \text{ odd,} \\ 0, & j-i \text{ even, } j\neq i, \\ N/2, & j=i, \end{cases} \tag{5.15}$$

where, for every integer m, x_m signifies $\cos(m\pi/N)$.

If we apply the general formula (5.11) to the simple case of a weighting function $w(y) = (1-y^2)^{\frac{1}{2}}$ (i.e. independent of x) we obtain

$$\int_{-1}^{1} (1-y^2)^{\frac{1}{2}} F(y)\,dy = \frac{\pi}{N}\sum_{j=1}^{N-1} (1-x_j^2)\, F(x_j), \tag{5.16}$$

which is simply the trapezium rule applied, with interval π/N, to the transformed integral

$$\int_0^{\pi} \sin^2\phi\, F(\cos\phi)\, d\phi.$$

In fact (5.16) is exact for polynomials F(y) of degree $\leq 2N-3$, being closely related to the Gauss-Chebyshev quadrature formula. The term in equation (4.8) involving the smooth kernel $K(x-y)$ is conveniently approximated with the aid of this formula.

It may be noted that a bridge between Methods A and B is provided by formulae (5.8) and (5.9), which enable the coefficients $a_r^{(N)}$ to be computed from function values $\psi(x_j)$ and vice versa. We note also that from the series form of the solution we can derive a corresponding approximation for the Fourier transform of p(x), namely

$$\bar{p}(s) = \sum_{r=1}^{N-1} \frac{\pi r a_r^{(N)}}{i^{r-1}} \frac{J_r(s)}{s} \tag{5.17}$$

(see [1], equation 11.4.25), which is needed in the computation of the expression (2.11) for the far-field pressure.

While the presentation and discussion of numerical results is beyond the scope of this paper, it may be remarked that the methods described have been successfully applied to the problem of Section 1 for a wide range of wavenumbers.

ACKNOWLEDGEMENTS

Thanks are due to the Admiralty Marine Technology Establishment, Teddington, for suggesting the problem, and to my colleague Dr A.G.P. Warham for his able assistance in implementing the methods herein described.

REFERENCES

1. ABRAMOWITZ, M. and STEGUN, A. (Ed.) *Handbook of mathematical functions*. Dover Publications, Inc., New York (1972).
2. ERDOGAN, F. and GUPTA, G.D. On the numerical solution of singular integral equations. *Q appl. Math.* 29, 525-535 (1972).
3. JUNGER, M.C. and FEIT, D. *Sound, Structures, and Their Interaction.* MIT Press: Cambridge, Mass. (1972).
4. KRENK, S. *Polynomial solutions to singular integral equations.* Riso National Laboratory, Dk-4000 Roskilde, Denmark (1981).
5. MILLER, G.F. Provision of library programs for the numerical solution of integral equations. pp. 247-256 of L.M. Delves and J. Walsh (Ed.) *The Numerical Solution of Integral Equations.* Clarendon Press, Oxford (1974).
6. NOBLE, B. *The Wiener-Hopf Technique.* Pergamon Press, New York (1958).

ITERATIVE REFINEMENT TECHNIQUES FOR THE EIGENVALUE PROBLEM OF COMPACT INTEGRAL OPERATORS

*Mario Ahués, *Filomena d'Almeida, †Françoise Chatelin, *Mauricio Telias

**IMAG, Université de Grenoble (France),* † *IBM T. J. Watson Research Center, Yorktown Heights (USA), on leave from IMAG*

1. INTRODUCTION

We are concerned with the numerical solution of the eigenvalue problem

$$T\varphi = \lambda\varphi,\ \varphi\neq 0,\ \varphi\in X \tag{1.1}$$

where T is the integral operator

$$x(t) \to (Tx)(t) = \int_\Omega k(t,s)x(s)ds$$

where Ω is a bounded domain in $\mathbb{R}^N$, $N\geq 1$, and X is often in practice $C(\Omega)$ or $L^2(\Omega)$. We assume that the kernel k is such that T is compact in X. Eq. (1.1) is approximated by a discretized version

$$T_n\varphi_n = \lambda_n\varphi_n,\ 0\neq\varphi_n\in X_n \tag{1.2}$$

where X_n is a finite dimensional space, most often a subspace of X. The approximation T_n is the result of a projection or an approximate quadrature method. The resulting matrix is *full* so that the solution of (1.2) may be expensive in computer time and storage. In Iterative Refinement methods, (1.2) is solved with a relatively small n and T (or rather T_M, $M>>n$, a finer approximation of T) is only used for evaluation, that is to compute Tx, $x\in X$.

When T_n corresponds to the Nyström method, such methods have been used on the second kind Fredholm equation $Tx-zx = f$ (cf. [4], [5]). In this paper we present and compare four iterative methods for the eigenvalue problem.

2. MATHEMATICAL BACKGROUND

We recall here some notions of spectral theory. For a more detailed treatment the reader is referred to Kato [8] or Chatelin [7]. Throughout the text T

and T_n are bounded linear operators defined on X, a separable Banach space over $\mathbb{C}$, $\mathcal{L}(X)$ denotes the algebra of such operators.

2.1 Convergence Notions

Let $\{T_n\}$ be a sequence in $\mathcal{L}(X)$ and $T\in\mathcal{L}(X)$, $\{T_n\}$ is *pointwise* convergent to T, denoted $T_n \xrightarrow{p} T$ if and only if for each x in X, $T_n x \to Tx$. A sequence $\{T_n\}$ converges in *norm* if and only if $\| T - T_n \| \to 0$. Set $B = \{x\in X; \|x\| \leq 1\}$, $\{T_n\}$ is a *collectively compact* approximation of T, denoted $T_n \xrightarrow{cc} T$ if and only if in addition to $T_n \xrightarrow{p} T$, the set $\bigcup_{n\in N}(T_n - T)B$ is relatively compact in X.

2.2 Spectral Definitions

The *spectral radius* of T is $r_\sigma(T) = \lim_k \inf \|T^k\|^{1/k}$. The *resolvent set* is $\rho(T) = \{z\in\mathbb{C};(T-z)^{-1}\in\mathcal{L}(X)\}$ where z stands for $z1$, 1 being the identity operator on X, the *spectrum* is $\sigma(T) = \mathbb{C}-\rho(T)$. For $z\in\rho(T)$, $R(z) = (T-z)^{-1}$ is the *resolvent operator.* Since T is compact, its spectrum consists of countably many isolated eigenvalues with finite multiplicity, plus zero. Let λ be isolated by the Jordan curve Γ, $P = (-1/2i\pi)\int_\Gamma R(z)dz$ is the *spectral projection* associated with λ and its range $M = PX$ is the corresponding *invariant subspace,* $m = \dim M$ is the *algebraic multiplicity of* λ. Set $N = (1-P)X$, $(T-\lambda)_{\restriction N}:N \to N$ is a bijection and $S = [(T-\lambda)_{\restriction N}]^{-1}(1-P)$ which belongs to $\mathcal{L}(X)$ is called the *reduced resolvent* operator at λ, it satisfies $(T-\lambda)S = S(T-\lambda) = 1-P$.

2.3 Convergence Notions Related to Spectral Concepts

For $z\in\rho(T_n)$, we set $R_n(z) = (T_n - z)^{-1}$. $\{T_n\}$ is a *stable* approximation of T at $z\in\rho(T)$, denoted $T_n - z \xrightarrow{s} T - z$, if and only if $T_n \xrightarrow{p} T$ and for n large enough, $z\in\rho(T_n)$ and $\|R_n(z)\|$ is uniformly bounded in n. Let Δ be the domain enclosed by the Jordan curve Γ around λ, we define accordingly $P_n = (-1/2i\pi))\int_\Gamma R_n(z)dz$ and $M_n = P_n X$. $\{T_n\}$ is a *strongly stable* approximation of T on Γ (resp. in Δ) denoted $T_n - z \xrightarrow{ss} T - z$ on Γ (resp. in Δ) if and only if $T_n - z \xrightarrow{s} T - z$ for all z on Γ (resp. in $\Delta - \{\lambda\}$) and $\dim M_n = \dim M$ for n large enough.

$\{T_n\}$ is a *radial* approximation of T at z, denoted $T_n - z \xrightarrow{\sigma} T - z$, if and only if $T_n - z \xrightarrow{s} T - z$ and $r_\sigma[(T - T_n)R(z)] \to 0$. The convergence is *uniformly radial* in $\rho(T)$ if $r_\sigma[(T-T_n)R(z)] \to 0$ uniformly in z for all z in each compact subset of $\rho(T)$, this is denoted $T_n - z \xrightarrow{u\sigma} T - z$ in $\rho(T)$.

2.4 Convergence of the Eigenelements

We recall the following results proved in [7]. c is a generic constant (independent of n).

- If $T_n - z \overset{ss}{\to} T - z$ on Γ then $P_n \overset{cc}{\to} P$ and there exist for n large enough exactly m eigenvalues of T_n (counting their algebraic multiplicities) which converge to λ.

- If T is compact, $T_n - z \overset{ss}{\to} T - z$ in $\mathbb{C}-\{0\}$ is equivalent to $T_n - z \overset{u\sigma}{\to} T - z$ in $\rho(T)$.

- If $\| T_n - T\| \to 0$ or $T_n \overset{cc}{\to} T$, then $T_n - z \overset{ss}{\to} T - z$ in $\rho(T)$.

- If $P_n \overset{cc}{\to} P$ then $P_{n\restriction M}$ is bijective for n large enough from M onto M_n.

- If $m = 1$, then, for n large enough, $\phi := (P_{n\restriction M})^{-1} \varphi_n$ is an eigenvector of T normalized by $P_n\phi = \varphi_n$, and the following bound holds

$$|\lambda-\lambda_n| + \|\phi-\varphi_n\| \leq c\,\|(T - T_n)P\|.$$

3. A LOOK AT THE LINEAR EQUATION $Tx - zx = f$.

For a fixed z in $\rho(T)$, we consider the equation $(T - z)x = f$ and its approximation $(T_n - z)x_n = f$, $x = R(z)f$ is approximated by $x_n = R_n(z)f$. If $T_n - z \overset{\sigma}{\to} T - z$, then for any fixed n, large enough, the following sequence converges to x as $k \to \infty$

$$x^0 := x_n,\ x^{k+1} := x^k + R_n(z)[f - (T - z)x^k],\ k>0 \tag{3.1}$$

at least as fast as a geometric progression with quotient arbitrarily close to $r_\sigma[(T_n - T)R(z)]$ (cf. [7]). Using the identity $R_n(z) = \frac{1}{z}(R_n(z)T_n - 1)$, (3.1) can be rewritten into the form

$$x^0 := x_n, x^{k+1} := \frac{1}{z}(Tx^k - f) + \frac{1}{z}R_n(z)T_n[f - (T - z)x^k],\ k\geq 0. \tag{3.2}$$

The form (3.1) is used in Atkinson [4] for the Nyström method and the form (3.2) is used in Lin Qun [10] in connection with the finite element method. A third equivalent form is given in Chatelin [7] in relation to perturbation theory and the Rayleigh-Schrödinger expansion of x in terms of the perturbation $T_n - T$. If T_M is used for the evaluation of Tx^k, and if either $\| T_n - T\| \to 0$ or $T_n \overset{cc}{\to} T$, then $x^k \to x_M = (T_M - z)^{-1}f$.

If, in the expression $\frac{1}{z}(R_n(z)T_n - 1)$ for $R_n(z)$, we replace T_n by T, this defines a second sequence

$$x^0 := x_n,\ x^{k+1} := \frac{1}{z}(Tx^k - f) + \frac{1}{z}R_n(z)T[f - (T - z)x^k],\ k\geq 0 \tag{3.3}$$

where the residual $f - (T - z)x^k$ is smoothed by T instead of being smoothed by T_n in (3.2) or remaining unchanged in (3.1). The form (3.3) is used in Brakhage [5] when the residual has an oscillatory behaviour. The approximate inverse for $T - z$ defined by $R_n^B(z) = \frac{1}{z}(R_n(z)T - 1)$ (where B stands for Brakhage) is such that $R(z) - R_n^B(z) = \frac{1}{z}(R(z) - R_n(z))T$, therefore the stable convergence $T_n - z \xrightarrow{s'} T - z$ is sufficient to ensure that $\| R(z) - R_n^B(z) \| \to 0$, hence the linear convergence of (3.3) in $\| (R_n(z) - R(z))T \|$, for n large enough.

We shall now present similar iterative schemes for the *eigenvalue* problem (1.2).

4. THE EIGENVALUE PROBLEM

The eigenvalue λ is supposed to be *simple* and nonzero. ϕ is the eigenvector of T normalized by $P_n\phi = \varphi_n$ for n large enough (note that ϕ depend on n, which will be fixed).

Let $\varphi_n{}^*$ be the eigenvector of the adjoint problem $T_n{}^*\varphi_n{}^* = \overline{\lambda}_n\varphi_n{}^*$ normalized by $<\varphi_n,\varphi_n{}^*> = 1$, where $<,>$ denotes the duality between X and its adjoint X^*, then $P_n = <\cdot,\varphi_n{}^*>\varphi_n$.

4.1 The Iterations

For a fixed n we define the four iterations for $k \geq 0$, in which the scalar μ^{k+1} stands for $<Tu^k,\varphi_n{}^*>$

$$u^0 := \varphi_n,\ u^{k+1} := u^k + S_n(\mu^{k+1}u^k - Tu^k) := F_1(u^k) \tag{4.1}$$

$$u^0 := \varphi_n,\ u^{k+1} := u^k + \frac{1}{\lambda_n}(S_nT - 1)(\mu^{k+1}u^k - Tu^k) := F_2(u^k) \tag{4.2}$$

$$u^0 := \varphi_n,\ u^{k+1} := \frac{Tu^k}{\mu^{k+1}} + S_nT_n(u^k - \frac{Tu^k}{\mu^{k+1}}) := G_1(u^k) \tag{4.3}$$

$$u^0 := \varphi_n,\ u^{k+1} := \frac{Tu^k}{\mu^{k+1}} + S_nT(u^k - \frac{Tu^k}{\mu^{k+1}}) := G_2(u^k). \tag{4.4}$$

In the next paragraph we prove under suitable assumptions that $u^k \to \phi$ and $\mu^k \to \lambda$ as $k \to \infty$. ϕ is the solution of $(T - \lambda)\phi = 0$. Therefore (4.1) (resp. (4.3)) for $(T - \lambda)\phi = 0$ (resp. $(1 - \frac{1}{\lambda}T)\phi = 0$) is a natural analogue of (3.1) for $(T - z)x = f$. By definition of μ^{k+1}, $P_n(Tu^k - \mu^{k+1}u^k) = <Tu^k,\varphi_n{}^*>\varphi_n - \mu^{k+1}\varphi_n = 0$. Then, by using the identity $\lambda_nS_n = S_nT_n - 1 + P_n$, we remark that (4.2) (resp. (4.4)) is deduced from (4.1) (resp. (4.3)) by replacing T_n by T, in the way (3.3) is deduced from (3.2). An equivalent form of (4.3) is $u^{k+1} = u^k + \lambda_nS_n(u^k - Tu^k/\mu^{k+1})$ which is very close to (4.1) written as

$u^{k+1} = u^k + \mu^{k+1} S_n(u^k - Tu^k/\mu^{k+1})$.

Iteration (4.1) is proposed in Chatelin [7] and related to multigrid methods; Lin Qun [9,10] has first proposed (4.3) for a selfadjoint operator in connection with finite element methods; (4.2) and (4.4) are given by Ahués et al. [3].

4.2 Convergence

We first prove the convergence of (4.1) and (4.3) under the assumption that $\| T_n - T \| \to 0$ or $T_n \overset{cc}{\to} T$. We remark that

$$\| (T - T_n)P \| \leq c \| (T - T_n)_{\restriction M} \| \leq c \| T - T_n \|.$$

Proposition 1. If $\| T_n - T \| \to 0$, *then for any fixed n, large enough,* $\lambda = \lim_k \mu^k$ *and* $\phi = \lim_k u^k$, *where* u^k *is defined* (i) *by* (4.1) *or* (ii) *by* (4.3). *And the convergence is at least linear*

$$| \mu^k - \lambda | + \| u^k - \phi \| \leq c \| T - T_n \|^{k+1}, \ k \geq 0.$$

Proof. We prove the convergence by induction. For $k = 0, | \lambda_n - \lambda | + \| \varphi_n - \phi \| \leq c \| T - T_n \|$ and we assume that $| \mu^k - \lambda | + \| u^k - \phi \| \leq c \| T - T_n \|^{k+1}$ holds.

Case (i) For (4.1) we write the identity

$$\begin{aligned}(T_n - \lambda_n)(u^{k+1} - \phi) &= (T_n - \lambda_n)u^k - (1 - P_n)(Tu^k - \mu^{k+1}u^k) - (T_n - \lambda_n)\phi \\ &= (T_n - T)(u^k - \phi) + (\mu^{k+1} - \lambda)u^k + (\lambda - \lambda_n)(u^k - \phi).\end{aligned}$$

Then

$$\mu^{k+1} - \lambda = <(T - T_n)(u^k - \phi), \varphi_n{}^*> \tag{4.5}$$

and

$$u^{k+1} - \phi = (\mu^{k+1} - \lambda)S_n u^k + (\lambda - \lambda_n)S_n(u^k - \phi) + S_n(T_n - T)(u^k - \phi) \tag{4.6}$$

since $P_n(u^{k+1} - \phi) = 0$. We deduce

$$\begin{aligned}| \mu^{k+1} - \lambda | &\leq c(| \lambda - \lambda_n | + \| T - T_n \|) \| u^k - \phi \| \\ \| u^{k+1} - \phi \| &\leq c[| \mu^{k+1} - \lambda | + (| \lambda - \lambda_n | + \| T - T_n \|) \| u^k - \phi \|] \\ &\leq c \| T - T_n \|^{k+2}.\end{aligned}$$

Case (ii) For (4.3) we write

$$(T_n-\lambda_n)(u^{k+1}-\phi) = (T_n-\lambda_n)u^k + \lambda_n(1-P_n)(u^k - \frac{1}{\mu^{k+1}}Tu^k) - T_n\phi + \lambda_n\phi$$
$$= T_n u^k - \frac{\lambda_n}{\mu^{k+1}}Tu^k - T_n\phi + \lambda_n\phi$$
$$= (\mu^{k+1}-\lambda)\frac{\lambda_n}{\mu^{k+1}\lambda}Tu^k - \frac{1}{\lambda}(\lambda_n-\lambda)T(u^k-\phi)-(T-T_n)(u^k-\phi).$$

$\mu^{k+1}-\lambda = \langle T(u^k-\phi),\varphi_n^*\rangle \to 0$, hence $\mu^{k+1}\neq 0$.

$$\mu^{k+1}-\lambda = \frac{\lambda_n-\lambda}{\lambda_n}\langle T(u^k-\phi),\varphi_n^*\rangle + \frac{\lambda}{\lambda_n}\langle (T-T_n)(u^k-\phi),\varphi_n^*\rangle,$$

$$u^{k+1}-\phi = (\mu^{k+1}-\lambda)\frac{\lambda_n}{\mu^{k+1}\lambda}[S_nT(u^k-\phi)+\lambda S_n\phi]$$
$$+ \frac{\lambda-\lambda_n}{\lambda}S_nT(u_k-\phi) + S_n(T_n-T)(u_k-\phi).$$

We conclude as in (i). □

Proposition 2. If $T_n \xrightarrow{cc} T$, *then for n fixed large enough and for* u^k *defined* (i) *by* (4.1) *or* (ii) *by* (4.3), *the convergence* $u^k \to \phi$, $\mu^k \to \lambda$ *is such that*

$$\max(|\mu^{2k+1}-\lambda| + \|u^{2k+1}-\phi\|,$$
$$|\mu^{2k}-\lambda| + \|u^{2k}-\phi\|) \le c\varepsilon_n^k \|(T-T_n)P\|, k\ge 0,$$

with $\varepsilon_n = \|(T-T_n)P\| + \|(T-T_n)S_n(T-T_n)\|$.

Proof. We first note that by definition

$$\mu^{2k+1}-\lambda = \langle T(u^{2k}-\phi),\varphi_n^*\rangle$$

and that by equation (4.6)

$$\|u^{2k+1}-\phi\| \le c[|\mu^{2k+1}-\lambda| + |\lambda-\lambda_n|\,\|u^{2k}-\phi\| + \|u^{2k}-\phi\|],$$

so

$$|\mu^{2k+1}-\lambda| \le c\|u^{2k}-\phi\|$$

and

$$\|u^{2k+1}-\phi\| \le c\|u^{2k}-\phi\|.$$

Hence, it suffices to prove that

$$|\mu^{2k}-\lambda| + \|u^{2k}-\phi\| \le c\varepsilon_n^k\|(T-T_n)P\|. \tag{4.7}$$

For $k = 0$ the result is clear. And $\varepsilon_n \to 0$ as $n\to\infty$ since $T_n \xrightarrow{cc} T$. Suppose now that (4.7) holds for a certain k arbitrarily chosen.

Case (i) From (4.5) we get $\mu^{2k+2}-\lambda = \langle (T - T_n)(u^{2k+1}-\phi), \varphi_n{}^* \rangle$ and from (4.6) and the fact that $S_n\varphi_n = 0$:

$$(T-T_n)(u^{2k+1}-\phi) = (\mu^{2k+1}-\lambda)(T-T_n)S_n[(u^{2k}-\phi)+(\phi-\varphi_n)] \\ +(\lambda-\lambda_n)(T-T_n)S_n(u^{2k}-\phi)+(T-T_n)S_n(T_n-T)(u^{2k}-\phi).$$

Then

$$|\mu^{2k+2}-\lambda| \leq c(\|u^{2k}-\phi\| + \|\phi-\varphi_n\| + |\lambda-\lambda_n| + \varepsilon_n)\|u^{2k}-\phi\| \\ \leq c\varepsilon_n \|u^{2k}-\phi\|.$$

Using again (4.6) we bound similarly

$$\|u^{2k+2}-\phi\| \leq c\varepsilon_n \|u^{2k}-\phi\|.$$

Case (ii) is treated the same way.□

Proposition 3. If $T_n - z \overset{ss}{\rightarrow} T - z$ *on* Γ *then* u^k *defined* (i) *by* (4.2) *or* (ii) *by* (4.4) *converges to* ϕ , *and*

$$|\mu^{k+1}-\lambda| + \|u^k-\phi\| \leq c\|(T-T_n)T\|^{k+1}, k \geq 0.$$

Proof. We first remark that $\|T-T_n)P\| \leq c\|(T-T_n)_{\restriction M}\| \leq c\|(T-T_n)T\|$. Then for $k = 0$, $\mu^1-\lambda = \langle T(\varphi_n-\phi), \varphi^*{}_n \rangle$ and $|\mu^1-\lambda| + \|u^0-\phi\| \leq c\|(T-T_n)T\|$. And we suppose that $|\mu^{k+1}-\lambda| + \|u^k-\phi\| \leq c\|(T-T_n)T\|^{k+1}$. It is clear that $|\mu^{k+1}-\lambda| = |\langle T(u^k-\phi), \varphi^*{}_n \rangle| \leq c\|u^k-\phi\|$.

Case (i) For (4.2) we write

$$(T_n-\lambda_n)(u^{k+1}-\phi) = (T_n-\lambda_n)(u^k-\phi) + \frac{1}{\lambda_n}(1-P_n)(T_n-T)(Tu^k-\mu^{k+1}u^k) \\ -(1-P_n)(Tu^k-\mu^{k+1}u^k)$$

$$= (T_n-T)(u^k-\phi) + \frac{1}{\lambda_n}(1-P_n)(T_n-T)T(u^k-\phi) + \frac{\lambda}{\lambda_n}(1-P_n)(T_n-T)\phi \\ -\frac{\mu^{k+1}}{\lambda_n}(1-P_n)(T_n-T)u^k + \mu^{k+1}u^k - Tu^k$$

$$= (T_n-T)u^k - \frac{\mu^{k+1}}{\lambda_n}(T_n - T)u^k - \frac{\mu^{k+1}}{\lambda_n}P_nTu^k + \frac{\mu^{k+1}}{\lambda_n}P_nT_nu^k$$

$$+ \lambda_n\phi + \mu^{k+1}u^k - T_n\phi + \frac{\lambda}{\lambda_n}(T_n-T)\phi - \frac{\lambda}{\lambda_n}P_nT_n\phi + \frac{\lambda^2}{\lambda_n}P_n\phi$$

$$+ \frac{1}{\lambda_n}(1 - P_n)(T_n-T)T(u^k-\phi)$$

$$= \frac{\lambda-\lambda_n}{\lambda_n}(T_n-\lambda_n-T)(\phi-u^k) + \frac{\lambda-\mu^{k+1}}{\lambda_n}[(\lambda-\lambda_n)\varphi_n + (T_n-\lambda_n)(u^k-\phi)$$

$$+ (T_n-\lambda_n)(\phi-\varphi_n) - T(u^k-\phi) + (\mu^{k+1}-\lambda)\phi + \mu^{k+1}(\varphi_n-\phi)]$$

$$+ \frac{1}{\lambda_n}(1 - P_n)(T_n - T)T(u^k-\phi).$$

We apply S_n and get

$$u^{k+1}-\phi = \frac{\lambda-\lambda_n}{\lambda_n}(1 - S_nT)(\phi-u^k) + \frac{\lambda-\mu^{k+1}}{\lambda_n}[(1 - S_nT)(u^k-\phi) + \phi-\varphi_n$$

$$+ (\mu^{k+1}-\lambda)S_n\phi + \mu^{k+1}S_n(\varphi_n-\phi)] + \frac{1}{\lambda_n}S_n(T_n-T)T(u^k-\phi).$$

Case (ii) For (4.4) we write

$$(T_n-\lambda_n)(u^{k+1}-\phi) = (T_n-\lambda_n)\frac{Tu^k}{\mu^{k+1}} + Tu^k - \mu^{k+1}\varphi_n + (P_n - 1)\frac{T^2u^k}{\mu^{k+1}} - (T_n-\lambda_n)\phi$$

$$= \frac{1}{\mu^{k+1}}(T_n - T)T(u^k-\phi) + \frac{\lambda-\mu^{k+1}}{\mu^{k+1}}(T_n - T)\phi - \lambda\phi + \lambda_n\phi + \frac{1}{\mu^{k+1}}P_nT^2u^k$$

$$+ Tu^k - \frac{\lambda_n}{\mu^{k+1}}Tu^k - \mu^{k+1}\varphi_n - (\lambda-\lambda_n)\phi + \frac{\mu^{k+1}-\lambda_n}{\mu^{k+1}}Tu^k = A$$

$$(\mu^{k+1}-\lambda_n)Tu^k = (\mu^{k+1}-\lambda)Tu^k + (\lambda-\lambda_n)T(u^k-\phi) + (\lambda-\lambda_n)\lambda\phi,$$

$$(1-P_n)A = \frac{1}{\mu^{k+1}}(T_n - T)T(u^k-\phi) + \frac{\lambda-\mu^{k+1}}{\mu^{k+1}}[(T_n-T)\phi - Tu^k]$$

$$+ \frac{\lambda-\lambda_n}{\mu^{k+1}}[T(u^k-\phi) + (\lambda-\mu^{k+1})\phi]$$

$$= \frac{1}{\mu^{k+1}}(T_n-T)T(u^k-\phi) + \frac{\mu^{k+1}-\lambda_n}{\mu^{k+1}}T(u^k-\phi) + \frac{\mu^{k+1}-\lambda}{\mu^{k+1}}\lambda\phi.$$

By multiplication by S_n we get

$$u^{k+1}-\phi = = \frac{1}{\mu^{k+1}} S_n(T_n - T)T(u^k - \phi) + \frac{\mu^{k+1}-\lambda_n}{\mu^{k+1}} S_n T(u^k - \phi) + \frac{\mu^{k+1}-\lambda}{\mu^{k+1}} \lambda S_n(\phi - \varphi_n).$$

In both cases we conclude easily that $\| u^{k+1} - \phi \| \le c \| (T - T_n)T \|^{k+2}$.□

In [3] the definition and proof of convergence of the four iterations under consideration are presented in the context of the Iterative Defect Correction method [12] on two different fixed point formulations of (1.1) which are mildly nonlinear equations in the eigenvector ϕ (see also [2]).

Other iterations can be thought of, for example in (4.1) and (4.2), λ_n may be replaced by μ^{k+1} and S_n by $\Sigma_n^{k+1} := [(T_n - \mu^{k+1})_{\restriction (1-P_n)X}]^{-1}(1 - P_n)$. The resulting iterations correspond to Newton's method with an approximate Jacobian [1,7]. Note that Σ_n^{k+1} varies over the iterations.

Clearly the computation of u^{k+1} depends on the condition of $(T_n - \lambda_n)_{\restriction (1-P_n)X}$, and in particular on the distance of λ_n to the rest of the spectrum of T_n. To deal with close or multiple eigenvalues, a somewhat different approach, based on analytic perturbation theory and Rayleigh-Schrödinger expansions in $T - T_n$ has been found useful. It is worth noticing that the two points of view give identical results on the *linear* equation $Tx - zx = f$ (see Chatelin [7]).

5. NUMERICAL EXPERIMENTS

We report the results of numerical experiments done on the three integral operators T^α, T^β, T^γ defined respectively by the kernels

$$\alpha) \quad k^\alpha(t,s) = \begin{cases} (1-s)t & \text{if } 0 \le t \le s \le 1, \\ (1-t)s & \text{if } 0 \le s < t \le 1. \end{cases}$$

$$\beta) \quad k^\beta(t,s) = \begin{cases} e(1-e)^{-1}(1-e^{s-1})(e^{-t}-1) & \text{if } 0 \le t \le s \le 1, \\ (1-e)^{-1}(1-e^{s})(e^{1-t}-1) & \text{if } 0 \le s < t \le 1. \end{cases}$$

$$\gamma) \quad k^\gamma(t,s) = 2 \,|\, \sin\, 10\pi t - \sin\, 10\pi s \,|^{1/2} \quad \textit{for } 0 \le t,\ s \le 1.$$

The dominant eigenvalue λ and its eigenvector ϕ are computed. For k^α (resp. k^β) the analytic expressions are $\lambda = \pi^{-2}$ (resp.$4(4\pi^2 + 1)^{-1}$, $\phi(t) = C \sin \pi t$ (resp. $Ce^{-t/2} \sin \pi t$) where C is such that $<\phi, \varphi_n{}^*> = 1$. (1.1) is set in $X^1 = L^2(0.1)$ or $X^2 = C(0,1)$. And the following subspaces and projections are used

* X_n^1 spanned by n Legendre polynomials defined on [0,1], π_n^1 is the orthogonal projection on X_n^1.
* X_n^2 spanned by n hat functions (i.e. piecewise linear) defined by the partition $\{\frac{i}{n}\}_0^n$ of [0,1], π_n^2 is the Lagrange interpolation at $\frac{i}{n}$.

The following approximations are used [7]:

* as projection methods, the Galerkin, projection and Sloan methods, respectively defined by $T_n^G = \pi_n T \pi_n$, $T_n^P = \pi_n T$ and $T_n^S = T\pi_n$
* as approximate quadrature methods, the Fredholm and Nyström approximations T_n^F and T_n^N with the trapezoidal rule.

Table 1 displays the different approximations used in each example, the corresponding discretization policies (values of n and M) and the number of iterations required to achieve a relative precision of 10^{-12}. The iteration is stopped when ℓ has been found such that $\ell > 120$ or $(|u^\ell - u^{\ell-1}|_\infty^{(M)} / |u^\ell|_\infty^{(M)}) < 10^{-12}$, where $|u|_\infty^{(M)} := \max(|u(t_i^{(M)})|, 1 \leq i \leq M)$ if the $t_i^{(M)}$ are the M points of the fine grid. We set $\tilde{\mu} := \langle Tu^\ell, \varphi_n^* \rangle$ and $\tilde{u} := u^\ell$; $\tilde{\mu}$ and $\tilde{u}$ are eigenelements of T_M, up to a precision of 10^{-12}. The error $\varepsilon = |\tilde{\mu} - \lambda|$ is given for the kernels k^α and k^β. The following notation is used in Table 1.

I example number

II kernel

III approximation (the subspace X_n^2 is used everywhere with the exception of example 1, where G^* means the Galerkin method on X_n^1)

IV type of convergence

> $\ell > 120$.

TABLE 1

Summary of results

I	II	III	IV	Policy n	Policy M	ε	Number of Iterations F_1	F_2	G_1	G_2
1		G^*	norm	2	8	2×10^{-4}	16	37	12	19
2	α	G	c.c	5	101	8×10^{-5}	>	17	8	18
3		F	c.c	5	101	9×10^{-5}	>	20	8	18
4		G	c.c			8×10^{-5}	>	17	20	18
5		P	norm			"	>	15	12	12
6	β	S	c.c	5	101	"	>	14	12	12
7		F	c.c			9×10^{-5}	>	21	19	19
8		N	c.c			"	113	12	13	14
9	γ	F	c.c	5	101	-	>	19	22	18
10	γ	N	c.c	5	101	-	>	12	21	15

All computations have been done in double precision on the HB68 at the University Computing Center of Grenoble. Displayed on Fig. 1 and Fig. 2 are graphs of the relative errors as functions of k in a decimal logarithmic scale, when the iterations F_2, G_1 or G_2 are used. For the eigenvalue, the graph of $\log_{10}|(\mu^k-\widetilde{\mu})/\widetilde{\mu}|$ versus k is plotted with crosses $\times$; for the eigenvector, the graph of $\log_{10}(|u^k-\widetilde{u}|_\infty^{(M)}/|\widetilde{u}|_\infty^{(M)})$ is plotted with a solid line ———.

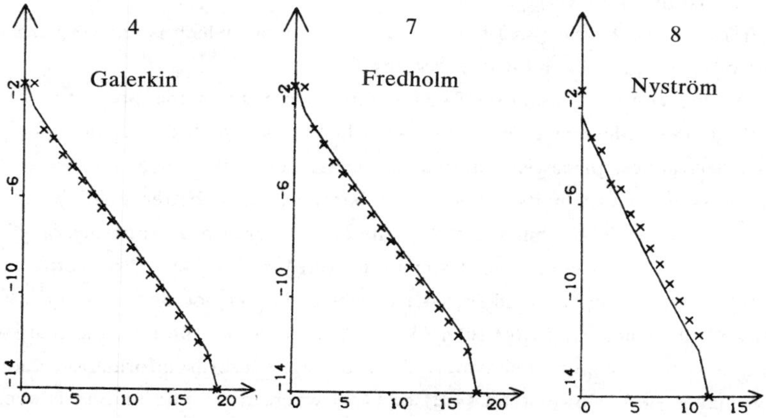

FIG. 1. Examples 4, 7 and 8 with iteration G_1.

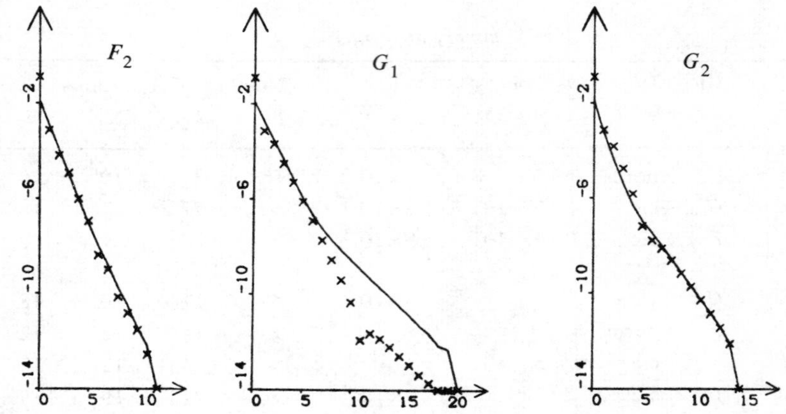

FIG. 2. Example 10 (Nyström) with iterations F_2, G_1 and G_2.

6. CONCLUSIONS

Iterations (4.1) and (4.3) are cheaper to perform than (4.2) and (4.4) because the latter require one evaluation of T^2u^k. On the other hand convergence of (4.2) and (4.4) is proved under weaker assumptions than those used in the proof of convergence of (4.1) and (4.3).

From the limited experiments performed we may draw the following conclusions:

(i) Iteration (4.1) is almost always not convergent for a policy (n,M) for which the other methods converge.

(ii) Iterations (4.2) and (4.4) have a similar behaviour which is not very sensitive to the different kernels and discretizations T_n.

(iii) On the contrary, iteration (4.3) seems to recognize the properties of the kernel, in the following sense. For k^α which is symmetric, has bounded first partial derivatives, piecewise continuous partial derivatives of any order and no oscillations nor picks in its shape, (4.3) becomes much better than the others. For k^β, symmetry is lost but the rest of the above mentioned properties of k^α still hold. Now (4.3) becomes very similar to (4.2) and (4.4). The kernel k^γ is symmetric, but has unbounded first derivatives and a very oscillatory shape. Then (4.2) and (4.4) become faster than (4.3). A possible interpretation is that, since k^γ has a very oscillatory behaviour, T_n loses some useful information about T which makes (4.3) slower than (4.2) or (4.4) where more information is kept by means of T_M.

(iv) The logarithmic graphs show that the theoretical error bounds are reasonably sharp in practice (with moderate constants).
(v) The effect of numerical quadrature seems to be marginal. We observe similar results for Sloan and Nyström methods, as well as for Galerkin and Freholm methods.

REFERENCES

1. AHUES, M., Raffinement des éléments propres d'un opérateur compact sur un espace de Banach pour des méthodes de type Newton à Jacobien approché, Rep. Université de Grenoble (1982).
2. AHUES, M.; CHATELIN, F., The use of Defect Correction to refine the eigenelements of compact integral operators. *SIAM J. Num. Anal.* (to appear).
3. AHUES, M.; D'ALMEIDA, F.; TELIAS, M., Two Defect Correction Methods for the Eigenvalue Problem of Compact Operators in Banach Spaces. *J. Int. Equ.* (submitted).
4. ATKINSON, K. E., Iterative variants of the Nyström Method for the Numerical Solution of Integral Equations. *Numer. Math. 22,* 17-31 (1973).
5. BRAKHAGE, H., Über die numerische Behandlung von Integralgleichungen nach der Quadraturformelmethode. *Numer. Math. 2,* 183-196 (1960).
6. CHATELIN, F., Iterative refinement for the Fredholm integral eigenvalue problem. In *Mini-Symposium on Iterative Refinement Techniques in Numerical Analysis,* SIAM Meeting, Stanford (1982).
7. CHATELIN, F., *Spectral Approximation of Linear Operators.* Academic Press, New York, (to appear, 1983).
8. KATO, T., *Perturbation Theory for Linear Operators.* Springer Verlag, Berlin, New York (1976).
9. LIN QUN, How to increase the accuracy of lower order elements in nonlinear finite element methods, pp 41-47 of R. Glowinski and J-L. Lions (eds.) *Computational Methods in Applied Science and Engineering.* North-Holland, Amsterdam (1980).
10. LIN QUN, Iterative refinement of finite elements approximations for elliptic problems. RAIRO Anal. Numér. 16, pp. 39-47 (1982).
11. SPENCE, A., Error bounds and estimates for eigenvalues of integral equations. *Numer. Math.* 29, 133-147 (1978) and 32, 139-146 (1979).
12. STETTER, H. H., The defect correction principle and discretization methods. *Numer. Math.* 29, 425-443 (1978).

SINGULAR INTEGRAL EQUATIONS AND MIXED BOUNDARY VALUE PROBLEMS FOR HARMONIC FUNCTIONS.

M.R. Razali and K.S. Thomas.

University of Southampton.

1. INTRODUCTION

In this paper, we discuss the application of singular integral equations to the numerical solution of mixed boundary value problems in potential theory. The problem we will study is the following (the Volterra problem).

Let $[a_j,b_j]$ $j = 1,\ldots,p$ be disjoint intervals on the real line ($-\infty < a_1 < b_1 < a_2 < \ldots < b_p < \infty$). We let

$$L_1 = \bigcup_{j=1}^{p} [a_j,b_j], \quad L_2 = \mathbb{R} \setminus L_1 .$$

We require a complex function $\Phi(z) = u(x,y) + iv(x,y)$ holomorphic in the upper half plane $y \geqq 0$ and bounded at infinity, and constants $C_1,\ldots,C_p$ such that

$$v(x,0) = v_0(x) + C_j \quad \text{if } x \in [a_j,b_j] \; j=1,\ldots p, \tag{1.1}$$

$$u(x,0) = u_0(x) \quad \text{if } x \in L_2. \tag{1.2}$$

$v_0(x)$ and $u_0(x)$ are two given functions assumed to be smooth (satisfy a Hölder condition).

Many authors have treated this problem and variants thereof. The problem is equivalent to a Riemann problem and the theory is described in [4,§95]. The paper of Homentcovschi [2] is worthy of mention. He evaluates the analytic solution of the Riemann problem by the use of Chebyshev polynomials, a technique akin to ours.

Mixed boundary problems involving Laplace's equation on simply connected regions can be formulated as Volterra problems when the conformal mapping of the region to the half plane is known.

The method of solution is to obtain a singular integral equation for u(x,0) on the *whole* of the real line. $\Phi(z)$ then satisfies [4,p112]

$$\Phi(z) = \frac{1}{\pi i} \int_{-\infty}^{\infty} \frac{u(t,0)dt}{t-z} + iC, \tag{1.3}$$

where C is an arbitrary constant. The integral equation is solved numerically.

The outline of the paper is as follows. In §2, we describe two familiar boundary value problems and show how they may be recast as a Volterra problem. In §3, we derive the basic integral equation. §4 contains a description of the numerical solution and §5 contains the results of some computations.

Other treatment of mixed boundary value problems may be found in [1], [3], [6], [7].

2. TWO MODEL PROBLEMS

We introduce two examples in which mixed boundary value problems are reduced to Volterra problems .

Problem A. Symm [5]

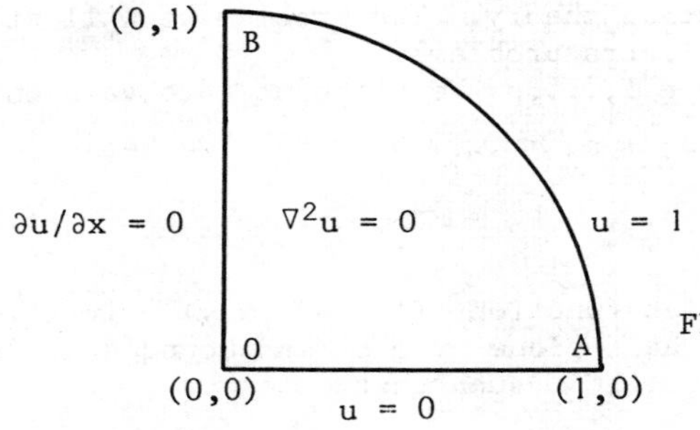

FIG.1 PROBLEM A

We require the solution of Laplace's equation in the first quadrant of the unit circle. The boundary conditions are given in Fig. 1. The problem has the analytic solution

$$u(x,y) = \frac{2}{\pi} \arctan\left(\frac{2y}{1 - x^2 - y^2}\right). \tag{2.1}$$

The conformal mapping used to transform the region into the upper half plane is the composition of two conformal mappings. These are

$$w_1 = \left((z^2 + 1)/(z^2 - 1)\right)^2 , \tag{2.2}$$

$$w_2 = (4w_1 - 3)/(3 - 2w_1). \tag{2.3}$$

The second of these is an automorphism of the upper half plane and is used to rearrange the images of O,A, and B from the first mapping.

The final "working" plane and Volterra problem are shown in Fig. 2.

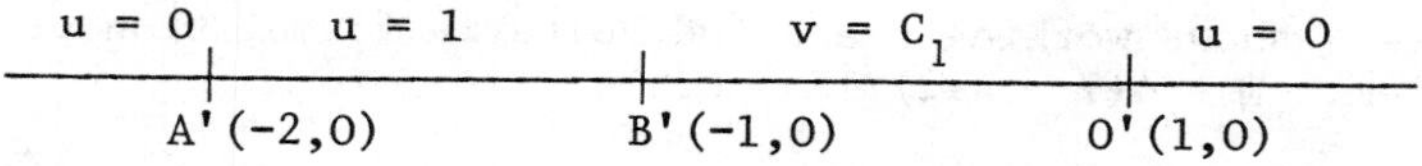

FIG. 2. Working plane and Volterra problem for Problem A.

The points O', A' and B' are the images of O, A and B resp. under the conformal mapping. The Neumann boundary condition is transformed to a boundary condition involving v by the Cauchy-Riemann conditions and integration, namely

$$\partial u/\partial x = 0 \Rightarrow \partial v/\partial y = 0 \Rightarrow v = C_1 \text{(constant)}.$$

Problem B. The Motz Problem.

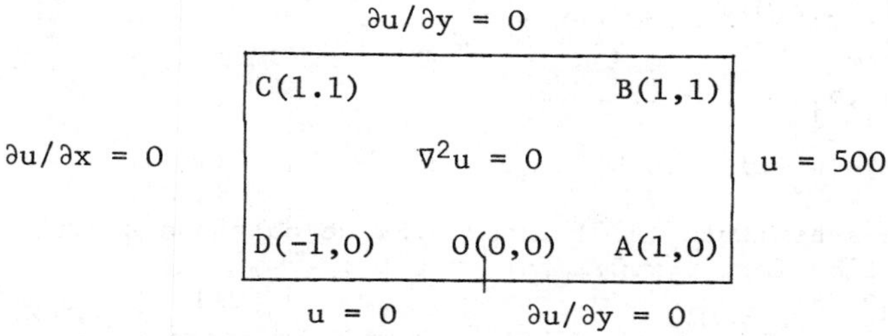

FIG. 3. The Motz Problem.

This problem has been solved by numerous authors. The usual boundary conditions are $u = 500$ on OD and $u = 1000$ on AB. Again the composition of two conformal mappings is used (cf. [7]). The first involves the elliptic sine function and the second is a half plane automorphism.

$$w_1 = \mathrm{sn}(Kz, 1/\sqrt{2}) \text{ where } K = \int_0^1 (1-t^2)^{-\frac{1}{2}}(1 - 0.5t^2)^{-\frac{1}{2}}dt. \quad (2.4)$$

$$w_2 = 4(w_1 - 1)/(3w_1 + 1). \quad (2.5)$$

Under this transformation the final working plane and Volterra problem are produced and shown in Fig. 4. The images of the points O,A,B,C,D are O', A', B', C' and D'.

We note in both problems the second conformal mapping ensures a boundary condition of the form $u = 0$ at infinity. This device is convenient but not necessary.

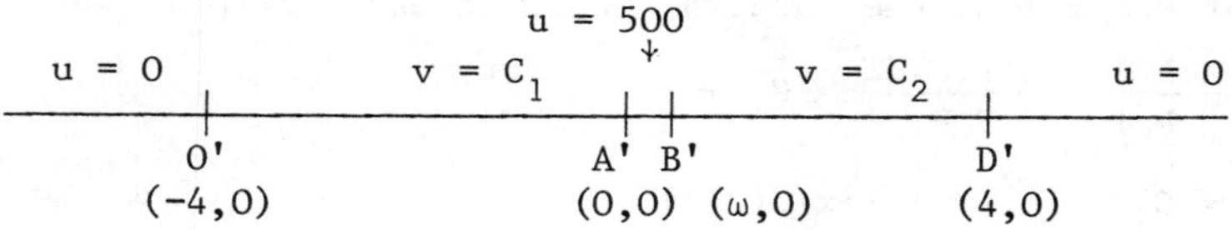

FIG. 4.

Figure 4 shows working plane and Volterra formulation of the Motz problem. $\omega = 4(7 - 4\sqrt{2})/17 \simeq 0.316$.

3. DERIVATION OF INTEGRAL EQUATION

One can easily derive a singular integral equation for $u(x,0)$ on those intervals $[a_j,b_j]$ $j = 1,\dots,p$ where it is unknown. We apply the Plemelj formula to (1.3) to get

$$\Phi^{+}(x,0) = u(x,0) + \frac{1}{\pi i}\int_{-\infty}^{\infty} \frac{u(t,0)dt}{t-x} + iC. \tag{3.1}$$

Taking the real and imaginary parts shows that in order that (1.1) and (1.2) be satisfied, $u(x,0)$ should satisfy

$$-\frac{1}{\pi}\int_{-\infty}^{\infty} \frac{u(t,0)dt}{t-x} = v_0(x) + C_j;$$

$$x \in [a_j,b_j]\ ,\ j = 1,\dots,p, \tag{3.2}$$

$$\text{and } u(x,0) = u_0(x),\ x \in L_2. \tag{3.3}$$

If we then substitute (3.3) into (3.2) we obtain a singular integral equation for $u(x,0)$ for $x \in L_1$:

$$-\frac{1}{\pi}\left(\sum_{j=1}^{p}\int_{a_j}^{b_j} \frac{u(t,0)dt}{t-x}\right) = v_0(x)+C_j + \frac{1}{\pi}\int_{L_2} \frac{u_0(t)dt}{t-x}\ ,$$

$$\text{for } x \in [a_j,b_j]\ ,\quad j=1,\dots,p. \tag{3.4}$$

Equation (3.4) is a singular integral equation on several disjoint arcs. A comprehensive account of the theory of such equations may be found in [4]. The function $u(x,0)$ is to be bounded at all the points a_j, b_j $j=1,\dots,p$. Hence the integral equation has a *negative* index. In fact the index is $-p$. There will thus be p compatibility conditions that the right hand side of (3.4) should satisfy. The approach of [2] is to use these to compute the constants C_j $j = 1,\dots,p$. (The arbitrary constant C in (3.1) may be absorbed into the C_j.) Our approach is to calculate the C_j *simultaneously* with the numerical solution of (3.4).

There is a strong connection between this derivative of the integral equation and the representation of $\Phi(z)$ as a double-layer potential.

For Problem A, the integral equation has the form

$$-\frac{1}{\pi}\int_{-1}^{1} \frac{u(t,0)dt}{t-x} = C_1 + \frac{1}{\pi}\int_{-2}^{-1} \frac{dt}{t-x}$$

$$= C_1 + \pi^{-1}\ \ln|(1+x)/(2+x)|. \tag{3.5}$$

There is only one arc in this case.

For Problem B, there are two arcs and the integral equation has the form:

$$-\frac{1}{\pi}\int_{-4}^{0}\frac{u(t,0)\,dt}{t-x}-\frac{1}{\pi}\int_{\omega}^{4}\frac{u(t,0)\,dt}{t-x}$$
$$= C_j + \frac{500}{\pi}\ \ln\left|\frac{\omega - x}{x}\right| , \tag{3.6}$$

where $j = 1$ if $x \,\varepsilon[-4,0]$, $j=2$ if $x\ \varepsilon\ [\omega,4]$.

4. NUMERICAL SOLUTION OF THE INTEGRAL EQUATION

Let us first consider the case of the integral equation on a single arc, typified by problem A,

$$-C_1 - \frac{1}{\pi}\int_{-1}^{1}\frac{u(t,0)\,dt}{t - x} = g(x) = \frac{1}{\pi}\ln\left|\frac{1 + x}{2 + x}\right| \tag{4.1}$$

in which C_1 and $u(t,0)$ are to be determined. The index of this equation is -1 and so we seek an approximation to $u(t,0)$

$$u(t,0) \simeq \surd(1 - t^2)\ \sum_{j=0}^{N}\alpha_j U_j(t)\ ;\ |t| \leqq 1\ , \tag{4.2}$$

namely a truncated series of Chebyshev polynomials of the second kind. On substituting (4.2) into (4.1) the left hand side of (4.1) becomes :

$$-C_1 + \sum_{j=0}^{N}\alpha_j T_{j+1}(x) \tag{4.3}$$

where $T_j(x)$ is the Chebyshev polynomial of the first kind. The calculation of the constant C_1 and the coefficients $\{\alpha_j\}_{J=0}^{N}$ may be carried out by collocation or Galerkin's method. Since (4.3) is a polynomial of degree $N+1$, the problem of determining the constant and the coefficients is equivalent to polynomial approximation of the right hand side $g(x)$. Collocation is thus equivalent to interpolation and Galerkin's method is equivalent to a least squares approximation of g. *Hence the efficient application of the Chebyshev series methods and the related quadrature methods depends on $g(x)$ being smooth.* In our case $g(x)$ has a logarithmic singularity at $x = -1$. Thus, the naive application of Chebyshev series methods leads to very slow convergence.

In fact, the form of the solution (4.2) is at fault. $u(t,0)$ must be continuous on the whole real line and the representation (4.2) produces a discontinuity in $u(t,0)$ at $t = -1$. Instead we put:

$$u(t,0) \simeq \surd(1-t^2)\ \sum_{j=0}^{N}\alpha_j U_j(t) + \tfrac{1}{2}(1-t);\ |t|\leqq 1\ . \tag{4.4}$$

The second term ensures the continuity of u(t,0) on the whole real line and is constructed from linear interpolation. The values of C_1 and $\{\alpha_j\}_{j=0}^{N}$ are obtained by substitution of (4.4) into (4.1) and collocation at the zeros $\{\eta_k\}_{k=1}^{N+2}$ of $T_{N+2}(x)$. This gives the linear equations:

$$-C_1 + \sum_{j=0}^{N} \alpha_j T_{j+1}(\eta_k) = \psi(\eta_k); \quad k = 1,\ldots,N+2, \tag{4.5}$$

where

$$\psi(x) = \frac{1}{\pi} \ln \left|\frac{1+x}{2+x}\right| + \frac{1}{2\pi} \int_{-1}^{1} \frac{1 - t}{t - x} dt. \tag{4.6}$$

The logarithmic singularity has been subtracted out and a marked improvement in the convergence is obtained

Having obtained u(t,0), one can compute the potentials in the half plane from the formula

$$u(x,y) = \frac{1}{\pi} \int_{-\infty}^{\infty} \frac{yu(t,0)dt}{(t-x)^2 + y^2} \quad . \tag{4.7}$$

Details of this are omitted from this paper.

Problem B involves a singular integral equation on two disjoint intervals. We solve this in a similar manner, namely using a series in Chebyshev polynomials of the second kind. To this end, we transform the ranges of integration on each arc to [-1,1] the "standard" range for Chebyshev polynomials. We introduce two functions

$$s_1(x) = \tfrac{1}{2}x + 1, \tag{4.8}$$

$$s_2(x) = 2(x - 2\ \omega/2)/(4 - \omega) \tag{4.9}$$

which map [-4,0] and [ω,4] resp. onto [-1,1]. (3.6) then becomes

$$-\frac{1}{\pi}\int_{-1}^{1} \frac{u_1(\tau)d\tau}{\tau - s_1(x)} - \frac{1}{\pi}\int_{-1}^{1} \frac{u_2(\tau)d\tau}{\tau - s_2(x)} = C_j + \frac{500}{\pi} \ln|1-\omega/x| \tag{4.10}$$

In (4.10),

$$u(t,0) = u_1(s_1(t)) \qquad t\ \varepsilon[-4,0] \tag{4.11}$$

$$u(t,0) = u_2(s_2(t)) \qquad t\ \varepsilon[\omega,4]. \tag{4.12}$$

We then calculate the approximations to u_1 and u_2 given by:

$$u_1(t) \simeq \surd(1 - t^2) \sum_{j=0}^{N} \alpha_j U_j(t) + 250(1+t), \tag{4.13}$$

$$u_2(t) \simeq \surd(1 - t^2) \sum_{j=0}^{N} \beta_j U_j(t) + 250(1-t). \tag{4.14}$$

Again, introduction of extra terms ensures continuity of the resulting $u(t,0)$. We have $2N + 4$ unknowns and these are again found by collocation. The collocation points are:

$$s_1^{-1}(\eta_k) \quad \text{and} \quad s_2^{-1}(\eta_k) \quad \text{where} \quad T_{N+2}(\eta_k) = 0.$$

We find the constants and coefficients from the linear equations

$$-C_1 + \sum_{j=0}^{N} \alpha_j T_{j+1}(\eta_k) + \sum_{j=0}^{N} \beta_j G_{j+1}(s_2(s_1^{-1}(\eta_k)))$$
$$= \zeta(s_1^{-1}(\eta_k)); \quad k = 1,\ldots,N+2. \tag{4.15}$$

$$-C_2 + \sum_{j=0}^{N} \alpha_j H_{j+1}(s_1(s_2^{-1}(\eta_k))) + \sum_{j=0}^{N} \beta_j T_{j+1}(\eta_k)$$
$$= \zeta(s_2^{-1}(\eta_k)); \quad k = 1,\ldots,N+2. \tag{4.16}$$

where

$$\zeta(x) = \frac{500}{\pi} \ln\left|\frac{\omega - x}{x}\right| + \frac{250}{\pi}\left((1 + s_1(x))\ln\left|\frac{1 - s_1(x)}{1 + s_1(x)}\right|\right.$$
$$\left. + (1 - s_2(x))\ln\left|\frac{1 - s_2(x)}{1 + s_2(x)}\right|\right). \tag{4.17}$$

The functions G_j and H_j are given in the appendix. Again, the logarithmic singularities have been subtracted out.

Having found $u(t,0)$, the potential in the upper half plane may be found using (4.7).

This method of solving singular integral equations on several arcs is of marked contrast to the more traditional approach of Homentcovschi [2] who makes use of the function

$$P(z) = \sqrt{\prod_{j=1}^{p} (z - a_j)(z - b_j)}.$$

Our method has the advantage of using the well known and extensively documented properties of Chebyshev polynomials.

5. NUMERICAL RESULTS

We give the solutions of Problems A and B using our technique and compare these with those of other authors.

Problem A.

Symm [5] gives values of the potential at the points P (0.1,0.1), Q (0.5,0.1) and R (0.9,0.1). In Table 1, N is the number of terms in the Chebyshev series (4.4).

TABLE 1.

Results for Problem A

	P	Q	R
N = 10	.1284	.1681	.5335
N = 22	.1282	.1681	.5335
Symm (n=64)	.1272	.1676	.5323
Symm (n=64*)	.1277	.1680	.5333
Analytic	.1282	.1680	.5335

Problem B.

In Table 2, we give values of the potential near the singularity at $z = 0$ and compare these with [3] and [7].

TABLE 2.

Results for the Motz Problem

Point	L=36	L=68	J & S	W & P
(1/28,0)	576.42	576.41	576.41	576.41
(1/14,0)	608.92	608.91	608.91	608.91
(3/28,0)	634.45	634.45	634.45	634.45
(0,1/28)	553.18	553.19	553.18	553.19
(1/28,1/28)	583.18	583.67	583.67	583.67
(1/14,1/28)	611.87	611.86	611.86	611.86
(3/28,1/28)	636.07	636.07	636.07	636.07

In Table 2, L denotes the *total* number of equations to be solved i.e. $2N + 4$. "J&S" are the results of [3] with $n=96$. "W & P" are the results of [7].

6. CONCLUSION

The use of singular integral equations in conjunction with conformal mapping is a useful technique for mixed boundary value problems. One could avoid the conformal mapping using a double layer representation of the potential. This leads to a pair of coupled integral equations which are more complicated.

7. APPENDIX

One can show that:

$$-\frac{1}{\pi}\int_{-1}^{1}\frac{\sqrt{(1-t^2)}U_j(t)}{t-s}\,dt = T_{j+1}(s), |s| \leq 1$$
$$= G_{j+1}(s),\ s < -1$$
$$= H_{j+1}(s),\ s > 1,$$

where

$$G_j(s) = (s - \sqrt{(s^2-1)})^{-j},$$
$$H_j(s) = (s + \sqrt{(s^2-1)})^{-j}.$$

The proof uses a change of variables to transform the integral to a contour integral around the unit circle. The calculus of residues is then applicable.

8. ACKNOWLEDGEMENT

One of us, M.R.R., wishes to express his gratitude to the University of Technology, Malaysia for the financial support during the tenure of his research at the University of Southampton.

REFERENCES

1. BREBBIA, C.A. *The Boundary Element Method for Engineers.* Pentech Press, Plymouth, England (1978).
2. HOMENTCOVSCHI, D., On the Mixed Boundary-value Problem for Harmonic Functions in Plane Domains. Zamp 31, 352-366 (1980).
3. JASWON, M.A. and SYMM, G.T. *Integral Equation Methods in Potential Theory and Elastostatics.* Academic Press, London (1977).
4. MUSKHELISHVILI, N.I. *Singular Integral Equations.* P. Noordhoff N.V. Groningen-Holland (1953).
5. SYMM, G.T. *The Robin Problem for Laplace's equation.* NPL Report DNACS 32/80 (1980.
6. WENDLAND, W.L., STEPHAN, E. and HSIAO, G.C., On the Integral Equation Method for the Plane Mixed Boundary Value Problem of the Laplacian. *Math. Meth. in the Appl. Sci.* 1, 265-321 (1979).
7. WHITEMAN, J.R. and PAPAMICHAEL, N. *Numerical Solution of Two Dimensional Harmonic Boundary Value Problems Containing Singularities by Conformal Transformation Methods.* Brunel University Report TR/2. (1971).

TRIAL FUNCTIONS WITH FAST ORTHONORMALIZATION

Eugen Schäfer

Mathematical Institute, University of Munich, FRG.

1. INTRODUCTION

Consider the integral equation eigenvalue problem

$$Kx = \lambda x, \quad x \in L_2(I), \; I = [0,1] .$$

Discretization of this problem by Galerkin's method has the disadvantage of producing a generalized matrix eigenvalue problem

$$C_h \xi_h = \lambda_h G_h \xi_h \quad ,$$

where C_h is the stiffness matrix and G_h is the Gramian matrix of the trial functions. If one discretizes the above integral equation eigenvalue problem by some kind of interpolating degenerate kernel method there results a standard matrix eigenvalue problem of the form

$$A_h \xi_h = \lambda_h \xi_h \quad ,$$

where $A_h = D_h G_h$. D_h is the result of applying the interpolation scheme to the kernel k of K, and G_h is again the Gramian matrix (cf. e.g. [2]) .

Of course both matrix eigenvalue problems can be brought into a symmetric form by Cholesky decomposition of the Gramian G_h, $G_h = L_h \times L_h^t$. The only essential drawback of symmetrization by Cholesky decomposition is the repeated square rooting.

The subject of this paper is the construction of special trial functions with the property that the Cholesky decomposition can be done in advance, once and for all. We consider polynomial spline spaces with B-splines as basis functions. If the given knot sequence has a regular structure then the Gramian matrix of the corresponding B-splines is periodic, apart from an irregular beginning block. In section 2 we introduce the necessary notations. In section 3 we show that one can modify a small left upper block such that there results a constant Cholesky decomposition of G_h. We realize this favourable modified matrix as the Gramian matrix of suitable B-splines. This will be done in section 4 by proper adaption of the underlying spline knotes. For the special case of (n-1)-fold knots we give a theoretical and numerical treatment for arbitrary spline order n. Numerical results are given also for simple knots.

With these adapted trial functions we thus avoid the necessity of repeated square rooting, and in addition reduce the amount of storage, in obtaining symmetric matrices. Of course the usual favourable properties of B-splines are unaltered. Let us finally remark that these trial functions can also be applied to second kind integral equations.

2. NOTATIONS

Given $n \geq 2$, the order of the splines, consider a knot mesh $\underline{s} = (s_i \mid i \in \mathbb{N})$ with

$$s_1 \leq \ldots \leq s_i \leq s_{i+1} \leq \ldots, \quad s_i < s_{i+n}, \quad \lim_{i\to\infty} s_i = \infty .$$

The corresponding B-splines (cf. e.g. [1]) are defined as divided differences of the truncated power function

$$B_i(\underline{s})(t) = (s_{i+n}-s_i)[s_i,\ldots,s_{i+n}](.-t)_+^{n-1} .$$

These B-splines are normalized by $\sum_i B_i = 1$ for $t \geq s_n$.

The Gramian matrix corresponding to $\underline{s}$ is $G(\underline{s}) = (\langle B_i(\underline{s}),B_j(\underline{s})\rangle)$, where we set

$$\langle B_i(\underline{s}),B_j(\underline{s})\rangle = \int_{s_n}^{\infty} B_i(\underline{s})(t)B_j(\underline{s})(t)dt .$$

<u>Definition:</u> A knot mesh $\underline{s}$ is called periodic if there exists $p \in \mathbb{N}$ and $H > 0$ such that for all $k \in \mathbb{N}$ there holds

$$s_{i+kp} = s_i + kH, \quad 1 \leq i \leq p .$$

For a periodic mesh $\underline{s}$ the Gramian matrix $G(\underline{s})$ is nearly periodic, i.e. we have

<u>Remark:</u> For periodic $\underline{s}$ let k_o be such that $k_o p \geq n-1$. Then there holds for all $k \geq k_o$, and all i,j with $1 \leq i,\ j \leq p$

$$\langle B_{i+kp}(\underline{s}),B_{j+kp}(\underline{s})\rangle = \langle B_{i+k_op}(\underline{s}),B_{j+k_op}(\underline{s})\rangle .$$

This property holds because of the identity

$$B_{i+kp}(\underline{s})(s_{i+kp}+t) = B_{i+k_op}(\underline{s})(s_{i+k_op}+t) ,$$

which is a consequence of the periodicity of $\underline{s}$. For general basis functions $(w_j \mid j \in \mathbb{N})$ which share this property above, and have compact support, we show in the next section that there exists a suitable left upper starting block with a resulting constant

Cholesky decomposition.

That it is sufficient to consider a standard knot mesh $\underline{s}$ can be seen as follows. Choose $h > 0$ and take $n(h) \in \mathbb{N}$ with $hs_{n(h)} < 1 \leq hs_{n(h)+1}$. Trial functions u_i to solve $Kx = \lambda x$ approximately, are given as

$$u_i = B_i(\underline{t}), \quad 1 \leq i \leq n(h) \ ,$$

where $\underline{t}$ is given by

$$\underline{t} = h(\underline{s-s_n}) \ , \quad \text{i.e. } t_i = h(s_i - s_n) \quad \text{for } i \in \mathbb{N} \ .$$

By the well-known B-spline properties,

$$B_i(\underline{t})(h\tau) = B_i(\underline{s})(s_n + \tau) \quad ,$$

$$\text{supp}(B_i(\underline{s})) = [s_i, s_{i+n}] \quad ,$$

we have

$$\int_0^1 B_i(\underline{t}) B_j(\underline{t}) d\tau = h \int_{s_n}^{\infty} B_i(\underline{s}) B_j(\underline{s}) d\tau, \quad n \leq i,j \leq n(h)-n \ .$$

Apart from irregular blocks in the left upper and right lower corners, the Gramian matrix $G(\underline{t})$ has a periodic structure similar to $G(\underline{s})$.

3. FIXED POINTS OF THE CHOLESKY MAP

In accordance with the Cholesky decomposition we define the following map φ, dependent on a given matrix A. Fix a symmetric matrix $A = (a_{ij}) \in \mathbb{R}^{2p,2p}$. For a given lower triangular matrix $L = (l_{ij}) \in \mathbb{R}^{p,p}$ with positive diagonal elements, we consider the map φ, $\varphi(L) = M$ with $M = (m_{ij}) \in \mathbb{R}^{p,p}$ a lower triangular matrix to be defined (if it exists) by

$$m_{ij} := l_{p+i,p+j} \ , \quad 1 \leq j \leq i \leq p \quad ,$$

where l_{ij} for $i > p$ is the solution of the system

$$\sum_{k=1}^{j} l_{ik} l_{jk} = a_{ij}, \quad 1 \le j \le i, \quad p+1 \le i \le 2p ,$$

and $l_{ij} = 0$ for $i < j$.

<u>Theorem:</u> Let A be the Gramian matrix of the 2p linear independent functions $w_i : \mathbb{R} \to \mathbb{R}$. Suppose that for all i, $1 \le i \le p$, there holds

$$\operatorname{supp}(w_i) \subset [0,b], \text{ and } w_{i+p}(t) = w_i(t-b) \text{ for } t \in \mathbb{R} .$$

Then there exists $q \in \mathbb{N}$ and a lower triangular matrix $\tilde{L} \in \mathbb{R}^{p,p}$ with positive diagonal elements such that

$$\varphi^q(\tilde{L}) = \tilde{L} .$$

Proof: In addition to $w_1,\ldots,w_{2p}$ we define functions w_j, $j \in \mathbb{N}$, by translation (in analogy to the already given functions B_j of section 2). For $k \in \mathbb{N}$, $k \ge 2$, define

$$w_{i+kp}(t) := w_{i+(k-1)p}(t-b), \quad 1 \le i \le p .$$

Consider the Cholesky decomposition (l_{ij}) of all finite sections of the infinite Gramian matrix

$$\left(\int_{\mathbb{R}} w_i(t) w_j(t)\,dt\right)_{i,j=1,2,\ldots} .$$

Fix (i,j) with $j \le i$. Then the sequence (c_k) ,

$$c_k = \sum_{m=j}^{i} l^2_{i+kp,m+kp} ,$$

is bounded and monotone decreasing, hence convergent. The monotonicity is implied by the well known property

$$\sum_{m=j+1}^{i} l^2_{i+kp,m+kp} = \operatorname{dist}^2(w_{i+kp};[w_1,\ldots,w_{j+kp}]) .$$

Here $[w_1,\ldots,w_m]$ is the linear hull of the elements $w_1,\ldots,w_m$ and

$$\mathrm{dist}^2(w_i;[w_1,\ldots,w_j]) = \min_{a_1,\ldots,a_j} \int_{\mathbb{R}} |w_i(t) - \sum_{m=1}^{j} a_m w_m(t)|^2 dt.$$

The convergence of (c_k) implies the convergence of $(|l_{i+kp,j+kp}|\,|k\in\mathbb{N})$ with limit, say $\bar{l}_{ij}$. Define the $p\times p$ matrices $L^k=(l_{i+kp,j+kp})_{1\le i,j\le p}$ and $\bar{L}=(\bar{l}_{i,j})_{1\le i,j\le p}$, and set $|C|=(|c_{ij}|)$ for a given matrix $C=(c_{ij})$. Then the convergence of $(|l_{i+kp,j+kp}|\,|k\in\mathbb{N})$ can be stated as $\lim_{k\to\infty}|L^k|=\bar{L}$. In addition the diagonal elements $l_{i+kp,i+kp}$ themselves, and not only their absolute values, are convergent to a positive limit for k tending to infinity. This fact is implied by the inequalities

$$l_{i+kp,i+kp} \ge \left(\min_{a_m} \int_0^\infty |w_i - \sum_{m=1}^{i-1} a_m w_m|^2\, dt\right)^{1/2}$$

for all $i\in\mathbb{N}$, and the linear independence of the $\{w_j\}$. Consider any sign choice $\varepsilon=(\varepsilon_{ij})\in\mathbb{R}^{p,p}$ with $\varepsilon_{ij}\in\{-1,+1\}$, if $\bar{l}_{ij}\ne 0$, and $\varepsilon_{ij}=0$ for $\bar{l}_{ij}=0$. Define the corresponding sign class

$$S(\varepsilon) = \{C\in\mathbb{R}^{p,p}\,|\,\mathrm{sign}\,c_{ij}=\varepsilon_{ij},\ \text{if } \bar{l}_{ij}\ne 0\}.$$

There exist only a finite number, say n_1, of sign choices ε. Consequently, there exist $q\in\mathbb{N}$, $1\le q\le n_1$, and a suitable sign choice $\tilde{\varepsilon}$ such that for some subsequence (k') of (k) there holds

$$L^{k'},\ L^{k'+q}\in S(\tilde{\varepsilon}),\quad \text{for all } k'.$$

Because of the fact $\lim_{k\to\infty}|L^k|=\bar{L}$, there follows

$$\lim_{k'\to\infty} L^{k'} = \lim_{k'\to\infty} L^{k'+q} = \tilde{\varepsilon}\bar{L}.$$

Since φ^q is continuous at $\tilde{\varepsilon}\bar{L}$, as $\bar{l}_{ii} > 0$, we finally obtain

$$\varphi^q(\tilde{\varepsilon}\bar{L}) = \tilde{\varepsilon}\bar{L} \ . \qquad ***$$

This theorem applies to the B-splines corresponding to a periodic mesh $\underline{s}$ by setting

$$w_i := B_i(\underline{s-s_n}) \ .$$

Even for the simplest case of spline order $n = 2$, the fixed point $\tilde{L}$ whose existence is assured by the above theorem, is not unique. In applications we take that fixed point $\tilde{L}$ which gives the best stability properties.

Efficiency requires q to be equal to 1. This is assured if $l_{ij} \geq 0$, $1 \leq j \leq i$, $i \in \mathbb{N}$.

Remark: The property $l_{ij} \geq 0$, $1 \leq j \leq i$, for all $i \in \mathbb{N}$ is implied by the inequalities

$$(\mathrm{dist}^2(w_i;[w_1,..,w_{i-p}])-\mathrm{dist}^2(w_i;[w_1,..,w_{j-1}]))\times$$
$$(\mathrm{dist}^2(w_j;[w_1,..,w_{i-p}])-\mathrm{dist}^2(w_j;[w_1,..,w_{j-1}]))$$
$$\leq (\int_{\mathbb{R}} w_i w_j dt)^2 \quad \text{for } i-p < j < i, \quad p < i \ .$$

Corollary: For the following knot meshes defined for $i = 1,2,...$ by

(i) $s_i = i$ (equidistant simple knots)

(ii) $s_i = \mathrm{entier}((i+n-3)/(n-1))$ (equidistant $(n-1)$-fold knots) we have $l_{ij} \geq 0$ for spline orders n with $2 \leq n \leq 5$.

4. SPLINES WITH CONSTANT CHOLESKY DECOMPOSITION OF THEIR GRAMIAN

Given a periodic knot sequence $\underline{s}$ we showed in the last section the existence of a fixed point L* of the

Cholesky map. Now we try to vary the first few knots in such a way that the beginning block of the corresponding Gramian matrix of B-splines equals L* whilst the remaining knots have the given periodic structure.

For equidistant (n-1)-fold knots we have the following existence result.

Theorem: Given the spline order $n \geq 2$, there exist τ_1^*, τ_2^* with $0 < \tau_1^* < \tau_2^*$ such that for the mesh $\underline{\tilde{s}}$ with

$$\tilde{s}_1 = -1,\ \tilde{s}_2 = 0,\ \tilde{s}_{(n-1)+2} = \tau_1^*,\ \tilde{s}_{i(n-1)+2} = \tau_2^* + i - 2, i = 2,3,\ldots$$

and $\tilde{s}_{i(n-1)+j} = \tilde{s}_{i(n-1)+2}$ for $2 < j \leq n$, the corresponding B-splines fulfill

$$(\langle B_i(\underline{\tilde{s}}), B_j(\underline{\tilde{s}})\rangle)_{i_o+kp+1 \leq i,j \leq i_o+(k+1)p} = \tilde{L}\tilde{L}^t\ .$$

Here we have set $p = n-1$, $i_o = 2n-1$, and $\tilde{L}$ is the fixed point of the Cholesky map according to the proof of the theorem of section 3.

Proof: Consider a mesh $\underline{t} = \underline{t}(\tau_1, \tau_2)$ of the form similar to $\underline{\tilde{s}}$ with $0 < \tau_1 < \tau_2$ not yet specified. Because of the special form of $\underline{t}$ it is sufficient for the theorem to hold, that there exist $0 < \tau_1^* < \tau_2^*$ such that

$$l_{2n-1,2n-1}(\tau_1^*, \tau_2^*) = \tilde{l}_{2n-1,2n-1}\ ,$$

where $L(\tau_1,\tau_2)L(\tau_1,\tau_2)^t = G(\underline{t}(\tau_1,\tau_2))$. But this is equivalent to have for $\underline{t} = \underline{t}(\tau_1^*,\tau_2^*)$

$$\mathrm{dist}^2_{]t_n,\infty[}(B_{2n-1}(\underline{t});[B_1(\underline{t}),\ldots,B_{2n-2}(\underline{t})])$$

$$= \lim_{k\to\infty} \mathrm{dist}^2_{]-\infty,\infty[}(B_{2n-1+kp}(\underline{s});[B_n(\underline{s}),\ldots,B_{2n-2+kp}(\underline{s})]),$$

with $\underline{s} = (s_i)$ defined by $s_i = \mathrm{entier}((i+n-3)/(n-1))$ (cp. the corollary of the last section).

Consider knots $\underline{t}(\tau)$ dependent on $\tau \in \mathbb{R}$ of the form

$$t_1 = \tau_- - 1 ,$$

$$t_j = \tau_- \quad \text{for } j = 2,\dots,n$$

$$t_j = \tau \quad \text{for } j = n+1,\dots,2n-1, \text{ and}$$

$$t_{2n} = 2 ,$$

where $\tau_- := \min(\tau,0)$. Then we consider $f :]-\infty,1[\to \mathbb{R}$ defined by

$$f(\tau) = \mathrm{dist}^2_{]t_n(\tau),\infty[}(B_{2n-1}(\underline{t}(\tau));[B_1(\underline{t}(\tau)),\dots,B_{2n-2}(\underline{t}(\tau))]).$$

The periodicity of $\underline{s}$, together with the fact that $\mathrm{supp}(B_j(\underline{s})) \cap]s_{n+kp},\infty[= \emptyset$ for $j \leq kp$, implies

$$f(1) < \tilde{I}^2_{2n-1,2n-1} \leq \mathrm{dist}^2_{]s_1,\infty[}(B_n(\underline{s});[B_1(\underline{s}),\dots,B_{n-1}(\underline{s})]),$$

by the definition of the minimal distance. As $B_j(\underline{t})|_{]t_{j+n-1},t_{j+n}[}$ is independent of t_j and similarly $B_j(\underline{t})|_{]t_j,t_{j+1}[}$ is independent of t_{j+n}, a closer look on the dependence of the B-splines relative to $\underline{t}(\tau)$ gives the identity

$$f(\tau) = \int_1^2 B_n(\underline{t}(\tau))^2 dt + (2-\tau)\,\mathrm{dist}^2_{]0,1[}(B_n(\underline{s});[B_1(\underline{s}),\dots,B_{n-1}(\underline{s})]).$$

Therefore $\lim_{\tau\to-\infty} f(\tau) = +\infty$ and consequently by the continuity of f there exists $\tau^* \in]-\infty,1[$ with $f(\tau^*) = \tilde{I}^2_{2n-1,2n-1}$.

The theorem follows with $\tau_1^* := \tau^* - \tau_-^*$ and $\tau_2^* := 2 - \tau_-^*$.

***

The continuity of f, though not obvious, can be proved elementary.

At least for spline orders $n \leq 5$ we have the additional periodicity property $\tau_2^* = 2$. Actual computation of τ_1^* shows that the modified meshes are nearly equidistant, compare table 1. For simple knots and spline order n = 3 table 2 shows the modified inner knots s_{n+i}^*, together with a row of the corresponding constant Cholesky matrix, multiplied by $((2n-1)!)^{1/2}$.

Approximation schemes, resp. degenerate kernels, based on quasiinterpolants can be constructed for our modified knots as in [3], and error estimates for the eigenvalue and eigenvector approximations can be given, cp. [4].

Finally let us mention another favourable property of these trial functions with respect to degenerate kernel methods. As the resulting matrices are symmetric this discretization is always stable regarding eigenvalues, if the given operator is selfadjoint.

Table 1

n	2	3	4	5
τ_1^*	.9879	.9982	.9995	.9998

Table 2 $n = 3$, $c_i = ((2n-1)!)^{1/2}\, l_{k,k-n+i}^*$;

i	1	2	3
s_{n+i}^*	0.891719	2.001625	2.999997
c_i	0.136221	3.477226	7.341004

REFERENCES

[1] De Boor, C.: Splines as linear combinations of B-splines. A survey. In: *Approximation Theory II*, G.G. Lorentz, C.K. Chui, L.L. Schumaker (eds.), New York e.a., Academic Press 1976.

[2] Hämmerlin, G.: Ersatzkernverfahren zur numerischen Behandlung von Integralgleichungen 2. Art. *ZAMM* 42, 439-463 (1962).

[3] Lyche, T., L.L. Schumaker: Local spline approximation methods. *J. Approximation Theory* 15, 294-325 (1975).

[4] Schäfer, E.: Spectral approximation for compact integral operators by degenerate kernel methods. *Numer. Funct. Anal. and Optimiz.* 2, 43-63 (1980).

ON COLLOCATION APPROXIMATIONS FOR VOLTERRA EQUATIONS WITH WEAKLY SINGULAR KERNELS

* Hermann Brunner

* *University of Fribourg, CH-1700 Fribourg (Switzerland)*

1. INTRODUCTION

For a given compact interval $I:=[0,T]$ let Δ_N denote the (uniform) partition $\{t_n = nh: n=0,1,\ldots, N\ (N \geqq 1),\ t_N = T\}$, and set $Z_N := \{t_n: n = 1,\ldots, N-1\}$, $\sigma_n := (t_n,t_{n+1}]$ $(n=1,\ldots,N-1)$, $\sigma_o := [t_o,t_1]$. Approximate solutions for various types of Volterra functional equations introduced below will be sought in certain polynomial spline spaces,

$$S_m^{(d)}(Z_N) := \{u: u|_{\sigma_n} = u_n \in \pi_m (n=0,\ldots,N-1);\ u_{n-1}^{(j)}(t_n) = u_n^{(j)}(t_n),\ t_n \in Z_N\ (j=0,\ldots,d)\}. \tag{1.1}$$

Here, $-1 \leqq d \leqq m-1$; the choice $d = m-1$ yields the classical spline space (full continuity), while for $d = -1$ we have polynomial splines possessing jump discontinuities at their knots Z_N. Clearly, $S_m^{(d)}(Z_N)$ is a linear space whose (finite) dimension is given by

$$\dim S_m^{(d)}(Z_N) = N(m-d) + (d+1); \tag{1.2}$$

hence,

$$\dim S_m^{(o)}(Z_N) = Nm + 1, \text{ and} \tag{1.2a}$$

$$\dim S_m^{(-1)}(Z_N) = N(m+1). \tag{1.2b}$$

Note that (1.2a) and (1.2b) imply that $\dim S_m^{(o)}(Z_N) = \dim S_{m-1}^{(-1)}(Z_N)+1$ $(m \geqq 1)$.

Once an approximating space of the above type (with $d \in \{-1,0\}$)

has been chosen we approximate the (smooth) solution of a given Volterra equation by requiring that this polynomial spline approximation satisfy the equation on a appropriately chosen finite subset of I whose cardinality is consistent with the dimension of the given approximating space. To be precise let

$$0 \leqq c_1 < \dots < c_{m-d} \leqq 1, \tag{1.3}$$

and set

$$X(N) := \bigcup_{n=o}^{N-1} X_n \text{ , with } X_n := \{t_n + c_i h:\ i=1,\dots,m-d\}. \tag{1.4}$$

The desired approximation is then generated recursively by calculating its restrictions on $\sigma_o, \sigma_1, \dots, \sigma_{N-1}$, using the sets of collocation points $X_o, X_1, \dots, X_{N-1}$.

In this note we investigate the application of collocation techniques to three classes of Volterra equations, namely,

$$y(t) = g(t) + \int_o^t (t-s)^{-\alpha} K(t,s)y(s)ds, \quad t \in I \ (0 \leqq \alpha < 1); \tag{1.5}$$

$$y'(t) = p(t)y(t)+q(t) + \int_o^t (t-s)^{-\alpha} K(t,s)y(s)ds, \quad t \in I (0 \leqq \alpha < 1); \tag{1.6}$$

and

$$\int_o^t (t-s)^{-\alpha} K(t,s)y(s)ds = g(t), \quad t \in I \ (0 \leqq \alpha < 1). \tag{1.7}$$

Our interest will focus on the construction of high-order methods (with the particular aim of generating high-order starting values for other methods), especially for the case where the exact solution is not smooth at $t = t_o$. Let $e(t) := y(t) - u(t)$ (with $e_n(t) := y(t) - u_n(t),\ t \in \sigma_n$) denote the collocation error associated with the approximation $u \in S_m^{(d)}(Z_N)(d \in \{-1,0\})$; then, ideally, we should like to have

$$\|e\| := \sup\{|e_n(t)|:\ t \in \sigma_n \ (n=0,\dots,N-1)\} = \mathcal{O}(h^p),$$

with $m \leqq p \leqq m+1$. It is of course well-known that the existence of such a p depends on the degree of smoothness of y on the closed interval. If $y \in C^{m+1}(I)$ we may write

$$e_n(t_n+\tau h) = \sum_{j=o}^{m} \left(\frac{h^j y^{(j)}(t_n)}{j!} - \alpha_{n,j}\right)\tau^j + h^{m+1}R_n(\tau), t_n+\tau h \in \sigma_n,$$

where

$$u_n(t_n+\tau h) = \sum_{j=o}^{m} \alpha_{n,j}\tau^j, \text{ and } R_n(\tau) := y^{(m+1)}(\xi_n)\cdot\tau^{m+1}/(m+1)! \quad (1.8)$$

$$(\tau = (t-t_n)/h).$$

If we now set

$$h^p\cdot\beta_{n,j} := h^j y^{(j)}(t_n)/j! - \alpha_{n,j} \quad (j=0,\ldots,m;\ n=0,\ldots,N-1) \quad (1.9)$$

then (1.8) mey be replaced by

$$e_n(t_n+\tau h) = h^p\left(\sum_{j=o}^{m} \beta_{n,j}\tau^j + h^{m+1-p}\cdot R_n(\tau)\right),\ t_n+\tau h \in \sigma_n \quad .(1.10)$$

Thus, if we can show that the vectors $\beta_n := (\beta_{n,o},\ldots,\beta_{n,m})^T \in \mathbb{R}^{m+1}$ satisfy $\|\beta_n\|_1 \leqq B < \infty$ uniformly as $h \to 0_+$ $(nh \leqq Nh = T)$, and if $M_{m+1} := \max\{|y^{(m+1)}(t)|/(m+1)! : t \in I\}$ then we shall have established the result

$$|e_n(t_n+\tau h)| \leqq h^p\cdot(B + h^{m+1-p}\cdot M_{m+1}),\ t_n+\tau h \in \sigma_n (n=0,\ldots,N-1),$$

and hence $\|e\| \leqq \text{const.}\ h^p$, $m \leqq p \leqq m+1$.

For the regular case $\alpha = 0$ smoothness of the given functions characterizing the respective equations (1.5), (1.6), and (1.7) (where $K(t,t) \neq 0$ for $t \in I$) will imply smoothness of y on the closed interval I; more generally, if $0 \leqq \alpha < 1$ and if the corresponding solution is smooth on I then the following results are known to hold:

Equation ($y \in C^{m+1}(I)$)	$S_m^{(d)}(Z_N)$	$\dim S_m^{(d)}(Z_N)$	p
ODE: (1.6), $K \equiv 0$	$d = 0$	$Nm + 1$	m
VIDE: (1.6), $0 \leqq \alpha < 1$	$d = 0$	$Nm + 1$	m
V2: (1.5), $0 \leqq \alpha < 1$	$d = -1$ ($d = 0$)	$N(m + 1)$	$m+1$
V1: (1.7), $\alpha = 0$	$d = -1$ ($c_{m+1}=1$)	$N(m + 1)$	$m+1$

Table 1.1

However, if $\alpha \neq 0$ $(0 < \alpha < 1)$ then, in general, (1.5) and (1.6) do not possess solutions $y \in C^{m+1}(I)$: there will be loss of smoothness at $t = t_o$. As will be made precise in the following section, equation (1.5) with $\alpha = \frac{1}{2}$, $g \in C^{m+1}(I)$, $K \in C^{m+1}(S)$ $(S := \{(t,s): t_o \leq s \leq t \leq T\})$, has a solution which may be written as

$$y(t) = v(t) + t^{\frac{1}{2}} \cdot w(t), \text{ with } v,w \in C^{m+1}(I) \tag{1.11}$$

(compare [4]). It has been shown ([8]) that $u \in S_m^{(-1)}(Z_N)$ yields $\|e\| \leq \text{const.}\ h^{\frac{1}{2}}$: the loss of order near $t = t_o$ (caused by the singularity of $y^{(j)}(t)$ at $t = t_o$, $j \geq 1$) leads to a corresponding loss in the global order of convergence on I.

Intuitively, we may expect to restore the previous high-order convergence $(m \leq p \leq m+1)$ if we fill the "gaps" $\{t^{1/2}, t^{3/2}, \ldots, t^{(2m-1)/2}\}$ in the polynomial spline space $S_m^{(-1)}(Z_N)$; i.e. if this space is replaced by a non-polynomial spline space of larger dimension. It is the purpose of this note to study this question; it will be shown that (i) for the second-kind integral equation (1.5) one can obtain $p = m+1$ for $\alpha \to 1_-$ only if the dimension of the augmented space tends to infinity; (ii) for the Volterra integro-differential equation (1.6) one has $p = m$ as $\alpha \to 1_-$ if the dimension of the augmented space tends to $(m+1)(m+2)/2$. These results are closely related to analogous results for Runge-Kutta methods for (1.5) derived recently by Lubich [5].

In order to analyze the above problem one needs to know the structure of the exact solutions of (1.5) and (1.6) (i.e. the expressions extending (1.11). These results will be given in the following section, mostly without detailed proofs. For the non-linear versions of (1.5), (1.6) (with $(t-s)^{-\alpha}$ replaced by $(t-s)^{\lambda}$, $\lambda > -1$) the behavior of solutions near $t = 0$ has recently been analyzed by Lubich [5], using techniques based on formal power series and the implicit function theorem.

2. BEHAVIOR OF NON-SMOOTH SOLUTIONS NEAR $t = 0$

Suppose that the nonhomogeneous term g in (1.5) has the form

$$g(t) = g_1(t) + t^{\beta}\cdot g_2(t), \quad \beta > 0, \ \beta \notin \mathbb{N},$$

with $g_i \in C^{m+1}(I)$ $(i = 1,2)$, and let $K \in C^{m+1}(S)$.

Theorem 2.1. Under the above hypotheses the (unique) solution of the Volterra integral equation of the second kind (1.5) $(0 < \alpha < 1)$ has the form

$$y(t)=g_1(t)+t^{\beta}\cdot g_2(t)+\sum_{n=1}^{\infty}\{\Psi_{n,1}(t;\alpha)+t^{\beta}\cdot\Psi_{n,2}(t;\alpha,\beta)\}\cdot(t^{1-\alpha})^n, \quad t \in [0,T]; \qquad (2.2)$$

here, the functions

$$\Psi_{n,1}(t;\alpha):= \int_0^1 (1-\tau)^{n(1-\alpha)-1}\cdot\Phi_n(t,\tau t;\alpha)\cdot g_1(\tau t)d\tau, \qquad (2.3a)$$

$$\Psi_{n,2}(t;\alpha,\beta):= \int_0^1 (1-\tau)^{n(1-\alpha)-1}\cdot\tau^{\beta}\cdot\Phi_n(t,\tau t;\alpha)\cdot g_2(\tau t)d\tau, \qquad (2.3b)$$

satisfy $\Psi_{n,i} \in C^{m+1}[0,T]$ $(i = 1,2;\ n \geqq 1)$, and Φ_n is given recursively by

$$\Phi_n(t,s;\alpha)= \int_0^1 (1-u)^{-\alpha}\cdot u^{(n-1)(1-\alpha)-1}\cdot K(t,s+(t-s)u)\cdot \Phi_{n-1}(s+(t-s)u,s;\alpha)du, \qquad (2.4)$$

with $\Phi_1(t,s;\alpha):= K(t,s)$.

The proof of this result is based on the classical construction of the iterated kernels for the given kernel $(t-s)^{-\alpha}\cdot K(t,s)$ and on the corresponding representation of the solution.

Corollary 2.1. Under the hypotheses of Theorem 2.1, and with $\alpha = p/q \in (0,1)$ rational (reduced to lowest terms; $p,q \in \mathbb{N}$) the solution of (1.5) may be written in the form

$$y(t)=v_0(t)+\sum_{s=1}^{q-1} v_s(t)\cdot(t^{1-\alpha})^s+t^{\beta}\cdot\{w_0(t)+\sum_{s=1}^{q-1} w_s(t)\cdot(t^{1-\alpha})^s\}, \quad t \in [0,T], \qquad (2.5)$$

where the functions v_s, w_s $(s = 0,\dots,q-1)$ are in $C^{m+1}[0,T]$.

If $g_2(t) \equiv 0$ (i.e. if the forcing function is smooth on $[0,T]$) then

$$y(t) = v_0(t) + \sum_{s=1}^{q-1} v_s(t)\cdot(t^{1-\alpha})^s, \quad t \in [0,T], \qquad (2.6)$$

while for $g_2(t) \not\equiv 0$, $\beta = 1-\alpha$, we have

$$y(t) = z_o(t) + \sum_{s=1}^{q-1} z_s(t)\cdot(t^{1-\alpha})^s,\ t \in [0,T], \tag{2.7}$$

with

$$z_o(t) := v_o(t) + t^{q-p}\cdot w_{q-1}(t),\ z_s(t) := v_s(t) + w_{s-1}(t)$$
$$(s = 1,\ldots,q-1).$$

The proof follows without difficulty from (2.2) by observing that $\alpha = p/q$ implies, setting $\ell' = rq+\ell\ (\ell < q)$, that $\ell'(1-\alpha) = r(q-p) + \ell(1-\alpha)$, with $r(q-p) \in \mathbb{N}$. We have

$$v_o(t) := g_1(t) + \sum_{\ell=1}^{\infty} \Psi_{q\ell,1}(t;\alpha)\cdot t^{\ell(q-p)},$$

$$v_s(t) := \sum_{\ell=o}^{\infty} \Psi_{q\ell+s,1}(t;\alpha)\cdot t^{\ell(q-p)}\ (s = 1,\ldots,q-1);$$

$$w_o(t) := g_2(t) + \sum_{\ell=1}^{\infty} \Psi_{q\ell,2}(t;\alpha,\beta)\cdot t^{\ell(q-p)},$$

$$w_s(t) := \sum_{\ell=o}^{\infty} \Psi_{q\ell+s,2}(t;\alpha,\beta)\cdot t^{\ell(q-p)}\ (s = 1,\ldots,q-1).$$

The above result contains (1.11) as a special case $(q-1 = 1)$.

We add two remarks. First, if g is not smooth near $t_o=0$ (i.e. $g_2(t) \not\equiv 0$), and if $K(t,s) \not\equiv 0$, then smooth solutions can exist. We mention one nontrivial example: the choice $K(t,s) = \lambda = -1\ ((t,s) \in S)$, $g_1(t) = \exp(-t)$, $g_2(t) = {}_1F_1(1;2-\alpha;-t)/(1-\alpha)$ (with ${}_1F_1(\cdot;\cdot;\cdot)$ denoting the confluent hypergeometric series), and $\beta = 1-\alpha$, leads to $y(t) = \exp(-t)$.

We now return our attention to the Volterra integro-differential equation (1.6); it is easily verified that any non-smooth behavior of the solution at $t = t_o$ is caused by the kernel $(t-s)^{-\alpha}\cdot K(t,s)$ (and by a nonsmoothness in q); we thus assume, without loss of generality, that $p(t) \equiv 0$, and we set

$$q(t) = q_1(t) + t^{\beta}\cdot q_2(t),\ \beta > 0,\ \beta \notin \mathbb{N}, \tag{2.8}$$

with $q_i \in C^{m+1}[0,T]$ $(i = 1,2)$. In addition let $K \in C^{m+1}(S)$.

Theorem 2.2. Under the above assumptions the (unique) solution of the Volterra integro-differential equation (1.6) $(0 < \alpha < 1,\ p(t) \equiv 0)$ has the form

$$y(t) = \{y_o + \int_o^t q_1(s)ds + \int_o^t s^\beta \cdot q_2(s)ds + \sum_{n=1}^\infty \gamma_{n,o}(t;\alpha)(t^{2-\alpha})^n \cdot y_o\} +$$
$$+t \sum_{n=1}^\infty \gamma_{n,1}(t;\alpha)\cdot(t^{2-\alpha})^n + t^{1+\beta}\cdot \sum_{n=1}^\infty \gamma_{n,2}(t;\alpha,\beta)(t^{2-\alpha})^n,$$
$$t \in [0,T]; \quad (2.9)$$

here, the functions $\{\gamma_{n,i}\}$ $(i=0,1,2)$ satisfy $\gamma_{n,i} \in C^{m+1}[0,T]$ for all $\alpha \in (0,1)$, $\beta > 0$.

This result can be proved along the lines of the proof of Theorem 2.1: the given Volterra integro-differential equation may be written as a integral equation of the second kind, namely,

$$y(t) = g(t) + \int_o^t G(t,s;\alpha)y(s)ds, \quad t \in I,$$

where

$$g(t) := y_o + \int_o^t q(s)ds, \quad G(t,s;\alpha) := p(s) + Q(t,s;\alpha),$$

with $Q(t,s;\alpha) := \int_s^t (u-s)^{-\alpha} K(u,s)du$, $(t,s) \in S$. For the special case under consideration, $p(t) \equiv 0$, the iterated kernels for $Q(t,s;\alpha)$ are then easily constructed.

Corollary 2.2. If $\alpha = p/q$ (reduced to lowest terms) then the solution of (1.6) $(0 < \alpha < 1,\ p(t) \equiv 0)$ is given by

$$y(t) = v_o(t) + \sum_{s=1}^{q-1} v_s(t)\cdot(t^{2-\alpha})^s + t^{1+\beta}\cdot\{w_o(t) + \sum_{s=1}^{q-1} w_s(t)(t^{2-\alpha})^s\},$$
$$t \in [0,T], \quad (2.10)$$

where v_s, $w_s \in C^{m+1}[0,T]$ $(s = 0,\dots,q-1)$ are defined in analogy to those in Corollary 2.1.

If $\beta = 1-\alpha$, with $\alpha = p/q$, then

$$y(t) = z_o(t) + \sum_{s=1}^{q-1} z_s(t)\cdot(t^{2-\alpha})^s, \quad t \in [0,T], \quad (2.11)$$

with $z_o(t) := v_o(t) + t^{2q-p}\cdot w_{q-1}(t)$, $z_s(t) := v_s(t) + w_{s-1}(t)$ $(s = 1,\dots,q-1)$.

If $\beta = 1-\alpha = 1/2$ then the solution reduces to

$$y(t) = z_o(t) + t^{3/2}\cdot z_1(t), \quad t \in [0,T], \quad (2.12)$$

in analogy to (1.11).

For Volterra equations of the first kind of the form (1.7) there emerges a completely different picture: if $g \in C^{m+2}(I)$, $K \in C^{m+2}(S)$, with $K(t,t) \neq 0$ for all $t \in I$, then $y \in C^{m+1}[0,T]$ for all $\alpha \in [0,1)$, provided $g^{(j)}(0) = 0$ $(j=0,\ldots,m)$.

3. CHOICE OF THE NON-POLYNOMIAL SPLINE SPACES

3.1. Integral equations of the second kind:

In order to illuminate the principal ideas we restrict our discussion to the case where $\beta = 1-\alpha$ (see (2.1)), and we assume that α is rational, $\alpha = p/q$ (reduced to lowest terms). Recall that under these assumptions the solution of (1.5) may be written in the form

$$y(t) = z_o(t) + \sum_{s=1}^{q-1} z_s(t)(t^{1-\alpha})^s, \quad z_s \in C^{m+1}(I).$$

Let m_s be given nonnegative integers satisfying $m_s \leqq m$ $(s = 0,\ldots,q-1)$. Then, for $t = t_o+\tau h(=\tau h) \in \sigma_o$ we may write

$$y(t_o+\tau h)= \sum_{s=o}^{q-1} \Big(\sum_{j=o}^{m_s} h^j z^{(j)}(0)\tau^j/j!+h^{m_s+1}\cdot R_{o,m_s}(\tau)\Big)\cdot((\tau h)^{1-\alpha})^s,$$

with $R_{o,m_s}(\tau):= z_s^{(m_s+1)}(\xi_o)\cdot\tau^{m_s+1}/(m_s+1)!$ $(0 < \xi_o < \tau h)$.

Consider now the function

$$u_o(t_o+ h):= \sum_{s=o}^{q-1} u_{o,s}(t_o+\tau h)\cdot(\tau h)^{1-\alpha})^s, \quad u_{o,s}(\tau h)= \sum_{j=o}^{m_s}\alpha_{o,j}^{(s)}\tau^j \in \pi_{m_s}.$$

Hence we have (3.1)

$$e_o(\tau h):= y(\tau h)-u_o(\tau h) = \sum_{s=o}^{q-1} h^{s(1-\alpha)}\Big\{ \sum_{j=o}^{m_s}\Big(\frac{h^j z_s^{(j)}(0)}{j!} -\alpha_{o,j}^{(s)}\Big)\tau^j + h^{m_s+1}\cdot R_{o,m_s}(\tau)\Big\}\cdot\tau^{s(1-\alpha)}, \tau\in[0,1].$$

Setting, in analogy to (1.9) (see also Table 1.1),

$$h^{m_s+1}\cdot\beta_{o,j}^{(s)}:=(h^j z_s^{(j)}(0)/j! - \alpha_{o,j}^{(s)})\cdot h^{s(1-\alpha)} \tag{3.2}$$

$(j = 0,\ldots,m_s;\ s = 0,\ldots,q-1)$,

we are led to the representation

$$e_o(\tau h)=\sum_{s=o}^{q-1} h^{m_s+1+s(1-\alpha)}\{\sum_{j=o}^{m_s}\beta_{o,j}^{(s)}\cdot\tau^j+R_{o,m_s}(\tau)\}\cdot\tau^{s(1-\alpha)},\tau\in[0,1]. \tag{3.3}$$

Suppose now that for a given $\alpha \in (0,1)$ and for a fixed $s \in \{0,...,q-1\}$ the degree $m_s \leqq m$ is the smallest integer with

$$m_s + s(1-\alpha) \geqq m \,. \tag{3.4}$$

Setting $M_{m_s+1} := \max\{|z_s^{(m_s+1)}(t)|/(m_s+1)!: t \in I\}$ we obtain

$$|e_o(\tau h)| \leqq h^{m+1} \sum_{s=o}^{q-1} (B_s + M_{m_s+1}),\ \tau h \in \sigma_o ,$$

provided we can show that the construction of the approximation u_o by collocation at $\{t_o+c_ih: 0 \leqq c_1<...<c_{M_q} \leqq 1\}$ $(M_q := \sum_{s=o}^{q-1} m_s+q)$ furnishes vectors $\beta_o^{(s)} := (\beta_{o,o}^{(s)},...,\beta_{o,m_s}^{(s)})^T$ for which $\|\beta_o^{(s)}\|_1 \leqq B_s < \infty$ for all sufficiently small $h > 0$.

The crucial condition is obviously (3.4): it will characterize the non-polynomial spline space containing the collocation approximation u whose restriction to the subinterval σ_o is the above function u_o .

Definition 3.1. Let $\alpha = p/q$ (reduced to lowest terms). Then we define

$$V_{m,\alpha}^{(-1)}(Z_N) := \{u: u|_{\sigma_o} = u_o(t_o+\tau h) = \sum_{s=o}^{q-1} h^{s(1-\alpha)}\cdot u_{o,s}(t_o+\tau h)\cdot\tau^{s(1-\alpha)};$$

$$u|_{\sigma_n} = u_n(t_n+\tau h) = \sum_{s=o}^{q-1} t_n^{s(1-\alpha)}\cdot u_{n,s}(t_n+\tau h)\cdot(1+\frac{\tau}{n})^{s(1-\alpha)}$$

$$(n = 1,...,N-1),\text{ with } u_{n,s} \in \pi_{m_s}\ (n=0,...,N-1)\}, \tag{3.5}$$

with m_s satisfying (3.4).

In order to determine the dimension of this linear space we introduce the following

Definition 3.2. Two lattice points (n_1,ℓ_1) and (n_2,ℓ_2) (i.e. ordered pairs in $\mathbb{N}_o \times \mathbb{N}_o$) are said to be $(1-\alpha)$-equivalent if

$$n_1(1-\alpha) + \ell_1 = n_2(1-\alpha) + \ell_2 \,.$$

The lattice points are called non-equivalent if they are not $(1-\alpha)$-equivalent.

Let $\nu_m^{(1)}(\alpha)$ denote the largest number of non-equivalent lattice points (n,ℓ) contained in $\{n(1-\alpha) + \ell \leqq m+1,\ \ell \leqq m\}$, and set $\mu_1 := q-p$.

Lemma 3.1.

$$\nu_m^{(1)}(\alpha) = \begin{cases} \nu_1^{(1)}(\alpha) + m + \sum_{k=2}^{m} \left[\frac{(k+1)q}{\mu_1}\right], & \text{if } 2 \leqq m < \mu_1 ; \\ (m+1)q - (\mu_1-1)(q-1)/2, & \text{if } \mu_1 \leqq m \end{cases} \tag{3.6}$$

(Compare [7, p.19] for a related problem, dealing essentially with the case of irrational α.)

As an immediate and crucial consequence we find that for $V_{m,\alpha}^{(-1)}(Z_N)$, with the degrees m_s $(s=0(1)q-1)$ being subject to $m_s + s(1-\alpha) \geqq m$, the following result holds.

Lemma 3.2.

$$\dim V_{m,\alpha}^{(-1)}(Z_N) \geqq N\cdot\nu_m^{(1)}(\alpha); \tag{3.7}$$

for $p = q-1$, we have $\nu_m^{(1)}(\alpha) = (m+1)q$, and hence for $q \to \infty$ (i.e. for $\alpha \to 1_-$),

$$\lim_{\alpha \to 1_-} \dim V_{m,\alpha}^{(-1)}(Z_N) = \infty .$$

3.2. Integro-differential equations

The modification of the arguments in the preceding section for the case of a Volterra integro-differential equation of the form (1.6) (with $p(t) \equiv 0$) is now clear: since the exact solution (again for the case $\beta = 1-\alpha$ in (2.8)) is given by

$$y(t) = z_o(t) + \sum_{s=o}^{q-1} z_s(t)\cdot(t^{2-\alpha})^s,\quad z_s \in C^{m+1}(I),$$

and in analogy to the remarks of Section 1 (see Table 1.1), the exponent (m_s+1) in (3.2) will be replaced by m_s (recall that, now, the approximating non-polynomial spline space will be a subspace of $C(I)$). Hence we define:

Definition 3.3. Let $\alpha = p/q$ (reduced to lowest terms). Then

$$V^{(o)}_{m,\alpha}(Z_N):=\{u \in C(I):\ u|_{\sigma_o}=u_o(t_o+\tau h)=\sum_{s=o}^{q-1} h^{s(2-\alpha)}\cdot u_{o,s}(t_o+\tau h)\cdot \tau^{s(2-\alpha)};$$

$$u|_{\sigma_n}=u_n(t_n+\tau h)=\sum_{s=o}^{q-1} t_n^{s(2-\alpha)}\cdot u_{n,s}(t_n+\tau h)\cdot(1+\frac{\tau}{n})^{s(2-\alpha)}$$

$$(n=1,\ldots,N-1),\ \text{with}\ u_{n,s}\in\pi_{m_s}\ (n=0,\ldots,N-1)\},$$

where m_s satisfies

$$m_s + s(2-\alpha) \geqq m \quad (m_s \leqq m),\ s = 0,\ldots,q-1. \tag{3.9}$$

Let $\nu^{(2)}_m(\alpha)$ denote the largest number of non-equivalent lattice points (Definition 3.2, with $1-\alpha$ now replaced by $2-\alpha$) contained in $\{n(2-\alpha) + \ell \leqq m+1,\ \ell \leqq m\}$, and set $\mu_2 := 2q-p$.

Lemma 3.2.

$$\dim V^{(o)}_{m,\alpha}(Z_N) \geqq N\cdot(\nu^{(2)}_m(\alpha) - 1) + 1\ ; \tag{3.10}$$

here,

$$\nu^{(2)}_m(\alpha) = \begin{cases} m + 1 + \sum_{k=1}^{m} [\frac{(k+1)q}{\mu_2}]\ , & \text{if } m < \mu_2-1\ ; \\ (m+1)q - (\mu_2-1)(q-1)/2, & \text{if } \mu_2-1 \leqq m. \end{cases} \tag{3.11}$$

If $p = q-1$ then $\nu^{(2)}_m(\alpha) = (m+1)(m+2)/2$ for all $q > m$, and hence

$$\lim_{\alpha\to 1_-} \dim V^{(o)}_{m,\alpha}(Z_N) = N\cdot(m^2+3m+1)/2 + 1 < \infty\ ,$$

if m_s is the smallest integer satisfying (3.9).

To conclude the discussion of the choice of the approximating non-polynomial spline spaces for (1.5) and (1.6) we mention, in view of practical applications, the most important special case, $\alpha = 1/2$.

Theorem 3.1.

(i) $\nu^{(1)}_m(1/2) = 2(m+1)$, and hence $\dim V^{(-1)}_{m,\frac{1}{2}}(Z_N) \geqq 2N(m+1)\ (m \geqq 0)$;

(ii) $\nu^{(2)}_m(1/2) = 2m+1$, and hence $\dim V^{(o)}_{m,\frac{1}{2}}(Z_N) \geqq 2Nm+1\ (m \geqq 1)$.

We note that (i) has been studied in [3](Theorem 1).

REFERENCES

1. BRUNNER, H., Non-polynomial spline collocation for Volterra equations with weakly singular kernels (to appear in *SIAM J. Numer. Anal.*).

2. BRUNNER, H. and NØRSETT, S.P., Superconvergence of collocation methods for Volterra and Abel integral equations of the second kind, *Numer. Math.* 36, 347-358 (1981).

3. BRUNNER, H. and TE RIELE, H.J.J., Volterra-type integral equations of the second kind with non-smooth solutions: high-order methods based on collocation techniques, Report NW 118/82, Mathematisch Centrum, Amsterdam (1982).

4. DE HOOG, F. and WEISS, R., High order methods for a class of Volterra integral equations with weakly singular kernels, *SIAM J. Numer. Anal.* 11, 1166-1180 (1974).

5. LUBICH, C., Runge-Kutta theory for Volterra and Abel integral equations of the second kind, Preprint Nr. 154, Sonderforschungsbereich 123, Universität Heidelberg (1982).

6. MILLER, R.K. and FELDSTEIN, A., Smoothness of solutions of Volterra integral equations with weakly singular kernels, *SIAM J. Math. Anal.* 2, 242-258 (1971).

7. PÓLYA, G. und SZEGÖ, G., *Aufgaben und Lehrsätze der Analysis*, Bd. 2. Dover Publications, New York (1945).

8. TE RIELE, H.J.J., Collocation methods for weakly singular second kind Volterra integral equations with non-smooth solution, Report NW 115/81, Mathematisch Centrum, Amsterdam (1981).

INTEGRAL EQUATIONS FOR BOUNDARY VALUE PROBLEMS EXTERIOR TO OPEN ARCS AND SURFACES

G.R. Wickham

Department of Mathematics, University of Manchester.

1. INTRODUCTION

One of the most important methods for solving exterior boundary value problems for partial differential equations is the so called "boundary integral equation method". This technique involves the use of Green functions to reduce the problem, which is formulated over an infinite domain, to finding the solution of an integral equation on the finite internal boundary. When that boundary is modelled by an open arc in two dimensions or an open surface in $\mathbb{R}_3$ the usual approach gives rise to a number of difficulties. As an example we shall consider the exterior Neumann problem of acoustics for an open arc; the boundary integral equation method is particularly useful here because the required solution of the partial differential equation can vary rapidly in the near field and only decays slowly at large distances:

PROBLEM $\mathcal{B}$ (ϕ_0): Let D be the domain exterior to an open boundary ∂D, where $\partial D \equiv L_+ \cup L_-$ and $L_\pm$ are the two sides of a twice continuously differentiable open arc L; we require the solution of

$$(\nabla^2 + k^2)\phi = 0, \qquad P \in D, \tag{1.1}$$

satisfying the boundary condition

$$\frac{\partial \phi}{\partial n}(p) = -\frac{\partial \phi_0}{\partial n}(p), \qquad p \in \partial D, \tag{1.2}$$

and the radiation condition

$$r^{\frac{1}{2}}\left(\frac{\partial}{\partial r} - ik\right)\phi \to 0, \qquad r \to \infty,\ 0 \leq \theta < 2\pi, \tag{1.3}$$

and which is bounded everywhere in $D \cup \partial D$. Here, (r,θ) are plane polar coordinates and $\underline{n}$ is the unit normal to L_+ pointing into D. Thus, ϕ is the field scattered from a rigid barrier immersed in an ideal compressible fluid when it is irradiated by time harmonic waves of wavenumber k and velocity potential ϕ_0; we shall take ϕ_0 to satisfy (1.1) in the whole plane. It may be shown using arguments similar to those employed in [10] that the only solution of the homogeneous problem $\mathcal{B}(0)$ is identically zero.

We shall seek a solution of $\mathcal{B}(\phi_0)$ in the form of a double layer potential, namely

$$\phi(P) = \int_L \mu(s) \frac{\partial}{\partial n_q} G(P,q)ds, \qquad P \in D, \tag{1.4}$$

where s denotes arc length measured along L, $q(s)$ is a point on L and the fundamental source or Green function $G(P,Q)$ is given by

$$G(P,Q) = \frac{i}{4} H_0^{(1)}(kR), \tag{1.5}$$

where $R(P,Q)$ is the distance between points P and Q. We shall suppose that if s is measured so that $\pm s_0$ are the ends of the arc L then the density function $\mu(s)$ has *PROPERTIES* $\mathcal{P}$: (i) $\mu(\pm s_0) = 0$
(ii) $\mu'(s)$ exists and is Hölder continuous in $(-s_0, s_0)$.

It is not too difficult to show that under these conditions (1.4) satisfies (1.1) and (1.3), is bounded everywhere in $D \cup \partial D$, and the potential difference across L is given by

$$[\phi(p)] = \phi(p_+) - \phi(p_-) = \mu(s) \tag{1.6}$$

where $p_\pm$ are corresponding points on $L_\pm$. It therefore follows that (1.4) solves $\mathcal{B}(\phi_0)$ provided the boundary condition is satisfied, i.e.

$$-\frac{\partial \phi_0}{\partial n_p} = \frac{\partial}{\partial n_p} \int_L \mu(s) \frac{\partial}{\partial n_q} G(p,q)ds, \qquad p \in L, \tag{1.7}$$

which is an *integro-differential equation of the first kind* for μ. Since

$$G(P,Q) = \frac{1}{2\pi} \log \frac{1}{R(P,Q)} + O(1)$$

as $P \to Q$, it follows that we cannot immediately interchange the order of differentiation and integration in (1.7). In any case, first-kind integral equations tend to be very difficult to solve even when it may be proven that a solution exists. In practice we are faced with the problem of either devising a special numerical method for treating (1.7) or finding an explicit regularisation which reduces the integro-differential equation to a Fredholm equation of the second kind. Whether we adopt the first or second approach, full account must be taken of the fact that the potential ϕ is not differentiable at the ends of the arc; in fact it turns out that

$$\mu(s) = O([s \pm s_0]^{\frac{1}{2}}) \qquad \text{as} \quad s \to \mp s_0. \tag{1.8}$$

When L is a straight line segment it is a reasonably straightforward matter to apply a numerical scheme due to Erdogan and Gupta [4]. Here, (1.8) is exploited in a rather ingenious way to derive a system of algebraic equations for approximating values of $\mu'(s_n)$, $n = 1,2,3, \ldots N$ where the s_n are the zeros of the Chebyshev polynomial $T_N(s)$. Again it is possible to apply integral transform techniques to convert (1.7) for a straight arc into a second kind equation, see for example Mal [5], but these methods cannot be applied to arcs of general shape.

Readers familiar with potential theoretic methods for closed boundaries will recall that the exterior Neumann problem of acoustics is usually solved using a single layer; this leads to a second kind equation soluble for all smooth boundaries and all values of the parameter k barring a certain discrete set, c.f. [8]. If in the present case, we apply Green's theorem to G and ϕ in the region D we simply reproduce (1.4). Thus, if we choose (1.5) as our fundamental solution and ϕ exists, ϕ *is necessarily a double layer potential* and an equation of the form (1.7) is unavoidable. In what follows we shall propose an alternative approach based on a new class of Green functions defined in the cut plane D. These give rise to Fredholm equations of the second kind uniquely soluble for μ irrespective of the geometry of L.

2. GREEN FUNCTIONS IN THE CUT PLANE

Let us suppose that we can find a Green function $G(P,q)$ satisfying (1.3), the partial differential equation (1.1) in the plane cut along L and which is such that G corresponds to equal and opposite sources at q_+ and q_-, where q_+ and q_- are opposite points on L_+ and L_- respectively, i.e.

$$\frac{\partial G}{\partial R} \pm \frac{1}{\pi R(P,q)} = o\left(\frac{1}{R(P,q)}\right), \qquad P \to q_{\pm}\ , \tag{2.1}$$

and

$$\lim_{P \to p_+} \frac{\partial G}{\partial n_p}(P,q) = \lim_{P \to p_-} \frac{\partial G}{\partial n_p}(P,q), \qquad p,q \in L. \tag{2.2}$$

We shall denote either side of (2.2) by $K(p,q)$ and it will be supposed that $K(p,q)$ is a continuous function of s_p and s_q on the square $L_p \times L_q$, where $L_p = [-s_0,s_0]$, expect possibly when $s_p = s_q$. Assuming such a G and the solution of problem

$\mathcal{B}(\phi_0)$ both exist, we can apply Green's theorem to any annular region $\mathcal{D} \in D$, where the inner boundary of $\mathcal{D}$ encloses a domain containing L. In the limit, when that contour is deformed into ∂D and the other boundary recedes to infinity, we obtain the equation

$$\mu(q) - \int_L K(p,q)\mu(p)ds_p = \int_L [G(p,q)] \frac{\partial\phi_0}{\partial n_p} ds_p, \quad p \in L, \tag{2.3}$$

where

$$[G(p,q)] = G(p_+,q) - G(p_-,q), \qquad p, q \in L, \tag{2.4}$$

and we have assumed that the contribution from the ends of L is zero in the limit. Thus, provided the associated kernel $K(p,q)$ is sufficiently smooth, *a Green function of the type described will give rise to a Fredholm integral equation of the second kind.*

To answer the question of how such a function may be constructed we recall that the double layer potential of density ρ, say, has a discontinuity equal to ρ and a continuous normal derivative across L. Thus, in view of the requirement (2.2), we shall look for a G in the form

$$G(P,q) = \int_L \rho(t,q) \frac{\partial G}{\partial n_t}(t,P)ds_t, \qquad P \in D, \tag{2.5}$$

where $\rho(t,q) \in \mathcal{P}(t)$ for each $q \in L$. However, it is evident that, if (2.1) is to be satisfied, $\rho(t,q)$ cannot be continuous at $t = q$; indeed (2.1) suggests that

$$\rho(t,q) = O(\log R(t,q)), \qquad t \to q. \tag{2.6}$$

Further, it may be shown that the required potential difference, $\mu(s_q)$, $(= [\phi(q)])$, satisfies (1.8) and so it would be desirable to choose ρ so that

$$\rho(t,q) = O((s_0 \mp s_q)^{\frac{1}{2}}), \qquad s_q \to \pm s_0, \quad t \in L, \tag{2.7}$$

for then every solution of (2.3) would have the same property. It therefore seems quite natural to look for a $\rho(t,q)$ which satisfies (2.6) and has properties $\mathcal{P}_{\frac{1}{2}}(t) \times \mathcal{P}_{\frac{1}{2}}(q) t \neq q$, where the subscript denotes that the required function vanishes according to (2.7). One such function is given by

$$\rho(t,q) = -\frac{2}{\pi}\{\log|t-q| - \log|q_0^2 - qt + (q_0^2-t^2)^{\frac{1}{2}}(q_0^2-q^2)^{\frac{1}{2}}| + \log|q_0|\}, \tag{2.8}$$

where t and q now denote the complex numbers $f(s_t) + ig(s_t)$ and $f(s_q) + ig(s_q)$ respectively,

$$\pm q_0 = f(\pm s_0) + i g(\pm s_0), \tag{2.9}$$

and

$$z = f(s) + ig(s) \tag{2.10}$$

is a parametric representation of the arc L. An expression of the form $(q^2 - q_0^2)^{\frac{1}{2}}$ is taken to be the boundary value on L_+ of the function $(z^2 - q_0^2)^{\frac{1}{2}}$ which is holomorphic in D and has power series $z - q_0^2/2z + \ldots$ about the point at infinity.

Having motivated the construction of a particular Green function G in what can only be regarded as a speculative manner we shall prove shortly that it does indeed have the desired properties. In particular we shall show that the normalisation constant $(-2/\pi)$ in (2.8) has been chosen so that (2.1) is satisfied. Before proceeding however, we note that if we add to ρ, given by (2.8), any function $\sigma(t,q) \in \mathcal{P}_{\frac{1}{2}}(t) \times \mathcal{P}_{\frac{1}{2}}(q)$ then the corresponding $G(P,q)$ given by (2.5) will also be of the required type; *in this way we can construct an infinite family of Green functions leading to second kind equations like (2.3)*.

THEOREM I: The Green function (2.5), where ρ is given by (2.8) satisfies (1.1), (1.3), (2.1), (2.2) and is such that

$$[G(p,q)] = \rho(p,q), \qquad p \neq q. \tag{2.11}$$

PROOF: We first note that $\rho(t,q)$ is an absolutely integrable function of t on L and since $G(P,q)$ satisfies

$$(\nabla^2 + k^2)G(P,q) = 0, \qquad P \in D,$$

and $$r^{\frac{1}{2}}\left(\frac{\partial}{\partial r}G - ikG\right) \to 0, \qquad r \to \infty,$$

then it follows immediately that G satisfies the Helmholtz equation and radiation condition in the independent variable P.

To prove (2.1), we observe that G may be written in the form

$$G(P,q) = \mathcal{R}\frac{1}{2\pi i}\int_L \frac{\rho(t,q)}{(t-z)}dt + G_1(P,q), \tag{2.12}$$

where

$$G_1(P,q) = \int_L \rho(t,q)\,\frac{\partial G_1}{\partial n_t}(t,P)ds_t, \tag{2.13}$$

and $$G_1(t,P) = G(t,P) + \frac{1}{2\pi}\log R(t,P). \tag{2.14}$$

It is easily shown that

$$\frac{\partial^2 G}{\partial R\partial n_t} = O(\log R(t,P)) \qquad \text{as } P \to t$$

and hence $\partial G_1/\partial R$ is bounded in the whole plane. On integrating by parts we find that the first term in (2.12) may be written as

$$G_0(P,q) = -\mathcal{R}\frac{1}{2\pi i}\int_L \ell n(t-z)\frac{\partial\rho}{\partial s}\,ds \qquad (2.15)$$

and hence

$$\frac{\partial G_0}{\partial R} = \mathcal{R}\,\frac{e^{i\theta(q,P)}}{2\pi i}\int_L \frac{1}{(t-z)}\frac{\partial\rho}{\partial s}\,ds,$$

where

$$\Theta(q,P) = \arg(P-q)\ .$$

Now, since ρ vanishes at the ends of the arc L, we have

$$⨍_L \frac{\partial\rho}{\partial s}\,ds = 0,$$

where $⨍_L$ denotes the Cauchy principal value, and so we may write

$$\frac{\partial G_0}{\partial R} = \mathcal{R}\,\frac{e^{i\theta(q,P)}}{2\pi i}\ ⨍_L\left\{\frac{1}{(t-z)} - \frac{1}{(q-z)}\right\}\frac{\partial\rho}{\partial s}\,ds$$

$$= \mathcal{R}\,\frac{1}{2\pi i R(q,P)}\,\Phi(z,q), \qquad (2.16)$$

where

$$\Phi(z,q) = \int_L \frac{\xi(t(s),q)}{(t-z)}\,dt, \qquad (2.17)$$

$$\xi(t(s),q) = (t-q)\,\frac{\partial\rho}{\partial s}\,e^{-i\beta(s)} \qquad (2.18)$$

and $\beta(s)$ is the angle which the positive tangent to L at the point t makes with a fixed direction in the plane (see fig.1).

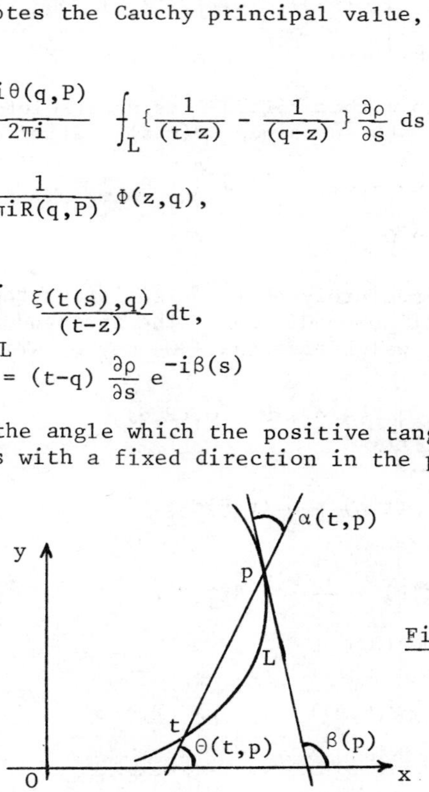

Fig.1

It follows from (2.8) that

$$\xi(t(s),q) = -\frac{2}{\pi} e^{-i(\beta(s)-\Theta(t,q))} \mathcal{R}\{\frac{(q_0^2-q^2)^{\frac{1}{2}} e^{i(\beta(s)-\Theta(t,q))}}{(q_0^2 - t^2)^{\frac{1}{2}}}\} \tag{2.19}$$

which is clearly a Holder continuous function of s in the open interval $(-s_0, s_0)$. It is known that integrals of the form (2.17) are continuous in D from either side of the arc L; in particular, it is readily shown, c.f. [7], ch.II, §22, that if q is an interior point of L, then there exists positive constants C, ε and μ such that

$$|\Phi(z,q) - \Phi(q^{\pm},q)| < C|z-q|^{\mu}$$

where $\mu \in (\varepsilon,1)$. But it follows from the Plemelj formulae and equations (2.18) and (2.19) that

$$\begin{aligned}\Phi(q^{\pm},q) &= \pm \pi i\xi(q,q) + \int_L \frac{\partial\rho}{\partial s} ds \\ &= \mp 2i,\end{aligned}$$

and so (2.1) follows immediately. It also follows from results given in [7], ch.IV that $\Phi(z,\pm q_0)$ is bounded in any neighbourhood of $z = \pm q_0$ respectively. Finally, the expression (2.11) may be deduced by applying the Plemelj formulae to (2.12) since G_1 is clearly continuous across the arc; this concludes the proof of theorem I. In the next section we shall investigate the properties of the associated kernel function $K(p,q)$.

3. THE KERNEL FUNCTION AND EXISTENCE

Consider first the function

$$\begin{aligned}K_0(p,q) &= \lim_{P\to p_{\pm}} \frac{\partial G_0}{\partial n_p}(P,q) \\ &= -\frac{1}{2\pi} \int_L \frac{\partial\Theta}{\partial n_p}(p,t) \frac{\partial\rho}{\partial s} ds\end{aligned} \tag{3.1}$$

by (2.15). But

$$\Theta(p,t) - \Theta(t,p) = \pm \pi$$

and so it follows from the Cauchy-Riemann relations that

$$\frac{\partial\Theta}{\partial n_p}(p,t) = \frac{\partial\Theta}{\partial n_p}(t,p) = \frac{\partial}{\partial s_p} \log R(t,p)$$

Hence

$$K_0(p,q) = -\frac{1}{2\pi}\int_L \frac{\cos\alpha(t,p)}{R(t,p)}\frac{\partial\rho}{\partial s}\,ds \tag{3.2}$$

where $\alpha(t,p)$ is the angle between the positive tangent to L at p and the directed line segment $\vec{tp}$ measured from the latter in the anticlockwise direction (see fig.1). Since

$$\alpha(t,p) = \beta(p) - \Theta(t,p)$$

we have

$$\frac{e^{i\alpha(t,p)}}{R(t,p)} = \frac{e^{i\beta(p)}}{(p-t)}$$

and so on substituting from (2.8) into (3.2) we find that

$$K_0(p,q)\,\frac{-1}{\pi^2}\,\mathcal{R}e^{i\beta(p)}(q_0^2-q^2)^{\frac{1}{2}}\int_L \frac{dt}{(t-q)(t-p)(q_0^2-t^2)^{\frac{1}{2}}}$$

$$\frac{-1}{\pi^2}\,\mathcal{R}i(q_0^2-q^2)^{\frac{1}{2}}\int_L \frac{\kappa(t,p)dt}{(t-q)(q_0^2-t^2)^{\frac{1}{2}}} \tag{3.3}$$

where $\kappa(t,p) = \sin\alpha(t,p)/R(t,p)$. The first term in this expression is easily shown to be zero using the residue theorem and, if L is a twice continuously differentiable curve, $\kappa(t,p)$ is a Hölder continuous function of both t and p. We conclude that $K_0(p,q)$ is a continuous kernel and hence the complete kernel

$$K(p,q) = \frac{-1}{\pi^2}\,\mathcal{R}i(q_0^2-q^2)^{\frac{1}{2}}\int_L \frac{\kappa(t,p)dt}{(t-q)(q_0^2-t^2)^{\frac{1}{2}}}$$

$$+\int_L \rho(t,q)\,\frac{\partial^2 G_1}{\partial n_t\,\partial n_p}(t,p)dt \tag{3.4}$$

is sufficiently smooth for the Fredholm alternatives to apply to (2.3). Further, it is a simple matter to show that any particular solution μ of (2.3) has properties $\mathcal{P}$ and hence when that solution is substituted into (1.4) we generate a bounded potential in D satisfying (1.1) and (1.3). It is also possible, though this is somewhat more difficult, to show that if μ satisfies (2.3) then the corresponding potential ϕ satisfies the boundary condition (1.2). It then follows from the uniqueness theorem for $\mathcal{B}$ that (+1) is not an eigenvalue of (2.3) and hence we have:

THEOREM II; there exists one and only one solution of problem $\mathcal{B}(\phi_0)$ for each value of the frequency parameter k. The details of the proof of this theorem will be reported elsewhere.

4. A SIMPLE EXAMPLE AND ITS NUMERICAL TREATMENT

The simplest example is when L is the straight line segment $|x| \leqslant 1$, $y = 0$; we then find that $\kappa(t,p) \equiv 0$ and (3.4) becomes

$$K(p,q) = \frac{ik^2}{4} \int_{-1}^{1} \rho(t,q) \left[\frac{H_1^{(1)}(k|t-p|)}{k|t-p|} + \frac{2i}{\pi k^2 (t-p)^2} \right] dt \tag{4.1}$$

with

$$\rho(t,q) = -\frac{2}{\pi} \{\log|t-q| - \log|1 - tq + \sqrt{(1-t^2)(1-q^2)}|\}$$

It is easily verified that (4.1) is a continuous kernel and is in fact uniformly $O(k^2 \log k)$ as $k \to 0$. It follows that, in this case, (2.3) may be solved by iteration when k is sufficiently small. For higher values the integral equation must be solved numerically. This of course provides no difficulty in practice providing the kernel function can be computed accurately. The factors in the integrand of (4.1) have logarithmic singularities at $t = q$ and $t = p$ respectively and so some care is required. A direct approach would be to split the range of integration into the three open intervals on which the integrand is continuous and differentiable and use, for example, an adaptive quadrature scheme in each interval. Such a procedure, while capable of attaining more or less any desired accuracy, is very expensive in computing time. Considerable savings can be achieved by noting that the function

$$\frac{dF}{dx} = \frac{H_1^{(1)}(|x|)}{|x|} + \frac{2i}{\pi x^2} - \frac{1}{2} - \frac{i}{\pi}(\gamma - \tfrac{1}{2} + \log \tfrac{1}{2}|x|) \tag{4.2}$$

where γ is Euler's constant is continuous and differentiable and $O(u^2 \log u)$ as $u \to 0$, [1]. Substituting from (4.2) into (4.1) and performing an integration by parts yields.

$$K(p,q) = \frac{ik^2}{2} \left[\frac{1}{2} + \frac{i}{\pi}(\gamma - \frac{1}{2} + \log \tfrac{1}{2}k)\right](1-q^2)^{\frac{1}{2}}$$

$$- \frac{1}{2\pi} k^2 \{2|p-q| \arctan(\frac{1 \pm q}{1 \mp q}) - (1 + \log 2)(1 - q^2)^{\frac{1}{2}}\}$$

$$+ \frac{ik^2}{2\pi}(1 - q^2)^{\frac{1}{2}} ⨍_{-1}^{1} \frac{F(k|t-p|)}{(t-q)(1-t^2)^{\frac{1}{2}}} dt \tag{4.3}$$

where we take the upper signs in the second term if $p > q$ and the lower signs otherwise. If we now assume that for each value of the parameter k and each p, $F(k|t-p|)$ may be sufficiently accurately approximated by the Chebyshev expansion

$$F(k|t-p|) = \sum_{n=0}^{N} B_n(k,p)T_n(t), \tag{4.4}$$

the remaining integral may be evaluated explicitly using the formula

$$\int_{-1}^{1} \frac{T_n(t)}{(1-t^2)^{\frac{1}{2}}(t-q)} dt = \begin{matrix} 0 & n = 0 \\ U_{n-1}(q), & n > 0 \end{matrix}, \quad q \in (-1,1). \tag{4.5}$$

see for example, [3].

Equations (4.3) through (4.5) provide an efficient quadrature rule for evaluating K and so we are now at liberty to choose from the various well tried numerical methods available for solving second kind equations with smooth kernels [2].

6. CONCLUSIONS

The usual treatment of boundary value problems exterior to open arcs and surfaces is to "reduce" them to the solution of singular integral equations of the first kind. Although there has been some progress, see for example [6], in the numerical solution of such equations using special quadrature formulae, such formulations still suffer from theoretical and practical disadvantages. In this article we have discussed a classical boundary value problem in diffraction theory in order to display a new approach which avoids these difficulties. We have shown how it is possible to construct an infinite family of Green functions which are analytic in the domain exterior to an open arc of general shape. Application of Green's theorem to any one of these functions and the solution of the original boundary value problem then leads to a Fredholm integral equation of the second kind with a continuous kernel. These equations also explicitly exhibit the singularities in the field quantities at the geometrical discontinuities so that their numerical solution may be effected by the most straightforward methods. Although we have only considered a relatively simple example it is hoped that the generalisation to more complicated problems is evident.

In [9] details of the numerical solution of (2.3) for the simple example considered in §5 and also the case when L is the arc of a circle are given. For the straight line segment a comparison is made with a direct numerical solution of (1.7) using the method of Erdogan & Gupta [4].

7. REFERENCES

1. ABRAMOWITZ, M. and STEGUN, I.A., (eds), *Handbook of mathematical functions*, National Bureau of Standards, Appl. Math. Series 55 (1964).
2. BAKER, C.T.H., *The numerical treatment of integral equations*. Oxford University Press, Oxford (1977).
3. ERDELYI, A., (ed.), *Higher transcendental functions, Vol.2*, McGraw-Hill, New York (1953).
4. ERDOGAN, F. and GUPTA, G.D., On the numerical solution of singular integral equations. *Q. Appl. Math.* 29, 525-534 (1972).
5. MAL, A.K., Interaction of elastic waves with a Griffith crack *Int. J. Eng. Sci.* 8, 763-776 (1970).
6. MILLER, G.F. (These proceedings).
7. MUSKHELISHVILI, N.I., *Singular integral equations*. P. Noordhoff, N.V., Groningen, Holland, (1946).
8. URSELL, F., On the exterior problem of acoustics. *Proc. Camb. Phil. Soc.* 74, 117-125, (1973).
9. WALKER, C.P., and WICKHAM, G.R. *Numerical analysis technical report No. (to appear)*, Dept. of Mathematics, University of Manchester.
10. WICKHAM, G.R. The diffraction of stress waves by a plane finite crack in two dimensions : uniqueness and existence. *Proc. Roy. Soc. Lond. A.* 378, 241-261 (1981).

EIGENVALUE PROBLEMS OF INTEGRO-DIFFERENTIAL EQUATIONS ON THE CONE OF NONDECREASING FUNCTIONS

E. Jäger, communicated by E. Bohl

University of Konstanz

1. INTRODUCTION

This paper is based on the Diploma Thesis of E. Jäger, prepared under the supervision of E. Bohl.

For certain matrices with nonnegative elements, Frobenius [1912] and Perron [1907] proved the existence of a positive eigenvalue λ and of an eigenvector with nonnegative components belonging to λ. Extensions of this result were made by Jentzsch [1912] to integral operators with positive kernel and by Krein-Rutman [1950] to linear operators leaving a cone invariant in a Banach space. In Bohl [1974], Theorem (III,3.3) shows moreover a possibility of iterative computing approximations to the spectral radius of a linear operator leaving a cone invariant in a partially ordered vector space.

As applications we consider eigenvalue problems

$$-(px')' = \lambda Bx, \quad \alpha x(0)-\beta x'(0) = x'(1) = 0 \qquad (1)$$

where $p\in C[0,1]$ is positive, $\alpha>0$, $\beta\geq 0$, and B is a linear operator on $C^1[0,1]$ to $C[0,1]$.

If B leaves the cone of nonnegative functions invariant, the above mentioned theory can obviously be applied for the treatment of (1). However, we do not need this strong monotonicity property of B to be able to apply Bohl's Theorem (III,3.3), and there are applications of this Theorem where B does not map the cone of all nonnegative functions into itself.

Let us for example consider the buckling of an elastic rod due to its dead weight (see Collatz [1945], Jäger [1982]). This problem yields the boundary value problem (see Collatz [1945])

$$-x''(s) = \lambda \int_0^s \frac{F(u)}{I(s)}(x(s)-x(u))\,du \quad (s\in[0,1]) \tag{2a}$$

$$\lambda = \frac{\gamma l^3}{E} \tag{2b}$$

$$x(0) = x'(1) = 0 \tag{2c}$$

where E is the modulus of elasticity, γ the specific gravity, I(s) the geometric moment of inertia and F(s) the cross-sectional area of the rod at the point $s\cdot l$. The function x(s) describes the buckling of the rod, transformed to the interval [0,1]. We suppose the functions F and I to be independent of l = length of the rod. (2) is an eigenvalue problem of the type (1), where the operator B, given by

$$(Bx)(s) = \int_0^s \frac{F(u)}{I(s)}(x(s)-x(u))\,du$$

($x\in C^1[0,1], s\in[0,1]$), does not map all nonnegative functions to nonnegative ones. However, the present property of B that the image of every nondecreasing function under B is nonnegative, along with a certain strong monotonicity property of Green's function will still allow application of Theorem (III,3.3) to (2). Hence, there exists a real number $\sigma>0$ such that (2) has a nontrivial solution for $\lambda=\sigma$ and only the trivial solution if $|\lambda|<\sigma$. Therefore, the rod remains straight as long as $(l^3\gamma)/E<\sigma$ and we have buckling if $(l^3\gamma)/E=\sigma$.

2. THE GENERAL FORM OF THE EIGENVALUE PROBLEM

Throughout this paper, C^i will denote the set of all functions on [0,1] with continuous derivatives up to order $i \in \mathbb{N}$, and C will be the set of all continuous functions on [0,1]. For the set of all linear operators on the real vector space X to the real vector space Y, we will use the symbol L[X,Y]. The symbol L[X] will be written for L[X,X].

Let $R_i : C^1 \to C$ $(i=0,1)$ be the boundary operators given by $R_o \in \{R_o^1, R_o^2\}$, where $R_o^1 x = x(0)$, $R_o^2 x = \alpha x(0) - x'(0)$ (α being a positive real number), and $R_1 x = x'(1)$. Let p be a continuous function on [0,1] such that $p(t) > 0$ for all $t \in [0,1]$. By L we denote the differential operator

$$L : W = \{x \in C^1 : px' \in C^1 \text{ and } R_t x = 0 \ (t=0,1)\} \longrightarrow C$$

$$x \mapsto -(px')'.$$

Let us consider now eigenvalue problems of the form

$$Lx = \lambda Bx, \tag{3}$$

where $B \in L[C^1, C]$. The operator L is invertible and $L^{-1} \in L[C,W]$ is given by $L^{-1} r = \int_0^1 G(\cdot,s) r(s) ds$ $(r \in C)$, where G(t,s) is Green's function for the differential operator $x \mapsto -(px')'$ and the boundary conditions $R_t x = 0$ $(t=0,1)$. Therefore, (3) has only the trivial solution for $\lambda = 0$. Our aim is to determine a real number λ such that (3) has a nontrivial solution and $|\lambda|$ is minimal. For this purpose, we first have to introduce some notations (see also Bohl [1970]).

3. THE SPECTRAL RADIUS OF A MONOTONE OPERATOR

Let X be a real vector space. We will assume $X \neq \{\Theta\}$ where Θ denotes the zero element in X. A subset $K \subset X$

is said to be a *cone*, if $K+K\subset K$, $\lambda K\subset K$ for all reals $\lambda>o$, and $K\cap(-K)=\{\Theta\}$. A cone is called *Archimedian* if $x\in X$, $z\in K$ and $z-nx\in K$ for all natural numbers n imply $-x\in K$. For every $e\in K$ we define a subspace X_e by setting

$$X_e = \{x\in X: re\pm x\in K \text{ for some } r\in\mathbb{R}\}.$$

An element $e\in K$ is said to be an *order unit of K*, if $X_e=X$. By oK we denote the set of all order units of K. By a *partially ordered vector space* we shall understand a real vector space endowed with an Archimedian cone K which contains order units.

Let X now be a partially ordered vector space. For every $e\in oK$ we define functionals $\|\ \|_e$ and $|\ |_e$ by setting

$$\|x\|_e = \inf\{r>o: re\pm x\in K\},\quad |x|_e=\sup\{r\in\mathbb{R}: x-re\in K\}\ (x\in X).$$

For every $e\in oK$, $\|\ \|_e$ is a norm on X, and two norms $\|\ \|_e$, $\|\ \|_{e'}$ $(e,e'\in oK)$ are equivalent.

3.1 Examples:

Let $\overline{X}$ be the set of all functions $x\in C^1$ such that $R_t x=0$ $(t=0,1)$. Then $\overline{K}=\{x\in\overline{X}: x'(t)\geq 0$ for all $t\in[0,1]\}$ is an Archimedian cone, but $o\overline{K}=\emptyset$. To obtain a partially ordered vector space, we consider $X=(\overline{X})_{\overline{e}}$, where $\overline{e}\in\overline{K}$ is given by

$$\overline{e}(t) = \begin{cases} 0.5t(2-t) & , \quad \text{if } R_o=R_o^1 \\ \frac{1}{\alpha}+0.5t(2-t), & \quad \text{if } R_o=R_o^2 \end{cases}.$$

X consists precisely of all elements $x\in\overline{X}$ such that $|x'(t)|\leq r(1-t)$ for some $r>0$ and all $t\in[0,1]$. $K=\overline{K}\cap X$ is an Archimedian cone with $oK=\{e\in K: e'(t)>0$ for all $t\in[0,1)$ and $M(1-t)\leq e'(t)$ for some $M>0$ and all $t\in[0,1]\}$. Furthermore, the functionals $\|\ \|_e$, $|\ |_e$ $(e\in oK)$ on X are given by

$\|x\|_e = \sup\{\frac{|x'(t)|}{e'(t)} : t \in [0,1)\}$, $|x|_e = \inf\{\frac{x'(t)}{e'(t)} : t \in [0,1)\}$.

3.2

Let $2 \leq M$ be a natural number and let $h = \frac{1}{M}$. We define a grid Ω_h by setting $\Omega_h = \{t_j : j=0,\ldots,M\}$, where $t_j = j \cdot h$. Let us consider the finite dimensional vector spaces $X_h^i = \{x \in \mathbb{R}^{\Omega_h} : i \cdot x(0) = 0\}$ $(i=0,1)$. The set $K_h^i = \{x \in X_h^i : x(0) \geq 0, x(t_{j+1}) - x(t_j) \geq 0 \text{ for } j=0,\ldots,M-1\}$ is an Archimedian cone, and we have $oK_h^i = \{e \in K_h^i : e(t_{j+1}) - e(t_j) > 0$ for all $j \in \{0,\ldots,M-1\}$ and $i=0 \Rightarrow e(0) > 0\}$ $(i=0,1)$. For $e \in oK_h^i$ the functionals $\|\ \|_e$ and $|\ |_e$ are given by

$$\|x\|_e = \max\{|m_x|, \max\{\frac{|x(t_{j+1}) - x(t_j)|}{e(t_{j+1}) - e(t_j)} : j=0,\ldots,M-1\}\},$$

$$|x|_e = \min\{m_x, \min\{\frac{x(t_{j+1}) - x(t_j)}{e(t_{j+1}) - e(t_j)} : j=0,\ldots,M-1\}\},$$

$$\text{where} \quad m_x = \begin{cases} \dfrac{x(0)}{e(0)} \quad , & \text{if } i=0 \\ \dfrac{x(t_1) - x(t_o)}{e(t_1) - e(t_o)}, & \text{if } i=1 \end{cases} .$$

A linear operator P on a partially ordered vector space is called *monotone* if $P(K) \subset K$. The symbol $L_+[X]$ denotes the set of all monotone elements of $L[X]$. Every $P \in L_+[X]$ is bounded with respect to every norm $\|\ \|_e$. The operator norm $\|P\|_e$ is given by $\|P\|_e = \|Pe\|_e$.

An operator $P \in L_+[X]$ is said to be *strongly monotonic* if for every $x \in K - oK$, $x \neq 0$ there exists a natural number $n = n(x) \geq 1$ such that $P^n x \in oK$.

For a strongly monotonic operator $P \in L_+[X]$ such that P^r is completely continuous (with respect to some norm $\|\ \|_e$) for some $r \in \mathbb{N}$, Bohl's Theorem (III,3.3) shows the existence of an eigenvector $z \in oK$ of P. The eigenvalue belonging to z coincides with the spectral radius $\sigma(P)$ of P. Furthermore, in this case, the equation

$$\sigma(P) = \max\{|\lambda| : \lambda \in \mathbb{R} \text{ is an eigenvalue of } P\} \quad (4)$$

and the inclusion

$$|Pe^n|_{e^n} \leq |Pe^{n+1}|_{e^{n+1}} \leq \sigma(P) \leq \|P\|_{e^{n+1}} \leq \|P\|_{e^n} \quad (n \in \mathbb{N}) \quad (5)$$

hold for every sequence $e^o \in oK$, $e^{n+1} = Pe^n$.

4. APPLICATION TO THE EIGENVALUE PROBLEM $Lx=\lambda Bx$

For every $x \in W$, there exists $\lim_{t\to 1} \frac{x'(t)}{1-t}$, hence $W \subset X$ (X being the vector space defined in example 3.1). Therefore, we get an operator $T \in L[X]$ by $Tx := L^{-1}Bx$ $(x \in X)$.

Theorem 1: Suppose that B satisfies the following conditions (i), (ii):

(i) $x \in K - \{\Theta\} \Rightarrow (Bx)(t) \geq 0$ for all $t \in [0,1]$ and $(Bx)(1) > 0$.

(ii) The image $B(Q)$ of every bounded subset $Q \subset X$ (with respect to some $\|\ \|_e$) is bounded in C (with respect to the maximum norm on C) and $B(Q)$ is equicontinuous at $t=1$.

Then T is monotone, strongly monotonic and completely continuous.

Suppose now that $Lx = \lambda Bx$ for some $\lambda \in \mathbb{R}$, $x \neq \Theta$. Since $\lambda \neq 0$, we have $\lambda^{-1}x = L^{-1}Bx$, hence λ^{-1} is an eigenvalue of T and by (4) we have $|\lambda^{-1}| \leq \sigma(T)$. On the other hand, $Tx = \sigma(T)x$ implies $x \in W$ and $Lx = \sigma(T)^{-1}Bx$, hence $\sigma(T)^{-1}$ is the minimal positive eigenvalue of (3). We now list a few examples of operators $B \in L[C^1, C]$ such that the conditions (i) and (ii) are satisfied.

4.1

Let $(Bx)(s)=\int_0^s K(s,u)(x(s)-x(u)du$ $(s\in[0,1],x\in C^1)$, where $K\in C([0,1]^2)$, $K(s,u)\geq 0$ for all $(s,u)\in[0,1]^2$ and $K(1,0)>0$.

The conditions are easy to verify. Hence, we can treat the problem (2) derived from the buckling of the the rod (there, we assume F and I to be continuous and $F(0)>0$, $I(1)>0$).

4.2

Let $(Bx)(s)=\int_0^1 K(s,\xi)x'(\xi)d\xi$ $(s\in[0,1],x\in C^1)$. Here, we make the assumptions $K\in C([0,1]^2)$, $K(s,\xi)\geq 0$ for all $(s,\xi)\in[0,1]^2$ and $K(1,\xi)>0$ for all $\xi\in(0,1)$.

4.3

Let $(Bx)(s)=(qx)'$, where $q\in C^1, q(t), q'(t)\geq 0$ for all $t\in[0,1]$ and $q'(1)>0$.

Remark: $q'(1)=0$ implies $(Bx)(1)=0$ for all $x\in X$, hence (i) is violated in this case. For example, the eigenvalue problem

$$-(px')' = \lambda x', \quad R_0x = R_1x = 0, \tag{6}$$

corresponding to $q\equiv 1$, has only the trivial solution for every $\lambda\in\mathbb{R}$.

5. THE DISCRETE EIGENVALUE PROBLEM

Our aim is now to compute approximations to $\sigma(T)$. For that purpose, we construct a grid Ω_h as described above and a finite dimensional vector space

$$X_h = \begin{cases} \mathbb{R}^{\Omega_h}, & \text{if } R_0=R_0^2 \\ \{x\in\mathbb{R}^{\Omega_h}: x(0)=0\}, & \text{if } R_0=R_0^1 \end{cases}$$

We define an operator $A_h \in L[X_h]$ by

$$(A_h x)(t_i) = h^{-2} \begin{cases} 0, & \text{if } i=0 \text{ and } R_0=R_0^1 \\ (p(0.5h)+p(-0.5h)(1+2h\alpha))x(0)- \\ \quad (p(0.5h)+p(-0.5h))x(h), & \text{if } i=0 \text{ and } R_0=R_0^2 \\ -p(t_i-0.5h)x(t_{i-1})+ \\ (p(t_i-0.5h)+p(t_i+0.5h))x(t_i)- \\ p(t_i+0.5h)x(t_{i+1}), & \text{if } i\in\{1,\dots,M-1\} \\ -(p(1-0.5h)+p(1+0.5h))x(1-h)+ \\ \quad (p(1-0.5h)+p(1+0.5h))x(1), & \text{if } i=M \end{cases}$$

(we make the assumption that p has a positive continuation in some neighborhood of $t=1$ and, if $R_0=R_0^2$, in some neighborhood of $t=0$ too). Now let $B_h \in L[X_h, \mathbb{R}^{\Omega_h}]$ and let $C_h \in L[\mathbb{R}^{\Omega_h}, X_h]$ be given by

$$(C_h x)(t_i) = \begin{cases} 0, & \text{if } i=0 \text{ and } R_0=R_0^1 \\ x(t_i) & \text{otherwise}. \end{cases}$$

We consider the eigenvalue problem

$$A_h x = \lambda C_h B_h x. \tag{7}$$

As in the previous chapter, we define an endomorphism $T_h \in L[X_h]$ by $T_h = A_h^{-1} C_h B_h$. By endowing X_h with the cone $K_h = \{x \in X_h : x(0) \geq 0, x(t_{i+1}) - x(t_i) \geq 0 \text{ for } i=0,\dots,M-1\}$, X_h becomes a partially ordered vector space (see example 3.2). It is clear that T_h is completely continuous on X_h.

Theorem 2: Let B_h satisfy the condition

$$x \in K_h - \{\Theta\} \Rightarrow (B_h x)(t) \geq 0 \text{ for all } t \in \Omega_h \text{ and } (B_h x)(1) > 0.$$

Then T_h is monotone and strongly monotonic.

Hence, $\sigma(T_h)^{-1}$ is the minimal positive eigenvalue of (7).

Example 5.1:

Let B be the operator considered in example 4.1. Using the trapezoidal rule, we get the approximation

$$h[\,0.5K(t_i,0)(x(t_i)-x(0))+\sum_{j=1}^{i-1} K(t_i,t_j)(x(t_i)-x(t_j))\,] \tag{8}$$

for $(Bx)(t_i)$ $(i=1,\dots,M)$.

Now let $x\in X_h$ and let $(B_hx)(t_i)$ be given by formula (8) for $i=0,\dots,M$. $K(t,s)\geq 0$ $(t,s\in[0,1])$ and $K(1,0)>0$ imply $(B_hx)(t_i)\geq 0$ $(i=0,\dots,M)$ and $(B_hx)(1)>0$ for $x\in K_h-\{\Theta\}$. This enables us to compute approximations to the critical value at which the rod buckles out.

6. CONVERGENCE

For $x\in C$ we denote the restriction of x to the grid Ω_h by x_h. Suppose now that the assumptions on B and B_h made in Theorem 1 and Theorem 2 may hold.

As a consequence of (5) we get for every $e\in oK$ an error estimate

$$|\sigma(T)-\sigma(T_h)| \leq C_e \,\|T_hz_h-(Tz)_h\|_{e_h}, \tag{9}$$

where z is an order unit of K such that $Tz=\sigma(T)z$ and C_e is a positive number independent of h.

Theorem 3: Suppose

$$p\in\begin{Bmatrix} C^2[0,1+\varepsilon], & \text{if } R_o=R_o^1 \\ C^2[-\varepsilon,1+\varepsilon], & \text{if } R_o=R_o^2 \end{Bmatrix}$$

for some $\varepsilon>0$ and $p(t)>0$ for all t in the domain of p. Let z be an order unit of K such that $Tz=\sigma(T)z$ and let $h<2\varepsilon$.

Then $Bz\in C^2$ and $d_h:=\max\{|(Bz)_h(t)-(B_hz_h)(t)|:t\in\Omega_h\}=O(h)$ imply $|\sigma(T)-\sigma(T_h)|=O(h)$.

Furthermore, if $R_o=R_o^1$, the conditions

$$Bz \in C^3,\ p'(1)=(Bz)'(1)=0,\ d_h=O(h^2) \tag{10}$$

imply $|\sigma(T)-\sigma(T_h)|=O(h^2)$.

If $R_o=R_o^2$, and if additional to (10) we have $p'(0)=0$, then $|\sigma(T)-\sigma(T_h)|=O(h^2)$.

7. SOME NUMERICAL RESTULTS

Now let us return to the buckling rod. Here we have the boundary condition $R_o x=x(0)=0$. We assume that the cross-section of the rod is a circular disk of radius $\overline{r}(s)$. Then we have

$$F(s)=\pi(\overline{r}(sl))^2,\ I(s)=0.25\pi(\overline{r}(s\cdot l))^4\quad (s\in[0,1]). \tag{11}$$

Now let $r(s)=\overline{r}(sl)$ for $s\in[0,1]$. To make F and I independent of l, we fix r independent of l. In the examples treated below, r is always a polynominal of degree ≤ 2. To be able to compare different statures of a rod of length l, we fix its volume to be $\pi\cdot l$. With

$$K(s,t)=\frac{r(t)^2}{r(s)^4}$$

we define B_h as in example 5.1 and we compute the minimal positive eigenvalue λ_h of $A_h x=\lambda C_h B_h x$, where A_h and C_h are defined as in Chapter 5 (here we have $p\equiv 1$).

To get approximations for the minimal value of $(l^3\gamma)/E$ such that the rod buckles out, we have to divide λ_h by 4.

The inclusion (5) implies

$$|e^{n-1}|_{e^n} \leq |e^n|_{e^{n+1}} \leq \sigma(T_h)^{-1} \leq \|e^n\|_{e^{n+1}} \leq \|e^{n-1}\|_{e^n} \quad (n\geq 1) \tag{12}$$

for every sequence $e^o \in oK_h$, $e^{n+1}=T_h e^n$ $(T_h=A_h^{-1}C_hB_h)$. We compute the series $|e^n|_{e^{n+1}}$, $\|e^n\|_{e^{n+1}}$ until

$$\|e^k\|^{-1}_{e^{k+1}}(\|e^k\|_{e^{k+1}}-|e^k|_{e^{k+1}})\leq 0.5\cdot 10^{-3}$$

for some $k\in\mathbb{N}$.

The following table shows the last required values $\|e^k\|_{e^{k+1}}$ for different statures of a rod of length 1 (the headline of the table shows the longitudinal views of the statures).

h	① 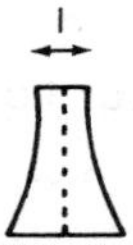	② 	③ 	④ 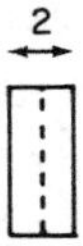
0.1	30.3592	29.9191	22.9716	7.7969
0.05	30.3666	29.6533	22.8571	7.8272
0.025	30.3681	29.5874	22.8283	7.8348
0.0125	30.3684	29.5710	22.8211	7.8367

For the examples 3 and 4 , Theorem 3 shows convergence of order 2: $|\sigma(T)-\sigma(T_h)|=O(h^2)$.

In the other cases we have convergence of order 1 (see Jäger [1982]).

REFERENCES:

1. BOHL, E., Linear operator equations on a partially ordered vector space. *Aequationes math.*, 4, (1970).

2. BOHL, E., *Monotonie: Lösbarkeit und Numerik bei Operatorgleichungen.* Springer tracts in natural philosophy 25, Berlin, Heidelberg, New York (1974).

3. COLLATZ, L., *Eigenwertaufgaben und ihre numerische Behandlung*. Akademische Verlagsgesellschaft Becker & Erler, Leipzig (1945).

4. FROBENIUS, G., Über Matrizen aus nicht negativen Elementen. *S.-B. Preuss.Akad.Wiss.Berlin* 456-477 (1912).

5. JENTZSCH, R., Über Integralgleichungen mit positivem Kern. *J.reine und angew. Math.* 141, 235-244 (1912).

6. JÄGER, E., *Ein numerisches Modell für die Stabknickung mit Berücksichtigung des Eigengewichts.* Diplomarbeit. Konstanz (1982).

7. KREIN-RUTMAN, Linear operators leaving invariant a cone in a Banach space. *Uspehi Math. Nauk SSSR* 3, 3-95 (1948). *American Math. Soc. Transl.* 29 (1950).

8. PERRON, O., Zur Theorie der Matrizen. *Math. Ann.* 64, 248-263 (1907).

THE FAST GALERKIN METHOD FOR THE SOLUTION OF CAUCHY SINGULAR INTEGRAL EQUATIONS

S. M. Hashmi and L. M. Delves

Department of Statistics and Computational Mathematics
University of Liverpool

1. INTRODUCTION

Cauchy type singular integral equations of Fredholm type are often encountered in problems of mathematical physics and their mathematical properties have been well investigated (see for example [8]). For numerical solution it is possible to reduce them to an equivalent Fredholm integral equation of the second kind, solving this by any numerical technique (see for example [7]).N.I. Ioakimidis and P.S. Theocaris [6] considered direct methods for the solution of Cauchy singular integral equations; after separating the dominant singular part, their equation may be expressed in the form:

$$A\phi(s) + \frac{B}{\pi}⨍_a^b \frac{\phi(t)}{t-s}\,dt + \int_a^b k(s,t)\,\phi(t)dt = h(s); \quad a \leqslant s \leqslant b \tag{1.1}$$

where h(s) and k(s,t) are known functions satisfying a Hölder condition on the interval [a,b], and A,B are real constants.

We present here a method for the numerical solution of Cauchy type singular integral equations of the form

$$\phi(s) + ⨍_a^b \frac{\phi(t)}{t-s}\,dt = h(s) \qquad a \leqslant s \leqslant b \tag{1.2}$$

which is a generalisation of the Fast Galerkin Method for normal integral equations [1,2]. We consider Fredholm, Volterra and inverse Volterra singular integral equations of Cauchy type; the extension of the method to equations of the more general form (1.1) is trivial.

1.1 *The finite part of an infinite integral*

The integral in (1.2) does not exist in the Riemann sense; for a Fredholm equation (a, b fixed) we interpret it here in the usual Cauchy sense:

$$⨍_a^b \frac{\phi(s)}{t-s} ds = \lim_{\varepsilon\to 0} \left[\int_a^{t-\varepsilon} + \int_{t+\varepsilon}^b \right] \frac{\phi(s)}{t-s} ds , \qquad a \leqslant t \leqslant b .$$

However, the Fast Galerkin method as extended in [2] and implemented as described in [3] treats Green's-function type operators (having kernels with a discontinuous derivative along the line t=s) as the sum of "Volterra" and "inverse-Volterra" operators; for Volterra and inverse-Volterra equations with Cauchy kernel, the singularity appears at one end of the range of integration, and the principal value integral does not exist. We give meaning to these integral equations in terms of the Hadamard finite part integral [5] for the integral

$$⨍_a^s \frac{A(t)}{s-t} dt \; ; \qquad a \leqslant s \leqslant b . \tag{1.3}$$

Hadamard definition:

The finite part of the infinite integral (1.3) is defined by adding a term $B(t)\log_e|s-t|$ and taking the limit as $t\to s$:

$$⨍_a^s \frac{A(z)}{(s-z)} dz = \lim_{t\to s} \left[\int_a^t \frac{A(z)}{(s-z)} dz + B(t) \log_e|s-t| \right] ,$$

where the function $A(z)$ is assumed to satisfy a Lipschitz condition, and $B(t)$ is a function satisfying the conditions:

(a) The limit must exist

(b) $B(t)$ has a continuous first derivative at least in the vicinity of $t=s$.

Here we make the consistent choice $B(t)$ = constant.

The following results then hold:

Lemma (1.1)

$$⨍_a^s \frac{dz}{z-s} = -\log_e|s-a| \quad ; \quad a \leqslant s \leqslant b.$$

$$⨍_s^b \frac{dz}{z-s} = \log_e|s-b| \quad ; \quad a \leqslant s \leqslant b$$

Proof: Follows from the definition with $A(z) = 1$.

We shall require within the formalism here to carry out a

change of variables; the Hadamard integral (of integral order) is not invariant under such a transformation (see reference [5] pages 137-138). Carrying out a linear transformation explicitly we find:

Lemma (1.2)

$$\int_{-1}^{x} \frac{dy}{y-x} = -\log_e \left|\frac{b-a}{2}(x+1)\right| \quad ; \quad -1 \leqslant x \leqslant 1$$

$$\int_{x}^{1} \frac{dy}{y-x} = \log_e \left|\frac{b-a}{2}(x-1)\right| \quad ; \quad -1 \leqslant x \leqslant 1.$$

Proof: Follows from Lemma (1.1) by mapping the variables onto the standard interval [-1, 1].

2. OUTLINE OF THE ALGORITHM

We consider the numerical solution of Cauchy type singular integral equations over a finite interval [a, b] of the form:

(a) Volterra:

$$\phi(s) + \int_{a}^{s} \frac{\phi(t)}{t-s} dt = h(s) \quad ; \qquad a \leqslant s \leqslant b \tag{2.1a}$$

(b) Inverse-Volterra:

$$\phi(s) + \int_{s}^{b} \frac{\phi(t)}{t-s} dt = h(s) \quad ; \qquad a \leqslant s \leqslant b \tag{2.1b}$$

(c) Fredholm:

$$\phi(s) + \int_{a}^{b} \frac{\phi(t)}{t-s} dt = h(s) \quad ; \qquad a \leqslant s \leqslant b. \tag{2.1c}$$

Let us map the variables (t,s) onto the finite interval [-1, 1] by setting

$$t = \frac{b-a}{2}y + \frac{b+a}{2} \; ; \qquad s = \frac{b-a}{2}x + \frac{b+a}{2} \; ; \qquad x, y \in [-1,1].$$

Substituting these in the above equations we get:

$$(a^1) \quad f(x) + \int_{-1}^{x} \frac{f(y)}{y-x} dy = g(x) \quad ; \quad -1 \leqslant x \leqslant 1 \tag{2.2a}$$

$$(b^1) \quad f(x) + \int_{x}^{1} \frac{f(y)}{y-x} dy = g(x) \quad ; \quad -1 \leqslant x \leqslant 1 \tag{2.2b}$$

$$(c^1)\quad f(x) + \fint_{-1}^{1} \frac{f(y)}{y-x}\,dy = g(x) \quad ; \quad -1 \leqslant x \leqslant 1 . \tag{2.2c}$$

We approximate the solution $f \in L_2$ by its truncated Chebyshev expansion:

$$f(x) \simeq F_N(x) = \sum_{j=0}^{N} a_j \, T_j(x) \qquad -1 \leqslant x \leqslant 1 . \tag{2.3}$$

The fast Galerkin method computes the coefficients a_j as the solution of the equations

$$(D+B)\,\underline{a} = \underline{g} \tag{2.4}$$

where D is the diagonal matrix with elements

$$D_{ij} = \int_{-1}^{1} \frac{T_i(x)}{\sqrt{1-x^2}} T_j(x)\,dx = \pi \begin{cases} 1 & i=j=0 \\ \frac{1}{2} & i=j>0 \\ 0 & i \neq j \end{cases} \tag{2.5}$$

$$i, j = 0(1)\ N$$

$$g_i = \int_{-1}^{1} \frac{T_i(x)}{\sqrt{1-x^2}}\, g(x)\,dx \; ; \qquad i = 0(1)\ N \tag{2.6}$$

$$B_{i,j}(\text{Volterra}) = \int_{-1}^{1} \frac{T_i(x)}{\sqrt{1-x^2}} \fint_{-1}^{x} \frac{T_j(y)}{y-x}\,dy\,dx \tag{2.7a}$$

$$B_{i,j}(\text{Inverse-Volterra}) = \int_{-1}^{1} \frac{T_i(x)}{\sqrt{1-x^2}} \fint_{x}^{1} \frac{T_j(y)}{y-x}\,dy\,dx \tag{2.7b}$$

$$B_{i,j}(\text{Fredholm}) = \int_{-1}^{1} \frac{T_i(x)}{\sqrt{1-x^2}} \fint_{-1}^{1} \frac{T_j(y)}{y-x}\,dy\,dx . \tag{2.7c}$$

We use the Fast Fourier Transform (FFT) techniques described in [1] for approximating the integral in (2.6); but for evaluating the matrix B, it is essential for accuracy to use analytic methods. We shall produce recurrence relations to evaluate the Volterra and inverse Volterra matrices; we could then calculate the Fredholm matrix as the sum of Volterra and inverse Volterra matrices, but it proves possible (and more accurate) to compute the Fredholm matrix directly.

3. FREDHOLM OPERATOR

From (2.7c) we have see [5] :

$$B_{i,j}=\int_{-1}^{1}⨍_{-1}^{1}\frac{T_i(x)}{\sqrt{1-x^2}}\frac{T_j(y)}{y-x}\,dy\,dx$$

$$= -\int_{-1}^{1}T_j(x)\,⨍_{-1}^{1}\frac{T_i(y)}{\sqrt{1-y^2}}\frac{dy}{y-x}\,dx$$

$$= -\int_{-1}^{1}T_j(x)⨍_{-1}^{1}\frac{T_i(y)}{\sqrt{1-y^2}}\frac{dy}{y-x}\,dx.$$

But (see for example [9, p.180])

$$⨍_{-1}^{1}\frac{T_i(y)}{\sqrt{1-y^2}}\frac{dy}{y-x} = \begin{cases}\pi U_{i-1}(x) & i>0\\ 0 & i=0\end{cases} \qquad |x|<1,$$

where $U_i(x)$ is a second kind Chebyshev polynomial.

Hence:

$$B_{ij} = -\pi\int_{-1}^{1}T_j(x)\,U_{i-1}(x)\,dx; \qquad i>0$$

$$= 0 \qquad\qquad i=0$$

$$= \begin{cases}\dfrac{2\pi i}{j^2-i^2} & |j-1|\text{ odd}\\ 0 & \text{otherwise.}\end{cases}$$

4. VOLTERRA AND INVERSE-VOLTERRA OPERATORS

For the Volterra operator we have from (2.7a)

$$B_{i,j} = \int_{-1}^{1}w(x)\,T_i(x)\,⨍_{-1}^{x}\frac{T_j(y)}{y-x}\,dy\,dx;\ w(x) = (1-x^2)^{-\frac{1}{2}}. \qquad (4.1)$$

Using the identity

$$T_j(y) = 2y\,T_{j-1}(y) - T_{j-2}(y) \quad ,\ j \geq 2 \qquad (4.2)$$

it follows that

$$B_{i,j} = 2\int_{-1}^{1} w(x)\ T_i(x) \fint_{-1}^{x} \frac{y}{y-x}\ T_{j-1}(y)\ dy\ dx$$

$$- \int_{-1}^{1} w(x)\ T_i(x) \fint_{-1}^{x} \frac{T_{j-2}(y)}{y-x}\ dy\ dx\ ; \qquad j \geq 2$$

$$= 2\int_{-1}^{1} w(x)T_i(x)\int_{-1}^{x} T_{j-1}(y)\ dy\,dx + 2\int_{-1}^{1} xT_i(x)\ w(x)\fint_{-1}^{x} \frac{T_{j-1}(y)}{y-x}\ dy\ dx$$

$$- \int_{-1}^{1} w(x)T_i(x) \fint_{-1}^{x} \frac{T_{j-2}(y)}{y-x}\ dy\ dx.$$

Using (4.2) then we have:

$$B_{i,j} = 2\int_{-1}^{1} w(x)\ T_i(x)\ dx \int_{-1}^{x} T_{j-1}(y)\ dy + B_{i+1,j-1} + B_{i-1,j-1}$$

$$- B_{i,j-2} \quad (i \geq 1,\ j \geq 2)$$

But

$$\int_{-1}^{x} T_{j-1}(y)\ dy = \tfrac{1}{2} \left[\frac{T_j(y)}{j} - \frac{T_{j-2}(y)}{j-2}\right]_{-1}^{x} \qquad j \geq 3 \qquad (4.3)$$

and hence

$$B_{i,j} = \frac{1}{j} \int_{-1}^{1} w(x)\ T_i(x)\ T_j(x)\ dx - \frac{1}{j-2} \int_{-1}^{1} w(x)\ T_i(x)\ T_{j-2}(x)\ dx$$

$$- \frac{(-1)^j}{j} \int_{-1}^{1} w(x)\ T_i(x)\ dx + \frac{(-1)^{j-2}}{j-2} \int_{-1}^{1} w(x)\ T_i(x)\ dx$$

$$+ B_{i+1,j-1} + B_{i-1,j-1} - B_{i,j-2} \qquad (i \geq 1,\ j \geq 3).$$

Now, using the identity

$$\int_{-1}^{1} w(x)\ T_i(x) = \begin{cases} \pi & i = 0 \\ 0 & i \neq 0 \end{cases} \qquad (4.4)$$

we end up with the general recurrence relation

$$B_{i,j} = \frac{1}{j} D_{i,j} - \frac{1}{j-2} D_{i,j-2} + B_{i+1,j-1} + B_{i-1,j-1}$$

$$- B_{i,j-2} \qquad (i \geq 1,\ j \geq 3) \quad (4.5)$$

where D is a diagonal matrix with elements as in (2.5).

Equations (4.2, 4.3 and 4.4) lead directly to a recurrence relation for computing the first row of the matrix B:

$$B_{0,j} = \int_{-1}^{1} \frac{1}{\sqrt{1-x^2}} ⨍_{-1}^{x} \frac{T_j(y)}{y-x}\, dy\, dx$$

$$= 2\pi \left(\frac{(-1)^{j-2}}{j(j-2)}\right) + 2\, B_{1,j-1} - B_{0,j-2} \quad (j \geq 3). \qquad (4.6)$$

Now for computing the first column of the matrix we use the Hadamard definition for evaluating the finite part of the infinite integral:

$$B_{i,0} = \int_{-1}^{1} \frac{T_i(x)}{\sqrt{1-x^2}} ⨍_{-1}^{x} \frac{dy}{y-x}\, dx\, dy\, .$$

Applying Lemma (1.2) we find:

$$B_{i,0} = ⨍_{-1}^{1} \frac{T_i(x)}{\sqrt{1-x^2}} \left(- \text{Log}_e \left| \frac{b-a}{2} (x+1) \right|\right) dx$$

$$= - \text{Log}_e \left| \frac{b-a}{2} \right| \int_{-1}^{1} \frac{T_i(x)}{\sqrt{1-x^2}}\, dx - ⨍_{-1}^{1} \frac{T_i(x)}{\sqrt{1-x^2}} \text{Log}_e \left| x + 1 \right| dx.$$

But

$$⨍_{-1}^{1} \frac{T_{i}(y)}{\sqrt{1-y^2}} \text{Log}_e \left| x - y \right| dy = -\pi \begin{cases} \text{Log}_e |2| & ; \quad i = 0 \\ \frac{1}{i} T_i(x) & ; \quad i > 0 \end{cases} \qquad (4.7)$$

From identity (4.4) and (4.7) we find

$$B_{i,0} = \pi \begin{cases} \text{Log}_e \left| \frac{4}{b-a} \right| & ; \quad i = 0 \\ \frac{(-1)^i}{i} & ; \quad i > 0 \end{cases} \qquad (4.8)$$

We can now calculate the second and third columns of the matrix from the recurrence relations:

$$B_{i,1} = \tfrac{1}{2}\,(B_{i+1,0} + B_{i-1,0}) + \frac{\pi}{2}\;\begin{matrix}1\\0\end{matrix}\begin{pmatrix}i=1\\i>1\end{pmatrix} \;;\quad i \geq 1 \tag{4.9}$$

$$B_{i,2} = B_{i+1,1} + B_{i-1,1} - B_{i,0} + \frac{\pi}{4}\;\begin{matrix}1\\0\end{matrix}\begin{pmatrix}i=2\\i\neq 2\end{pmatrix} \;;\quad i \geq 1 . \tag{4.10}$$

Hence we can summarize the algorithm for computing the matrix elements $B_{i,j}$ as follows:

step (1) $$B_{i,0} = \pi \begin{cases} \mathrm{Log}_e \left|\dfrac{4}{b-a}\right| ; & i = 0 \\ \dfrac{(-1)^i}{i} ; & i > 0 \end{cases}$$

step (2) $$B_{0,1} = \pi + B_{1,0} ;$$

$$B_{i,1} = \tfrac{1}{2}\,(B_{i+1,0} + B_{i-1,0}) + \frac{\pi}{2}\;\begin{matrix}1\\0\end{matrix}\begin{pmatrix}i=1\\i>1\end{pmatrix} \;;\quad i \geq 1$$

step (3) $$B_{0,2} = 2B_{1,1} - B_{0,0} - \frac{\pi}{2} ;$$

$$B_{i,2} = B_{i+1,1} + B_{i-1,1} - B_{i,0} + \frac{\pi}{4}\;\begin{matrix}1\\0\end{matrix}\begin{pmatrix}i=2\\i\neq 2\end{pmatrix} ;\quad i \geq 1$$

step (4) $$B_{0,j} = 2\pi \left(\frac{(-1)^{j-2}}{j(j-2)}\right) + 2B_{1,j-1} - B_{0,j-2} \;;\quad j \geq 3$$

step (5) $$B_{i,j} = \frac{1}{j}\, D_{i,j} - \frac{1}{j-2}\, D_{i,j-2} + B_{i+1,j-1} + B_{i-1,j-1} - B_{i,j-2} \;;\quad (i \geq 1,\ j \geq 3)$$

where

$D_{i,j}$ and $D_{i,j-2}$ are given by (2.5) .

Recurrence relations for the Inverse-Volterra matrix (2.7b) follow similarly or on using the obvious identity:

$$B_{i,j}\ (\text{Inverse-Volterra}) = (-1)^{i+j+1}\, B_{i,j}\ (\text{Volterra})$$

which follows on inserting the change of variables $x=-s, y=-t$ and using the identity

$$T_i(-s) = (-1)^i\, T_i(s)$$

5. NUMERICAL EXAMPLES

These analytic results for the matrix B are all that is required to extend the Fast Galerkin formalism; all other aspects of the algorithm go through unchanged, including error estimates - for details, see [2]. Within the implementation (Fag_1) described in [3] it is necessary only to add these matrices to the "singular-kernel" library. We report here numerical results obtained from the package Fag 1 in this way, on an IBM4341 computer. The problems presented are of two types:

(i) Problems with smooth solution

(ii) Problems whose solution, although analytic, has singularities close to the integration region.

For each problem, we take an interval $a \leqslant x \leqslant b$ and report results for each of the three integral equations: Fredholm (F), Volterra (V), Inverse-Volterra (IV). The problems are then defined by equation (2.2c, a, b) respectively; and by the stated solutions f(x) and driving term g(x). The numerical solution is defined by the degree (N-1) of the approximate polynomial used.

Problem (1)

$$f(x) = x \; ; \qquad [a, b] = [-1, 1]$$

$$g(x) = (2x+1) - x \operatorname{Log}_e |x+1| \qquad (V)$$

$$= 1 + x \operatorname{Log}_e |x-1| \qquad (IV)$$

$$= (x+2) + x \operatorname{Log}_e \left| \frac{1-x}{1+x} \right| \qquad (F)$$

For this problem we expect to obtain the exact solution, apart from round-off errors, provided that $N \geqslant 2$. We obtain in fact 16-17 significant figure accuracy on the IBM 4341 computer using 64 bit reals.

Problem (2)

$$f(x) = (x+\alpha)^{-1} \; ; \qquad [a,b] = [-1,1]$$

$$g(x) = (x+\alpha)^{-1} \left(1 + \operatorname{Log}_e \left| \frac{\alpha-1}{\alpha+x} \right| \right) - (x+\alpha)^{-1} \operatorname{Log}_e |x+1| \quad (V)$$

$$= (x+\alpha)^{-1} \left(1 + \operatorname{Log}_e \left| \frac{x+\alpha}{\alpha+1} \right| \right) + (x+\alpha)^{-1} \operatorname{Log}_e |x-1| \quad (IV)$$

$$= (x+\alpha)^{-1}\left(1 - \mathrm{Log}_e\left|\frac{\alpha+1}{\alpha-1}\right|\right) + (x+\alpha)^{-1}\,\mathrm{Log}_e\left|\frac{1-x}{1+x}\right| \quad (F)$$

For this problem we assume that α lies outside the range $[-1,1]$, and expect to obtain "rapid" convergence provided that α is not close to this interval, with convergence being slower as $|\alpha|$ approaches 1. The results obtained are shown graphically in Figures (1-3), which shows that the convergence is, as expected, representable in the form

$$||\, f - f_N \,|| \sim CA^N$$

apart from an odd-even effect in the Fredholm case, with the convergence parameter A approaching 1 as α approaches the integration region. The results were obtained on the IBM 4341 (64 bit reals).

Problems (3, 4)

As problem (2), but with

$$[a,b] = [-2,\tfrac{1}{2}] \qquad \text{problem (3)}$$

$$= [0,1] \qquad \text{problem (4)}.$$

For these problems we present results obtained on the IBM 4341 computer for the Fredholm operator. These are given in table (1) for values of α close to, and well away from, the integration region. It is clear that the accuracy achieved is not sensitive to the interval used, but only to the distance from the nearest singularity of the solution.

6. CONCLUSIONS

These results show how easily the Fast Galerkin framework handles even strongly singular problems; the accuracy achieved is independent of the strength of the singularities in the equation, and depends only on the smoothness of the solution. An additional advantage of the formalism, not shown here but discussed in [2] is that of handling problems whose kernel contains a singular factor together with an additive or multiplicative smooth term. The Volterra and Inverse Volterra results are perhaps amusing rather than of immediate practical significance, but kernels of Fredholm type are of common occurrence, and the technique used seems to be well suited to these.

Table 1

Computed error = $||f - fN||_{\infty}$

for Problems 3 and 4

N	Problem 3				Problem 4			
	$\alpha = 2.1$		$\alpha = 9.0$		$\alpha = 1.1$		$\alpha = 9.0$	
3	4.3208	10^{-1}	3.0979	10^{-4}	6.9934	10^{-3}	3.2784	10^{-5}
4	9.2606	-1	4.2196	-5	2.2973	-3	1.4320	-6
5	1.6922	-1	1.6536	-6	1.8003	-4	1.9748	-8
6	3.1608	-1	1.8061	-7	4.3501	-5	7.2544	-10
7	7.7263	-2	9.1731	-9	4.5240	-6	1.2917	-11
8	1.2345	-1	9.2333	-10	9.9007	-7	4.4056	-13
9	3.4902	-2	5.1559	-11	1.1327	-7	8.6217	-15
10	4.7674	-2	4.8700	-12	2.3119	-8	3.3961	-16
11	1.4038	-2	2.8295	-13	2.7411	-9	5.5485	-17
12	1.7069	-2	2.5037	-14	5.2492	-10	8.0938	-17
13	6.1197	-3	1.4839	-15	6.6564	-11	9.5895	-17
14	7.2625	-3	2.2365	-16	1.2784	-11	1.2167	-16
15	3.5254	-3	8.6017	-17	1.7909	-12	7.1339	-17
16	3.7019	-3	1.1297	-16	3.4573	-13	1.2329	-16
17	1.5954	-3	1.1207	-16	4.5651	-14	1.2993	-16
18	1.3569	-3	1.3438	-16	8.6908	-15	1.6655	-16

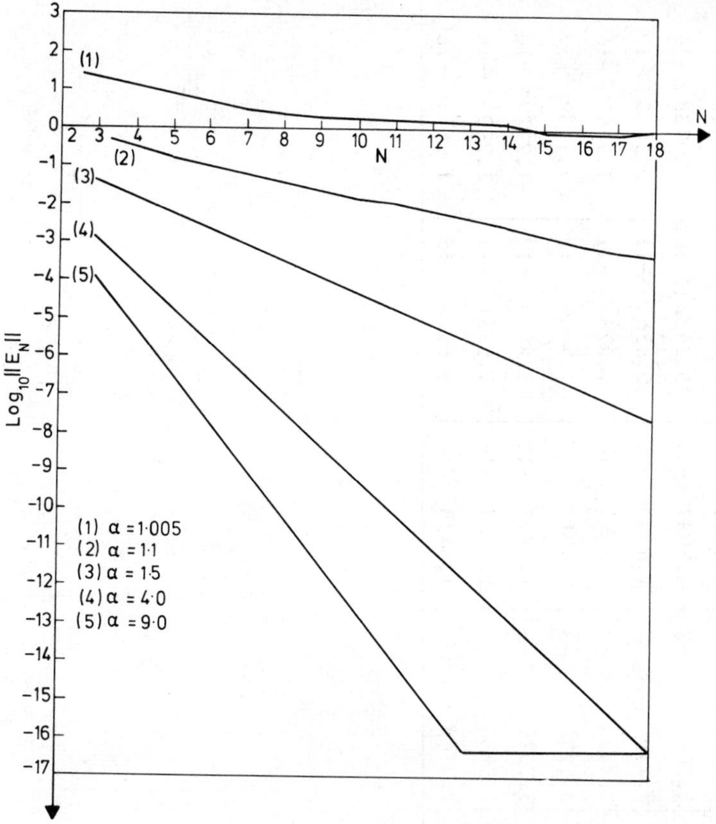

FIG. 1. Problem 2(V)

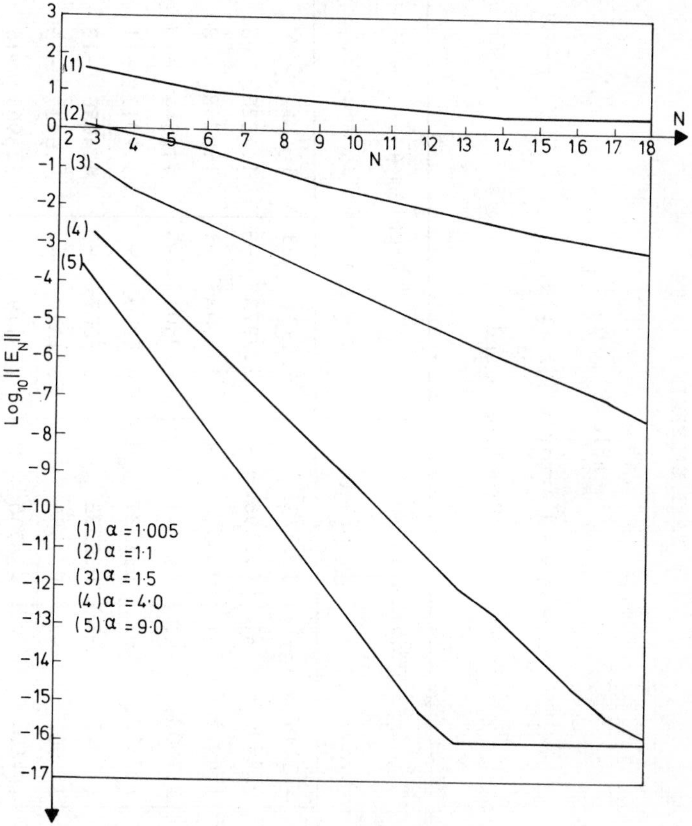

FIG. 2. Problem 2(IV)

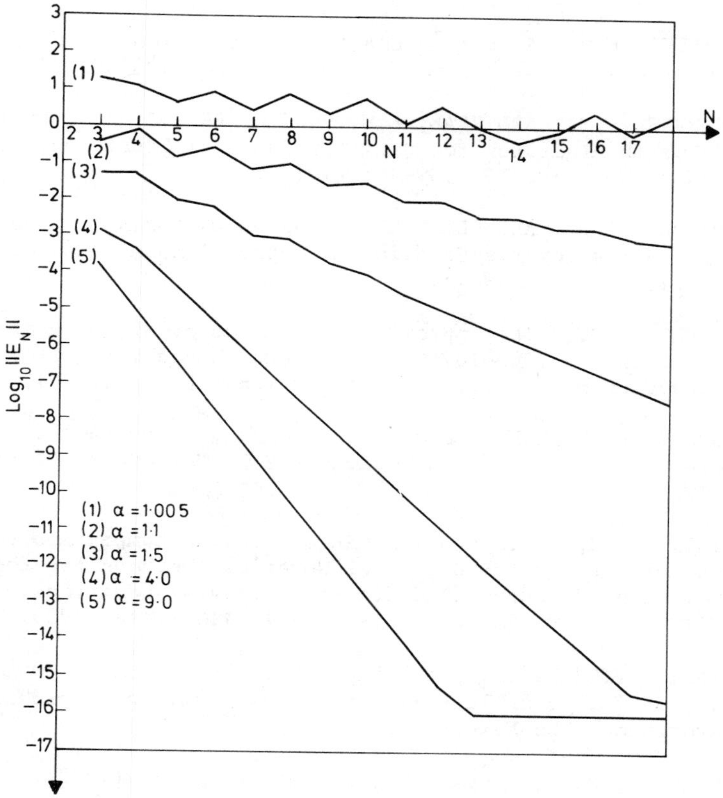

FIG. 3. Problem 2(F)

REFERENCES

[1] DELVES, L.M., A Fast Method for the Solution of Integral Equations. *J. Inst. Maths. Applics.* 20, 173-182 (1977a).

[2] DELVES, L.M., ABD-ELAL, L.F., and HENDRY, J.A., A Fast Galerkin Algorithm for Singular Integral Equations. *J. Inst. Maths. Applics.* 23, 139-166 (1979).

[3] DELVES, L.M., ABD-ELAL, L.F., and HENDRY, J.A., A Set of Modules for the Solution of Integral Equations. *Computer J.* 24, 184-190 (1981).

[4] ELLIOTT, D., The Approximate Solution of Singular Integral Equations. pp. 83-107 of M.A. GOLDBERG (ed), *Solution Methods for Integral Equations.* Plenum Press (1978).

[5] HADAMARD, J.LL.D., *Lectures on Cauchy's Problems in Partial Differential Equations.* Yale University Press (1923).

[6] IOAKIMIDIS, N.I., and THEOCARIS, P.S., A Comparison between the Direct and the Classical Numerical Methods for the Solution of Cauchy type Singular Integral Equations. *SIAM J. Numer. Anal. Vol.* 17 *No.* 1, 115 - 118 (1980).

[7] KANTOROVICH, L.V., and KRYLOV, V.I., *Approximate Methods of Higher Analysis,* Inter-science, New York, and Noordhoff, Groningen, The Netherlands (1958).

[8] MIKHLIN, S.G., *Integral Equations.* Pergamon (1957).

[9] TRICOMI, F.G., *Integral Equations.* Inter-science, London (1957).

[10] ZABREYKO, P.P., et. al, *Integral Equations* - a reference text, (trans. R.S. Anderson et al). Noordhoff International, (1975).

THE STABILITY OF A NUMERICAL METHOD FOR A SECOND KIND ABEL EQUATION

D. Kershaw

University of Lancaster

The analysis of the numerical procedure given above in [2] will be completed by showing that when it is applied to the test equation

$$f_0(x) = 1 - \frac{\lambda}{\Gamma(\alpha)}\int_0^x \frac{f_0(t)}{(x-t)^{1-\alpha}}\,dt, \quad 0 < \alpha < 1, \lambda > 0 \tag{1}$$

then $\lim_{n\to\infty} f_n = 0$ where f_n is the approximation to $f_0(x_n)$. In [2] equ.(4.6) set $g(x) = 1$ and $K[x,t,f] = -\lambda f$ (note that λ here should not be confused with the one which occurs in the final section of [2]), then

$$\left[1 + \frac{\lambda h^\alpha}{\Gamma(\alpha+2)}\right] f_n = 1 - \frac{\lambda h^\alpha}{\Gamma(\alpha+2)}\left[(\alpha+1)n^\alpha+(n-1)^{\alpha+1}-n^{\alpha+1}\right] -$$

$$- \frac{\lambda h^\alpha}{\Gamma(\alpha+2)}\sum_{r=1}^{n-1} f_r\left[(n-r+1)^{\alpha+1}-2(n-r)^{\alpha+1}+(n-r-1)^{\alpha+1}\right],\ n=1,2,.... \tag{2}$$

Define a generating function ϕ formally by

$$\phi(z) = \sum_{n=1}^{\infty} f_n z^{n-1}; \tag{3}$$

the remainder of this note will be devoted to a proof that $\phi(e^{i\theta})$ is integrable on $(0,2\pi)$, $\phi(z)$ is regular for $|z| <1$, and possesses but one integrable singularity on $|z|= 1$. For such a function, the Fourier coefficients decay, *that is*, $f_n \to 0$ as $n\to\infty$.

Define the function F_α by

$$F_\alpha(z) = \frac{1}{\Gamma(\alpha)}\sum_{n=1}^{\infty} n^{\alpha-1}z^{n-1}, \tag{4}$$

this is a special case of the Φ function described in [1] p.27 etc. The series converges for $|z| < 1$. The following results will be needed:

$$F_\alpha(z) = \frac{\sin\pi\alpha}{\pi}\int_0^\infty \frac{t^{-\alpha}}{(e^t-z)}\,dt, \quad 0 < \alpha < 1 \tag{5}$$

and

$$zF_\alpha(z) = (-\log z)^\alpha + \frac{1}{\Gamma(\alpha)} \sum_{r=0}^{\infty} \zeta(1-r-\alpha\ ,1)\frac{(\log z)^r}{r!}, \alpha \neq 0,-1,-2,\ldots \quad (6)$$

The plane is cut along the real axis from z=1 to z=∞ in order to make F_α analytic. It is not difficult to show with the aid of (2) that

$$\phi(z) = \frac{1-\lambda h^\alpha(1-z)[F_{\alpha+1}(z)-(1-z)F_{\alpha+2}(z)]}{(1-z)[1+\lambda h^\alpha(1-z)^2F_{\alpha+2}(z)]} . \quad (7)$$

We require integral representations of $F_{\alpha+1}$ and $F_{\alpha+2}$, both for $0 < \alpha < 1$. To this end note that

$$\alpha . F_{\alpha+1}(z) = \frac{d}{dz} z\, F_\alpha(z) = F_\alpha(z) + z\frac{d}{dz} F_\alpha(z),$$

and so from (5) we have

$$\alpha . F_{\alpha+1}(z) = \frac{\sin\pi\alpha}{\pi} \int_0^\infty \left[\frac{t^{-\alpha}}{e^t-z} + \frac{zt^{-\alpha}}{(e^t-z)^2}\right] dt .$$

Integrate by parts to give the result that

$$(1-z)F_{\alpha+1}(z) = \frac{\sin\pi\alpha}{\pi} \int_0^\infty t^{-(\alpha+1)} . \frac{e^t-1}{e^t-z} . dt, \quad 0 < \alpha < 1. \quad (8)$$

In a similar fashion we can show that

$$(1-z)^2 F_{\alpha+2}(z) = \frac{\sin\pi\alpha}{\pi} \int_0^\infty t^{-(\alpha+2)}\left[t-(1-z) \frac{e^t-1}{e^t-z}\right] dt. \quad (9)$$

Consequently

$$F_{\alpha+1}(z)-(1-z)F_{\alpha+2}(z) = \frac{\sin\pi\alpha}{\pi} \int_0^\infty t^{-(\alpha+1)}\left[\frac{e^t-1}{t} - 1\right] \frac{dt}{e^t-z}. \quad (10)$$

Clearly this function is continuous on the unit circle with the possible exception of the point z=1. However the representation (6) can be used to show that

$$\lim_{z\to 1}(1-z)^\alpha[F_{\alpha+1}(z)-(1-z)F_{\alpha+2}(z)] = \frac{1}{2}, \quad 0 < \alpha < 1, \quad (11)$$

and so we can assert that the numerator of (7) is continuous on $|z| = 1$.

Similarly from (9) we see that $(1-z)^2F_{\alpha+2}(z)$ is also continuous on the unit circle with the possible exceptions of the point z = 1, but again with the aid of (6) we can show that

$$\lim_{z\to 1} (1-z)^{\alpha+2}F_{\alpha+2}(z) = 1. \quad (12)$$

We now assert that if $\lambda>0$ then

$$1 + \lambda h^{\alpha}(1-z)^2 F_{\alpha+2}(z) \tag{13}$$

cannot vanish on the unit circle. We see from (9) that the imaginary part of this expression is a constant multiple of

$$y \int_0^{\infty} t^{-(\alpha+2)} \frac{(1-e^t)^2}{|e^t-z|^2} dt \tag{14}$$

which can vanish only if $y=0$.

Consider the integrand of (9) when $z=x$. Remove the positive factor $t^{-(\alpha+2)}/(e^t-x)$ to give for examination

$$g(t) = t(e^t-x) - (1-x)(e^t-1). \tag{15}$$

Now $g(0) = 0$, and

$$g'(t) = te^t + x(e^t-1) = t(1+x) + \frac{t^2}{1!}(1 + \frac{x}{2}) + \ldots$$

it follows that if $|x| < 1$ then $g'(t) > 0$ and so the integrand is positive for $t > 0$ if $|x| \leqslant 1$ which proves the assertion.

It will have been noticed that (12) shows that (13) is unbounded at $z = 1$, however we can write the denominator of ϕ as

$$(1-z)^{1-\alpha}[(1-z)^{\alpha} + \lambda h^{\alpha}(1-z)^{2+\alpha} F_{\alpha+2}(z)]. \tag{16}$$

The expression inside the square brackets is continuous on the unit circle and also does not vanish there. Hence we can write

$$\phi(z) = (1-z)^{\alpha-1}. \ F(z) \tag{17}$$

where F is continuous on $|z| = 1$; since $(1-z)^{\alpha-1}$ is integrable on $|z| = 1$ we have proved what was required, namely that ϕ is integrable on the unit circle and has the other properties required.

ACKNOWLEDGEMENT

The author is indebted to Dr. P.L. Walker of the University of Lancaster and to Christopher T. H. Baker for discussions on the underlying principles.

REFERENCES

1. ERDELYI, A. et al. *Higher Transcendental Functions*, volume 1. McGraw-Hill, New York (1953).

2. KERSHAW, D., Some results for Abel-Volterra integral equations of the second kind. *These proceedings*.

ON APPLICATIONS AND THE CONVERGENCE OF BOUNDARY INTEGRAL METHODS

Wolfgang L. Wendland

Technische Hochschule Darmstadt, Federal Republic Germany

1. INTRODUCTION

Although the reduction of elliptic boundary value problems to equivalent integral equations on the boundary represents historically the earliest method of corresponding mathematical analysis,its numerical exploitation has been developed only more recently creating many activities in computational mathematics and engineering from several different fields to implement boundary integral methods. Therefore the corresponding integral equations form now a much larger class than the classical Fredholm integral equations of the second kind with weakly singular kernels. They contain singular integral equations with Cauchy respectively Giraud kernels in elasticity and thermoelasticity, Fredholm integral equations of the first kind with weakly singular kernels as in elasticity, flow problems, electrostatics and conformal mapping and integrodifferential operators with nonintegrable kernels as in acoustics or elasticity. Whereas in classical analysis these types of equations have been treated differently, modern Fourier analysis for pseudodifferential operators allows us to formulate unifying properties which provide also an analysis of numerical methods for their approximate solution. We shall consider here the class of strongly elliptic boundary integral equations which provide coerciveness in form of Gårding inequalities. In special cases coercivity can also be obtained with potential theoretic methods as in [26],[31],[41],[42] but here we prefer the more general approach via pseudodifferential operators For the numerical treatment of these equations we formulate asymptotic error estimates in terms of the diminishing mesh width h of families of boundary elements. These asymptotic error estimates are in rather good agreement with numerical experiments and can also be used for making decisions concerning the choice of numerical quadrature formulas [4],[24],[57] or the choice of orders of finite element spaces approximating the boundary and the desired charges [19],[39],[58].

In Section 2 we give a brief survey of the asymptotic error results. As in [51] we have for strongly elliptic equations quasi-optimality for Galerkin's method in the energy space and superapproximation due to Aubin-Nitsche duality [27]. In two dimensions, i.e. equations on curves we present corresponding results for collocation methods with odd degree splines [3] and even degree splines [49]. In Section 3 we present four different types of strongly elliptic boundary integral equations belonging to engineering applications.

The equations considered are systems of the form

$$Au + B\omega = f\ ,\quad \Lambda u = b\ ,\quad \text{in short}\ \ \mathcal{A}(u,\omega) = (f,b) \tag{1.1}$$

where A is a $p\times p$ matrix of linear operators mapping the p-vector valued functions u on Γ into p-vector valued functions. B is a given $p\times q$ matrix of suitable smooth functions on Γ ,and Λ is a given $q\times p$ matrix of linear functionals. f , a p-vector valued function and $b\in\mathbb{R}^q$ are given, u and $\omega\in\mathbb{R}^q$ are the unknowns. Γ is a given compact $n-1$ dimensional sufficiently smooth manifold in $\mathbb{R}^n$, $n=2$ or 3 .

2. BOUNDARY ELEMENT METHODS

The numerical solution of equations (1.1) by using finite element approximations on Γ yields the boundary element methods. Here we shall consider Galerkin's method as well as classical collocation. In both methods u is to be approximated by a finite linear combination

$$u_h = \sum_{\ell=1}^{N} \gamma_\ell\ \mu_\ell \tag{2.1}$$

of the basis trial functions μ_ℓ forming the N-dimensional trial space H_h where $h \sim N^{-(n-1)}$ will denote the meshwidth. Γ is given by local representations $\Gamma : x = \underset{\sim}{x}(t)$ such that partitions in the parameter domains are mapped onto corresponding partitions of Γ . On partitions in the parameter domains we use a $(m+1,m)$-system of finite elements [8]. Then the local representations of Γ transplant these finite element functions onto Γ . Boundary integrals can then be evaluated in the local coordinates in which these finite elements appear as simple functions. However, in higher dimensions as for $n = 3$ the parameter representations of Γ can be rather involved. Hence in this case an additional approximation of the surface representation is mostly used leading to isoparametric elements on Γ as well as to a boundary approximations [19], [39], [41]. For the finite element $(m+1,m)$-system we have the

Approximation property [8],[21] : *If* $\infty < t \leq s \leq m+1$ *and* $t \leq m$ *then there exists a constant* c *depending only on* t,s,m *such that*

$$\inf_{\chi \in H_h} \| u-\chi \|_t \leq ch^{s-t} \| u \|_s \quad \text{for} \quad u \in H^s . \tag{2.2}$$

In case n=2 (2.2) *holds also for* $t < m + \frac{1}{2}$ [3].

The well known Galerkin procedure with (1.1), (2.1) now is to find the coefficients $\gamma_1,\dots,\gamma_N$ and ω_h by solving the finite system of linear equations

$$\begin{aligned} &\sum_{\ell=1}^{N} \gamma_\ell (A\mu_\ell,\mu_k) + (B\omega_h,\mu_k) = (f,\mu_k) \quad \text{and} \\ &\sum_{\ell=1}^{N} \gamma_\ell \Lambda\mu_\ell = b \quad \text{for} \quad k = 1,\dots,N \end{aligned} \tag{2.3}$$

where $(f,g) = \int_\Gamma f\bar{g}\, ds$ denotes the L_2-scalar product. By P_h we shall denote the $L_2 \times \mathbb{R}^q$-orthogonal projection onto $H_h \times \mathbb{R}^q$.

For the ordinary collocation method let $x_k \in \Gamma$ be N appropriately chosen collocation points forming a unisolvent set $\tilde{\Delta} = \{x_k\}$ with respect to H_h. The corresponding interpolation operator let us denote by I_h. Then the collocation method is to find $u_h \in H_h$ and $\omega_h \in \mathbb{R}^q$ by solving

$$\begin{aligned} &\sum_{\ell=1}^{N} \gamma_\ell A\mu_\ell(x_k) + B(x_k)\omega_h = f(x_k) \quad \text{and} \\ &\sum_{\ell=1}^{N} \gamma_\ell \Lambda\mu_\ell = b \quad \text{for} \quad k = 1,\dots,N . \end{aligned} \tag{2.4}$$

2.1 *Strongly elliptic equations on* Γ

For A we assume that the Sobolev-Slobodetski spaces $H^{\kappa+\alpha}(\Gamma)$ (see [1]) are continuously mapped into $H^{\kappa-\alpha}(\Gamma)$ and that A can be written as $A = A_o + C_o$ where $C_o : H^{\kappa+\alpha} \to H^{\kappa-\alpha}$ is compact and A_o is a strongly elliptic pseudodifferential operator of real order 2α [51],[53] with

$$Au(x) = \int_{\xi=-\infty}^{+\infty} \int e^{i(\tau-t)\cdot\xi} a_o(\tau,\xi)\Psi_\tau(t)u(\underset{\sim}{x}(t))\,dt\,d\xi + C_1 u(x) . \tag{2.5}$$

Here $x = \underset{\sim}{x}(\tau)$ and $y = \underset{\sim}{x}(t)$ are given by the local parameter representation of Γ. $\Psi_\tau(t)$ is a C^∞ cut-off function with compact support and $\Psi_\tau(t) \equiv 1$ for $|t-\tau| \leq \delta$ with a suitable $\delta > 0$. The principal symbol $a_o(\tau,\xi)$ is a C^∞ p×p-matrix with respect to ξ and sufficiently smooth with respect to τ satisfying

$$a_o(\tau,\lambda\xi) = \lambda^{2\alpha} a_o(\tau,\xi) \quad \text{for all} \quad \lambda\geq 1 \text{ and } |\xi|\geq 1 .$$

A is called *strongly elliptic of order* 2α *if* there exist a positive constant κ_1 and a complex valued C^∞ matrix $\Theta(x)$ such that

$$\text{Re } \zeta^T\Theta(\underset{\sim}{x}(\tau))\, a_o(\tau,\xi)\overline{\zeta} \geq \kappa_1|\zeta|^2$$

$$\text{for all} \quad |\xi| = 1 , \text{ all } \quad x\in\Gamma \quad \text{and all} \quad \zeta\in\mathbb{C}^p . \tag{2.6}$$

Theorem 2.1 [23],[29] *: If* A *is strongly elliptic of order* 2α *then for* ΘA *there holds Gårding's inequality, i.e. to any chosen real* κ *there exist a positive constant* κ_o *and a compact linear operator* $C : H^{\kappa+\alpha} \to H^{\kappa-\alpha}$ *such that*

$$\text{Re}(\Theta Av,v)_\kappa \geq \kappa_o\| v \|^2_{\kappa+\alpha} - \text{Re}(Cv,v)_\kappa \quad \text{for all } v\in H^{\kappa+\alpha}. \tag{2.7}$$

Here $(\, , \,)_\kappa$ denotes the H^κ-scalar product. For Galerkin's method *we first multiply* (1.1) *by* Θ .

2.2 Galerkin's method

If Galerkin's method is executed with a (m+1,m)-system then we have

Theorem 2.2 [6],[22],[39],[51]: *Let* A *be strongly elliptic,* $\kappa = 0$ *and assume uniqueness for* (1.1). *Let* $\alpha \leq m$ *, respectively* $\alpha < m+1/2$ *for* $n = 2$ *. Then there is an* $h_o > 0$ *such that for any* $0 < h \leq h_o$ *the Galerkin equations are uniquely solvable providing*

$$|\omega-\omega_h| + \| u-u_h \|_\alpha \leq c \inf_{\chi\in H_h} \| u-\chi\|_\alpha \tag{2.8}$$

where c *is independent of* h,u,u_h .

Now, if in addition $u \in H^s$, $s > \alpha$, then (2.8) with (2.2) implies optimal order convergence of Galerkin's procedure in H^α . This can further be improved with the Aubin-Nitsche duality argument.

Theorem 2.3 [27]: *Let* $2\alpha-m-1 \leq t \leq \alpha \leq s \leq m+1$ *and in addition to the assumptions of Theorem* 2.2 *let also* $A : H^t \times \mathbb{R}^q \to H^{t-2\alpha} \times \mathbb{R}^q$ *be an isomorphism. Then for* $u\in H^s$ *we have super approximation*

$$|\omega-\omega_h| + \| u-u_h \|_t \leq ch^{s-t} \| u \|_s . \tag{2.9}$$

Remarks: 1) If in addition the (m+1,m)-family is quasi-regular providing an inverse assumption [8] then (2.9) holds also for $\alpha < t \leq s \leq m+1$, $t \leq m$. In this case we have [3],[57], [58 (2.8)]

$$|\omega_h| + \| u_h \|_o \leq c(h^{2\min\{0,\alpha\}} \| P_h f \|_o + |b|) \quad \text{and}$$
$$\| P_h A u_h \|_o \leq c\, h^{2\min\{0,-\alpha\}} \| u_h \|_o \tag{2.10}$$

for the conditioning of (2.3) which is basic for estimates of the effects due to numerical integration [4],[39],[57]. 2) Most of the above results also hold for systems of equations with *different orders* $\alpha \in \mathbb{R}^p$ [51],[58]. 3) For some of the above equations and (2.3) one can also prove L_∞-estimates,

$$\| u-u_h \|_{L_\infty} \leq c_\varepsilon h^{s-\varepsilon} \|u\|_{W^{s,\infty}}, \; 0<s\leq m+1 \text{ and any } \varepsilon>0 \quad [47].$$

4) With augmentation techniques the above approach can be extended to mixed boundary value problems and domains with corners including direct computations of stress intensity factors [11],[12],[13],[14],[32],[59]. 5) If H_h is defined by trigonometric polynomials then (2.3) corresponds to "spectral methods" which converge for a much larger class of equations (see for $\alpha=0$ in [20],[46]).

2.3 Collocation method

In contrary to Galerkin's procedure, for the collocation (2.4) convergence results for arbitrary n are known yet only in the special case of Fredholm integral equations of the second kind [2],[5],[9],[10],[28],[43],[54],[55] and for more general equations only if the trial functions are specifically chosen depending on A [38]. Convergence results for simple boundary element collocation are yet to be found if $n \geq 3$.

For $n = 2$ (i.e. Γ a Jordan curve) however, some general results have been obtained recently. Here let $\underset{\sim}{x}(t)$ be a 1-periodic regular parameter representation of Γ . Let $H_h = S_m(\Delta)$ denote the space of 1-periodic C^{m-1} splines of degree m (respectively piecewise constant functions for $m=0$) subject to the set Δ of break points with maximal meshwidth h . For collocation we require

$$m > 2\alpha \; .$$

For *odd* splines let $j = 1/2(m+1)$ and choose for the collocation points x_k the break points, i.e. $\tilde{\Delta}=\Delta$. For *even* splines we further require a *regular* meshwidth with $t_\ell = \ell\cdot h$, $\ell = 1,\dots,N = 1/h$ for the break points Δ but *collocate at the midpoints* $\tilde{\Delta}$ given by $x_k = \underset{\sim}{x}((k-1/2)h)$. In this case we also must further restrict us to operators A with *convolutional* principal part, i.e with $a_o = a_o(\xi)$ in (2.5) not depending on τ .

The collocation equations (2.4) now can be written as *modified Galerkin* equations. To this end we introduce the functionals

$$Ju := \int_\Gamma u ds \ , \quad J_{\tilde\Delta} u := \sum_{k=1}^{N} \delta_k u(x_k)$$

where δ_k denotes the weight of the trapezoidal rule on $\tilde\Delta$.

Theorem 2.4 [3] *: The collocation equations* (2.4) *are equivalent to the Galerkin-Petrov equations to find* $u_h \in S_m(\Delta)$, ω_h *satisfying*

$$((I-J + J_{\tilde\Delta})(A(u_h-u) + B(\omega_h-\omega)), \chi)_j = 0$$

for all test functions $\chi \in \begin{cases} S_m(\Delta) & \text{for } m \text{ odd} , \\ S_{m+1}(\tilde\Delta) & \text{for } m \text{ even} . \end{cases}$

For estimates we write (2.4) in short as

$$I_n AP_h(u_h,\omega_h) = I_h A(u,\omega) = I_h(f,b)$$

where I_h is extended to $H^s \times \mathbb{R}^q$ with $I_h|_{\mathbb{R}^q}$ = identity on $\mathbb{R}^q$.

Theorem 2.5*:* ([3] for odd m , [49] for even m) . *Let* $\kappa = (m+1)/2$. *Then there exist positive constants* h_o *and* c *such that for all* $0 < h \le h_o$ (2.4) *is uniquely solvable and stable, i.e.*

$$\|(I_h AP_h)^{-1} I_h A(u,\omega)\|_{H^{\kappa+\alpha}\times\mathbb{R}^q} = \|u_h\|_{\kappa+\alpha} + |\omega_h|$$

$$\le c(\|u\|_{\kappa+\alpha} + |\omega|) \quad \text{and}$$

$$|\omega - \omega_h| + \|u-u_h\|_{\kappa+\alpha} \le (c+1) \inf_{\chi\in S_m(\Delta)} \|u-\chi\|_{\kappa+\alpha} .$$

For odd m the proof in [3] rests on (2.7) whereas for even m the proof in [49] is based on a rather tedious Fourier series analysis and (2.6). Again the Aubin-Nitsche duality argument provides super approximation.

Theorem 2.6 : ([3] for odd m , [49] for even m) . *Let* $2\alpha \le t \le \kappa+\alpha \le s \le m+1$ *and in addition to the assumptions of Theorem* 2.5 *let also* $A : H^t \times \mathbb{R}^q \to H^{t-2\alpha} \times \mathbb{R}^q$ *be an isomorphism. Then for* $u \in H^s$ *we have*

$$|\omega-\omega_h| + \|u-u_h\|_t \le ch^{s-t}\|u\|_s . \qquad (2.11)$$

Remarks: 1) If the (m+1,m)-family is quasiregular then (2.10) holds also for $\kappa+\alpha \le t \le s$, $t < m+1/2$ and (2.10) is valid with I_h instead of P_h . 2) Piecewise linear spline collocation (2.4) on *regular* meshes for singular integral equations with Cauchy kernel (α=0,m=1) has been analyzed in [44],[45] to obtain error estimated in Hölder-spaces. There it is also shown that in this case strong ellipticity of A is even *necessary* for the convergence of the collocation method. (From [50] it follows that here strong ellipticity is also necessary for the convergence of Galerkin's method.) 3) In [50] for singular integral equations ($\alpha=0$) even degree splines and collocation at the *break points* are used and it is shown that strong ellipticity of SA with S the Hilbert transform is necessary and sufficient for asymptotic convergence.

3. EXAMPLES AND APPLICATIONS

3.1 Fredholm integral equations of the second kind

Many problems of classical mathematical physics as stationary ideal flows around obstacles [10],[55], stationary electromagnetic fields [33] and classical scattering problems [17], [28],[58] can be modelled by Fredholm integral equations of the second kind having smooth ($n=2$) or weakly singular kernels ($n=3$). For instance the pressure amplitude u of the acoustic field of the scattered plane wave e^{ikx_1} by the hard obstacle with boundary Γ solves on Γ

$$Au(x) = u(x) - \frac{1}{(n-1)\pi} \int_\Gamma u(y)\frac{\partial}{\partial \nu_y}\gamma(x,y;k)ds_y = e^{ikx_1} \quad \text{for } x\in\Gamma \qquad (3.1)$$

where $\frac{\partial}{\partial \nu_y}$ normal derivative, $\nu(y)$ outer normal at $y\in\Gamma$,

$$\gamma = \begin{cases} \frac{\pi}{2i} H_o^{(1)}(kr) & \text{for } n = 2 , \\ \frac{1}{r}\exp(ikr) & \text{for } n = 3 , \end{cases} \qquad r = |x-y| ,$$

$A = I-C$, $\alpha = 0$, $a_o = 1$, I identity, $C : H^\kappa \to H^\kappa$ compact for any $\kappa \in \mathbb{R}$ provided smooth enough Γ [58 §3]. (2.7) is trivially satisfied. For k not critical (3.1) is uniquely solvable.

3.2 Singular integral equations with Cauchy kernels [35],[36]

3.2.1 Plane problems, $n = 2$. Identify $\mathbb{R}^2$ with the complex plane $\mathbb{C}$ and set $x = x_1 + ix_2 \in \Gamma$, $y = \underset{\sim}{x}_1(t) + \underset{\approx}{x}_2(t)$. Here

$$Au(x) = a(x)u(x) + \frac{1}{\pi i}\, \text{p.v.} \int_\Gamma \frac{b(x,y)u(y)}{y - x}\, dy + C_1u(x) \quad \text{with } \alpha=0 .$$

C_1 is compact, $a_o = a(x) + b(x,x)\xi \,/\, |\xi|$.

Now strong ellipticity (2.6) is equivalent to the condition

$$\det(a(x) + \lambda\, b(x,x)) \neq 0 \quad \text{for all } \lambda\in[-1,1] \quad [44], [45] .$$

Many examples of *strongly elliptic* singular integral equations can be found in plane elasticity [16],[30],[37],[48]. Following [3] we present the equations of the second fundmental problem. Here the tractions g on Γ are given and the displacement $u + \begin{pmatrix} 1,0, x_2 \\ 0,1,-x_1 \end{pmatrix}\omega$, $p = 2$, $q = 3$ on Γ is to be computed where ω gives the rigid motion. From Somigliana's identity one finds on Γ

$$\begin{aligned}
&Au(x)+B(x)\omega = \varepsilon u(x) + \int_\Gamma S(x,y)u(y)ds_y + \begin{pmatrix}1,0,x_2\\0,1,-x_1\end{pmatrix}\omega = f(x) ,\\
&\Lambda u = 0 \sim \int_\Gamma u_1(y)ds_y = \int_\Gamma u_2(y)ds_y = \int_\Gamma (u_1dy_1 + u_2dy_2) = 0
\end{aligned} \tag{3.2}$$

where $\varepsilon = -1$ for the interior and $\varepsilon = +1$ for the exterior problem,

$$\begin{aligned}
S_{jk}(x,y) &= \frac{1}{\pi}\frac{1}{(\lambda+2\mu)}\{(\mu\delta_{jk} + 2(\lambda+\mu)L_{jk})\frac{\partial}{\partial\nu_y}\log r\\
&\quad + (-1)^j\mu(1-\delta_{jk})\frac{d}{ds_y}\log r\},\\
L_{jk}(x,y) &= (x_j - y_j)(x_k - y_k)/r^2 , \; j,k = 1,2 .
\end{aligned} \tag{3.3}$$

$\lambda > -\mu$ and $\mu > 0$ are the Lamé constants. Λ excludes rigid motions from u. The right hand side f is given by

$$\begin{aligned}
f(x) &= \frac{\lambda+3\mu}{2\mu(\lambda+2\mu)}\{\frac{1}{\pi}\int_\Gamma g(y)\log r\, ds_y - \frac{\tilde q}{\pi}\int_\Gamma (L+\frac{1}{2})g(y)ds_y\} ,\\
\tilde q &= (\lambda+\mu)/(\lambda+3\mu) .
\end{aligned} \tag{3.4}$$

The principal symbol of (3.2) is given by

$$a_o(\xi) = \begin{pmatrix}\varepsilon & ,-i\nu\frac{\xi}{|\xi|}\\ i\nu\frac{\xi}{|\xi|}, & \varepsilon\end{pmatrix} \quad \text{where } 0<\nu = \frac{\mu}{\lambda+2\mu} < 1 \text{ and } \alpha=0 .$$

Obviously, a_o is Hermitian satisfying (2.6) with $\Theta = \varepsilon I$. Thus (3.2) is strongly elliptic and all our assumptions are verified.

3.2.2 Spacial problems, $n = 3$. Singular integral operators of Giraud's type on the surface Γ have the form

$$Au(x) = a(x)u(x) + \frac{1}{2\pi}\text{ p.v. }\int_{y\in\Gamma}\frac{b(x,y)}{r^2}u(y)ds_y + C_1u(x) \text{ with } \alpha=0$$

where $b(x,y)$ is given with respect to local surface polar coordinates (r,ϕ) [33, p. 76 ff.] by the Fourier expansion

$$b(x,y) = \sum_{\ell\neq 0} b_\ell(x)\, e^{i\ell\phi} .$$

Then with $\cos\phi = \xi_1/|\xi|$, $\sin\phi = \xi_2/|\xi|$ the symbol a_o is given by [35]

$$a_o(x,\xi) = a(x) + \sum_{\ell\neq 0} i^{|\ell|}\frac{2\pi}{|\ell|}b_\ell(x)\, e^{i\ell\phi} .$$

Strong ellipticity (2.6) can be checked for many problems of thermoelasticity from the symbols in [30],[36].

3.3 Some Fredholm integral equations of the first kind

3.3.1 Plane problems, $n = 2$. Many interior and exterior plane boundary value problems can be reduced to equations [16]

$$-\frac{1}{\pi}\int_\Gamma \log r\, u(y)ds_y + \frac{\tilde{q}}{\pi}\int_\Gamma (L+\frac{1}{2})u(y)ds_y + B(x)\omega = \Psi(x) \text{ on } \Gamma,\ \Lambda u = b \tag{3.5}$$

where the kernel $L(x,y)$ is a smooth matrix function. Here

$$a_o = \frac{1}{|\xi|} I \text{ for } |\xi| \geq 1 \text{ and } \alpha = -\frac{1}{2},$$

so A is strongly elliptic. In case $p=q=1$, $B=1$, $L=0$, $\Lambda u = \int uds$ (3.5) is Symm's integral equation in conformal mapping [52],[56], in electrostatics [31], in torsion problems and with smooth L in soft scattering [58].

For the *first fundamental problem of plane elasticity* in (3.2), (3.4) the boundary displacement $\phi|_\Gamma$ is given and the boundary tractions $g|_\Gamma = u$ are to be computed [3],[16], [48]. Here in (3.5) $L, \tilde{q}$ are given by (3.3), (3.4), $p=2$, $q=3$, $B(x)\omega = \begin{pmatrix}1,0,x_2\\0,1,-x_1\end{pmatrix}\omega$ rigid motion,

$$\Lambda u = 0 \sim \int_\Gamma u_1(y)ds_y = \int_\Gamma u_2(y)ds_y = \int_\Gamma (y_2u_1 - y_1u_2)ds_y = 0$$

is the equilibrium condition. Ψ in (3.5) is given by

$$\Psi(x) = -\frac{2\mu(\lambda+2\mu)}{\lambda+3\mu}\{\varepsilon\,\phi(x) + \int_\Gamma S(x,y)\phi(y)ds_y\}.$$

(3.5) with $\tilde{q} = -1$, L (3.3) : plate equation, interior and exterior Stokes flows and slow viscous flows [13],[16],[24],[25], [26].

3.3.2 Spacial problems, $n = 3$. On the surface Γ,

$$Au(x) = \frac{1}{2\pi}\int_\Gamma \frac{1}{r}u(y)ds_y + C_1 u \tag{3.6}$$

with $a_o = \frac{1}{|\xi|}$ for $|\xi| \geq 1$ and $\alpha = -\frac{1}{2}$ was used in electrostatics [42] and in soft scattering [58]. In [18] the Stokes flow around Γ yields a system with $p=3$, $q=1$,

$$Au(x)+B(x)\omega = \int_\Gamma E(x,y)u(y)ds_y + \omega\nu(x) = f(x) \text{ on } \Gamma,$$

$$\Lambda u = \int_\Gamma \nu(y)\cdot u(y)ds_y = b,$$

$$E(x,y)_{jk} = \frac{1}{8\pi}\{\frac{1}{r}\delta_{jk} + (x_j-y_j)(x_k-y_k)/r^3\}.$$

Using local coordinates at x with the x_3-axis in the direction of the outer normal $\nu(x)$ and local surface polar coordinates [33, p. 76 ff.] we find that $E(x,y)$ defines a pseudohomogeneous kernel and

$$a_o = \frac{1}{\Psi|\xi|^3}\begin{pmatrix} \xi_1^2+2\xi_2^2, & -\xi_1\xi_2, & 0\\ -\xi_1\xi_2, & 2\xi_1^2+\xi_2^2, & 0\\ 0, & 0, & \xi_1^2+\xi_2^2\end{pmatrix} \text{ for } |\xi| \geq 1.$$

Thus A is strongly elliptic.

3.4 Equations with hypersingular nonintegrable kernels

Diffraction by a hard open screen $\Gamma_o \subset \Gamma \subset \mathbb{R}^n$, $n = 2$ or 3 has been reduced in [15],[17] to

$$Au(x) = \frac{1}{2(n-1)\pi} \int_\Gamma (u(y)-u(x)) \frac{\partial}{\partial\nu_x} \frac{\partial}{\partial\nu_y} \gamma_o(x,y)ds_y + C_1u(x) \qquad (3.7)$$
$$= -\frac{k}{\eta} e^{ikx_1} \frac{\partial}{\partial\nu_x} x_1|_\Gamma$$

where

$$C_1u(x) = \frac{i}{\eta}u(x) + \frac{1}{2(n-1)\pi}\int_\Gamma u(y)\frac{\partial}{\partial\nu_x}[(\frac{\partial}{\partial\nu_y} - \frac{i}{\eta})\{\gamma(x,y;k)-\gamma_o\} + \frac{i}{\eta}\gamma_o]ds_y ,$$

$$\gamma_o = \begin{cases} -\log r & \text{for } n = 2 , \\ 1/r & \text{for } n = 3 , \end{cases} \qquad \eta > 0 .$$

Here u is to be found in $H^{1/2}(\Gamma)$ with $u|_{\Gamma\setminus\Gamma_o} = 0$. Equations of the same type without the latter ristriction for u can be found for scattering and potential problems in [6],[7],[19],[40],[60], for elasticity in [41] and in the form of Prandtl's wing equation in [3],[36]. With the Gauss-Cartan theorem or local Taylor expansion the principal part of A in (3.7) becomes either a composition of the Hilbert transform with $\frac{du}{ds}$ [3, (2.3.11)] , [58 §1] or of (3.6) with appropriate covariant differential operators on Γ [41],[60]. These provide us with

$$a_o = |\xi| \quad \text{for } |\xi| \geq 1 , \ \alpha = \frac{1}{2} .$$

Hence A in (3.6) is strongly elliptic satisfying our requirements.

References

1. ADAMS, R.A. *Sobolev Spaces*. Academic Press, New York (1975).
2. ANSELONE, P.M. *Collectively Compact Operator Approximation Theory*. Prentice Hall, London (1971).
3. ARNOLD, D., WENDLAND, W.L., On the asymptotic convergence of collocation methods. *Math. Comp.*, to appear.
4. ARNOLD, D., WENDLAND, W.L., Collocation versus Galerkin procedures for boundary integral methods. *Boundary Element Meth. Eng.* (ed. Brebbia) Springer, Berlin, 18-33 (1982).
5. ATKINSON, K.E. *A Survey of Numerical Methods for the Solution of Fredholm Integral Equations of the Second Kind*. SIAM, Philadelphia (1976).
6. AZIZ, A.K., DORR, M.R., KELLOGG, R.B., Calculation of electromagnetic scattering by a perfect conductor. *Navel Surface Weapons Center* TR80-245 (1980).

7. AZIZ, A.K., KELLOG, B., Finite element analysis of a scattering problem. *Math. Comp.* 37, 261-272 (1981).
8. BABUSKA, I., AZIZ, K.A., Survey lectures on the mathematical foundations of the finite element method. *The Math. Found. Finite Element Meth.* (ed. Aziz) Academic Press, N.Y., 3-359 (1972).
9. BEN NOBLE, Error Analysis of Collocation Methods for Solving Fredholm Integral Equations. *Topics Num. Anal.* (ed. Miller) Academic Press, London, 211-232 (1972).
10. BRUHN, G., WENDLAND, W., Über die näherungsweise Lösung von linearen Funktionalgleichungen. *Int. Ser. Num. Math.* 7, 136-164 Birkhäuser, Basel (1967).
11. COSTABEL, M., STEPHAN, E., Boundary integral equations for mixed boundary value problems in polygonal domains and Galerkin approximation. *Banach Center Publ.*, Warsaw. To appear.
12. COSTABEL, M., STEPHAN, E., Curvature terms in the asymptotic expansions for sulutions of boundary integral equations on curved polygons. To appear.
13. COSTABEL, M., STEPHAN, E., WENDLAND, W.L., On boundary integral equations of the first kind for the bi-Laplacian in a polygonal plane domain. To appear.
14. DJAOUA, M., A method of calculation of lifting flows around 2-dimensional corner spaced bodies. Centre Math. Appl., Ecole Polytechnique, 34 (1978).
15. DURAND, M., Diffraction d'ondes acoustiques par un écran mince. *Publ. de Math. Appl.* Marseille-Toulon 80/3, Université de Provence, France (1980).
16. FICHERA, G., Linear elliptic equations of higher order in two independent variables and singular integral equations. *Proc. Diff. Cont. Mech.* Univ. Wisconsin Press, Madison (1961).
17. FILIPPI, P., Potentiels de couche pour les ondes mécaniques scalaires. *Revue du Cethedec* 51, 121-175 (1977).
18. FISCHER, T., An integral equation procedure for the exterior three-dimensional viscous flow. *Integral Equations and Operator Theory* 5, 480-505 (1982).
19. GIROIRE, J., Integral equation methods for exterior problems for the Helmholtz equation. Centre de Math. Appl., Ecole Polytechnique, 40 (1978).
20. GOHBERG, I.C., FELDMAN, I.A. *Convolution Equations and Projection Methods for their Solution.* AMS. Transl., Providence (1974).
21. HELFRICH, H.P., Simultaneous approximation in negative norms of arbitrary order. *R.A.I.R.O. Num. Analysis* 15, 231-235 (1981).
22. HILDEBRANDT, ST., WIENHOLTZ, E., Constructive proofs of representation theorems in separable Hilbert space. *Comm. Pure Appl. Math.* 17, 369-373 (1964).

23. HÖRMANDER, L., Pseudo-differential operators and non-elliptic boundary problems. *Annals Math.* 83, 129-209 (1966).
24. HSIAO, G.C., KOPP, P., WENDLAND, W.L., A Galerkin collocation method for some integral equations of the first kind. *Computing* 25, 89-130 (1980).
25. HSIAO, G.C., MAC CAMY, R.C., Solution of boundary value problems by integral equations of the first kind. *SIAM Rev.* 15, 687-705 (1973).
26. HSIAO, G.C., WENDLAND, W.L., A finite element method for some integral equations of the first kind. *J.M.A.A.* 58, 449-481 (1977).
27. HSIAO, G.C., WENDLAND, W.L., The Aubin-Nitsche lemma for integral equations. *J. Integral Equations* 3, 299-315 (1981).
28. KLEINMAN, R., WENDLAND, W.L., On Neumann's method for the exterior Neumann problem for the Helmholtz equation. *J.M.A.A.* 57, 170-202 (1977).
29. KOHN, J.J., NIRENBERG, L., On the algebra of pseudo-differential operators. *Comm. Pure Appl. Math.* 18, 269-305 (1965).
30. KUPRADZE, V.D., GEGELIA, T.G., BASHELEISHVILI, M.O., BURCHULADZE, T.V. *Three-Dimensional Problems of the Mathematical Theory of Elasticity and Thermoelasticity.* North Holland, Amsterdam (1979).
31. LE ROUX, M.N., Résolution numérique du problème du potential dans le plan par une méthode variationelle d'élements finis. Thèse, Rennes (1974).
32. LAMP, U., SCHLEICHER, T., STEPHAN, E., WENDLAND, W.L., Theoretical and experimental asymptotic convergence of the boundary integral method for a plane mixed boundary value problem. *Boundary Element Meth. Eng.* (ed. Brebbia) Springer, Berlin, 3-17 (1982).
33. MARTENSEN, E. *Potentialtheorie.* B.G. Teubner, Stuttgart (1968).
34. MIKHLIN, S.G. *Variationsmethoden der Mathematischen Physik.* Akademie-Verlag, Berlin (1962).
35. MIKHLIN, S.G. *Multidimensional Singular Integrals and Integral Equations.* Pergamon, Oxford (1965).
36. MIKHLIN, S.G., PRÖSSDORF, S. *Singuläre Integraloperatoren.* Akademie-Verlag, Berlin (1980).
37. MUSKHELISHVILI, N.I. *Some Basic Problems of Mathematical Theory of Elasticity.* Noordhoff, Groningen (1953).
38. NASHED, M.Z., WAHBA, G., Generalized inverses in reproducing kernel spaces. *SIAM J. Math. Anal.* 5, 974-987 (1974).
39. NEDELEC, J.C., Curved finite element methods for the solution of singular integral equations on surfaces in $\mathbb{R}^3$. *Comp. Math. A. M. E.* 8, 61-80 (1976).

40. NEDELEC, J.C. Approximation par potentiel de double cuche du problème de Neumann extérieur. *C.R. Acad. Sci.* Paris, Sér. A 286, 616-619 (1977).
41. NEDELEC, J.C., Formulations variationelles de quelques equations intégrales faisant intervenir des parties finies. *Innovative Num. Anal. Eng. Sc.* (ed. Shaw). Univ. Press Virginia, Charlottesville, 517-524 (1980).
42. NEDELEC, J.C., PLANCHARD, J., Une méthode variationelle d'éléments finis pour la résolution numérique d'un problème extérieur des $\mathbb{R}^3$, 105-129 (1973).
43. PRENTER, P.M., A collocation method for the numerical solution of integral equations. *SIAM J. Num. Math.* 10, 570-581 (1973).
44. PRÖSSDORF, S., SCHMIDT, G., A finite element collocation method for singular integral equations. *Math. Nachr.* 100, 33-60 (1981).
45. PRÖSSDORF, S., SCHMIDT, G., A finite element collocation method for systems of singular integral equations. *Math. Nachr.* To appear.
46. PRÖSSDORF, S., SILBERMANN, B. *Projektionsverfahren und die näherungsweise Lösung singulärer Gleichungen.* B.G. Teubner, Leipzig (1977).
47. RANNACHER, R., WENDLAND, W.L., On the order of pointwise convergence of some boundary element methods. In preparation.
48. RIZZO, F.J., An integral equation approach to boundary value problems of classical elastostatics. *Quart. Appl. Math.* 25, 83-95 (1967).
49. SARANEN, J., WENDLAND, W.L., On the asymptotic convergence of collocation methods with spline functions of even degree. To appear.
50. SCHMIDT, G., On spline collocation for singular integral equations. *Math. Nachr.* To appear.
51. STEPHAN, E., WENDLAND, W.L., Remarks to Galerkin and least squares methods with finite elements for general elliptic problems. *Springer Lecture Notes Math.* 564, 461-471 (1976).
52. SYMM, G.T., Integral equation methods in potential theory, II. *Proc. Royal Soc. London* A275 , 33-46 (1963).
53. TREVES, F. *Introduction to Pseudodifferential and Fourier Integral Operators* I. Plenum Press, New York (1980) .
54. VAINIKKO, G. *Funktionalanalysis der Diskretisierungsmethoden.* G.B. Teubner, Leipzig (1976).
55. WENDLAND, W.L., Die Behandlung von Randwertaufgaben im $\mathbb{R}_3$ mit Hilfe von Einfach- und Doppelschichtpotentialen. *Num. Math.* 11, 380-404 (1968).
56. WENDLAND, W.L., On Galerkin collocation methods for integral equations of elliptic boundary value problems. In Numerical Treatment of Integral Equations. *Int. Ser. Num. Math.* 53, 224-275 Birkhäuser, Basel (1980).

57. WENDLAND, W.L., Asymptotic accuracy and convergence. In: *Progress in Boundary Element Methods* (ed. Brebbia), Pentech Press, London 1 , 289-313 (1981).
58. WENDLAND, W.L.,Boundary element methods and their asymptotic convergence. *Theoretical Acoustics and Numerical Techniques* (ed. Filippi). Springer, Lecture Notes in Physics. To appear.
59. WENDLAND, W.L., STEPHAN, E., HSIAO, G.C., On the integral equation method for the plane mixed boundary value problem of the Laplacian. *Math. Meth. in the Appl. Sci.* 1, 265-321 (1979).
60. STEPHAN, E., Solution procedures for interface problems in acoustics and electromagnetics. *Theoretical Acoustics and Numerical Techniques* (ed. Filippi). Springer, Lecture Notes in Physics. To appear.

INCLUSION OF SOLUTIONS OF CERTAIN TYPES OF INTEGRAL EQUATIONS

L. Collatz

Hamburg

Summary This survey describes methods for getting rigorous lower and upper bounds for solutions of certain types of integral equations. The numerical procedure uses approximation and optimization; the theoretical background is given by topological fixed point theorems and monotonicity principles. Numerical examples for linear and nonlinear integral equations and critical comparisons of different methods are given; further applications of inclusion methods are described, f.i. for mixed boundary value problems, integro-differential equations and for oscillations of systems with delay.

1. Introduction

Many different methods have been developed for the approximate numerical solution of integral equations (compare f.i. Kantorowitsch-Akilow [64], Collatz [6o], Colton [76], the comprehensive book of Chr. Baker [78], a.o.). Many of these methods give excellent numerical results and one has also considerations about error-estimations; but usually it is difficult to get error-bounds for the approximate solutions. Sometimes one may be interested to get rigorous error bounds one can guarantee, for instance: The computer gives numerical results with a certain number of decimals, and one wishes to know, how many of these decimals can be guaranted. In simple cases one can give such error bounds by using principles of monotonicity.

This survey begins with some well-known applications of monotonicity and fixed point theorems, which have had success also in nonlinear problems. Than we add some recent applications, also for integro-functional-equations.

Many problems with integral equations can be written in the form

(1.1) $u = Tu,$

where $u(x)=u(x_1,\ldots,x_n)$ is an unknown element of a Banach space R and T a given linear or nonlinear operator, which maps a given subset D of R into itself. In the following B may be a given closed bounded connected domain in the space $\mathbb{R}^n$ of real points $(x_1,\ldots,x_n)$ and D may be a set of realvalued functions defined on B; sometimes one supposes u(x) as contiuous or integrable, or differentiable a.o. In our case we suppose T as integral-operator which may contain also derivatives of u, a.o.

2. The classical contraction mapping theorem.

One of the possible descriptions of this well-known theorem is as follows. We use an iteration procedure

(2.1) $u_{n+1} = Tu_n, \ (n=0,1,\ldots)$

starting with an element $u_o \epsilon D$, for which also $u_1 \epsilon D$. We suppose, that T admits a Lipschitz constant $K<1$ with

(2.2) $||Tf-Tg|| \leq K||f-g||$ for all $f,g \epsilon D$

and that the sphere S of all elements h with

(2.3) $||h-u_1|| \leq \frac{K}{1-K}||u_1-u_o||$

belongs to D. Then one can prove (compare for all details f.i. Collatz [68], Bohl [74] a.o.), that there exists a uniquely determined fixed point u of (1.1) and that this fixed point u lies in S. Therefore we have an error bound.

Numerical examples are given f.i. in Collatz [68]. The theorem has had great success for existence theorems even in rather complicated nonlinear problems. Disadvantages are, that one needs the knowledge of a Lipschitzconstant K and that $K<1$; in cases, in which K is near to 1, the sphere S with (2.3) becomes very large and the error bound is not useful; Nr.4 gives an example in which the Schauder-theorem of Nr.3 is more powerful than the contracting mapping theorem.

3. Monotone iterations and Schauder's fixed point theorem

Now we suppose that the Banach space R is partially ordered. The operator T is called syntone (resp. antitone), if

(3.1) $\begin{cases} \text{from } f\leq g \text{ follows } Tf\leq Tg \text{ (resp. } Tf\geq Tg) \\ \text{for all } f,g\in D. \end{cases}$

and T is called "monotonically decomposssible" (J.Schröder [62],[8o], Bohl [74], Collatz [68]) if T has the form

(3.2) $T=T_1+T_2$, T_1 syntone, T_2 antitone.

In this case one calculates two sequences $v_o,v_1,\ldots$ and $w_o,w_1,\ldots$ of iterates with

(3.3) $\begin{cases} v_{n+1}=T_1v_n+T_2w_n \\ w_{n+1}=T_1w_n+T_2v_n \end{cases}$ $(n=0,1,\ldots)$

and which satisfy the initialconditions

(3.4) $v_o\leq v_1\leq w_1\leq w_o$.

Then is follows

(3.5) $v_o\leq v_1\leq v_2\leq\ldots\leq v_n\leq\ldots\leq w_n\leq\ldots\leq w_2\leq w_1\leq w_o$ $(n\geq 2)$

(For numerical calculations one let usually v_o depend on certain parameters a_ν and w_o depend on parameters $\overline{a_\nu}$, and determines the parameters a_ν, $\overline{a_\nu}$ from an optimization problem, see (4.4). This is in principle an approximation problem, mostly a simultaneous approximation. About approximation theory and its practical treatment compare f.i. Bredendiek [69], Meinardus[67], Meinardus-Merz [79], Werner-Schaback [79], Grothkopf [81] and many others.) If T is furthermore compact (J. Schröder [8o], Collatz [68], a.o.) one proves, that there exists at least one fixed point u between v_n and w_n for every n:

(3.6) $v_n\leq u\leq w_n$ for $n=1,2,\ldots$.

Hammerstein's Integral-Equations

We consider the Hammerstein Equations (compare Bohl [74]):

(3.7) $Tu(x)=f(x)+\lambda\int_B K(x,t)\Phi(u(t))dt=u(x),$

where K(x,t) is a given on BxB defined function (continuous or L_2-integrable), f(x) a given, f.i. continuous function, λ a given constant and Φ a given function of bounded variation; this means: ϕ admits a representation

$$(3.8)\begin{cases}\phi(z)=\phi_1(z)+\phi_2(z)\\ \text{for all } z \text{ of a certain real interval,}\end{cases}$$

where $\phi_1(z)$ is syntone and $\phi_2(z)$ antitone. We write K in the form

$$(3.9)\quad K(x,t)=K_1(x,t)+K_2(x,t)$$

with $K_1=\mathrm{Max}(K,0)$, $K_2=\mathrm{Min}(K,0)$. Then the operator T is monotonically decompossible:

$$T=T_1+T_2 \text{ with } T_1=f(x)+\lambda\int_B\{K_1\phi_1(u)+K_2\phi_2(u)\}dt,$$
$$T_2=\lambda\int_B\{K_1\phi_2(u)+K_2\phi_1(u)\}dt.$$

This holds for $\lambda>0$; for $\lambda<0$ one has to change T_1 and T_2.

4. Examples

A. Linear Equation (Love[49]). The linear Integral-Equation of Love for an electrostatic potential y(t) has the form (Sloan [82])

$$(4.1)\quad y(t)=\int_0^1\frac{y(s)ds}{A^2+(t-s)^2}. \text{ We choose } A=2.$$

First step: Here the operator T is syntone and (3.3) reduces to

$$(4.2)\quad v_{n+1}=Tv_n,\ w_{n+1}=Tw_n\quad (n=0,1,\dots)$$

We take the starting elements asconstants: $v_0=\underline{a}$, $w_0=\bar{a}$ and get

$$(4.3)\begin{cases}v_1=1+\underline{a}\phi_1(t),\ w_1=1+\bar{a}\phi_1(t)\\ \text{with } \phi_1(t)=\int_0^1\frac{ds}{A^2+(t-s)^2}=\frac{1}{A}\{\arctan(\frac{t}{A})-\arctan(\frac{t-1}{A})\}\end{cases}$$

The parameter $\underline{a},\bar{a}$ are determined from the optimization

$$(4.4)\quad \delta=\mathrm{Min},\ w_1-v_1\leq\delta \text{ and } (3.4).$$

The best values are $\underline{a}=\{1-\phi_1(0)\}^{-1}\simeq1.3018$, $\bar{a}=\{1-\phi_1(1/2)\}^{-1}\simeq1.3245$; fig.1 shows the graphs of v_0,v_1,w_1,w_0.

The solution $y(t)$ lies in the strip $v_1 \leq y \leq w_1$; we have for the approximate solution

(4.5) $\eta(t)=\frac{1}{2}(w_1(t)+v_1(t))$

the error bound

(4.6) $|\eta(t)-y(t)| \leq$

$\leq \frac{1}{2}\delta \simeq 0.00278.$

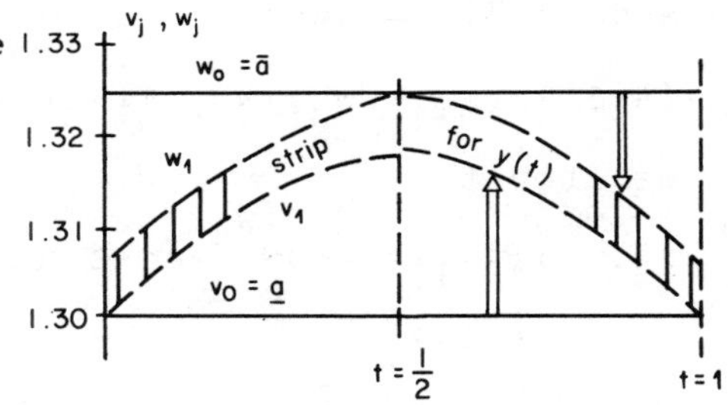

FIG. 1

Second approximation: We can improve the bound by using polynomials:

$$(4.7)\begin{cases} v_o=\underline{a}+\underline{b}(t-t^2), & v_1=1+\underline{a}\phi_1(t)+\underline{b}\phi_2(t) \\ w_o=\bar{a}+\bar{b}(t-t^2), & w_1=1+\bar{a}\phi_1(t)+\bar{b}\phi_2(t) \end{cases}$$

$$\text{with } \phi_2(t)=\int_o^a \frac{(s-s^2)ds}{4+(t-s)^2}=-1+(\frac{1}{2}-t)\,\ell n\frac{4+(t-1)^2}{4+t^2} +$$

$$+ \frac{4+t-t^2}{2}\{\arctan(\frac{t}{2})-\arctan(\frac{t-1}{2})\}.$$

Then one gets an approximate solution $\eta(t)$ with

$$|\eta(t)-y(t)| \leq 0.000036 \text{ for} \begin{cases} \underline{a}=1.30529, & \bar{a}=1.30500 \\ \underline{b}=0.069304, & \bar{b}=0.06931. \end{cases}$$

B. Nonlinear Equation (nonlinear oscillations).

We suppose that the following functions $y(t)$, $v_j(t)$, $w_j(t)$, $(j=0,1)$ have the period 2π. The differential equation

(4.8) $4\ddot{y}(t)+y(t)+\rho e^{y(t)}+\cos t=0$

for forced oscillations can be transformed into an Hammerstein-Equation

(4.9) $y(t)=Ty(t)=\int_{-\pi}^{\pi} G(t,s)\{-\rho e^{y(t)}-\cos s\}ds$

by using Greens function $G(t,s)$:

$$(4.10)\left.\begin{cases} 4\ddot{y}+y=r(t) \\ y,r \text{ period } 2\pi \end{cases}\right\} \Leftrightarrow y(t)=\int_{-\pi}^{\pi} G(t,s)r(s)ds.$$

Then $G(t,s) \geq 0$, therefore the operator T is antitone and (3.3) reduces to

(4.11) $v_{n+1}=Tw_n,\ w_{n+1}=Tv_n\,(n=0,1,\ldots)$

We iterate

(4.12) $\begin{cases} 4v_1+v_1=-\rho e^{w_0}-\cos t \text{ and correspondingly with} \\ w_1, v_0. \end{cases}$

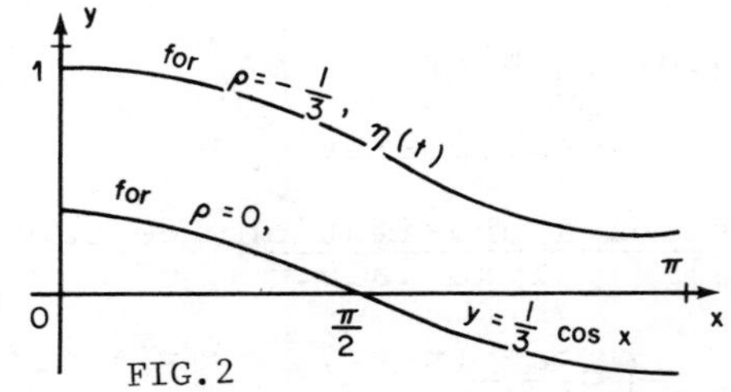

FIG. 2

We take

$v_1=\sum_{j=0}^{p} a_j\cos(jt)$ and

calculate w_0 from (4.11), and analogeously

$w_1=\sum_{j=0}^{p} \bar{a}_j\cos(j,t)$.

The parameters $a_j, \bar{a}_j$ are again determined from (4.4). Fig. 2 shows the approximate solution $\eta(t)$, given by (4.5), for $\rho=-1/3$; increasing p reduces the error bound δ corresponding the table

p	$w_1-v_1\leq\delta$
1	0.07216
2	0.00287

and therefore we have the solution for p=2

$$|\eta(t)-y(t)|\leq\frac{1}{2}\delta\simeq 0.00144.$$

The case $\rho=1/3$ is treated in Collatz [81].

C. Nonlinear Equation of Urysohn-Type.

(4.13) $Ty(x)=\lambda\int_0^1 \frac{dt}{1+x+y(t)}=y(x)$

Here we consider at first the case $\lambda=1$.

First Step: Starting with constant a we get $Ta=\frac{\lambda}{1+x+a}$. We wish to calculate y(1). The operator T is antitone. The best values of the constants are

$$\begin{cases} v_0=(1/2)(\sqrt{3}-1)\approx 0.366 \\ w_0=\sqrt{3}-1\approx 0.732 \end{cases}$$

Fig. 3 shows the graphs of v_o, v_1, w_1, w_o. We get as approximate value

$$y(1) \simeq \eta(1) = \frac{1}{4} + \frac{1}{12}\sqrt{3} \simeq 0.39434$$

with the error bound

$$|\eta(1) - y(1)| \leq \frac{3}{4} - \frac{5}{12}\sqrt{3} \simeq 0.0283.$$

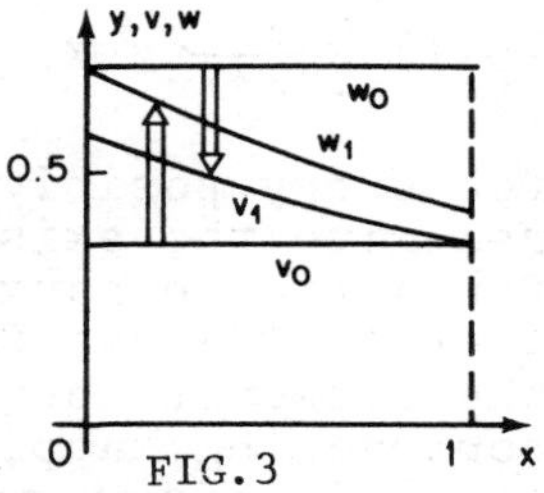

FIG.3

Second approximation: We use v_o, w_o as functions of the form $\alpha+\beta t$ and have the formula

$$\int_0^1 \frac{dt}{\alpha+\beta t} = \frac{1}{\beta}\ln(1+\frac{\beta}{\alpha}).$$

With $w_o = 0.68 - 0.27x$, $v_o = 0.61 - 0.25t$ we get

$$v_1 = \frac{100}{27}\ln\frac{1.68+x}{1.41+x},\quad w_1 = 4\ln\frac{1.61+x}{1.36+x},\ \text{fig. 4}$$

and for the approximate value

$$\eta(1) = \frac{1}{2}\{v_1(1) + w_1(1)\} \simeq 0.39802 \text{ and}$$

the error bound

$$|\eta(1) - y(1)| \leq 0.00473.$$

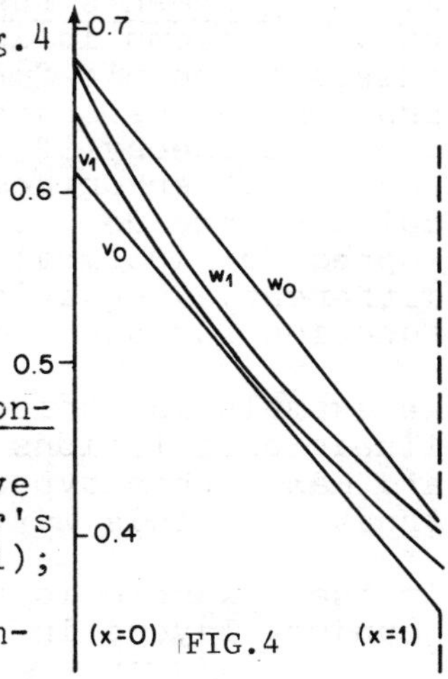

FIG.4

5. Comparison of the theorems of contracting mapping and of Schauder

Now let be λ in (4.13) any positive number. We apply at first Schauder's theorem, using the iteration (4.11); starting with $w_o = \infty$ (formally as limit for increasing positive constants) we get $v_1(x) = T(\infty) = 0$ for every λ. Taking $v_o(x) = 0$ we get

$$w_1(x) = \lambda\int_0^1 \frac{dt}{1+x} = \frac{\lambda}{1+x},\ \text{fig. 5.}$$

The initial conditions (3.4) are satisfied and we get the existence of a solution $y(x)$ for every positive λ and the inclusion $0 \leq y(x) \leq \lambda/(1+x) \leq \lambda$; therefore we can improve the bounds by taking $\hat{w}_o = \lambda$ instead of $w_o = \infty$, and we get $\hat{v}_1 = \lambda/(1+\lambda+x)$ as better lower bound, this means, we have the inclusion

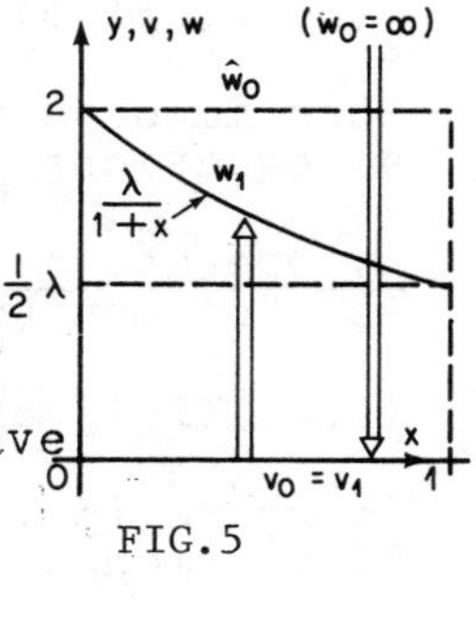

FIG.5

(5.1) $\frac{\lambda}{1+\lambda+x} \leq y(x) \leq \frac{\lambda}{-1+x}$

for every positive λ;but the contracting mapping theorem give the existence and bounds for a solution only for a finite range of λ, for $0<\lambda<\lambda_0$; if the operator T of (4.13) has for $\lambda=1$ the Lipschitzconstant K, then the Lipschitz constant of T in (4.13) is $K\cdot\lambda$ and the contracting mapping theorem is applicable only for $\lambda<1/K$. In this example the Schauder-theorem is highly superior to the contracting mapping theorem, an experience, which is known also for other cases (f.i. Collatz [66], p.360).

6. Application to ordinary and partial differential equations

A. Using Green's function. For many types of boundary value problems it is sufficient to know that there exists a Green's function, with the aid of which the problem can be transformed into an integral equation. Then the theory of integral equations can be applied for to get error bounds; but for the numerical calculation one rewrites the procedures (which are developped for integral equations) into procedures for differential equations, and one avoids the Green's function for the numerical calculation.

An example for this method is described for the nonlinear oscillations with (4.8) as special equation, and many other types of nonlinear vibrations are treated in this way.

Another example is the famous Prandtl-equation for boundary layers in hydrodynamics

(6.1) $y''' + y''y = 0$

$y(0)=0$, $y'(0)=a$, $y''(0)=1$ (Collatz [81])p.215).

More important are applications to partial differential equations.

B. Operators of monotonic type. An operator T is called "of monotonic type", (Collatz [52]) if

(6.2) $Tf \leq Tg$ has the consequence $f \leq g$ for all $f,g \in D$

(compare (3.1), these operators are also: called "inversmonotone", Schröder [62],[80]). It is known for wide classes of boundary value problems even for partial differential equations, that there exists a nonnegative Green's function so that(as in A) the theory of integral equations gives the theoretical background for the numerical procedure for inclusion of solutions. The Green's function itself in usually not known and not used in the numerical method.

Example: Mixed boundary value problem.
A function $u(x,y)$ should satisfy the potential equation

$$(6.2)\quad \Delta u=\frac{\partial^2 u}{\partial x^2}+\frac{\partial^2 u}{\partial y^2}=0 \text{ in } B=\{(x,y),\ -\infty<x<+\infty\ ,\ 0<y<\ell\}$$

and the boundary conditions

$$(6.3)\quad u=0 \text{ on } \Gamma_1\ (y=0),\ \frac{\partial u}{\partial y}=f(x)>0 \text{ on } \Gamma_2(y=\ell), \lim_{x\to\pm\infty} u(x,y)=0.$$

Here $f(x)$ is a given continuous positive function. $u(x,y)$ may be interpretad as steady distribution of temperature in an infinite strip, one side is hold on the temperature $u=0$, on the other side $y=\ell$ one is introducing heat corresponding to $f(x)$. Also a hydrodynamical interpretation is possible. We have the monotonicity principle:

$$(6.4)\quad \text{From } \Delta w\geq 0 \text{ in } B,\ w\geq 0 \text{ on } \Gamma_1,\ \frac{\partial w}{\partial y}\geq f(x) \text{ on } \Gamma_2$$

follows $w\geq u$ in B; analogeuously one can get a lower bound for u. Numerical Example: $\ell=1$; $f(x)=1/(1+x^2)$

Approximate solution:

$$u\ w=\sum_{\nu=1}^{p} c_\nu w_\nu \text{ with } w_\nu=\ell n\{x^2+(y-d_\nu)^2\}-\ell n\{x^2+\&y-d_\nu)^2\}$$

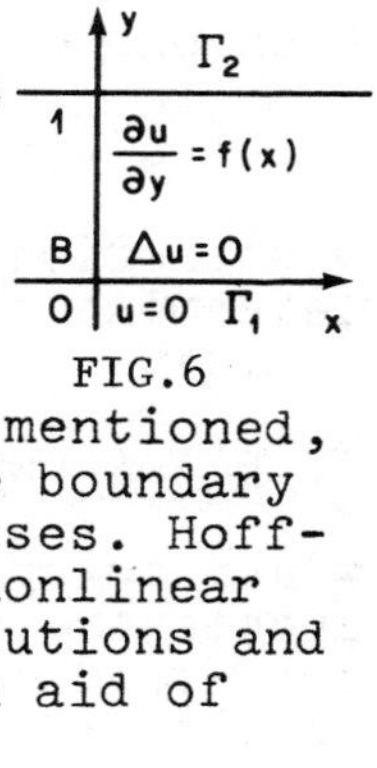

FIG.6

w_ν satisfies $\Delta w_\nu=0$ and $w_\nu=0$ on T_1; one has to determine c_ν, d_ν such that w satisfies the boundary condition on T_2 as good as possible. Mr. Kuhn calculated for $p=3$, $d_1=1.8$, $d_2=2.4$, $d_3=2.5$ an error bound $|\varepsilon|\leq 0.00218$.

C. Free Boundary value Problems It may be mentioned, that monotonicity principles also for free boundary value problems can be applied in simple cases. Hoffmann [78] transforms a Stefan problem to a nonlinear integral equations, and inclusions for solutions and for the free boundary can be obtained with aid of monotonicity (Collatz [8o]).

7. Equations with delay-term

Here we mention some types of equations for which monotonicity-principles have been proved recently.

A. Integro-differential equation.

We consider the equation for a function u(x,t):

$$(7.1)\quad Lu=\frac{\partial u(x,t)}{\partial t}-\frac{\partial^2 u(x,t)}{\partial x^2}-\int_0^t \frac{s}{t}\frac{\partial^2 u(x,s)}{\partial x^2}ds=q(x,t)$$

in $B=\{(x,t),\ |x|<a,\ t>0\}$

and prescribe boundary valuesu along the boundary ∂B; have q(x,t) is given in B; u(x,t) may be interpreted as temperature or as concentration of a gas at the place x at the time t, and it is supposed, that there is an influence of $\frac{\partial^2 u}{\partial x^2}$ not only at the time t (as in the classical case of heat-conduction-equation) but also of $\frac{\partial^2 u}{\partial x^2}$ at earlier times as it is indicated by the integral in (7.1). Earlier times have a smaller influence sorresponding to the factor s/t.

We have the monotonicity principle:

(7.2) From $Lw\geq q(x,t)$ in B, $w\geq u$ on ∂B follows $w\geq u$ in B.

This gives the possibility to include the solution u as in Nr. 6 B.

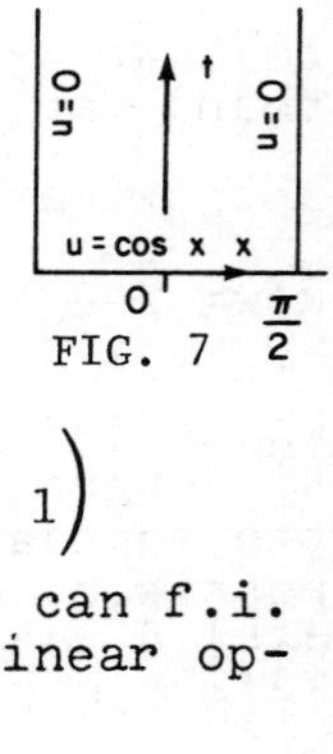

FIG. 7

For instance, it may be given

$$(7.3)\quad a=\frac{\pi}{2},\ q(x,t)=0,\ u(x,o)=\cos x \text{ for } |x|\leq\frac{\pi}{2},$$

$$u(\pm\frac{\pi}{2}t)=0 \text{ for } t\geq 0;$$

then one can take as approximate solution

$$(7.4)\quad u\ w(x,t)=(\cos x)\sum_{\nu=1}^{p} a_\nu e^{-b_\nu t};\quad \left(\sum_{\nu=1}^{p} a_\nu = 1\right)$$

w satisfies the boundary conditions and one can f.i. determine the parameter a_ν, b_ν from the nonlinear optimization

(7.5) $-\delta\leq Lw\leq\delta$, δ=Min; or one calculates a lower bound

$\underline{w}$ and an upper bound $\overline{w}$ for u from two optimization problems: $-\underline{\delta}<L\underline{w}<0$, $\underline{\delta}$= Min and $0<L\overline{w}<\overline{\delta}$, $\overline{\delta}$=Min. one would consider the restrictions only in a finite domain $|x|<\pi/2$, $o<t<t_o$ with fixed t_o.

B. Oscillations in a time-delay-system

Models for physical systems with time delay use often the difference-differential equation

(7.6) $Lu(t)=u^{(n)}(t)+au(t)+bu(t-\tau)=r(t)$ with $r(t)= r(t+T)$

with given real constants a,b,τ,T (Bellen-Zennaro [82]). One asks for periodic solutions $u(t)$ of period T: $u(t)=u(t+T)$. Bellen-Zennaro prove a maximum principle (under further assumptions on a,b) and monotonicity by using the inverse operator L^{-1} which is an integral operator with a resolvent kernel Γ; they show, that $\Gamma\geq 0$ in certain cases and this gives the possibility for inclusion of periodic solutions; it is intended to present numerical results (Collatz [82]).

I thank Mr. Th. Kuhn/Hamburg for numerical calculations on a computer.

Prof. Dr. Lothar Collatz
Institut für Angewandte Mathematik
Bundesstraße 55
2ooo Hamburg 13
Germany

References

Baker, Chr.T.H. [78] The numerical treatment of Integral Equations, Clarendon Press 1978, 1o34 p.

Bellen, A. - M. Zennaro [82] Maximum Principles for periodic solutions of Linear delay differential equations, Lecture at Oberwolfach, June 1982, to appear ISNM, Birkhäuser ed.

Bohl, E. [74] Monotonie, Lösbarkeit und Numerik bei Operatorgleichungen, Springer 1974, 255p.

Bredendiek, E. [69] Simultan Approximation, Arch.Rat. Mech.Anal. 33 (1969), 3o7-33o.

Collatz, L. [52] Aufgaben monotoner Art, Arch.Math. Anal.Mech 3 (1952) 366-376.

Collatz, L. [6o] Numverical treatment of differential equations, Springer 196o, 568p.

Collatz, L. [66] Functional Analysis and Numerical Mathematics, Acad.Press 1966, 473p.

Collatz, L. [8o] Monotonicity and free boundary value problems, Lect. Notes Math. Vol. 773, Springer (198o) 33-45.

Collatz, L. [81] Anwendung von Monotoniesätzen zur Einschließung der Lösungen von Gleichungen, Jahrbuch Überblicke Mathematik 1981, 189-225.

Collatz, L. [82] Inclusion of oscillations for some delay differential equations; lecture at Oberwolfach, June 1982, to appear ISNM, Birkhäuser ed.

Colton, D.L. [76] Solution of boundary value problems by the method of integral operators, Pitman 1976, 148 p.

Grothkopf, U. [81] Anwendungen der nichtlinearen Optimierung auf Randwertaufgaben bei partiellen Differentialgleichungen, Internat.Ser.Numer.Math. 56 (1981) 73-82.

Hoffmann, K.H. [78] Monotonie bei nichtlinearen Stefan Problemen. Intern.Ser.Num.Math. Vol. 39 (1978) 162-190.

Kantorowitsch, L.W. - G.P. Akilow [64] Funktionalanalysis in normierten Räumen. Akad. Verl. Berlin 1964, 622 p.

Love, E.R. [49] The electrostatic field of two equal circular co-axial conducting disks, Quart.J.Mech. Applied Math. 2 (1949), 428-451

Meinardus, G. [67] Approximation of functions, Theory and numerical methods, Springer 1967, 198p.

Meinardus, G. - G. Merz [79] Praktische Mathematik, I; B.I. Wissenschaftsverlag, 1979, 337p.

Schröder, J. [62] Invers-monotone Operatoren, Arch. Rat.Mech.Anal. 10 (1962), 276-295.

Schröder, J. [80] Operator Inequatilies, Acad.Press 1980, 367p.

Sloan, I.H. [82] Superconvergence to the Galerkin method for integral equations of the second kind, these proceedings, Durham 1982.

Walter, W. [70] Differential and Integral Inequalities, Springer 1970, 352 p.

Werner, H. - R. Schaback [79] Praktische Mathematik II, Springer 1979, 388p.

SUBJECT INDEX